K. Doksum

Institute of Mathematical Statistics

LECTURE NOTES–MONOGRAPH SERIES

Volume 16

Topics in Statistical Dependence

Henry W. Block, Allan R. Sampson,
and Thomas H. Savits, Editors
University of Pittsburgh

Institute of Mathematical Statistics
Hayward, California

Institute of Mathematical Statistics

Lecture Notes–Monograph Series

The production of the *IMS Lecture Notes–Monograph Series* is managed by the IMS Business Office: Jessica Utts, IMS Treasurer, and Jose L. Gonzalez, IMS Business Manager.

Library of Congress Catalog Card Number: 91-71759

International Standard Book Number 0-940600-23-4

Printed in the United States of America

PREFACE

In early 1985, after having witnessed the development of the area of "positive dependence" for a sufficiently long time, we decided to organize the first symposium dedicated *solely* to this research area. The dramatic growth of this realm of research began with the seminal works of Lehmann (1966) and Esary, Proschan and Walkup (1967). Partial summaries of this research area had been given by authors such as Barlow and Proschan (1981), Marshall and Olkin (1979), and Tong (1980). Individual sessions at national meetings had been organized around subtopics in this area, and a few conferences (e.g., the Tong conference on inequalities) had overlapped in part this area.

To our knowledge, however, no conference or book had yet focused purely on the burgeoning research in the area of positive and negative dependence for the modeling and analysis of multivariate data. We wanted to bring together as many researchers in this broad and diverse field as possible. Our goal was interaction and synthesis, as well as developing directions for future research in this area. This led to the Symposium on Dependence in Probability and Statistics which was held August 1–5, 1987.

To facilitate interaction, we chose a secluded conference site in Pennsylvania's Laurel Mountains. The Hidden Valley Conference Center provided a relaxed setting in which papers were presented, accompanied by vigorous discussion, in both organized and informal working sessions. In fact, at least one paper in this volume began during an evening discussion session at the conference. To provide some breaks from the intense program, several diversionary activities were planned. These included a trip to see nearby Fallingwater, Frank Lloyd Wright's masterpiece, and also an evening's musical entertainment based on old Pennsylvania music, instruments, and folklore. To give a more complete description of the Symposium, we have included an exact copy of the program on pages 497–504.

From our admittedly biased viewpoints, there was tremendous enthusiasm both in anticipation of, and especially during, the Symposium. Our colleagues in the field were all extremely supportive and helpful. Various federal agencies, including AFOSR, NSF and ARO, were early enthusiastic financial supporters of the Symposium, and eventually also became active participants.

Throughout the development of the Symposium, we set as a goal the production of a high quality volume reflecting the content of the proceedings. Our intent was to have the papers in such a volume represent the depth and diversity of current knowledge in the field, as well as point to directions for future research. The IMS Lecture Notes-Monograph Series represents for us an ideal forum to accomplish this.

To assure high standards, we required that all papers submitted be subjected to intensive refereeing. With the help of many colleagues we succeeded in having every paper refereed by one or more referees. Based upon their input, and space considerations, not all papers submitted were accepted for publication. We owe a debt of gratitude to the following for their conscientious refereeing:

Name	**Affiliation**
Elja Arjas	University of Oulu, Finland
Barry C. Arnold	University of California at Riverside
Asit P. Basu	University of Missouri
Philip Boland	University College, Ireland
Vanderlei Costa Bueno	Universidade de Sao Paulo, Brazil
Timothy Costigan	The Ohio State University
Ernest Enns	University of Calgary, Canada
Emad El-Neweihi	University of Illinois at Chicago
Stephen E. Fienberg	Carnegie Mellon University
Zvi Gilula	Hebrew University, Israel
Joseph Glaz	University of Connecticut
Z. Govindarajulu	University of Kentucky
William Griffith	University of Kentucky
Myles Hollander	Florida State University
F.W. Huffer	Florida State University
Satish Iyengar	University of Pittsburgh
Donald R. Jensen	Virginia Polytechnic Institute and State University
Harry Joe	University of British Columbia, Canada
George Kimeldorf	University of Texas at Dallas
Samuel Kotz	University of Maryland
Naftali Langberg	Haifa University, Israel
Nicholas T. Longford	Educational Testing Service
James Lynch	University of South Carolina
Albert W. Marshall	University of British Columbia, Canada
Jie Mi	University of Pittsburgh
Yashaswini D. Mittal	Virginia Polytechnic Institute and State University
Magdy Metry	Ain Shams University, Egypt
Charles M. Newman	University of Arizona
Truc T. Nguyen	Bowling Green State University
Ingram Olkin	Stanford University
Shixian Qian	University of Pittsburgh
M.B. Rao	North Dakota State University
Yosef Rinott	Hebrew University, Israel
Tim Robertson	University of Iowa
Marco Scarsini	Universita Degli Studi Di Roma "La Sapienza", Italy
Moshe Shaked	University of Arizona
David Stoffer	University of Pittsburgh
Yung L. Tong	Georgia Institute of Technology
Richard A. Vitale	University of Connecticut
Lyn Whitaker	University of California at Santa Barbara
Takemi Yanagimoto	The Institute of Statistical Mathematics, Japan

Additionally, we would like to note that a majority of the talks were videotaped to form a permanent record of the conference. Arrangements are being made to suitably archive these videotapes.

The editors wish to acknowledge the support of the Air Force Office of Scientific Research, the National Science Foundation (Grant No. DMS–8618897) and the Army Research Office (Grant No. AAL–03–87–G–0106) for their generous support. Also the University of Pittsburgh provided administrative support for the Symposium as well as for the preparation of this volume, and the Department of Statistics of Carnegie Mellon University provided computer support towards the production of this volume. Our thanks also to the staff of Hidden Valley Conference Center for their organization and their graciousness. We would also like to thank our local arrangements committee consisting of David Stoffer, Devendra Chhetry, and Zhaoben Fang for their dedicated assistance. Finally, we owe a great debt of gratitude to Diane Hall for coordinating the many articles required to produce this volume, and to Margie Smykla and Barbara Gastgeb for their outstanding TEX skill in the production.

H.W. Block, A.R. Sampson, T.H. Savits
November 1990

Table of Contents

INTRODUCTION

Concepts of positive dependence are becoming increasingly important in probability, statistics and their applications. While these concepts are traditionally viewed as focusing on positive and negative dependence for random vectors, they also are related to broader issues in the modeling and the analysis of multivariate data, and, in particular, ordinal data.

Historically, positive dependence for the multivariate normal distribution had been synonymous with positive correlations. Other subsequently developed multivariate distributions were often interpreted with this perspective. It was eventually realized that positive correlations can have substantially different meanings for other multivariate distributions than they have for the multivariate normal. In fact, it has been more recently demonstrated that several different important positive dependence concepts which are equivalent for multivariate normal distributions are not equivalent, in general, for multivariate distributions. In particular the concept of association is stronger than positive orthant dependence which is stronger than positive correlations. Thus, in a certain sense, many of the positive dependence concepts discussed or referenced in this volume, are outgrowths of original attempts to nonparametrically capture and extend certain properties of the multivariate normal. Additionally, other types of dependence came about from extending certain univariate properties, such as the memoryless property of the exponential distribution.

From the point of view expressed above, the theoretical origins of this research area include the fundamental works of Lehmann (1966) concerning orthant dependence, Esary, Proschan and Walkup (1967) dealing with the concept of association, and Marshall and Olkin (1967) modeling multivariate distributions. These three papers drew upon a rich historical stream and, in turn, have spawned numerous applications and inspired other related dependence concepts. The sources of this historical stream range broadly from reliability and mathematical inequalities to nonparametric statistical modeling and measures of association. Some of the researchers involved in these pioneering efforts include Goodman, Hardy-Littlewood-Polya, Hoeffding, Karlin, Kendall, Kruskal, Lancaster, Šidák, Sobel, and Tukey.

Barlow and Proschan (1981) further developed positive dependence concepts in their book (first printed in 1975) particularly in the bivariate case, and Tong (1980) provides additional material and development. A parallel development occurred independently in the mathematical physics literature (see Fortuin, Kastelyn and Ginibre (1971)). For a long time it was felt that negative dependence concepts were the mirror image of positive dependence. That this was not the case was demonstrated by Karlin and Rinott (1980), Block, Savits, and Shaked (1982), and by Joag-dev and Proschan (1983) among others. Related review articles on inequalities and dependence are those by Eaton (1982) and by Block and Sampson

(1983).

Currently, techniques and approaches are being developed specifically for modeling positive dependence for discrete multivariate distributions, with a perspective for applications to multivariate ordinal contingency tables. These techniques draw upon a diverse body of research including log-linear contingency table modeling (Bishop, Fienberg and Holland (1975)) and canonical representations of probability distributions (Lancaster (1969)). Due in part to natural constraints upon the discrete probabilities imposed by positive and negative dependence concepts, techniques from order-restricted inference have been fruitful in dealing with problems related to estimation (e.g., Robertson, Wright and Dykstra (1989)). The book by Agresti (1984) synthesizes and summarizes some of the basic connections between positive dependence and ordinal contingency tables, and the applications of order restricted inference to this area.

This broad field of research concerning positive and negative dependence can be categorized into four areas. These areas are currently in different stages of development.

(a) *Fundamental Concepts of Positive Dependence.* This area consists of the probabilistic modeling of concepts for positive and negative dependence. The primary foci are properties of dependence, orderings, related inequalities, measures and their interrelationships.

(b) *Applications to Probability Theory, Reliability Theory, Mathematical Physics, and Other Areas.* This area applies the theoretical developments in the preceding item, (a), in developing results in other research areas. Examples include (i) probability limit theorems under positive dependence conditions; (ii) system reliability bounds assuming certain dependence structures for component life lengths; (iii) models for causal relationships in the social sciences; and (iv) repeated testing schemes.

(c) *Statistical Considerations Involving Positive and Negative Dependence.* This relatively new area concerns estimation, hypothesis testing and distribution theory based upon data taken from multivariate distributions having certain positive or negative dependence structures, or having certain monotone order relationships. Instances of such research are: (i) testing for component life lengths having certain distributional structures such as quadrant dependence; and (ii) testing whether all odds ratio in an $r \times c$ contingency table are greater than or equal to 1 (which is the dependence concept called TP_2).

(d) *Inter-Relationships of Positive Dependence Concepts With Other Aspects of Statistics.* This area concerns the connections between the ideas developed in the theory of positive and negative dependence, with other areas of statistics. A large body of research is concerned with the interpretation of positive and negative dependence concepts for the multivariate normal distribution and other parametric families of distributions. Examples include: (i) conditions for certain positive dependence concepts to hold for elliptically symmetric distributions; and (ii) results for when certain parametric families of distributions are well-ordered by the parameter with respect to various positive dependence orderings. An emerging research area is the connection between various probability models for ordinal contingency tables and positive dependence concepts. For instance, certain types of odds-ratios properties have a natural interpretation as positive dependence concepts.

Bibliography of Pertinent Literature

AGRESTI, A. (1984). *Analysis of Ordinal Categorical Data.* John Wiley, New York.

BARLOW, R.E., BARTHOLOMEW, D.J., BREMNER, J.M., and BRUNK, H.D. (1972). *Statistical Inference Under Order Restrictions.* John Wiley, New York.

BARLOW, R. and PROSCHAN, F. (1981). *Statistical Theory of Reliability and Life Testing.* To Begin With, Silver Spring, Md.

BISHOP, Y.M., FIENBERG, S.E., and HOLLAND, P.W. (1975). *Discrete Multivariate Analysis.* MIT Press, Cambridge.

BLOCK, H.W. and SAMPSON, A.R. (1983). Inequalities on distributions: bivariate and multivariate. *Encyclopedia of Statistical Sciences, Volume 4* (S. Kotz and N.L. Johnson, eds.), John Wiley and Sons, Inc.

BLOCK, H.W., SAVITS, T.H., and SHAKED, M. (1982). Some concepts of negative dependence. *Ann. Prob.* **10**, 765–772.

EATON, M.L. (1982). A review of selected topics in multivariate probability inequalities. *Ann. Statist.* **10** 11–43.

ESARY, J.D., PROSCHAN, F., and WALKUP, D.W. (1967). Association of random variables with applications. *Ann. Math. Statist.* **38** 1466–1474.

FORTUIN, C., KASTELYN, P., and GINIBRE, J. (1971). Correlation inequalities on some partially ordered sets. *Comm. Math. Phys.* **22** 89–103.

GOODMAN, L.A. and KRUSKAL, W.H. (1979). *Measures of Association for Cross-classifications.* Springer-Verlag, New York.

JOAG-DEV, K. and PROSCHAN, F. (1983). Negative association of random variables with applications. *Ann. Statist.* **11** 286–295.

KARLIN, S. and RINOTT, Y. (1980). Classes of orderings of measures and related correlation inequalities. II. Multivariate reverse rule distributions. *J. Multi. Anal.* **10** 499–516.

LANCASTER, H.O. (1969). *The Chi-squared Distribution.* John Wiley and Sons, New York.

LEHMANN, E.L. (1966). Some concepts of dependence. *Ann. Math. Statist.* **37** 1137–1153.

MARSHALL, A.W. and OLKIN, I. (1967). A multivariate exponential distribution. *J. Amer. Statist. Assoc.* **62** 30–49.

ROBERTSON, T.J., WRIGHT, F.T., and DYKSTRA, R.L. (1988). *Order Restricted Statistical Inference.* John Wiley and Sons, New York.

TONG, Y.L. (1980). *Probability Inequalities in Multivariate Distribution.* Academic Press, New York.

SHOULD MINIMAL REPAIR DEPEND ON INFORMATION?

BY ELJA ARJAS AND ILKKA NORROS

University of Oulu and Technical Research Centre of Finland

The notion of minimal repair with respect to a history is defined in terms of a general filtration and a completely unpredictable stopping time. An inequality relating compensator transformations with respect to the minimal history of a one-point process and a richer history is proven. Applied to minimal repair, this result shows that the modeling of minimal repair in a "black box" sense always gives a stochastically longer life length than a more realistic model.

1. Introduction. The notion of minimal repair was introduced in reliability theory by Barlow and Hunter (1960). Its intuitive meaning is putting the system back to operation when it fails in such a way that the situation immediately preceding the failure is restored. The traditional probabilistic model is the following. Consider a nonnegative random variable S (the life length of the system) with a continuous distribution function F. When the system fails, say at time $S = s$, it is given an additional lifetime S' with conditional distribution

$$P(S' > t \mid S = s) = P(S > s + t \mid S > s) = (1 - F(s+t))/(1 - F(s)).$$

Equivalently, the minimal repair model can be defined in terms of the cumulative hazard function

$$R(t) = -\ln(1 - F(t)) = \int_0^t \frac{dF(s)}{1 - F(s)}$$

as follows. The original failure point $S(\omega)$ is "erased" and the hazard of the additional life time S' at age $t - S(\omega)$ is given the same value as the original hazard would have had at time t had there been no failure, that is, $dR(t)$. If minimal repairs are made repeatedly, the sequence of repair times is a nonhomogeneous Poisson process with integrated intensity $R(t)$.

This simple notion of minimal repair has obvious intuitive appeal. However, it may not be a realistic description of any actual repair done on a failed system.

AMS 1980 subject classifications. 90B25, 60G55.

Key words and phrases. Minimal repair, imperfect repair, compensator transformation, stochastic order.

In a sense, it treats the system as a black box, without any reference to what caused the failure and what repair was needed to put the system back to operation. Bergman makes this point in a review paper (Bergman, 1985), distinguishing between "statistical" and "physical" minimal repair. The comments on minimal repair on page 51 in Ascher and Feingold (1984) should also be mentioned.

As a very simple example illustrating this problem in the definition of minimal repair, consider a system consisting of two components in parallel, with independent $Exp(1)$ distributed life lengths. The system fails when the component with the longer life fails. A natural concept of minimal repair of the system would be the restoration of the working condition of this component when it fails, leaving the component which failed earlier in the down state. The additional lifetime obtained by this kind of repair is clearly independent of the failure time and $Exp(1)$ distributed. The black box model would consider a system life length S with distribution $P(S \leq t) = F(t) = (1-e^{-t})^2$ and an additional lifetime S' with conditional distribution

$$P(S' > t \mid S = s) = \frac{1 - F(s+t)}{1 - F(s)} = e^{-t}\frac{2 - e^{-s-t}}{2 - e^{-s}},$$

which is stochastically larger than $Exp(1)$ for every s.

In this example, the more realistic model gives a less optimistic estimate of the total life length than the black box model. We shall show below in Section 3 that this is always the case. Section 2 is devoted to the definition of minimal repair with respect to a general history. A more detailed exposition of Sections 2 and 3 can be found in Norros (1987) and Arjas and Norros (1989), respectively.

2. Minimal Repair with Respect to a General History. Consider a probability space $(\Omega, \mathcal{F}, P)$, a history (filtration) $\mathbf{F} = (\mathcal{F}_t)_{t \geq 0}$ of sub-σ-fields of $\mathcal{F}$ and an $\mathbf{F}$-stopping time S, satisfying the following conditions:

(i) Ω is a Polish space, that is, a complete separable metric space;

(ii) $\mathcal{F}$ is the completion w.r.t. P of $\mathcal{B}(\Omega)$, the Borel σ-field of Ω;

(iii) $\mathbf{F}$ satisfies Dellacherie's "usual conditions", that is, it is right continuous and $\mathcal{F}_0$ contains all P-null sets;

(iv) S is completely unpredictable and a.s. finite.

We denote by N the simple point process $N_t = 1_{\{t \geq S\}}$ and by A the $\mathbf{F}$-compensator of N. By (iv), A is a continuous process.

In this section we show how the stopping time S can be "minimally repaired". In other words, we define probability measures Q_n, $n = 1, 2, \ldots$, such that under Q_n, things behave as if S had been n times minimally repaired before its final failure. In our construction we need the following well-known result and some facts about the prediction process.

LEMMA 2.1. *For any finite stopping time T, the conditional distribution of $A_S - A_T$ given $\mathcal{F}_T$ is the $Exp(1)$ distribution on the set $\{S > T\}$.*

The notion of a *prediction process* was introduced by Knight (1975). Aldous

(1981) developed a somewhat different approach, which was applied in Norros (1985). The definition given below differs slightly from the above-mentioned ones since we are considering an abstract history instead of a process.

We denote by $\mathcal{P}(\Omega)$ the space of all probability measures on $\mathcal{B}(\Omega)$, endowed with the topology of weak convergence. $\mathcal{P}(\Omega)$ is in turn a Polish space, and $\mathcal{P}(\mathcal{P}(\Omega))$ can be defined in a similar manner.

THEOREM 2.2. *There exists a $\mathcal{P}(\Omega)$-valued cadlag process μ such that for any stopping time T, μ_T is a regular version of the conditional probability $P(\,\cdot \mid \mathcal{F}_T)$.*

The proof can be found in Aldous (1981), and it is reproduced in Norros (1985). In this paper, we call the process μ of Theorem 2.2 simply *the prediction process.*

PROPOSITION 2.3. *Let Y be a bounded $\mathcal{B}(\Omega)$-measurable random variable defined on $(\Omega, \mathcal{F}, P)$. Denote by M^Y a cadlag version of the martingale $E[Y \mid \mathcal{F}_t]$. Then the process $\int Y\, d\mu_{t-}$ is indistinguishable from M^Y_{t-}. In particular, $\int Y\, d\mu_{T-} = M^Y_{T-}$ a.s. for every stopping time T.*

For a proof, see, for example, Norros (1985). Note that, although $\mu_T = P[\,\cdot \mid \mathcal{F}_T]$, μ_{T-} is *not* $P[\,\cdot \mid \mathcal{F}_{T-}]$. For example, T is $\mathcal{F}_{T-}$-measurable, but if T is completely unpredictable, then the random measure μ_{T-} gives T a continuous distribution a.s. The random measure μ_{T-} is not a conditional distribution with respect to any σ-field, but it tells what the prediction was immediately before T occurred.

Intuitively, the difference between the σ-fields $\mathcal{F}_{S-}$ and $\mathcal{F}_S$ is that in $\mathcal{F}_{S-}$, it is known when S occurs, but it is not known what else happens at time S. For example, if S is a point in a marked point process, then S is known in $\mathcal{F}_{S-}$, but the mark is $\mathcal{F}_S$-measurable and may not be $\mathcal{F}_{S-}$-measurable.

Now we proceed to the construction of the "minimal repair" of S. Suppose that we choose an ω with distribution P and start proceeding at time 0. Suddenly S occurs. In order to make a minimal repair, we have to change our ω to another, say ω', which is indistinguishable from ω strictly before the time $S(\omega)$ and satisfies $S(\omega') > S(\omega)$. Moreover, ω' should be chosen according to an appropriate distribution among the candidates satisfying these conditions. This reasoning can be formalized by means of the prediction process.

Indeed, $\mu_{S-}(\omega)$ gives the conditional distribution with respect to the history strictly before $S(\omega)$ when it was not yet known that S would appear at time $S(\omega)$. Thus, if we choose ω' according to the distribution $\mu_{S-}(\omega)$, we may proceed further "as if nothing had happened". Intuitively, this means that at any time prior to the ultimate system failure it is not even "known" whether there has been a minimal repair or not.

In order to be more rigorous, let κ be a $\mathcal{B}(\Omega)$-measurable $\mathcal{P}(\Omega)$-valued random variable such that $\kappa = \mu_{S-}$ a.s. (κ can be constructed by means of the regular conditional distribution of μ_{S-} w.r.t. $\mathcal{B}(\Omega)$). Now define successively the probability measures Q_0, Q_1, $Q_2, \ldots$ on $\mathcal{B}(\Omega)$ by

$$Q_0 = P, \; Q_{n+1}(A) = \int_\Omega \kappa(\omega)(A) Q_n(d\omega).$$

Q_n is the measure which is obtained when S is deferred n times in the sense of minimal repair. The following theorem shows that the measures Q_n have a density (w.r.t. P) which has a very simple expression.

THEOREM 2.4. *For any n, the probability measure Q_n is absolutely continuous w.r.t. P, with the Radon-Nikodym derivative*

$$\frac{dQ_n}{dP} = \frac{1}{n!} A_S^n.$$

Moreover, for any stopping time T,

$$\left(\frac{dQ_n}{dP}\right)_{\mathcal{F}_T} = e^{A_T} 1_{\{S>T\}} \left(1 - \frac{1}{n!}\int_0^{A_T} x^n e^{-x} dx\right) + \frac{1}{n!} A_S^n 1_{\{S \le T\}}.$$

PROOF. We prove the first assertion by induction. It holds trivially for $n = 0$. Suppose that it holds for some fixed value of n. Let Y be any bounded $\mathcal{B}(\Omega)$-measurable random variable. We have to show that

$$\int Y dQ_{n+1} = EY \frac{1}{(n+1)!} A_S^{n+1}.$$

Denote by M^Y a cadlag version of the martingale $E[Y \mid \mathcal{F}_t]$. Now, by the definition of κ, Proposition 2.3, the rules of Stieltjes stochastic calculus and Dellacherie's integration formula (Dellacherie (1972), IV T 47),

$$\int_\Omega Y dQ_{n+1} = \int_\Omega Q_n(d\omega) \int_\Omega \kappa(\omega)(d\omega') Y(\omega') = \int_\Omega M_{S-}^Y dQ_n = E\frac{1}{n!} A_S^n M_{S-}^Y$$

$$= E\frac{1}{n!}\int_0^\infty A_t^n M_{t-}^Y dN_t = E\int_0^\infty M_{t-}^Y \frac{1}{n!} A_t^n dA_t = E\int_0^\infty M_t^Y \frac{1}{(n+1)!} dA_t^{n+1}$$

$$= EM_\infty^Y \frac{1}{(n+1)!} A_\infty^{n+1} = EY\frac{1}{(n+1)!} A_S^{n+1}.$$

The proof of the second equation is based on Lemma 2.1.:

$$\left(\frac{dQ_n}{dP}\right)_{\mathcal{F}_T} 1_{\{S>T\}} = 1_{\{S>T\}} \frac{1}{n!} E[A_S^n \mid \mathcal{F}_T]$$

$$= 1_{\{S>T\}} \frac{1}{n!} \int_0^\infty (A_T + x)^n e^{-x} dx = 1_{\{S>T\}} \frac{1}{n!} \int_{A_T}^\infty x^n e^{-(x-A_T)} dx$$

$$= e^{A_T} 1_{\{S>T\}} \left(1 - \frac{1}{n!}\int_0^{A_T} x^n e^{-x} dx\right). \quad \|$$

Theorem 2.4 gives the justification for the following definition.

DEFINITION 2.5. With notation as above, the probability measure corresponding to an *n-fold $\mathcal{F}$-minimal repair of the stopping time S* is the measure Q_n defined by

$$\frac{dQ_n}{dP} = \frac{1}{n!} A_S^n.$$

We close this section with a remark concerning the situation where the stopping time S is eliminated completely by repeating "minimal repairs" indefinitely. Since S is a.s. finite, a measure describing this operation in the sense of Theorem 2.4 can not in general be defined on $\mathcal{F}_\infty$, and if it can, it will be concentrated on the P-null set $\{S = \infty\}$. However, such a measure can be defined on each sub-σ-field $\mathcal{F}_T$, where T is a stopping time such that A_T is bounded. Letting n go to infinity in the second assertion of Theorem 2.4, it is seen that this measure, say, Q_∞^T, is absolutely continuous w.r.t. P on $\mathcal{F}_T$ and

$$\frac{dQ_\infty^T}{dP|_{\mathcal{F}_T}} = e^{A_T} 1_{\{S>T\}}.$$

3. Stochastic Comparison of Transformed Distributions. Let $(\Omega, \mathcal{F}, P)$, $\mathbf{F} = (\mathcal{F}_t)_{t\geq 0}$, S and N be as in the previous section. Let $\mathbf{G} = (\mathcal{G}_t)_{t\geq 0}$ be the history generated by the one point counting process N. Let $A^{\mathbf{F}}$ and $A^{\mathbf{G}}$ be the $\mathbf{F}$- and $\mathbf{G}$-compensators of N, respectively. Since S is assumed to be completely unpredictable w.r.t. $\mathbf{F}$, both compensators are continuous.

We consider the following kind of transformations of the compensators. Let $g : [0, \infty] \to [0, \infty]$ be an increasing differentiable function such that $g(0) = 0$ and $g(\infty) = \infty$. Consider the continuous increasing process $B^{\mathbf{F}}$ defined by

$$B_t^{\mathbf{F}} = g(A_t^{\mathbf{F}}),\ t \geq 0.$$

By the next well-known Girsanov type theorem, we can modify the probability P in an absolutely continuous way so that $B^{\mathbf{F}}$ is the compensator of N with respect to the new measure. For a proof, see Jacod (1975), Proposition 4.3 and Theorem 4.5.

THEOREM 3.1. *Denote by L the process*

$$L_t = g'(A_t^{\mathbf{F}})^{N_t} e^{A_t^{\mathbf{F}} - B_t^{\mathbf{F}}}.$$

L is a uniformly integrable martingale with expectation 1. *Define the probability measure Q on $\mathcal{F}$ by*

$$\frac{dQ^{\mathbf{F}}}{dP} = L_\infty = g'(A_S^{\mathbf{F}}) e^{A_S^{\mathbf{F}} - B_S^{\mathbf{F}}}.$$

Then the $(Q^{\mathbf{F}}, \mathbf{F})$-compensator of S is $B^{\mathbf{F}}$.

We consider two special cases of this transformation. The first is minimal repair. Indeed, choosing $g(x) = x - \ln(1+x)$, we have

$$L_\infty = \frac{A_S^{\mathbf{F}}}{1+A_S^{\mathbf{F}}} \exp(A_S^{\mathbf{F}} - (A_S^{\mathbf{F}} - \ln(1+A_S^{\mathbf{F}}))) = A_S^{\mathbf{F}},$$

which is the density corresponding to an $\mathcal{F}$-minimal repair of S (Definition 2.5).

The other case is the linear transformation

$$g(x) = \alpha x, \ \alpha \in (0,1),$$

which could be called *proportional improvement* since the hazard is reduced by a fixed percentage. One also quickly concludes that this is equivalent to the *imperfect repair* of Brown and Proschan (1983), where the device is repeatedly minimally repaired with probability $(1-\alpha)$ up to the first unsuccessful repair attempt. See also Shaked and Shanthikumar (1986).

We can now prove the main result of this paper.

THEOREM 3.2. *With the notation as above, suppose that the function g is such that $e^{-g(-\ln x)}$ is concave for $x \in (0,1)$. Then S is stochastically smaller under $Q^{\mathbf{F}}$ than under $Q^{\mathbf{G}}$.*

PROOF. Let T be an $\mathbf{F}$-stopping time such that $A_T^{\mathbf{F}}$ is bounded. We first observe that (cf. Norros (1986), Proposition 5.1)

$$E1_{\{S>t\wedge T\}} e^{A_{t\wedge T}^{\mathbf{F}}} = 1.$$

Thus, $\exp(A_{t\wedge T}^{\mathbf{F}})$ can be viewed as a density function on the set $\{S > t \wedge T\}$. Denote

$$f(x) = e^{-g(-\ln x)}, \ x \in (0,1),$$

which by assumption is a concave function. Now, by Jensen's inequality,

$$Q^{\mathbf{F}}(S > t\wedge T) = E1_{\{S>t\wedge T\}} e^{A_{t\wedge T}^{\mathbf{F}} - g(A_{t\wedge T}^{\mathbf{F}})} = E1_{\{S>t\wedge T\}} e^{A_{t\wedge T}^{\mathbf{F}}} f(e^{-A_{t\wedge T}^{\mathbf{F}}})$$

$$\leq f(E1_{\{S>t\wedge T\}} e^{A_{t\wedge T}^{\mathbf{F}}} e^{-A_{t\wedge T}^{\mathbf{F}}}) = f(P(S > t\wedge T)).$$

Applying this for $T = T_n = \inf\{t : A_t^{\mathbf{F}} \geq n\}$ and letting $n \to \infty$ we obtain, since $S \leq \sup T_n$ a.s.,

$$Q^{\mathbf{F}}(S > t) \leq f(P(S > t)).$$

But

$$f(P(S > t)) = e^{-g(-\ln P(S>t))} = e^{-g(A_t^{\mathbf{G}})} = e^{-B_t^{\mathbf{G}}} = Q^{\mathbf{G}}(S > t). \quad \|$$

It is easy to see that the transformations corresponding to minimal repair and imperfect repair both satisfy the conditions of Theorem 3.2. Thus we have the following corollaries, which are of certain practical interest.

COROLLARY 3.3. *Consider the change of distributions which corresponds to exactly one successful minimal repair on a failed device. Then the* **F***-hazard transformation leads to a stochastically shorter life length than the corresponding* **G***-hazard transformation.*

COROLLARY 3.4. *Consider the change of distributions which corresponds to a fixed proportional improvement, or, equivalently, imperfect repair with a constant probability for successful minimal repair. Then the* **F***-hazard transformation leads to a stochastically shorter life length than the corresponding* **G***-hazard transformation.*

We conclude with some remarks. First, we return to the distinction between "physical" and "statistical" minimal repair. Recall that the history **F** can be completely general, as long as S remains completely unpredictable. On the other hand, one could argue that if the minimal repair is an actual physical operation performed on a failed device, which returns it to the state immediately preceding the failure, the history should be one "giving a full description of the internal state".

It is also interesting that our inequalities hold irrespective of all dependencies that a conditioning on a history might reveal, whether positive or negative. This is not obvious since usually stochastic comparison results involving conditioning require some form of stochastic monotonicity.

Finally, we mention two open problems. First, Proposition 5.4 in Norros (1986) shows that in the case of proportional improvement, the stochastic comparison in Corollary 3.4 is reversed if **F** is the internal history of a set of component life lengths and if the transformation is made on the compensators of *all* components. Would a similar result hold for the minimal repair also, or for a more general class of compensator transformations? Second, does Theorem 3.2 hold when **G** is larger than the minimal history but smaller than **F**?

REFERENCES

ALDOUS, D. (1981). Weak convergence and the general theory of processes. Incomplete draft of a monograph.

ARJAS, E. and NORROS, I. (1989). Change of life distribution via a hazard transformation: An inequality with application to minimal repair. *Math. Oper. Res.* **14** 355–361.

ASCHER, H. and FEINGOLD, H. (1984). *Repairable Systems Reliability.* Marcel Dekker, New York and Basel.

BARLOW, R.E. and HUNTER, L. (1960). Optimum preventive maintenance policies. *Operations Research* **1** 90–100.

BROWN, M. and PROSCHAN, F. (1983). Imperfect repair. *J. Appl. Prob.* **20** 851–859.

BERGMAN, B. (1985). On reliability theory and its applications (with discussion). *Scand. J. Statist.* **12** 1–41.

DELLACHERIE, C. (1972). Capacités et processus stochastiques. Springer Verlag, Berlin.

JACOD, J. (1975). Multivariate point processes: Predictable projection, Radon-Nikodym derivatives, representation of martingales. *Z. Wahrsch. Verw. Geb.* **34** 235–253.

KNIGHT, F. (1975). A predictive view of continuous time processes. *Ann. Prob.* **3** 573–596.

NORROS, I. (1985). Systems weakened by failures. *Stoch. Proc. Appl.* **20** 181–196.

NORROS, I. (1986). A compensator representation of multivariate life length distributions, with applications. *Scand. J. Statist.* **13** 99–112.

NORROS, I. (1987). The "minimal repair", elimination and externalization of a totally inaccessible stopping time. Reports of Dep. of Math., Univ. of Helsinki.

SHAKED, M. and SHANTHIKUMAR, J.G. (1986). Multivariate imperfect repair. *Operations Research* **34** 437–448.

DEPARTMENT OF APPLIED MATHEMATICS
AND STATISTICS
UNIVERSITY OF OULU
LINNANMAA
SF–90570 OULU, FINLAND

TELECOMMUNICATIONS LABORATORY
TECHNICAL RESEARCH CENTRE OF
FINLAND
OTAKAARI 7 B
SF–02150 ESPOO, FINLAND

DEPENDENCE IN CONDITIONALLY SPECIFIED DISTRIBUTIONS

By Barry C. Arnold

University of California, Riverside

Suppose that (X,Y) is a two-dimensional random variable whose joint density is conditionally specified. If for each x, the conditional distribution of Y given $X = x$ is a member of a particular exponential family and for each y, the conditional distribution of X given $Y = y$ is a member of a possibly different exponential family, it is possible to determine sufficient conditions for negative (or positive) dependence of (X,Y). Analogous results are obtainable for certain conditionally specified distributions which do not involve exponential families.

1. Introduction. Castillo and Galambos (1987a,b) considered a spectrum of joint distributions for which the conditional distributions belonged to specified families. Subsequently Arnold (1987) and Arnold and Strauss (1987, 1988) considered broad classes of such conditionally specified distributions. Specific attention was focused on situations where the conditional distributions were posited to be members of particular exponential families of distributions. The initial example in this genre involved bivariate distributions in which both sets of conditional densities were negative exponential. Such joint distributions were shown to have densities of the form:

$$f_{X,Y}(x,y) = \beta\gamma\theta(\delta)\exp\{-[\beta x + \gamma y + \beta\gamma\delta xy]\}, \quad x > 0, y > 0, \tag{1}$$

where

$$\theta(\delta) \triangleq \left[\int_0^\infty e^{-u}(1+\delta u)^{-1}du\right]^{-1}. \tag{2}$$

In connection with the development of method of moments estimates for such a joint distribution, it was necessary to compute the correlation between X and Y. The resulting expression

$$\rho(X,Y) = \frac{\delta + \theta(\delta) - \theta^2(\delta)}{\theta(\delta)(1+\delta-\theta(\delta))} \tag{3}$$

AMS 1980 subject classifications. Primary 62E10, 62H05; secondary 60E05.

Key words and phrases. Negative dependence, positive dependence, exponential families, total positivity.

was evaluated numerically for various values of δ. It was consequently evident that $\rho(X,Y) \leq 0$ for every $\delta \leq 0$. Retrospectively this could have easily been obtained by observing that, since for the density (1) we have

$$P(X > x \mid Y = y) = \exp[-(1 + \gamma\delta y)\beta x], \tag{4}$$

it follows that X is stochastically decreasing in Y. Consequently using Theorem 5.4.2 of Barlow and Proschan (1981), X and Y are negatively quadrant dependent so that $\rho(X,Y) \leq 0$.

It is reasonable to ask to what extent does this negative dependence carry over to the broader classes of conditionally specified distributions discussed in Castillo and Galambos (1987a) and Arnold and Strauss (1987, 1988). The issue is addressed in Section 2. Certain conditionally specified distributions not of the exponential form are discussed in Section 3. The final section outlines certain multivariate versions of the results in the earlier sections. Since it is relatively easy to get experimenters to model conditional distributions and it is relatively easy to get them to discuss the signs of correlations between variables, it is of interest to be able to identify certain modelling inconsistencies which might result. For example we cannot have $X \mid Y = y$ exponential with parameter dependent on y, $Y \mid X = x$ exponential with parameter dependent on x and (X,Y) positively correlated. If we replace the word "exponential" with "normal," such a model is of course possible.

2. Bivariate Distributions Whose Conditionals Are in Prescribed Exponential Families. As much as possible we will use the notation of Arnold and Strauss (1987). Let $\{f_1(x;\underline{\theta}) : \underline{\theta} \in \Theta\}$ denote a k parameter exponential family of densities of the form

$$f_1(x;\underline{\theta}) = r_1(x)\beta_1(\underline{\theta}) \exp\left\{\sum_{i=1}^{k} \theta_i q_i^{(1)}(x)\right\} \tag{5}$$

where Θ is the natural parameter space and the densities are defined with respect to some convenient dominating measure on $\mathbf{R}$ (usually Lebesgue measure or counting measure). Analogously let $\{f_2(y;\underline{\tau}) : \underline{\tau} \in T\}$ denote an ℓ-parameter exponential family of the form

$$f_2(y;\underline{\tau}) = r_2(y)\beta_2(\underline{\tau}) \exp\left\{\sum_{j=1}^{\ell} \tau_j q_j^{(2)}(y)\right\} \tag{6}$$

where T is the natural parameter space. Assume that the $\{q_i^{(1)}\}$ are functionally independent, and similarly for the $\{q_i^{(2)}\}$.

Suppose that $f(x,y)$ is a bivariate density for which

(i) For every y for which $f(x \mid y)$ is defined, this density is a member of the family (5) for some $\underline{\theta}$ which may depend on y.

(ii) For every x for which $f(y \mid x)$ is defined, this conditional density is a member of the family (6) for some $\underset{\sim}{\tau}$ which may depend on x.

Arnold and Strauss show that if (i) and (ii) hold, then $f(x, y)$ must be of the form

$$(7) \qquad f(x,y) = r_1(x)r_2(y)\exp\{\underset{\sim}{q}^{(1)}(x)'M\underset{\sim}{q}^{(2)}(y) + \underset{\sim}{a}'\underset{\sim}{q}^{(1)}(x) + \underset{\sim}{b}'\underset{\sim}{q}^{(2)}(y) + c\}$$

for suitable choices of $M, \underset{\sim}{a}, \underset{\sim}{b}$, and c (subject to the requirement that $\int_{-\infty}^{\infty}\int_{-\infty}^{\infty} f(x,y)dxdy = 1$). Here

$$\underset{\sim}{q}^{(1)}(x) = (q_1^{(1)}(x), \ldots, q_k^{(1)}(x))$$

and

$$\underset{\sim}{q}^{(2)}(y) = (q_1^{(2)}(y), \ldots, q_\ell^{(2)}(y))$$

and the dimensions of M, $\underset{\sim}{a}$ and $\underset{\sim}{b}$ are appropriately selected. (Note that the case of independence is included; it corresponds to $M \equiv 0$. The parameters included in M, $\underset{\sim}{a}$ and $\underset{\sim}{b}$ are constrained to be such that

$$\psi(M, \underset{\sim}{a}, \underset{\sim}{b}) = \int_{S_x}\int_{S_y} e^{-c} f(x,y)dxdy < \infty.$$

The remaining constant c is then necessarily given by

$$e^c = \frac{1}{\psi(M, \underset{\sim}{a}, \underset{\sim}{b})}$$

so that the joint density integrates to 1. Here $S_x = \{x : r_1(x) > 0\}$ and $S_y = \{y : r_2(y) > 0\}$.

Under what conditions can we determine that $f(x, y)$ given by (7) is positive quadrant dependent and hence that $\rho(X, Y) \geq 0$? A convenient sufficient condition for such positive dependence is that the density be totally positive of order 2, i.e., that

$$(8) \qquad \begin{vmatrix} f(x_1, y_1) & f(x_1, y_2) \\ f(x_2, y_1) & f(x_2, y_2) \end{vmatrix} \geq 0$$

for every $x_1 < x_2$, $y_1 < y_2$ in S_x and S_y respectively. (This is part of Barlow and Proschan's (1981) Theorem 5.4.2). However the determinant (8) assumes a particularly simple form if the joint density is of the form (7). Substitution into (8) yields the following sufficient condition for total positivity of order 2:

$$(9) \qquad [\underset{\sim}{q}^{(1)}(x_1) - \underset{\sim}{q}^{(1)}(x_2)]'M[\underset{\sim}{q}^{(2)}(y_1) - \underset{\sim}{q}^{(2)}(y_2)] \geq 0$$

for every $x_1 < x_2$ in S_x and $y_1 < y_2$ in S_y. For example if $\underline{q}^{(1)}(x) \uparrow$ as $x \uparrow$ and $\underline{q}^{(2)}(y) \uparrow$ as $y \uparrow$, then a sufficient condition for TP_2 and consequently for nonnegative correlation is $M \geq 0$. If $\underline{q}^{(1)}$ and $q^{(2)}$ are not monotone, it is unlikely that any choice for M will yield a TP_2 density. If the determinant in (9) is always ≤ 0, then negative correlation is guaranteed. Consider the following examples abstracted from Arnold and Strauss (1987).

EXAMPLE 2.1. Bivariate exponential conditionals distribution. Here $k = \ell = 1$, $r_1(t) = r_2(t) = I(t > 0)$, $q^{(1)}(t) = q^{(2)}(t) = -t$. The joint densities are of the form

$$f(x, y) = \exp[m_{11}xy - ax - by + c], \quad x > 0, y > 0.$$

For convergence we require $m_{11} \leq 0$, $a > 0$, $b > 0$. This guarantees that expression (9) is ≤ 0 and consequently $\rho(X, Y) \leq 0$.

EXAMPLE 2.2. Bivariate Poisson conditionals distribution. Here

$$f(x, y) = \exp(m_{11}xy + ax + by + c)/x!y!$$

for $x = 0, 1, 2, \ldots, y = 0, 1, 2, \ldots$. For convergence we need $m_{11} \leq 0$. Again (9) is ≤ 0 and negative correlation is assured.

EXAMPLE 2.3. Bivariate geometric conditionals distribution. Here

$$f(x, y) = \exp(m_{11}xy + ax + by + c)$$

for $x = 0, 1, 2, \ldots, y = 0, 1, 2, \ldots$. For convergence we require $a < 0$, $b < 0$ and $m_{11} \leq 0$. Consequently (9) is ≤ 0 and negative correlation is encountered.

EXAMPLE 2.4. Normal conditionals with variance 1. In this case we find

$$f(x, y) = e^{-(x^2+y^2)/2} \exp(m_{11}xy + ax + by + c)$$

which yields a valid density provided $| m_{11} | \leq 1$. It is the classical bivariate normal. The correlation is determined by the sign of m_{11}.

EXAMPLE 2.5. Bivariate normal conditionals distribution. Here $k = \ell = 2$, $r_1(t) = r_2(t) = 1$, $\underline{q}^{(1)}(t) = \underline{q}^{(2)}(t) = \binom{t^2}{t}$. The resulting bivariate density is

$$f(x, y) = \exp\left\{\begin{pmatrix} x^2 \\ x \end{pmatrix}' M \begin{pmatrix} y^2 \\ y \end{pmatrix} + \underline{a}' \begin{pmatrix} x^2 \\ x \end{pmatrix} + \underline{b}' \begin{pmatrix} y^2 \\ y \end{pmatrix} + c\right\}.$$

Since $\underline{q}^{(1)}$ and $\underline{q}^{(2)}$ are not monotone, we cannot expect total positivity. The exceptional case occurs when $M_{11} = M_{12} = M_{21} = 0$ in which case the density reduces to a standard bivariate normal with correlation determined by the sign of M_{22}.

3. Other Conditionally Specified Bivariate Distributions.

(a) **Bivariate Pareto** (α) **conditionals.** Suppose $f(x,y)$ is such that each $f(x \mid y)$ and each $f(y \mid x)$ is a member of the Pareto(α) ($\alpha > 0$) family of densities, i.e., of the form

$$f(u) = \frac{\alpha}{\sigma}\left(1 + \frac{u}{\sigma}\right)^{-(\alpha+1)}, \quad u > 0. \tag{10}$$

Arnold (1987) showed that, necessarily, we then have

$$f(x,y) = [\lambda_{00} + \lambda_{10}x + \lambda_{01}y + \lambda_{11}xy]^{-(\alpha+1)}, \quad x > 0,\ y > 0. \tag{11}$$

In order to have a valid density we must have $\lambda_{00} > 0$, $\lambda_{01} > 0$, $\lambda_{10} > 0$, and $\lambda_{11} \geq 0(\lambda_{11} > 0$ if $0 < \alpha \leq 1)$. For such a density we find

$$P(X > x \mid Y = y) = \left(1 + \frac{\lambda_{10} + \lambda_{11}y}{\lambda_{00} + \lambda_{01}y}\, x\right)^{-\alpha}.$$

If we compute $\frac{d}{dy}P(X > x \mid Y = y)$ we find that its sign depends on the sign of $\lambda_{00}\lambda_{11} - \lambda_{10}\lambda_{01}$ and so X is either stochastically increasing or decreasing in Y. Consequently, when $\rho(X,Y)$ exists,

$$\operatorname{sign} \rho(X,Y) = \operatorname{sign}(\lambda_{00}\lambda_{11} - \lambda_{01}\lambda_{10}).$$

(b) **Bivariate uniform conditionals.** Suppose $f(x,y)$ is such that $f(x \mid y)$ is uniform $(0, \phi(y))$ for each $y \in (0,1)$ and $f(y \mid x)$ is uniform $(0, \psi(x))$ for each $x \in (0,1)$. By writing $f(x,y)$ as a product of marginal and conditional densities in both possible manners, we conclude that $\phi(y) = \psi^{-1}(y)$ and that ψ is an arbitrary nonincreasing function defined on (0,1) with $\psi(0) = 1$ and $\psi(1) = 0$. For such a function ψ, the joint density assumes the form

$$f(x,y) = I(0 < x < 1, 0 < y < \psi(x))/c(\psi), \tag{12}$$

where

$$c(\psi) = \int_0^1 \psi(x)dx.$$

For such a bivariate distribution we have $P(Y > y \mid X = x)$ nonincreasing in x for every y, so that (X,Y) are negatively correlated.

4. Multivariate Conditionally Specified Distributions.

There is in principle no difficulty in extending the discussion of Sections 2 and 3 to higher dimensions. Arnold and Strauss (1987, 1988) provide several examples. Two such examples will be described here.

(1) Multivariate exponential conditionals distributions. Here we assume X is a k-dimensional random variable with nonnegative coordinate random variables.

For $i = 1, 2, \ldots, k, X^{(i)}$ denotes the vector X with the ith coordinate deleted. If for each i and for each $x^{(i)} \in \mathbf{R}_+^{(k-1)}$, the conditional distribution of X_i given $X^{(i)} = x^{(i)}$ is exponential $\mu_i(x^{(i)}))$ for some functions $\mu_i(\cdot)$, it may be verified that the joint density of X must be of the form

$$f(\underline{x}) = \exp\left[-\sum_{s\in\xi_k} \lambda_s \left(\prod_{i=1}^{k} x_i^{s_i}\right)\right], \quad \underline{x} > \underline{0} \tag{13}$$

where ξ_k is the set of all vectors of 0's and 1's of dimension k. The parameters $\lambda_s(s \not\equiv \underline{0})$ are nonnegative, those for which $\sum_{i=1}^k s_i = 1$ are positive, and $\lambda_{\underline{0}}$ is such that the density integrates to 1. It is evident that X_1 is stochastically decreasing in $(X_2, \ldots, X_k)$ (more generally X_i is stochastically decreasing in $X^{(i)}$). Negative correlations are thus assured.

(2) Multivariate uniform conditionals distibution. Suppose that $X = (X_1, \ldots, X_k)$ is such that each X_i has the interval (0,1) as its set of possible values. Assume that for each i, the conditional distribution of X_i given $X^{(i)} = x^{(i)}$ is uniform $(0, \psi_i(x^{(i)}))$. In such a case the joint density must be uniform over a region in the k-dimensional positive orthant under some surface passing through the k points

$$(1, 0, \ldots, 0), (0, 1, 0, \ldots), \ldots, (0, \ldots, 0, 1).$$

Negative dependence and negative correlation are consequently present.

REFERENCES

ARNOLD, B.C. (1987). Bivariate distributions with Pareto conditionals. *Statist. and Prob. Letters* **5** 263–266.

ARNOLD, B.C. and STRAUSS, D. (1987). Bivariate distributions with conditionals in prescribed exponential families. Technical Report 151, University of California, Riverside, CA.

ARNOLD, B.C. and STRAUSS, D. (1988). Bivariate distributions with exponential conditionals. *J. Amer. Statist. Assoc.* **83** 522–527.

BARLOW, R.E. and PROSCHAN, F. (1981). *Statistical Theory of Reliability and Life Testing: Probability Models.* To Begin With, Silver Spring.

CASTILLO, E. and GALAMBOS, J. (1987a). Bivariate distributions with normal conditionals. *Proceedings of the International Association of Science and Technology for Development*, International Symposium on Simulation, Modeling and Development, Cairo, Egypt, Acta Press, Anaheim, CA, 59–62.

CASTILLO, E. and GALAMBOS, J. (1987b). Bivariate distributions with Weibull conditionals (to appear).

DEPARTMENT OF STATISTICS
UNIVERSITY OF CALIFORNIA
RIVERSIDE, CA 92521

CONDITIONAL INDEPENDENCE AND PROBABILISTIC INFLUENCE DIAGRAMS

BY RICHARD E. BARLOW[1] AND CARLOS ALBERTO DE BRAGANCA PEREIRA[1]

University of California, Berkeley, and Universidade de Sao Paulo, Brazil

A graphical approach to conditional independence is discussed. Some well known results concerning conditional independence are proved using simple influence diagram arguments. This material is, in part, from a book in progress tentatively titled *Applied Bayesian Statistics*, by the present authors.

1. Introduction. Influence diagrams with decision nodes were invented in 1976 by Miller et al. [cf. Howard and Matheson (1984)]. Shachter (1986) further developed methods for analyzing influence diagrams. S. Wright (1934) used diagrams to aid in understanding his "method of path coefficients." Although his diagrams pictorially resemble Gaussian influence diagrams [cf. Shachter and Kenley (1988)], they are not based on the Bayesian paradigm. They are not in any sense influence diagrams. I.J. Good (1961) invented "causal nets" that resemble influence diagrams. He used them to illustrate his ideas of causality and conditional independence. In this respect they are similar to influence diagrams. However he did not develop a comparable methodology for analyzing the diagrams. His diagrams are not influence diagrams as we define them below.

Influence diagrams are useful for **modeling** statistical problems. Construction of the diagram is helpful in understanding the problem and communicating the interdependencies to others. In the process of constructing the influence diagram, a representation of the joint distribution of random quantities related to the problem of interest is developed. Usually one does not start with the joint distribution but uses the influence diagram model to determine a useful representation of the joint distribution. In the case of decision influence diagrams, the diagram can be used to help solve the decision problem(s) of interest. Examples of the use of influence diagrams can be found in Barlow and Zhang (1987) and Lauritzen and Spiegelhalter (1988).

[1]This research was partially supported by the U. S. Army Research Office under Contract DAAG29-85-K-0208 with the University of California and by CNPq, Brasilia, Brazil under Contract No. 20771/85-MA.

AMS 1980 subject classifications. 60–02, 62A15, 62B15.

Key words and phrases. Conditional independence, conditional probability, graphs, influence diagrams.

1.1. Definitions and Basic Results. An influence diagram is, first of all, a **directed graph.** A **graph** is a set, V, of nodes or vertices together with a set, A, of arcs joining the nodes. It is said to be directed if the arcs are arrows (directed arcs). Let $V = \{v_1, \ldots, v_n\}$ and let A be a set of ordered pairs of elements of V, representing the directed arcs. That is, if $[v_i, v_j] \in A$ for $1 \leq i, j \leq n$, then there is a directed arc (arrow) from vertex v_i to vertex v_j (the arrow is directed from v_i to v_j). If $[v_i, v_j] \in A$, v_i is said to be an **adjacent predecessor** of v_j and v_j is said to be an **adjacent successor** of v_i. The direction of arcs is meant to denote influence (or possible dependence).

Circles (or ovals) represent random quantities which may, at some time, be observed and consequently may change to data. Circle nodes are called **probabilistic nodes.** Attached to each circle node is a conditional probability (density) function. This function is a function of the state of the node and also of the states of the adjacent predecessor nodes.

A double circle (or double oval) denotes a **deterministic node** which is a node with only one possible state, given the states of the adjacent predecessor nodes; i.e., it denotes a deterministic function of all adjacent predecessors. Thus, to include the background information, H, in the graph, we would have to use a double circle around H.

The following concepts formalize the ideas used in drawing the diagrams of this paper.

DEFINITION 1.1. A directed graph is cyclic, and is called a **cyclic directed graph,** if there exists a sequence of ordered pairs in A such that the initial and terminal vertices are identical; i.e., there exists an integer $k \leq n$ and a sequence of k arcs of the following type:

$$[v_{i_1}, v_{i_2}], [v_{i_2}, v_{i_3}], \ldots, [v_{i_{k-1}}, v_{i_k}], [v_{i_k}, v_{i_1}].$$

DEFINITION 1.2. An **acyclic directed graph** is a directed graph that is not cyclic.

DEFINITION 1.3. A **root node** is a node with no adjacent predecessors. A **sink node** is a node with no adjacent successors. Note that any acyclic directed graph must have at least one root and one sink node.

DEFINITION 1.4. A **Probabilistic Influence Diagram** is an acyclic directed graph in which

i) nodes represent random quantities while directed arcs indicate possible dependence; and

ii) attached to each node is a conditional probability function (for the node) which depends on the states of adjacent predecessor nodes.

Given a directed acyclic graph together with node conditional probabilities (i.e., a probabilistic influence diagram), there exists a *unique joint probability function* corresponding to the random quantities represented by the nodes of the graph. This is because a directed graph is acyclic if and only if there exists a list ordering of the nodes such that any successor of a node x in the graph follows node x in the list as well. Consequently, following the list ordering and taking the product of all node conditional probabilities we obtain the joint probability of the random quantities corresponding to the nodes in the graph. Note that in a cyclic graph the product of the conditional probability functions attached to the nodes *would not determine* the joint probability function.

The following basic result shows that the absence of an arc connecting two nodes in the influence diagram denotes the judgment that the unknown quantities associated with these nodes are conditionally independent given the states of all *adjacent* predecessor nodes.

REMARK 1.5. Let x_i and x_j represent two nodes in a probabilistic influence diagram. If there is no arc connecting x_i and x_j, then x_i and x_j are conditionally independent given the states of the adjacent predecessor nodes; i.e.,

$$p(x_i, x_j \mid w_i, w_j, w_{ij}) = p(x_i \mid w_i, w_j, w_{ij})p(x_j \mid w_i, w_j, w_{ij})$$

where $w_i(w_j)$ denotes the set of adjacent predecessor nodes to only $x_i(x_j)$ while w_{ij} denotes the set of adjacent predecessor nodes to both x_i and x_j.

REMARK 1.6. In a probabilistic influence diagram, if two nodes, x_i and x_j, are root nodes then they are independent.

EXAMPLE 1.7. (Forensic Science). A robbery has been committed and a suspect, a young man, is on trial. In the course of the robbery, a window pane was broken. The robber had apparently cut himself and a blood stain was left at the scene of the crime. Let x represent the blood type of the suspect, y the blood type of the blood stain found at the scene of the crime, and θ the quantity of interest, "the state of culpability" (guilt or innocence) of the suspect. Formally, and before using the actual values of the observable quantities, we have:

$$x = \begin{cases} 1 & \text{if the suspect's blood type is } A, \\ 0 & \text{otherwise.} \end{cases} \quad y = \begin{cases} 1 & \text{if the blood stain type is } A, \\ 0 & \text{otherwise.} \end{cases} \quad \theta = \begin{cases} 1 & \text{if the suspect is guilty,} \\ 0 & \text{otherwise.} \end{cases}$$

The following diagram is a probability model constructed for this case. Note that the actual values of x and y that are known at the time of the analysis are not yet used. In fact, the diagram describes the dependence relations among the quantities and the conditional probabilities to be used.

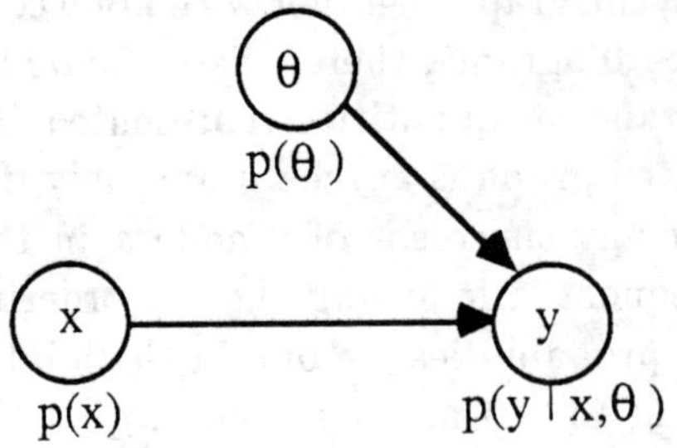

Figure 1.1. Influence Diagram for a Problem in Forensic Science

If p represents the proportion of people in the population with blood type A and if, for a jury member that happens to be interested in probability, q represents his probability that the suspect is guilty before the juror has learned about the blood evidence, then a reasonable probability model is:

$$p(\theta) = \begin{cases} q & \text{if } \theta = 1 \\ 1-q & \text{if } \theta = 0. \end{cases} \quad p(x) = \begin{cases} p & \text{if } x = 1 \\ 1-p & \text{if } x = 0. \end{cases}$$

$$p(y \mid x, \theta) = \begin{cases} p & \text{if } \theta \neq y = 1 \\ 1-p & \text{if } \theta = y = 0 \\ 1 & \text{if } \theta = 1 \text{ and} \\ & y = x \\ 0 & \text{otherwise.} \end{cases}$$

The objective of the jury member is to obtain the probability of guilt ($\theta = 1$) after observing the evidence ($x = y = 1$) namely that the blood type of the suspect is the same as that of the stain. That is, the jury member needs to obtain $p(\theta \mid x, y)$ evaluated at $\{\theta = x = y = 1\}$.

2. Probabilistic Influence Diagram Operations. The Bayesian approach to statistics is based on probability judgments and as such follows the laws of probability. You are said to be **coherent** if i) you use probability to measure your uncertainty about quantities of interest and ii) you do not violate the laws of probability when stating your measurements (probabilities). Probabilistic influence diagrams (and influence diagrams in general) are helpful in assuring coherence. Clearly, from coherence, any operation to be performed in a probabilistic influence diagram must not violate the laws of probability. The three basic probabilistic influence diagram operations that we discuss next are based on the addition and product laws. These operations are: 1) Splitting Nodes, 2) Merging Nodes, and 3) Arc Reversal.

2.1. Splitting Nodes. In general a node in a probabilistic influence diagram can denote a vector random quantity. It is always possible to split such a node into other nodes corresponding to the elements of the vector random quantity. To illustrate ideas, suppose that a node corresponds to a vector of two random quantities, x and y, with joint probability function $p(x, y)$. From the product law we know that

$$p(x, y) = p(x)p(y \mid x) = p(x \mid y)p(y).$$

Hence, Figure 2.1 presents the 3 possible probabilistic influence diagrams that can be used in this case showing the two ways of splitting node (x, y).

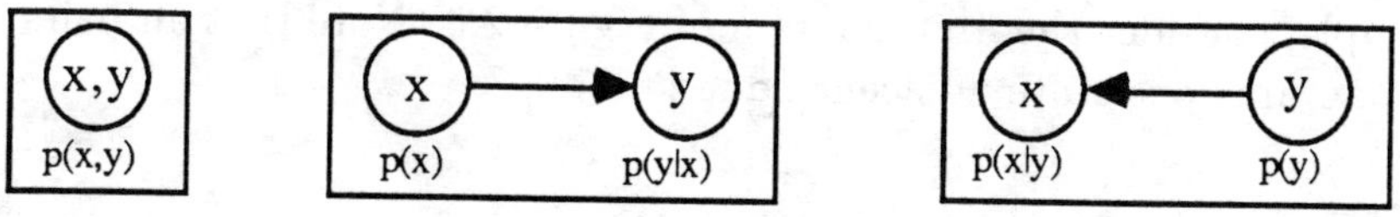

Figure 2.1.
Probabilistic Influence Diagrams for Two Random Quantities

The following property is also a direct consequence of the laws of probability and it is of special interest for statistical applications.

PROPERTY 2.1. Let x be a random quantity represented by a node of a probabilistic influence diagram and let $f(x)$ be a (deterministic) function of x. Suppose we connect to the original diagram a deterministic node representing $f(x)$ using a directed arc from x to $f(x)$. Then, the joint probability distributions for the two diagrams are equal. (See Figure 2.2 for illustration.)

PROOF. Let w and y represent the sets of random quantities that precede and succeed x, respectively, in a list ordering. Note that $p(f(x) \mid w, x) = p(f(x) \mid x) = 1$ and consequently from the product law $p(x, f(x) \mid w) = p(x \mid w)$. That is, node x may be replaced by node $(x, f(x))$ without changing the joint probability of the graph nodes. Using the splitting node operation in node $(x, f(x))$ with x preceding $f(x)$, we obtain the original graph with the additional deterministic node $f(x)$ and a directed arc from x to $f(x)$. Note also that no other arc is necessary since $f(x)$ is determined by x and $p(y \mid w, f(x), x) = p(y \mid w, x)$. ||

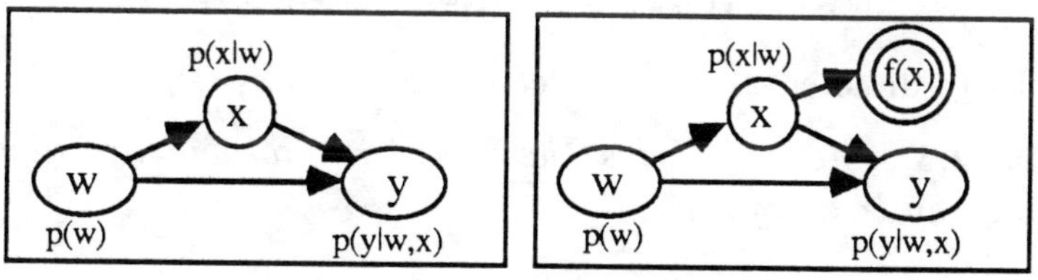

Figure 2.2. Addition of a Deterministic Node

2.2. Merging Nodes. The second probabilistic influence diagram operation is the merging of nodes. Consider first a probabilistic influence diagram with two nodes, x and y, with a directed arc from x to y. The product law states that $p(x,y) = p(x)p(y \mid x)$. Hence, without changing the joint probability of x and y, the original diagram can be replaced by a single node diagram representing the vector (x,y). The first two diagrams of Figure 2.1 in the reverse order illustrate this operation. In general, two nodes, x and y, can be replaced by a single node, representing the vector (x,y), if there is a list ordering such that x is an immediate predecessor or successor of y.

It is not always possible to merge two adjacent nodes in a probabilistic influence diagram. Note that two adjacent nodes may not be neighbors in any list ordering. For example, consider the first diagram of Figure 2.3. Note that all pairs of nodes in this diagram constitute adjacent nodes.

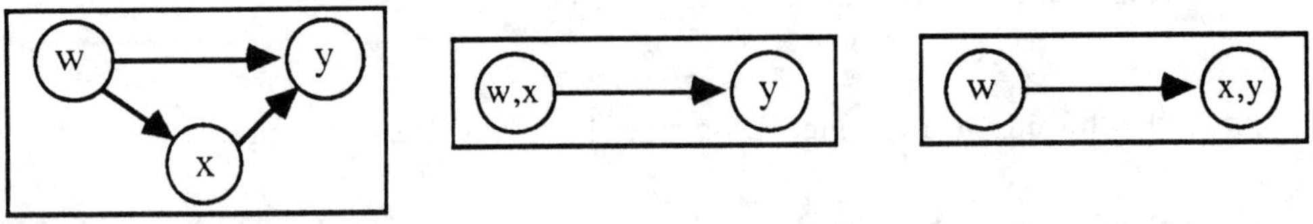

Figure 2.3. Diagram with Adjacent Nodes, w and y, Not Allowed to Be Merged

However, w and y cannot be merged into a node representing (w,y). Clearly the only list ordering here is $w < x < y$ and w and y are not immediate neighbors in this ordering. The problem here is that to merge w and y we would need an arc from (w,y) to x and another from x to (w,y). The reason for this is the existence of arcs $[w,x]$ and $[x,y]$ in the original graph. If we were to have arcs in both directions between (w,y) and x, we would not obtain, in general, the joint probability function from the diagram since $p(w,x,y) \neq p(x \mid w,y)p(w,y \mid x)$. Also it can be seen from the first diagram of Figure 2.3 that there exist two paths from w to y. This is the graphical way to see that w and y cannot be merged into a single node. To construct a graphical technique to check if two nodes can be merged, we need the following definition and theorem.

DEFINITION 2.2. A **directed path** from node x_i to node x_j is a chain of ordered pairs

$$([x_i, x_{k_1}], [x_{k_1}, x_{k_2}], \ldots, [x_{k_{t-1}}, x_{k_t}], [x_{k_t}, x_j])$$

corresponding to directed arcs which lead from x_i to x_j.

THEOREM 2.3. *(Merging Nodes Theorem) In a probabilistic influence diagram, nodes x and y can be merged if either*

1) the only directed path between x and y is a directed arc connecting x and y; or

2) there is no directed path connecting x and y.

PROOF. To be definite, suppose that x precedes y in an associated list ordering corresponding to a probabilistic influence diagram. Let $w_x(w_y)$ be the set of adjacent predecessors of $x(y)$ but not of $y(x)$ and let w_{xy} be the set of node which are adjacent predecessors of both x and y. Since there is no directed path from x to y except, possibly, for a directed arc from x to y, we may add arcs from each node in w_x to y and from each node in w_y to x without creating any cycles. This is possible because directed arcs indicate *possible dependence* not necessarily strict dependence. We have of course lost some graph information as a result of these arc additions.

In the associated list ordering of nodes for our modified diagram, the family of nodes $\{w_x, w_y, w_{xy}\}$ precede both x and y. Since there is no other directed path from x to y other than possibly a directed arc from x to y, there exists an associated list ordering of nodes for which x is an immediate predecessor of y in this list ordering. The product

$$p(x \mid w_x, w_y, w_{xy})p(y \mid x, w_x, w_y, w_{xy})$$

must appear in the representation for the joint probability function for all probabilistic nodes based on the list ordering. Since

$$p(y, x \mid w_x, w_y, w_{xy}) = p(x \mid w_x, w_y, w_{xy})p(y \mid x, w_x, w_y, w_{xy})$$

by the product law, we can merge x and y.

Finally, suppose that there is a directed path from x to y other than a directed arc from x to y. In this case it is not difficult to see that merging x and y would create a cycle which is not allowed. ||

The above result is related to arc reversal, an important operation discussed next.

2.3. Reversing Arcs. The probabilistic influence diagram operation corresponding to Bayes' formula is that of arc reversal. Consider the diagram on the left in Figure 2.4. Using the merging nodes operation we obtain the single node diagram in the center where the probability function of the node (x, y) is obtained from the first diagram as $p(x, y) = p(x)p(y \mid x)$. Using the splitting nodes operation we can obtain the diagram on the right of Figure 2.4. Note that to obtain the corresponding probability functions we use

1) the theorem of total probability for $p(y) = \Sigma_x p(y \mid x)p(x)$, where Σ_x is the sum (or integral) over all possible values of x, and

2) the multiplication law for $p(x \mid y) = p(x, y)/p(y)$ since $p(y)p(x \mid y) = p(x, y)$.

By substituting the appropriate expression in $p(x \mid y)$ we obtain Bayes' formula. That is,

$$p(x \mid y) = \{p(x)p(y \mid x)\}/\{\Sigma_x p(x)p(y \mid x)\}.$$

Hence, by using the theorem of total probability and Bayes' formula when performing an arc reversal operation, we can go directly from the left diagram to the right one in Figure 2.4 without having to consider the one in the center.

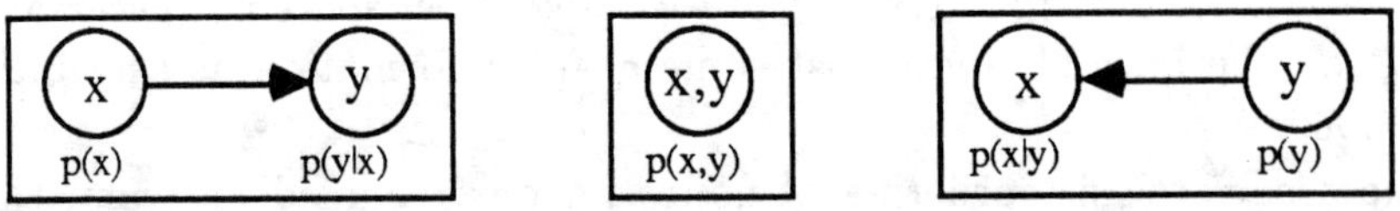

Figure 2.4. Reversing Arc Operation in a Two Node Probabilistic Influence Diagram

Although the diagrams are different they have the same joint probability function for node random quantities. This fact is formalized in the following definition.

DEFINITION 2.4. Two probabilistic influence diagrams are said to be **equivalent in probability** if they have the same joint probability function for node random quantities.

Consider the diagram of Figure 2.5 where w_x, w_y, and $w_{x,y}$ are sets of adjacent predecessors of x and (or) y as indicated by the figure. If arc $[x, y]$ is the *only* directed path from node x to node y, we may add arcs $[w_x, y]$ and $[w_y, x]$ to the diagram without introducing any cycles. (See left diagram of Figure 2.6.) Remember that a directed arc only indicates *possible* dependence.

The following result introduces the conditions under which arc reversal operations can be performed.

THEOREM 2.5. *(Reversing Arcs Theorem) Suppose that arc $[x, y]$ connects nodes x and y in a probabilistic influence diagram. $[x, y]$ can be reversed to $[y, x]$, without changing the joint probability function of the diagram if*

1) there is no other dirccted path from x to y,

2) all the adjacent predecessors of $x(y)$, in the original diagram, become also adjacent predecessors of $y(x)$, in the modified diagram, and

3) the conditional probability functions attached to nodes x and y are also modified in accord with the laws of probability.

PROOF. Let $w_x(w_y)$ be the set of adjacent predecessors of $x(y)$ but not of $y(x)$ and $w_{x,y}$ be the set of adjacent predecessors of both x and y. Since arcs represent possible dependence, we can add arcs to the diagram in order to make the set $(w_x, w_x, w_{x,y})$ an adjacent predecessor of both x and y. Since there is no other directed path connecting x and y, there is a list ordering such that x is an immediate predecessor of y in the list. Note also that the elements of the set $(w_x, w_x, w_{x,y})$ are all predecessors of both x and y in the list ordering. To obtain the joint probability function corresponding to the first diagram we consider the product, following the list ordering, of all node conditional probability functions. As a factor of this product we have

$$p(x \mid w_x, w_{x,y})p(y \mid x, w_y, w_{x,y}) = p(x \mid w_x, w_y, w_{x,y})p(y \mid x, w_x, w_y, w_{x,y}) = \\ p(x, y \mid w_x, w_y, w_{x,y}) = p(y \mid w_x, w_y, w_{x,y})p(x \mid y, w_x, w_y, w_{x,y}).$$

The first equality is due to the fact that x and w_y are conditionally independent given $(w_x, w_{x,y})$ and y and w_x are conditionally independent given $(w_y, w_{x,y})$. [See Figure 2.5.] The other two equalities follow from the product law.

Replacing $p(x \mid w_x, w_{x,y})p(y \mid x, w_y, w_{x,y})$ in the product of the conditional probability functions for the original diagram by $p(y \mid w_x, w_y, w_{x,y})p(x \mid y, w_x, w_y, w_{x,y})$ we obtain the product of the conditional probability functions for the second diagram. This proves that the joint probability functions of the two diagrams are equal. Finally, we notice that if there were another directed path from x to y, we would create a cycle by reversing arc $[x, y]$, which is not allowed. $\|$

In general, reversing an arc corresponds to applying Bayes' formula and the theorem of total probability. However, it may also involve the addition of arcs and such arcs, in some cases represent only pseudo dependencies. In this sense, some relevant information may have been lost after arc reversal.

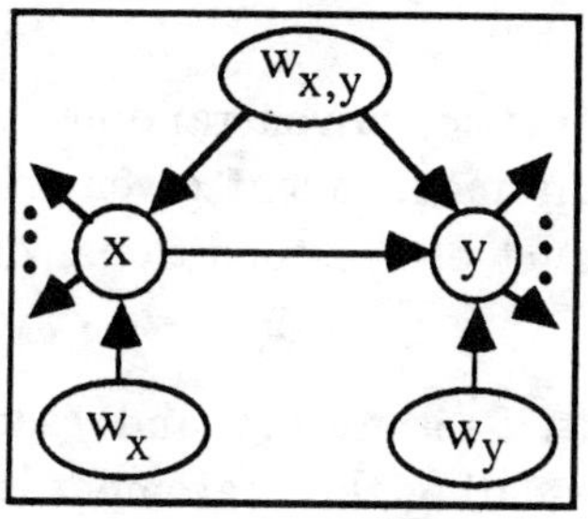

Figure 2.5.

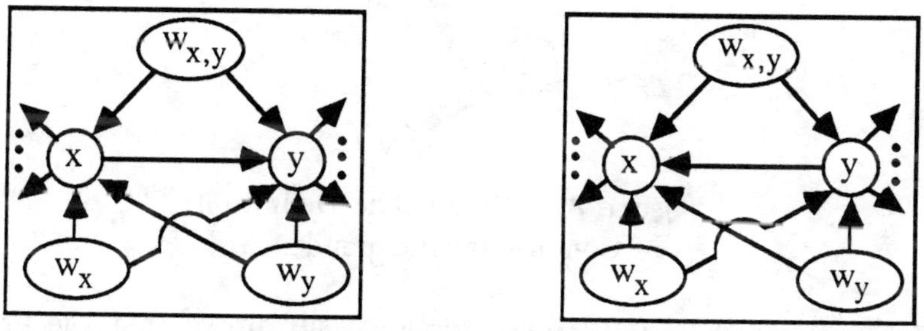

Figure 2.6. Equivalent Probabilistic Influence Diagrams. Probability Nodes in the Right Diagram are Obtained From the Left Diagram by Using Bayes' Formula and the Theorem of Total Probability

3. Conditional Independence. The objective of this section is to study the concept of conditional independence and introduce its basic properties. We believe that the simplest and most intuitive way that this study can be performed is by using all the visual force of the probabilistic diagram.

We now introduce the two most common definitions of conditional independence.

DEFINITION 3.1. (Intuitive) Given random quantities x, y, and z, we say that y is conditionally independent of x given z if the conditional distribution of y given (x, z) is equal to the conditional distribution of y given z.

The interpretation of this concept is that, if z is given, no additional information about y can be extracted from x. The influence diagram representing this statement is presented in Figure 3.1.

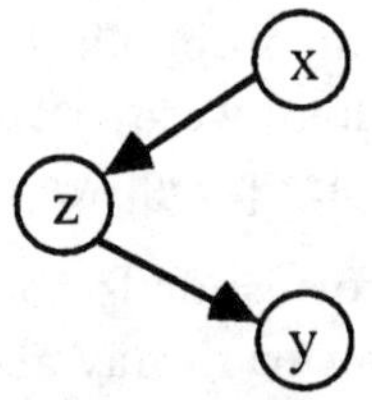

Figure 3.1. Intuitive Definition of Conditional Independence

DEFINITION 3.2. (Symmetric) Given random quantities x, y, and z, we say that x and y are conditionally independent given z if the conditional distribution of (x, y) given z is the product of the conditional distributions of x given z and that of y given z.

The interpretation is that, if z is given, x and y share no additional information. The influence diagram representing this statement is displayed in Figure 3.2.

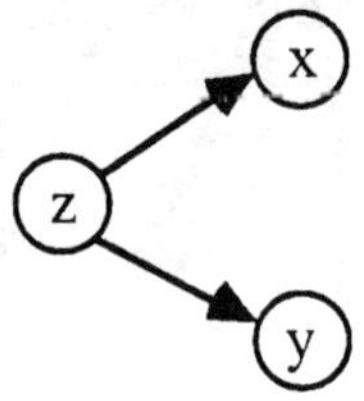

Figure 3.2. Symmetric Definition of Conditional Independence

Using the arc reversal operation, we can easily prove that the probabilistic influence diagrams in Figures 3.1 and 3.2 are equivalent. Thus, Definitions 3.1 and 3.2 are equivalent, which means that in a specific problem we can use either one. To represent the conditional independence described by both Figures 3.1 and 3.2

we can write either $x \perp\!\!\!\perp y \mid z$ or $y \perp\!\!\!\perp x \mid z$. This is a very general notation since x, y, and z are general random quantities (scalars, vectors, events, etc.). If in place of $\perp\!\!\!\perp$ we use $\not\!\perp\!\!\!\perp$, then x and y are said to be strictly dependent given z. We obtain independence (dependence) and write $x \perp\!\!\!\perp y$ ($x \not\!\perp\!\!\!\perp y$) if z is an event which occurs with probability one. It is important to notice that the symbol $\perp\!\!\!\perp$ corresponds to the absence of an arc in a probabilistic influence diagram. However, the existence of an arc only indicates possible dependence. Although $\not\!\perp\!\!\!\perp$ is the negation of $\perp\!\!\!\perp$, the "absence of an arc" is included in the "presence of an arc."

The following proposition introduces the essence of the DROP/ADD principles for conditional independence which are briefly discussed in the sequel.

PROPOSITION 3.3. *If $x \perp\!\!\!\perp y \mid z$ then, for every $f = f(x)$, we have:*

(i) $f \perp\!\!\!\perp y \mid z$; and

(ii) $x \perp\!\!\!\perp y \mid (z, f)$.

The proof of this property is the sequence of diagrams of Figure 3.3. First note that (by Property 2.1) to obtain the second diagram from the first we can connect to x a deterministic node f using arc $[x, f]$ without changing the joint probability function. Consequently, by reversing arc $[x, f]$ we obtain the third diagram. To obtain the last diagram from the third we use the merging nodes operation. Relations i) and ii) of Proposition 3.3 are represented by the second and the third diagrams of Figure 3.3.

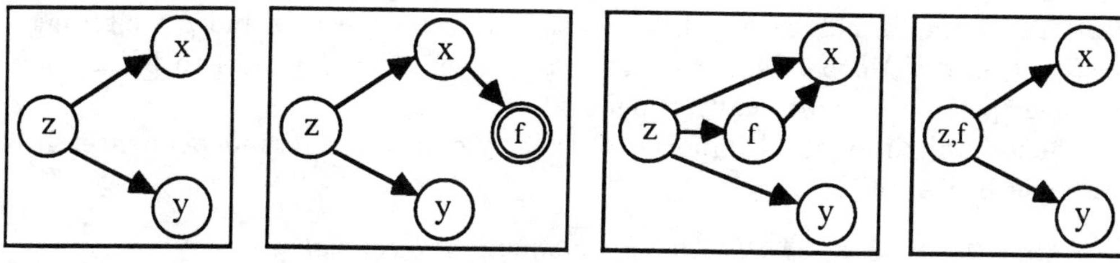

Figure 3.3. Proof of Proposition 3.3

As direct consequences of Proposition 3.3 we have:

$C1-$ If $g = g(z)$ then $x \perp\!\!\!\perp y \mid z$ if and only if $x \perp\!\!\!\perp (y, g) \mid z$.

$C2-$ Let $f = f(x, z)$ and $g = g(y, z)$. If $x \perp\!\!\!\perp y \mid z$ then, $f \perp\!\!\!\perp g \mid z$ and $x \perp\!\!\!\perp y \mid (z, f, g)$.

The concept of conditional independence gives rise to many questions. Among them are the ones involving the DROP/ADD principles that we describe next. Suppose that x, y, z, w, f, and g are random objects such that $x \perp\!\!\!\perp y \mid z$, $f = f(x)$ and $g = g(z)$. What can be said about the relation $\perp\!\!\!\perp$ if f is substituted for x, g for z, (y, w) for y or (z, w) for z? In other words, can x, y, and z be reduced or enlarged without destroying the $\perp\!\!\!\perp$ relation? In general, the answer is no. However, for special kinds of reductions or enlargements the conditional independence relation is preserved.

First we present two simple examples to show that arbitrary enlargements of x, y, or z may destroy the $\perp\!\!\!\perp$ relation. The forensic science example shows that $\theta \perp\!\!\!\perp y$ but $\theta \not\perp\!\!\!\perp (x, y)$ or, in the present notation, considering z a sure event and $w = x$, $\theta \perp\!\!\!\perp y \mid z$ but $\theta \not\perp\!\!\!\perp (y, w) \mid z$. Consider now that w_1 and w_2 are two independent standard normal random variables; i.e., $w_1 \sim w_2 \sim N(0, 1)$, and $w_1 \perp\!\!\!\perp w_2$. If $x = w_1 - w_2$ and $y = w_1 + w_2$, then $x \perp\!\!\!\perp y$ but certainly $x \not\perp\!\!\!\perp y \mid w_2$. Note that if z is a constant and $w = w_1$, we conclude that $x \perp\!\!\!\perp y \mid z$ but $x \not\perp\!\!\!\perp y \mid (z, w)$.

Secondly, we present an example to show that an arbitrary reduction of z, the conditioning quantity, can destroy the $\perp\!\!\!\perp$ relation. Let w_1, w_2, and w be three mutually independent standard normal random quantities; i.e., $w_1 \perp\!\!\!\perp (w_2, w)$, $(w_1, w_2) \perp\!\!\!\perp w$, $w_2 \perp\!\!\!\perp (w_1, w)$, $w_1, \perp\!\!\!\perp w_2$, $w_1 \perp\!\!\!\perp w$, $w_2 \perp\!\!\!\perp w$, and $w_1 \sim w_2 \sim w \sim N(0, 1)$. Define $x = w_1 - w_2 + w$ and $y = w_1 + w_2 + w$, and note that $x \perp\!\!\!\perp y \mid w$ but $x \not\perp\!\!\!\perp y$. As before, if z is a constant we can conclude that $x \perp\!\!\!\perp y \mid (z, w)$ but $x \not\perp\!\!\!\perp y \mid z$.

The destruction of the $\perp\!\!\!\perp$ relation by reducing or enlarging its arguments is known as Simpson's paradox (for more details, see Lindley and Novick, 1981). The paradox, however, is much stronger since highly positively correlated random variables could be highly negatively correlated after some Drop/Add operations. For instance, let z and w be two independent normal random variables with zero means. Define $x = z + w$ and $y = z - w$ and note that the correlation between x and y is given by correlation$(x, y) = (l - r)(l + r)^{-1}$ where r is equal to the variance of w divided by the variance of z. Also, if z is given it is clear that the conditional correlation is -1. In order to make cor(x, y) close to 1 we can consider r arbitrarily small. This shows that we can have cases where x and y are strongly positive (negative) dependent but, when z is given, x and y turn to be strongly negative (positive) conditionally dependent.

The following is another important property of conditional independence. It is presented in Dawid (1979).

PROPOSITION 3.4. *The following statements are equivalent:*

(i) $x \perp\!\!\!\perp y \mid z$ *and* $x \perp\!\!\!\perp w \mid (y, z)$*;*

(ii) $x \perp\!\!\!\perp (w, y) \mid z$*; and*

(iii) $x \perp\!\!\!\perp w \mid z$ *and* $x \perp\!\!\!\perp y \mid (w, z)$.

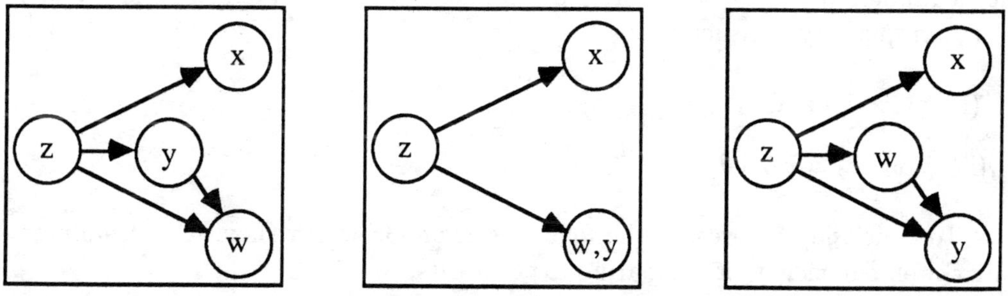

Figure 3.4. Proof of Proposition 3.4

Figure 3.4 is the proof of Proposition 3.4. Again, only the basic probabilistic influence diagrams operations are used. The second graph is obtained from the first by merging nodes w and y. The third graph is obtained from the second by splitting node (w, y) and the first is obtained from the third by reversing arc $[w, y]$.

The above simple properties are very useful in some statistical applications and they are related to the concept of sufficient statistic. In the context of comparisons of experiments a very general concept of sufficiency was introduced by Blackwell (1953). We next discuss Blackwell's concept of sufficiency using probabilistic influence diagrams.

3.1. Blackwell Sufficiency. Suppose that we can perform either one of two experiments to learn about a random quantity θ. In the first experiment, we observe x, knowing $p(x \mid \theta)$. In the second experiment, we observe y, knowing $p(y \mid \theta)$. If, furthermore, there exists a random quantity x' such that $\theta \perp\!\!\!\perp x' \mid y$ and $p(x' \mid \theta) = p(x \mid \theta)$, then we say that y is **Blackwell sufficient** for x relative to θ.

In terms of probabilistic influence diagrams, we construct two diagrams, the first with nodes θ and x connected by arc $[\theta, x]$ and the second with three nodes θ, y, and x' connected by arcs $[\theta, y]$ and $[y, x']$. If in the second diagram, after eliminating node y, we obtain a diagram having only two nodes, θ and x', equivalent to the first diagram, then we have Blackwell's concept of sufficiency. See Figure 3.5. In this sense x' is a "garbling" of y. If we cannot observe both x and y, it is better to observe y and use $p(y \mid \theta)$ to make inferences about θ.

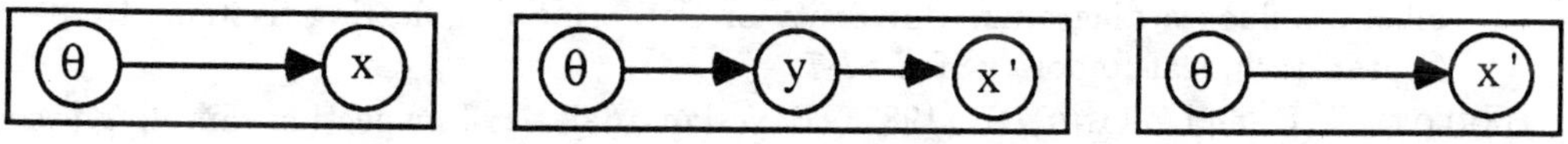

Figure 3.5. Blackwell Sufficiency When the Right and Left Diagrams are Equivalent

DEFINITION 3.5. (Blackwell Sufficiency). A random quantity, y, is sufficient for a random quantity, x, relative to a random quantity, θ, if there exists another random quantity, x', such that

(i) $\theta \perp\!\!\!\perp x' \mid y$ and

(ii) $p(x' \mid \theta) = p(x \mid \theta)$.

To conclude, we present the following example which shows the usefulness of Blackwell sufficiency in comparing experiments.

EXAMPLE 3.6. Let x and y be two Bernoulli quantities such that, given a parameter $\theta, \Pr\{x = 1 \mid \theta\} = \theta/2$ and $\Pr\{y = 1 \mid \theta\} = \theta$. Suppose that we want to learn more about the parameter θ, but we can only observe one of the random quantities x or y, but not both. The question of which one to observe involves the cost of observation and other considerations. For the moment let us suppose they have the same cost. If we can prove that y is Blackwell sufficient for x relative to θ, we must prefer y since it is at least as good as x for learning about θ. We now prove that y is in fact Blackwell sufficient for x.

Suppose that we toss a fair coin and record $r = 1$ if we obtain a tail and $r = 0$ if we obtain a head. Define now the random quantity $x' = yr$. Figure 3.6 shows, on the left, a diagram relating θ, y, r, and x'. After eliminating node r we obtain the diagram in the center of Figure 3.6. The right diagram of Figure 3.6 is obtained after the elimination of node y. This last diagram is equivalent to the probabilistic diagram relating x and θ. Hence, y is Blackwell sufficient for x relative to θ.

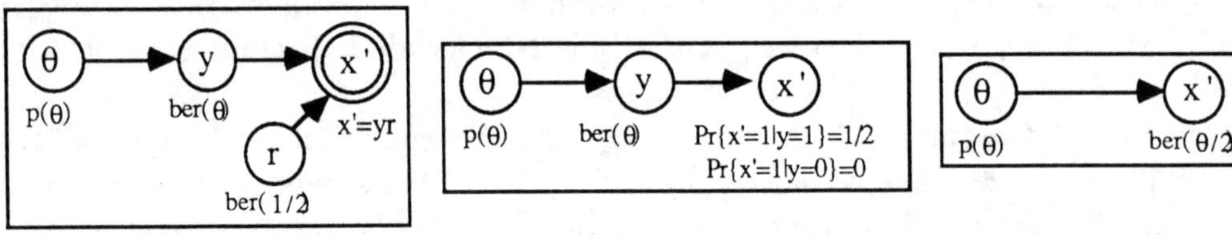

Figure 3.6. Proof of Blackwell Sufficiency

REFERENCES

BARLOW, R.E. and PEREIRA, C.A. DE B. (1987). The Bayesian operation and probabilistic influence diagrams. University of California Engineering Systems Research Center Technical Report ESRC 87-7.

BARLOW, R.E. and ZHANG, X. (1987). Bayesian analysis of inspection sampling procedures discussed by Deming. *J. of Stat. Planning and Inference* **16** 285–296.

BASU, D. (1975). Statistical information and likelihood (with discussion). *Sankhyā* (Ser. A) **37** 1–71.

BASU, D. and PEREIRA, C.A. DE B. (1983). Conditional independence in statistics. *Sankhyā* (Ser. A) **45** 324–337.

BLACKWELL, D. (1953). Equivalent comparisons of experiments. *Ann. Math. Statist.* **24** 265–272.

DAWID, A.P. (1979). Conditional independence in statistical theory. *J.R. Statist. Soc.* B **41** 1–31.

GOOD, I.J. (1961). A causal calculus I and II. *Br. J. Phil. Sci.* **XI**, 305–318 and **XII**, 43–51.

HOWARD, R.A. and MATHESON, J.E. (1984). Influence diagrams. In *Readings in the Principles and Applications of Decisions Analysis.* Two volumes. Howard and Matheson, eds., Strategic Decision Group, Menlo Park, CA.

LAURITZEN, S.L. and SPIEGELHALTER, D.J. (1988). Local computations with probabilities on graphical structures and their application to expert systems (with discussion). *J. Roy. Statist. Soc.* B **50** (to appear).

LINDLEY, D.V. and NOVICK, M.R. (1981). The role of exchangeability in inference. *Ann. Statist.* **9** 45–58.

SHACHTER, R. (1986). Evaluating influence diagrams. *Oper. Res.* **34** 871–882.

SHACHTER, R. and KENLEY, C.R. (1988). Gaussian influence diagrams. *Man. Sci.* To appear.

WRIGHT, S. (1934). The method of path coefficients. *Ann. Math. Statist.* **5** 161–215.

4177 ETCHEVERRY HALL
UNIVERSITY OF CALIFORNIA, BERKELEY
BERKELEY, CA 94720

INSTITUTO DE MATHEMATICA
E ESTATISTICA
UNIVERSIDADE DE SAO PAULO
CAIXA POSTAL 20570 (AG. IGUATEMI)
CEP 05508 SAO PAULO, S.P.
BRASIL

A SURVEY OF SOME INFERENCE PROBLEMS FOR DEPENDENT SYSTEMS

BY ASIT P. BASU[1]

University of Missouri–Columbia

In this paper a survey of some inference problems relating to dependent systems are considered. A number of multivariate models, both parametric and nonparametric, are given and related tests of dependence and tests of exponentiality are considered.

1. Introduction. The univariate exponential distribution with density function

$$f(x) = \lambda \exp(-\lambda x), \quad x \geq 0,\ \lambda > 0$$

and distribution function

$$F(x) = 1 - \exp(-\lambda x), \quad x \geq 0$$

is well known as the most important model in reliability theory. Here the survival function is given by

$$\bar{F}(x) = 1 - F(x) = \exp(-\lambda x), \tag{1}$$

and the failure rate function

$$r(x) = f(x)/\bar{F}(x)$$

for $F(x) < 1$, is λ, a constant. A random variable X with survival function (1) will be denoted by $X \sim e(\lambda)$.

The exponential distribution has a number of interesting properties. Some of these are given below.

P1. $F(x)$ is absolutely continuous.

[1]Research sponsored by the Air Force Office of Scientific Research, Air Force Systems command, USAF, under grant number AFOSR–87–0139. The U.S. Government is authorized to reproduce and distribute reprints for governmental purposes notwithstanding any copyright notation thereon.

AMS 1980 subject classifications. Primary 62N05; secondary 62E15, 62G10.

Key words and phrases. Reliability, exponential distribution, tests of independence, nonparametric tests, tests for aging.

The author would like to thank the editors and a referee for their constructive comments.

P2. $F(x)$ possesses the loss (or lack) of memory property (LMP). That is $\bar{F}(x+t) = \bar{F}(x)\bar{F}(t)$ for all $x, t \geq 0$.

P3. The failure rate $r(x)$ is a constant.

P4. Let X_1 and X_2 be independently distributed with $X_i \sim e(\lambda_i), i = 1, 2$. Then $\min(X_1, X_2) \sim e(\lambda_1 + \lambda_2)$.

Because of the usefulness of the univariate exponential distribution, it is natural to consider multivariate exponential distributions as models for multicomponent systems. However, unlike the normal distribution, there is no unique natural extension and a number of multivariate exponential distributions have been proposed. For a survey of useful multivariate exponential distributions, see Basu (1988). Section 2 describes a few of these multivariate distributions and related tests of independence are given in Section 3. Finally, some tests for multivariate exponentiality against distributions with specific types of multivariate failure rates (to be defined later) are given in Section 4.

2. Dependent Models.

A. *Multivariate exponential distributions.* A number of multivariate distributions have been derived where multivariate analogues of some of the properties for the univariate distribution have been considered. For simplicity, and without loss of much generality, we shall primarily consider bivariate exponential distributions. For example, Marshall and Olkin (1967) consider the following bivariate analogue of property P2:

$$(2) \qquad \bar{F}(x_1 + y_1, x_2 + y_2) = \bar{F}(x_1, x_2)\bar{F}(y_1, y_2), \text{ for all } x_1, x_2, y_1, y_2 \geq 0.$$

Here $\bar{F}(x, y) = P(X > x, Y > y)$, is the bivariate survival function. Marshall and Olkin showed that the only solution of (2) with univariate exponential marginals is

$$(3) \qquad \bar{F}(x, y) = \exp\{-\theta_1 x - \theta_2 y\},$$

for some $\theta_1, \theta_2 > 0$. That is, (2) implies that X and Y have independent exponential distributions.

By relaxing (2) we obtain the following definition of bivariate loss of memory property (BLMP).

DEFINITION 2.1. The random vector (X, Y) is said to have the BLMP if

$$(4) \qquad \bar{F}(x_1 + y, x_2 + y) = \bar{F}(x_1, x_2)\bar{F}(y, y), \text{ for } x_1, x_2, y \geq 0.$$

Assuming (4) and exponential marginals Marshall and Olkin (1967) obtain the following class of distributions, to be denoted by the BVE.

(5) $\bar{F}(x,y) = \exp\{-\lambda_1 x - \lambda_2 y - \lambda_{12}\max(x,y)\}, \quad \lambda_1, \lambda_2 > 0, \lambda_{12} \geq 0, x, y \geq 0.$

Note that $\bar{F}(x,y)$ is not absolutely continuous.

Similarly Basu (1971) considered an extension of property P_3 and defined the bivariate failure rate as

$$r(x,y) = f(x,y)/\bar{F}(x,y). \tag{6}$$

It is shown in Basu (1971) that, except for the case of independence, there does not exist any absolutely continuous bivariate exponential distribution with constant bivariate failure rate and marginal exponential distributions. Brindley and Thompson (1972) consider a more general definition of bivariate failure rate and obtained the BVE as the class of distributions with constant bivariate failure rate and marginal exponential distributions.

The BVE has a number of interesting properties. Let $(X,Y) \sim$ BVE with parameters λ_1, λ_2, and λ_{12}. Denote this by $(X,Y) \sim \text{BVE}(\lambda_1, \lambda_2, \lambda_{12})$. Then $X \sim e(\lambda_1 + \lambda_{12}), Y \sim e(\lambda_2 + \lambda_{12})$, the BLMP is satisfied, $\min(X,Y) \sim e(\lambda_1 + \lambda_2 + \lambda_{12})$. Note that $P(X = Y) \neq 0$, and the BVE is not absolutely continuous. The correlation coefficient $\rho = \rho_{XY} = \lambda_{12}/\lambda$, where $\lambda = \lambda_1 + \lambda_2 + \lambda_{12}$. Thus $\lambda_{12} = 0 (\rho_{XY} = 0)$ implies independence. If X and Y denote the lifetimes of a two component series system then it follows that the system lifetime will follow the exponential distribution if (X,Y) follows the BVE.

A related model is proposed by Block and Basu (1974). We shall denote this by the ACBVE. Here the survival function is given by

$$\begin{aligned} \bar{F}(x,y) &= \frac{\lambda}{\lambda_1 + \lambda_2} \exp[-\lambda_1 x - \lambda_2 y - \lambda_{12}\max(x,y)] \\ &\quad - \frac{\lambda_{12}}{\lambda_1 + \lambda_2} \exp[-\lambda \max(x,y)], \\ &\quad \lambda_1, \lambda_2 > 0, \lambda_{12} \geq 0, x, y \geq 0, \end{aligned} \tag{7}$$

where, as before, $\lambda = \lambda_1 + \lambda_2 + \lambda_{12}$. The ACBVE is absolutely continuous, satisfies the BLMP and here also $\min(X,Y) \sim e(\lambda)$. However, the marginals are not univariate exponential distributions. The ACBVE is the absolutely continuous part of the BVE. However, it is not a special case of the BVE since the BVE distributions are not absolutely continuous. As in the case of the BVE $\lambda_{12} = 0 (\rho_{xy} = 0)$ implies independence.

Multivariate extensions of the BVE and the ACBVE have been considered by Marshall and Olkin (1967) and Block (1975) respectively.

Note that min $(X,Y) \sim e(\lambda)$ for both BVE and ACBVE. Esary and Marshall (1974) consider the general class of distributions with exponential minima. As a special case, consider the following class of bivariate distributions, to be called the EM, with survival function

$$(8)\quad \bar{F}(x,y) = \exp[-\lambda_1 x - \lambda_2 y - \max(\lambda_3 x, \lambda_4 y)], \quad \lambda_1, \lambda_2, \lambda_3, \lambda_4 \geq 0,\ x, y \geq 0$$

The EM reduces to the BVE if $\lambda_3 = \lambda_4$.

B. *Multivariate distributions based on the notions of aging.* Although the exponential distribution is the most useful model, other models have also been found useful. Because of the nonrobustness of inference procedures based on the exponential distribution, a number of classes of nonparametric distributions, based on the notions of aging, have been proposed as models. The most commonly studied classes of life distributions in the univariate case, based on the notions of aging, are the following:

1) Increasing failure rate class (IFR);

2) Decreasing failure rate class (DFR);

3) Increasing failure rate in average class (IFRA);

4) Decreasing failure rate in average class (DFRA);

5) New better that used class (NBU);

6) New worse than used class (NWU);

7) Harmonic new better than used in expectation class (HNBUE);

8) Harmonic new worse than used in expectation class (HNWUE).

See Barlow and Proschan (1981) and Basu and Ebrahimi (1986a) for a description of these classes.

Multivariate versions of these and other classes have been defined and their properties have been developed by Basu and Ebrahimi (1988), Basu, Ebrahimi, and Klefsjö (1983), Block and Savits (1980, 1981), Buchanan and Singpurwalla (1977), Ghosh and Ebrahimi (1981) and others. An important problem is to see if a class of distributions is closed under convolution. That is, let $\underline{X} = (X_1, \ldots, X_p)$ and $\underline{Y} = (Y_1, \ldots, Y_p)$ both belong to the class of distributions G and let $\underline{X}$ and $\underline{Y}$ be independent. Then, under what condition will $\underline{X} + \underline{Y} \sim G$?

Block and Savits (1980) proved the closure under convolution for the class of multivariate IFRA distributions defined by them. El-Neweihi (1984) and El-Neweihi and Savits (1987) have proved the closure property of a multivariate NBU distribution under convolution. Similarly, Basu and Ebrahimi (1986b, 1988) have proved closure under convolution of the class of multivariate NBUE and the class multivariate HNBUE distributions.

Other properties have been discussed in the papers mentioned above.

3. Tests for Independence. Since independently distributed component lifetimes are easier to analyze, a number of tests for independence have been considered. For the BVE and the ACBVE, testing for independence is equivalent to testing the null hypothesis

$$H_0 : \lambda_{12} = 0, \tag{9}$$

against the alternative hypothesis

$$H_1 : \lambda_{12} > 0.$$

For the BVE, Bemis et al. (1972) derive the UMP test for the above hypothesis when λ_1 and λ_2 are known; and Bhattacharyya and Johnson (1973) derive the UMP test when $\lambda_1 = \lambda_2$ but the common value is unknown. Similarly for the ACBVE, Gupta, Mehrotra, and Michalek (1984) derive a test for (9) when the marginal distributions are equal.

Let $P_{\rm MO}(P_{\rm BB})$ denote the power of a test T when the BVE (ACBVE) is the underlying model. Since, $P(X = Y) = \lambda_{12}/\lambda = 0$, under the null hypothesis (9) for the BVE, H_0 is rejected if $X_i = Y_i$ for some i. Weier and Basu (1981) show that, based on a random sample $(X_1, Y_1), (X_2, Y_2), \ldots, (X_n, Y_n)$,

$$P_{\rm MO} = 1 - (1 - \rho)^n + (1 - \rho)^n P_{\rm BB}. \tag{10}$$

It is thus enough to consider tests for the ACBVE, which are easier to derive because of absolute continuity. Assuming $\lambda_1 = \lambda_2$, Weier and Basu (1981) have also considered nonparametric tests of independence of Kendall and Spearman. For a definition of Kendall's test for independence and that due to Spearman see, for example, Lehmann (1975). Assume $\lambda_1 = \lambda_2$. Let U denote the UMP test, M denote the test based on MLE of λ_{12}, T = Kendall's tau, R = Spearman's ρ based on a random sample of size n. Then the following result concerning Pitman asymptotic relative efficiency (ARE) is obtained by Weier and Basu (1981).

$$\begin{gathered} \text{ARE(T,U)} = \text{ARE(R,U)} \\ = \text{ARE (T,M)} = \text{ARE(R,M)} = .5, \\ \text{ARE(M,U)} = 1. \end{gathered}$$

The above ARE results hold for the BVE also. Weier and Basu (1980) have also considered tests for independence for a special trivariate distribution with exponential marginals. This model is an extension of the BVE. A similar trivariate extension of the ACBVE is also given. However, like the bivariate case, here the marginals are not exponential distributions. This ACBVE extension is a special case of Block's (1975) more general extension. The case for general multivariate exponential distribution is open.

4. Tests for Multivariate Exponentiality. In this section we consider tests for multivariate exponentiality. Without much loss of generality, we consider the bivariate case.

4.1. A Test for Bivariate Exponentiality Against BNBU. Basu and Ebrahimi (1984) have considered statistical procedures to test whether a bivariate distribution follows the Marshall-Olkin bivariate exponential distribution (BVE) against the alternative that it is nonexponential bivariate new better than used (BNBU). Hollander and Proschan (1972) have considered the univariate case. Throughout we assume $F(0,0)=0$.

DEFINITION 4.1. F is said to be BNBU-I if

$$\bar{F}(x+t,y+t) \leq \bar{F}(x,y)\bar{F}(t,t), x,y \geq 0, t \geq 0, \tag{11}$$

and similar inequalities hold for both marginal survival functions.

DEFINITION 4.2. F is said to be BNBU-II if $\bar{F}(x+t,x+t) \leq \bar{F}(x,x)\bar{F}(t,t)$, for all $x,t \geq 0$, and similar inequalities hold for the marginal survival functions.

The boundary members of BNBU-I obtained by insisting on equalities in (11), are the family of Marshall-Olkin bivariate exponential distributions (BVE).

Basu and Ebrahimi (1984) have considered testing

$$H_0 : \bar{F}(x,y) = \exp(-\lambda_1 x - \lambda_2 y - \lambda_{12}\max(x,y)), x,y,\lambda_1,\lambda_2 > 0, \lambda_{12} \geq 0 \tag{12}$$

versus

$$H_1 : F \text{ is BNBU-I and not BVE}, \tag{13}$$

on the basis of a random sample $(X_1,Y_1),(X_2,Y_2)\ldots(X_n,Y_n)$ from F. Consider the functionals

$$\Delta(\bar{F}) = \int\int\int\{\bar{F}(x,y)\bar{F}(t,t) - \bar{F}(x+t,y+t)\}dF(x,y)dF(t,t)$$

and

$$\Gamma(\bar{F}) = \int\int\int\{\bar{F}(x,y)\bar{F}(t,t) - \bar{F}(x+t,y+t)\}dxdydt$$

Under $H_0 : \Delta(\bar{F}) = 0$ and $\Gamma(\bar{F}) = 0$. If the H_0 is not true, both $\Delta(\bar{F})$ and $\Gamma(\bar{F})$ will be large. Estimates of $\Delta(\bar{F})$ and $\Gamma(\bar{F})$ will also be expected to be large under H_1. Thus estimates of $\Gamma(\bar{F})$ and $\Delta(\bar{F})$ or, quantities which are asymptotically equivalent to these estimates, could be used as test statistics.

Two test statistics have been proposed for testing the above hypothesis. These are given below.

a) Reject H_0 in favor of H_1 if $\Delta(\bar{F}_n)$ is too large. Here

$$\begin{aligned} \Delta(\bar{F}_n) &= \frac{1}{2n^2}\sum_{i=1}^{n}\sum_{j=1}^{n} t(X_j, X_i)t(Y_j, Y_i) - \frac{1}{n^3} \\ &\sum_{i=1}^{n}\sum_{j=1}^{n}\sum_{k=1}^{n}\{t(X_i, X_j + Z_j)t(Y_i, Y_j + Z_k)\} \end{aligned} \tag{14}$$

where $Z_k = \min(X_k, Y_k)$ and

$$t(a,b) = \begin{cases} 1, & \text{if } a > b, \\ 0, & \text{otherwise.} \end{cases}$$

b) Reject H_0 in favor of H_1 if $\hat{\Gamma}(\bar{F})$ is too large, where

$$\begin{aligned} \hat{\Gamma}(\bar{F}) = \frac{1}{n^2}\{\sum_{i=1}^{n} X_iY_i\}\{\sum_{i=1}^{n} Z_i\} \quad &+ \quad \{\frac{1}{2n}\sum_{i=1}^{n}(X_i + Y_i)Z_i^2\} - \\ &\{\frac{1}{n}\sum_{i=1}^{n} X_iY_iZ_i + \frac{1}{3n}\sum_{i=1}^{n} Z_i^3\} \end{aligned} \tag{15}$$

Properties of $\Delta(\bar{F}_n)$ and $\hat{\Gamma}(\bar{F})$ are given in Basu and Ebrahimi (1984). In particular both $\Delta(\bar{F}_n)$ and $\hat{\Gamma}(\bar{F})$ are consistent and asymptotically unbiased estimates of $\Delta(\bar{F})$ and $\hat{\Gamma}(\bar{F})$ respectively and are asymptotically normally distributed.

Note that the above tests can be considered bivariate extensions of the univariate Hollander-Proschan test (1972).

4.2. Testing for Bivariate Increasing Failure Rate Average (BIFRA). Basu and Habibullah (1987) have considered a test for bivariate exponentiality against the BIFRA alternative. Esary and Marshall (1979) and Block and Savits (1980) have studied properties of BIFRA distributions.

DEFINITION 4.3. (X, Y) is said to have a BIFRA distribution if and only if

$$\bar{F}^{\alpha}(x, y) \leq \bar{F}(\alpha x, \alpha y) \tag{16}$$

for all $x, y > 0$ and all $\alpha, 0 < \alpha < 1$.

Note that equality in (16) implies a bivariate distribution with exponential minimum. An example of a bivariate distribution with exponential minimum, which includes the BVE as a special case, is given by (8).

Note that the BIFRA distribution is also the BNBU-I. We consider testing the null hypothesis

(17) $$H_0 : \bar{F}^\alpha(x,y) = \bar{F}(\alpha x, \alpha y) \text{ for all } \alpha \epsilon (0,1) \text{ and all}, x > 0,\ y > 0$$

against the alternative hypothesis

(18) $$H_1 : \bar{F}^\alpha(x,y) \leq \bar{F}(\alpha x, \alpha y), \alpha\epsilon(0,1), x > 0,\ y > 0,$$

where the inequality holds for some $(x,y)\epsilon R_2^+$ and for at least some α. Define

$$\Delta_\alpha(F) = \int_0^\infty \int_0^\infty [\bar{F}^{1/\alpha}(\alpha x, \alpha y) - \bar{F}(x,y)] dF(x,y), (0 < \alpha < 1).$$

Under $H_0, \Delta_\alpha(F) = 0$ and under $H_1, \Delta_\alpha(F) > 0$. Thus $\Delta_\alpha(F)$ meaures the deviation from H_0. Basu and Habibullah (1987) propose a test for the above hypothesis based on an estimator of $\Delta_\alpha(F)$ where α is a fixed constant.

The following lemma due to Basu and Habibullah (1987) shows that if equality (17) holds for a particular $\alpha_0\epsilon(0,1)$ then it will imply that F has a bivariate exponential distribution with exponential minimum. To this end it is sufficient to prove that the distribution of $Z = \min(X,Y)$ is univariate exponential.

LEMMA 4.1. *Let* $\bar{F}^\alpha 0(t,t) = \bar{F}(\alpha_0 t, \alpha_0 t)$ *where* α_0 *is fixed,* $0 < \alpha_0 < 1, t > 0$ *and* $\bar{F}(t,t) = P(Z > t)$. *Then* $\bar{F}(t,t) = e^{-\lambda t}$ *for some* $\lambda > 0$.

From Lemma 4.1 it is clear that H_0 does not hold if (17) is not true for some fixed α. Let $(X_1,Y_1),(X_2,Y_2),\ldots,(X_n,Y_n)$ be a random sample from F. For testing the above hypothesis Basu and Habibullah (1987) propose rejecting H_0 in favor of H_1 if the statistic

$$J_{.5}^{(n)} = U_1 - U_2$$

is large. Note that $\alpha = .5$ is taken to simplify computations. Here

$$U_1 = \frac{2}{n(n-1)\binom{n-2}{1}} \sum_c h_1\{(X_{\alpha_1}, Y_{\alpha_1}), (X_{\alpha_2}, Y_{\alpha_2}); (\frac{1}{2}X_{\alpha_3}, \frac{1}{2}Y_{\alpha_3})\},$$

and the sum $\sum_c$ extends over all combinations $1 \leq \alpha_i \leq n$, $i = 1,2,3$, $\alpha_1 < \alpha_2$, $\alpha_1 \neq \alpha_3$, $\alpha_2 \neq \alpha_3$, and

$$h_1\{(a_1,b_1),(a_2,b_2);(a_3,b_3)\} = \begin{cases} 1 & \text{if } a_1, a_2 > a_3 \text{ and } b_1, b_2 > b_3 \\ 0, & \text{otherwise.} \end{cases}$$

Also

$$U_2 = \frac{1}{\binom{n}{2}} \sum_{\alpha_1 \neq \alpha_2} h_2\{(X_{\alpha_1}, Y_{\alpha_1}), (X_{\alpha_2}, Y_{\alpha_2})\},$$

where

$$h_2\{(a_1,b_1),(a_2,b_2) = \begin{cases} 1 & \text{if } (a_1 > a_2, b_1 > b_2) \\ 0, & \text{otherwise.} \end{cases}$$

$J_{.5}^{(n)}$ is a difference of two U-statistics and therefore is asymptotically normally distributed, details are given in Basu and Habibullah (1987).

5. Concluding Remarks. In this paper we present a survey of some inference problems relating to dependent systems. Some other related problems namely, the problem of competing risks and that of converting dependent models to independent ones have been described in Basu and Klein (1981).

REFERENCES

BARLOW, R.E. and PROSCHAN, F. (1981). *Statistical Theory of Reliability and Life Testing: Probability Models.* To Begin With, Silver Spring.

BASU, A.P. (1971). Bivariate failure rate. *J. Amer. Statist. Assoc.* **66** 103–104.

BASU, A.P. (1988). Multivariate exponential distributions and their applications in reliability. *Handbook of Statistics* **7** (P.R. Krishnaiah and C.R. Rao, eds.), North Holland, 467–477.

BASU, A.P. and EBRAHIMI, N. (1984). Testing whether survival function is bivariate new better than used. *Commun. Statist. Theory Methods* **13**(15), 1839–1849.

BASU, A.P. and EBRAHIMI, N. (1986a). HNBUE and HNWUE distributions—A survey. *Reliability and Quality Control* (A.P. Basu, ed.), North Holland, 33–46.

BASU, A.P. and EBRAHIMI, N. (1986b). Multivariate new better than used in expectation distributions. *Statist. Prob. Lett.* **4** 295–301.

BASU, A.P. and EBRAHIMI, N. (1988). Multivariate harmonic new better than used in expectation distributions. *J. Statist. Plan.* **20** 181–190.

BASU, A.P., EBRAHIMI, N., and KLEFSJŐ, B. (1983). Multivariate harmonic new better than used in expectation distributions. *Scand. J. Statist.* **10** 19–25.

BASU, A.P. and HABIBULLAH, M. (1987). A test for bivariate exponentiality against BIFRA alternatives. *Calcutta Statist. Assoc. Bull.* **36** 79–85.

BASU, A.P. and KLEIN, J.P. (1981). Some recent results in competing risks theory. Survival Analysis **2**, IMS Lecture notes–monograph series (R.A. Johnson and J. Crowley, eds.), 216–229.

BEMIS, B.M., BAIN, L.J., and HIGGINS, J.J. (1972). Estimation and hypothesis testing for the parameters of a bivariate exponential distribution. *J. Amer. Statist. Assoc.* **67** 927–929.

BHATTACHARYYA, G.K. and JOHNSON, R.A. (1973). On a test of independence in a bivariate exponential distribution. *J. Amer. Statist. Assoc.* **68** 704–706.

BLOCK, H.W. (1975). Continuous multivariate exponential extensions. *Reliability and Fault Tree Analysis* (R.E. Barlow, J.B. Fussell, and N.D. Singpurwalla, eds.), SIAM, Philadelphia, 285–306.

BLOCK, H.W. and BASU, A.P. (1974). A continuous bivariate exponential extension. *J. Amer. Statist. Assoc.* **69** 1031–1037.

BLOCK, H.W. and SAVITS, T.H. (1980). Multivariate IFRA distributions. *Ann. Probab.* **8** 793–801.

BLOCK, H.W. and SAVITS, T.H. (1981). Multivariate classes in reliability theory. *Math. Oper. Res.* **6** 453–461.

BRINDLEY, E.C., JR. and THOMPSON, W.A., JR. (1972). Dependence and aging aspects of multivariate survival. *J. Amer. Statist. Assoc.* **67** 822–830.

BUCHANAN, W.B. and SINGPURWALLA, N.D. (1977). Some stochastic characterizations of multivariate survival. *The Theory and Applications of Reliability* **1** (C.P. Tsokos and I.N. Shimi, eds.), 329–348. Academic Press, New York.

EL-NEWEIHI, E. (1984). Characterizations and closure under convolution of two classes of multivariate life distributions. *Statist. Prob. Lett.* **2** 333–335.

EL-NEWEIHI, E. and SAVITS, T.H. (1987). Convolution of the IFRA scaled-mins class. *Ann. Prob.* **15** 423–427.

ESARY, J.D. and MARSHALL, A.W. (1974). Multivariate distributions with exponential minimums. *Ann. Stat.* **2** 84–98.

ESARY, J.D. and MARSHALL, A.W. (1979). Multivariate distributions with increasing hazard rate average. *Ann. Prob.* **7** 359–370.

GHOSH, M. and EBRAHIMI, N. (1981). Multivariate NBU and NBUE distributions. *Egypt Statist. J.* **25** 36–55.

GUPTA, R.C., MEHROTRA, K.G., and MICHALEK, J.E. (1984). A small test for an absolutely continuous bivariate exponential model. *Comm. Statist. Theory and Meth.* **13** 1735–1740.

HOLLANDER, M. and PROSCHAN, F. (1972). Testing whether new is better than used. *Ann. Math. Statist.* **43** 1136–1146.

LEHMANN, E.L. (1975). *Nonparametrics: Statistical Methods Based on Ranks.* Holden-Day, Inc., San Francisco.

MARSHALL, A.W. and OLKIN, I. (1967). A multivariate exponential distribution. *J. Amer. Statist. Assoc.* **62** 30–44.

WEIER, D.R. and BASU, A.P. (1980). Testing for independence in multivariate exponential distributions. *Austral. J. Statist.* **22** 276–288.

WEIER, D.R. and BASU, A.P. (1981). On tests of independence under bivariate exponential models. STATISTICAL DISTRIBUTIONS IN SCIENTIFIC WORK **5** (C. Taillie, G.P. Patil, and B.A. Baldessari, eds.), 169–180, Reidel, Dordrecht.

DEPARTMENT OF STATISTICS
UNIVERSITY OF MISSOURI
COLUMBIA, MO 65211

PARTIAL ORDERINGS ON PERMUTATIONS

BY H.W. BLOCK,[1] D. CHHETRY[1], Z. FANG[1], A.R. SAMPSON[1]

University of Pittsburgh, Tribhuvan University, University of Science and Technology of China, and University of Pittsburgh

Results concerning partial orderings defined on permutations and their applications in statistics are surveyed.

1. Introduction. There are a variety of partial orderings on S_n, the set of all permutations of $\{1, \ldots, n\}$, which have applications in statistics. These orderings have been considered among others by Sobel (1955), Savage (1957), Lehmann (1966), Yanagimoto and Okamoto (1969), Hǎjek (1969), Hollander, Proschan, and Sethuraman (1977), hereafter referred to as HPS, and Schriever (1985), (1987a), (1987b). These partial orderings have been defined in various ways with different names and notations. They have been used to define monotone functions on S_n, and the monotonicity properties of such functions have been used in many areas including stochastic comparisons of rank order statistics. For example, using a particular ordering HPS introduced an important class of nondecreasing functions, called decreasing in transposition (DT) functions, and discussed their importance in statistics. Recently, Block, Chhetry, Fang, and Sampson (1990), hereafter referred to as BCFS(a), considered three well-known and one new partial orderings on S_n through a unified approach and characterized them by four important positive dependence orderings on bivariate empirical rank distributions.

In addition to partial orderings the notion of metrics on $S_n \times S_n$ plays an important role in many areas. The metrics arise naturally in a variety of situations, such as in the analysis of sorting algorithms (Knuth (1973)), in measuring association in bivariate rank data (Kendall (1970) and Diaconis and Graham (1977)), and in the study of ranking models (Fligner and Verducci (1986)). Recently, Block, Chhetry, Fang, and Sampson (1987) considered in a unified fashion three well-known and two new metrics on $S_n \times S_n$, and discussed their applications.

The main objective of this paper is to survey important results and applications of partial orderings defined over permutations. In Section 2 we review basic results concerning five partial orderings. Their applications are considered in Section 3 and Section 4. Also in Section 3 we introduce five classes of nondecreasing or arrangement increasing functions on S_n and discuss their basic implications.

[1]Research sponsored by the AFOSR Grant No. AFOSR–84–0113. The U.S. Government is authorized to reproduce and distribute reprints for governmental purposes notwithstanding any copyright notation thereon.

AMS 1980 subject classification. Primary 62E10; secondary 62G99.

Key words and phrases. Partial ordering, metrics, decreasing in transposition, arrangement increasing, dependence orderings.

Throughout this paper, permutations on S_n are denoted by lower case bold face letters, such as $\mathbf{i}, \mathbf{j}, \mathbf{r}, \mathbf{s}$, etc., and their components are denoted by the corresponding lower case letters. The permutations $(1, \ldots, n)$ and $(n, \ldots, 1)$ are denoted by $\mathbf{e}$ and $\mathbf{e}^*$, respectively. The composition or product of $\mathbf{r} = (r(1), \ldots, r(n))$ and $\mathbf{s} = (s(1), \cdots, s(n))$ is denoted by $\mathbf{r} \cdot \mathbf{s}$, where $\mathbf{r} \cdot \mathbf{s} = (r(s(1)), \cdots, r(s(n)))$. The inverse of $\mathbf{r}$ is denoted by $\mathbf{r}^{-1}$, where $\mathbf{r}^{-1} = (r^{-1}(1), \ldots, r^{-1}(n))$. An inversion of $\mathbf{r} = (r(1), \cdots, r(n))$ is a pair $(r(k), r(\ell))$ such that $(k - \ell)(r(k) - r(\ell)) < 0$.

2. Partial Orderings On S_n: In order to define several partial orderings on S_n through a unified approach, BCFS(a) considered the following definition.

Definition 2.1. An interchange of two components $i(k)$ and $i(\ell)$ of $\mathbf{i}$ is said to be: (1) a correction of type 1 inversion if $\Delta_{k\ell} < 0$; (2) a correction of type 2 inversion if $\Delta_{k\ell} < 0$ and $|i(k) - i(\ell)| = 1$; (3) a correction of type 3 inversion if $\Delta_{k\ell} < 0$ and $|k - \ell| = 1$, where $\Delta_{k\ell} = (k - \ell)(i(k) - i(\ell))$.

Definition 2.1 is used to define three well-known ordering relations for permutations as follows.

Definition 2.2. For $t = 1, 2, 3$ a permutation $\mathbf{i}$ is said to be better ordered in the sense of ordering b_t than $\mathbf{j}$, denoted by $\mathbf{i} \overset{b_t}{\geq} \mathbf{j}$, if $\mathbf{i} = \mathbf{j}$ or if $\mathbf{i}$ is obtainable from $\mathbf{j}$ in a number of steps each of which consists of correcting a type t inversion.

The following ordering relation is described by BCFS(a).

Definition 2.3. A permutation $\mathbf{i}$ is said to be better ordered in the sense of ordering b_4 than $\mathbf{j}$, denoted by $\mathbf{i} \overset{b_4}{\geq} \mathbf{j}$, if $\mathbf{i} = \mathbf{j}$ or if $\mathbf{i}$ is obtainable from $\mathbf{j}$ in a number of steps each of which consists of correcting a type 2 or type 3 inversion.

One more ordering relation whose application has been considered by Savage (1957) is as follows.

Definition 2.4. A permutation $\mathbf{i}$ is said to be better ordered in the sense of ordering b_0 than $\mathbf{j}$, denoted by $\mathbf{i} \overset{b_0}{\geq} \mathbf{j}$, if $\sum_{k=1}^{m} i(k) \leq \sum_{k=1}^{m} j(k)$ for all $m = 1, \ldots, n$.

Each of the above ordering relations is a partial ordering on S_n, in the sense that they are reflexive, transitive, and anti-symmetric. The implications among these orderings are as follows:

$$\begin{array}{lll} \mathbf{i} \overset{b_2}{\geq} \mathbf{j} & \searrow & \\ & & \mathbf{i} \overset{b_4}{\geq} \mathbf{j} \Rightarrow \mathbf{i} \overset{b_1}{\geq} \mathbf{j} \Rightarrow \mathbf{i} \overset{b_0}{\geq} \mathbf{j}. \\ \mathbf{i} \overset{b_3}{\geq} \mathbf{j} & \nearrow & \end{array}$$

The above implications are strict (BCFS(a) and Savage (1957)), and the ordering b_2 neither implies nor is implied by the ordering b_3 (Lehmann (1966)).

For theoretical as well as practical purposes it is natural to seek one-to-one functions $f\colon S_n \to S_n$ such that the order relation between $f(\mathbf{i})$ and $f(\mathbf{j})$ would be easily demonstrable once we know the order relation between $\mathbf{i}$ and $\mathbf{j}$. The results for two such functions are as follows.

THEOREM 2.5. *Let* $\mathbf{i}$ *and* $\mathbf{j}$ *be two permutations in* S_n. *Then*

(a) $\mathbf{i} \overset{b_1}{\geq} \mathbf{j} \Longleftrightarrow \mathbf{i}^{-1} \overset{b_1}{\geq} \mathbf{j}^{-1}$

(b) $\mathbf{i} \overset{b_2}{\geq} \mathbf{j} \Longleftrightarrow \mathbf{i}^{-1} \overset{b_3}{\geq} \mathbf{j}^{-1}$

(c) $\mathbf{i} \overset{b_4}{\geq} \mathbf{j} \Longleftrightarrow \mathbf{i}^{-1} \overset{b_4}{\geq} \mathbf{j}^{-1}$

(d) $\mathbf{i} \overset{b_t}{\geq} \mathbf{j} \Longleftrightarrow \bar{\mathbf{i}} \overset{b_t}{\geq} \bar{\mathbf{j}}$, *for* $t = 0, 1, 2, 3, 4$,

where for an arbitrary permutation $\mathbf{r} \in S_n$, $\bar{\mathbf{r}} = \mathbf{e}^* \cdot \mathbf{r} \cdot \mathbf{e}^*$ *and* $\bar{\mathbf{r}}$ *is known as the complement of* $\mathbf{r}$.

Note that $\mathbf{i} \overset{b_0}{\geq} \mathbf{j}$ does not necessarily imply that $\mathbf{i}^{-1} \overset{b_0}{\geq} \mathbf{j}^{-1}$. To see this choose $\mathbf{i} = (1432)$ and $\mathbf{j} = (3241)$. Consequently the comparison $\mathbf{i}^{-1} \overset{b_0}{\geq} \mathbf{j}^{-1}$ leads to an ordering different than $\mathbf{i} \overset{b_0}{\geq} \mathbf{j}$. One might label this companion ordering $\mathbf{i} \overset{b_0'}{\geq} \mathbf{j}$.

In the remainder of this section we state several theorems concerning the formulations of the above five orderings. These formulations are easier to handle than the original definitions. The following theorem characterizes the b_0 ordering.

THEOREM 2.6. $\mathbf{i} \overset{b_0}{\geq} \mathbf{j}$ *if and only if* $\sum_{k=1}^n a_k i(k) \geq \sum_{k=1}^n a_k j(k)$ *for every choice of* $a_1, \ldots, a_n$ *provided* $a_1 \leq \cdots \leq a_n$.

The following theorem which characterizes the b_1-ordering is stated in Yanagimoto and Okamoto (1969). A simple intuitive proof is given in Metry and Sampson (1988a).

THEOREM 2.7. *For any positive integer* $m \leq n$, *let* $i(1,m) < \cdots < i(m,m)$ *be the increasing rearrangement of the last* m *components of* $\mathbf{i}$, *and* $j(1,m) < \cdots < j(m,m)$ *be the increasing rearrangement of the last* m *components of* $\mathbf{j}$. *Then* $\mathbf{i} \overset{b_1}{\geq} \mathbf{j}$ *if and only if* $i(k,m) \geq j(k,m)$ *for any* k *and* m *such that* $1 \leq k \leq m \leq n$.

The following theorem is due to Hăjek (1969).

THEOREM 2.8. $\mathbf{i} \overset{b_2}{\geq} \mathbf{j}$ *if and only if the following holds*

$$k < \ell,\ j(k) < j(\ell) \Rightarrow i(k) < i(\ell).$$

The following two theorems are due to BCFS(a).

THEOREM 2.9. $\mathbf{i} \overset{b_3}{\geq} \mathbf{j}$ *if and only if the following holds*

$$k < \ell,\ j(k) < j(\ell) \Rightarrow i^{-1}(j(k)) < i^{-1}(j(\ell))$$

$$i^{-1}(j(k)) < i^{-1}(j(\ell)),\ j(k) > j(\ell) \Rightarrow k < \ell.$$

THEOREM 2.10. $\mathbf{i} \overset{b_4}{\geq} \mathbf{j}$ *if and only if there exists a permutation* $\mathbf{r}$ *such that*

$$k < \ell,\ j(k) < j(\ell) \Rightarrow r(k) < r(\ell),\ i(r(k)) < i(r(\ell)),$$

$$r(k) < r(\ell),\ i(r(k)) > i(r(\ell)) \Rightarrow k < \ell,\ j(k) > j(\ell).$$

An important consequence of Theorem 2.10 is as follows: $\mathbf{i} \overset{b_4}{\geq} \mathbf{j}$ if and only if there exists a permutation $\mathbf{t}$ such that $\mathbf{i} \overset{b_2}{\geq} \mathbf{t}$ and $\mathbf{t} \overset{b_3}{\geq} \mathbf{j}$ or $\mathbf{i} \overset{b_3}{\geq} \mathbf{t}$ and $\mathbf{t} \overset{b_2}{\geq} \mathbf{j}$.

3. Partial Orderings and Monotone Functions. Throughout the paper $x_1, \ldots, x_n$ denote n distinct observations and $x_{(1)} < \cdots < x_{(n)}$ denote the increasing rearrangement of $x_1, \ldots, x_n$. Also, $\mathbf{y} \cdot \mathbf{r}$ denotes the vector $(y_{r(1)}, \ldots, y_{r(n)})$ in R^n, for all $\mathbf{y} \in R^n$ and $\mathbf{r} \in S_n$. Unless otherwise stated we adopt the following definition of rank orders.

DEFINITION 3.1. The rank order of $x_1, \ldots, x_n$ is the permutation $\mathbf{r}$ in S_n such that

$$(x_{(1)}, \ldots, x_{(n)}) \cdot \mathbf{r} = (x_1, \ldots, x_n).$$

Another definition of rank order mentioned in Savage (1957) is as follows.

DEFINITION 3.2. The rank order of $x_1, \ldots, x_n$ is the permutation $\mathbf{r}'$ in S_n such that

$$(x_1, \ldots, x_n) \cdot \mathbf{r}' = (x_{(1)}, \ldots, x_{(n)}).$$

From the above two definitions it is clear that $\mathbf{r}' = \mathbf{r}^{-1}$, and in the literature $\mathbf{r}^{-1}$ is called the antirank of the rank order $\mathbf{r}$. In view of this fact throughout this paper antiranks may be viewed as the rank orders in the sense of Definition 3.2. In the above definition when the x values are replaced by the corresponding continuous random variables we replace $\mathbf{r}(\mathbf{r}')$ by the random vector $\mathbf{R}$ $(\mathbf{R}^*)$ of ranks. We now present an example of Savage (1957) to show that some results are valid only in terms of the rank orders of Definition 3.2.

EXAMPLE 3.3. Let $X_1, \ldots, X_n$ be n independent random variables such that each X_k has continuous density $g(k\lambda)h(x)e^{k\lambda x}$, where g and h are nonnegative functions and $\lambda > 0$. Assume that the possible rank orders $\mathbf{r}'$, $\mathbf{s}'$, etc. and the random vector $\mathbf{R}^*$ of ranks of $X_1, \ldots, X_n$ are defined as in the Definition 3.2. Then, Savage (1957, Theorem 3) proved that

$$(1) \qquad \mathbf{r}' \overset{b_0}{\geq} \mathbf{s}' \Rightarrow \text{Prob}(\mathbf{R}^* = \mathbf{r}') \geq \text{Prob}(\mathbf{R}^* = \mathbf{s}').$$

We also present an example of Savage (1957) to show that there are situations where it does not matter which definition of rank orders have been used.

EXAMPLE 3.4. Let $X_1, \ldots, X_n$ be n independent random variables such that each X_k has continuous TP_2 density $f(x; \lambda_k)$ where $\lambda_1 \leq \cdots \leq \lambda_n$. Assume that the rank orders $\mathbf{r}$, $\mathbf{s}$, etc. and the random vector $\mathbf{R}$ of ranks of $X_1, \ldots, X_n$ are defined as in Definition 3.1. Then, Savage (1957, Corollary 1.1) proved that

$$(2) \qquad \mathbf{r} \overset{b_1}{\geq} \mathbf{s} \Rightarrow \text{Prob}\,(\mathbf{R} = \mathbf{r}) \geq \text{Prob}\,(\mathbf{R} = \mathbf{s})$$

provided $\lambda_{r(k)}$ corresponding to these k for which $\mathbf{r}(k) \neq \mathbf{s}(k)$ are not all equal. The result in (2) remains valid even if we replace $\mathbf{r}$, $\mathbf{s}$ and $\mathbf{R}$ by $\mathbf{r}'$, $\mathbf{s}'$ and $\mathbf{R}^*$, respectively, where $\mathbf{r}'$, $\mathbf{s}'$ and $\mathbf{R}^*$ are defined in Example 3.3.

In the above examples two ordering relations of Section 2 were used to define nondecreasing functions in S_n. Such functions are desirable in statistics since their monotonicity properties may be used to establish important results concerning rank statistics. For example, Savage (1957) used the monotonicity property (1) in testing the hypothesis H_0: there exists a c.d.f. $F(x)$ such that Prob $(X_k \leq x) \equiv F_k(x) = F(x)$ for $k = 1, \ldots, n$ against $H_1 : X_k$ has density as described in Example 3.3 for $k = 1, \ldots, n$. Savage (1957, Corollary 1.1) gives a necessary criterion for a rank test of H_0 against H_1 to be admissible. For the application of the monotonicity property in (2) see Savage (1957, Corollary 1.2).

To study the properties and application of a class of monotone functions on S_n, HPS introduced the notion of DT (decreasing in transposition) functions as follows.

DEFINITION 3.5. A function $f : S_n \to R^1$ is said to be DT on S_n if $\mathbf{r} \overset{b_1}{\geq} \mathbf{s} \Rightarrow f(\mathbf{r}) \geq f(\mathbf{s})$.

The notion of DT functions on $M^n \times N^n$, where M and N are subsets of R^1, due to HPS, is as follows.

DEFINITION 3.6. Let $g : M^n \times N^n \to R^1$ be a function. Then g is said to be DT on $M^n \times N^n$ if

(a) $g(\boldsymbol{\lambda}, \mathbf{x}) = g(\boldsymbol{\lambda} \cdot \mathbf{r}, \mathbf{x} \cdot \mathbf{r})$ for all $\boldsymbol{\lambda} \in M^n$, $\mathbf{x} \in N^m$ and $\mathbf{r} \in S_n$;

(b) $\mathbf{r} \overset{b_1}{\geq} \mathbf{s} \Rightarrow g(\boldsymbol{\lambda}, \mathbf{x} \cdot \mathbf{r}) \geq g(\boldsymbol{\lambda}, \mathbf{x} \cdot \mathbf{s})$ whenever $\lambda_1 \leq \cdots \leq \lambda_n$ and $x_1 \leq \cdots \leq x_n$.

HPS in Lemma 2.2 proved that the DT class of functions includes as special cases other well-known classes of functions, such as Schur convex (concave) functions, TP_2-functions, and L-superadditive functions. Also HPS (Section 3) considered the preservation properties under mixtures, compositions, products and integral transformations of DT functions, and in their Example 3.10, proved that many multivariate density functions are DT functions. One of their important results, which has direct applications in statistics, is as follows.

THEOREM 3.7. *Let $X_1, \ldots, X_n$ have joint density function $h(\boldsymbol{\lambda}, \mathbf{x})$, where h is a DT function on $R^n \times R^n$ with vector parameter $\boldsymbol{\lambda}$. Let $\mathbf{R} = (R_1, \ldots, R_n)$ be the random vector of ranks of $X_1, \ldots, X_n$ and $g(\boldsymbol{\lambda}, \mathbf{r}) = Prob_{\boldsymbol{\lambda}}(\mathbf{R} = \mathbf{r})$. Then $g(\boldsymbol{\lambda}, \mathbf{r})$ is a DT function on $R^n \times S_n$.*

The above theorem with a slight modification in the definition of rank order (see HPS) is applicable even when the underlying multivariate density is discrete.

A DT function is essentially a nondecreasing function defined on S_n when the partial ordering on the domain of the function is b_1. Besides the b_1 ordering, other orderings have been employed to define nondecreasing functions on S_n. For example, see Savage (1957, Theorem 3, or Example 3.3) for the b_0 ordering; Lehmann (1966, Sections 6 and 7); and Hăjek (1969, Section 5) for the b_2 ordering; Yanagimoto and Okamoto (1969, Section 6) and Boland, El-Neweihi and Proschan (1989) for the b_3 ordering. Some of these results will be considered in the next section.

In view of the above facts the following definition which is a simple extension of Definition 3.5, seems imperative.

DEFINITION 3.8. Let $\overset{b}{\geq}$ be a partial ordering on S_n. Then, a function $f : S_n \to R^1$ is said to be arrangement increasing with respect to b on S_n, denoted AI(b) if $\mathbf{r} \overset{b}{\geq} \mathbf{s} \Rightarrow f(\mathbf{r}) \geq f(\mathbf{s})$.

4. Partial Orderings and Dependence Orderings. Several dependence orderings for bivariate distributions were considered among others by Yanagimoto and Okamoto (1969), Cambanis, Simons and Stout (1976), Ahmed, Langberg, Léon and Proschan (1979), Tchen (1980), Karlin and Rinott (1980), Whitt (1982), Block and Sampson (1988), Schriever (1985), (1987a), Kimeldorf and Sampson (1987), and Yanagimoto (1990). We describe some of these. Let the bivariate random variable $(X^{(1)}, Y^{(1)})$ and $(X^{(2)}, Y^{(2)})$ have joint c.d.f.'s $H_1(x,y)$ and $H_2(x,y)$, respectively. Denote the corresponding pairs of marginals by $F_1(x)$, $G_1(y)$ and $F_2(x)$, $G_2(y)$. Denote the supports of $X^{(1)}$, $Y^{(1)}$, $X^{(2)}$ and $Y^{(2)}$ by $D_x^{(1)}$, $D_y^{(1)}$, $D_x^{(2)}$ and $D_y^{(2)}$, respectively. By convention, we have all the marginal distributions are strictly increasing on their corresponding supports. The definitions of our orderings of positive dependence are as follows:

DEFINITION 4.1. (Tchen (1980)): H_2 is said to be more concordant than H_1, denoted by $H_2 \overset{c}{\geq} H_1$, if

(i) $F_1(x) = F_2(x)$ and $G_1(y) = G_2(y)$ for all x, y;

(ii) $H_2(x,y) \geq H_1(x,y)$ for all x, y.

DEFINITION 4.2. (Schriever (1987a)): H_2 is said to be more associated than H_1, denoted by $H_2 \overset{a}{\geq} H_1$, if there exist functions

$$\phi_1 : D_x^{(1)} \times D_y^{(1)} \to D_x^{(2)} \text{ and } \phi_2 : D_x^{(1)} \times D_y^{(1)} \to D_y^{(2)}$$

such that for all x_1, $x_2 \in D_x^{(1)}$ and y_1, $y_2 \in D_y^{(1)}$.

(i) $x_1 \leq x_2,\ y_1 \leq y_2 \Rightarrow \phi_1(x_1, y_1) \leq \phi_1(x_2, y_2),\ \phi_2(x_1, y_1) \leq \phi_2(x_2, y_2)$

(ii) $\phi_1(x_1, y_1) < \phi_1(x_2, y_2),\ \phi_2(x_1, y_1) > \phi_2(x_2, y_2) \Rightarrow x_1 < x_2,\ y_1 > y_2,$

(iii) $(X^{(2)}, Y^{(2)}) \sim (\phi_1(X^{(1)}, Y^{(1)}),\ \phi_2(X^{(1)}, Y^{(1)})),$
where $\sim$ means "distributed as."

DEFINITION 4.3. (Schriever (1985)): H_2 is said to be more row regression dependent than H_1, denoted by $H_2 \overset{rr}{\geq} H_1$, if there exists a function $\phi_2 : D_x^{(1)} \times D_y^{(1)} \to D_y^{(2)}$ such that

(i) $x_1 \leq x_2,\ y_1 \leq y_2 \Rightarrow \phi_2(x_1, y_1) \leq \phi_2(x_2, y_2)$;

(ii) $(X^{(2)}, Y^{(2)}) \sim (X^{(1)}, \phi_2(X^{(1)}, Y^{(1)}))$.

DEFINITION 4.4. (Schriever (1985)): H_2 is said to be more column regression dependent than H_1, denoted by $H_2 \overset{cr}{\geq} H_1$, if there exists a function $\phi_1 : D_x^{(1)} \times D_y^{(1)} \to D_x^{(2)}$ such that

(i) $x_1 \leq x_2,\ y_1 \leq y_2 \Rightarrow \phi_1(x_1, y_1) \leq \phi_1(x_2, y_2)$;

(ii) $(X^{(2)}, Y^{(2)}) \sim (\phi_1(X^{(1)}, Y^{(1)}), Y^{(1)})$.

The orderings "more concordant" and "more row regression dependent" were introduced by Yanagimoto and Okamoto (1969) under the additional assumptions that the distributions of both pairs have the same continuous marginals.

DEFINITION 4.4(A). H_2 is said to be more weakly column ordered than H_1, denoted by $H_2 \overset{w-c}{\geq} H_1$, if

(i) $F_1(x) = F_2(x)$ and $G_1(y) = G_2(y)$ for all x, y;

(ii) $E_{H_2}[\phi(X)Y] \geq E_{H_1}[\phi(X)Y]$ for all non-decreasing functions ϕ.

The notion of *more weakly row ordered*, denoted by $\overset{w-r}{\geq}$, is analogously defined. We summarize some of the basic consequences of these orderings below.

THEOREM 4.5.

(1) $H_2 \overset{c}{\geq} F_2 \cdot G_2$ *if and only if* $X^{(2)}$ *and* $Y^{(2)}$ *are PQD in the sense of Lehmann (1966).*

(2) $H_2 \overset{a}{\geq} F_2 \cdot G_2$ *implies* $X^{(2)}$ *and* $Y^{(2)}$ *are associated.*

(3) $H_2 \overset{rr}{\geq} F_2 \cdot G_2$ *if and only if* $Y^{(2)}$ *is positively regression dependent on* $X^{(2)}$ *in the sense of Lehmann (1966), provided* F_2 *and* G_2 *are continuous.*

(4) $H_2 \overset{cr}{\geq} F_2 \cdot G_2$ *if and only if* $X^{(2)}$ *is positively regression dependent on* $Y^{(2)}$ *in the sense of Lehmann (1966), provided* F_2 *and* G_2 *are continuous.*

(5)
$$\begin{array}{lcl} H_2 \overset{rr}{\geq} H_1 & \searrow & \\ & & H_2 \overset{a}{\geq} H_1. \\ H_2 \overset{cr}{\geq} H_1 & \nearrow & \end{array}$$

(6) $H_2 \overset{a}{\geq} H_1 \Rightarrow H_2 \overset{c}{\geq} H_1$, *provided* $F_1(x) = F_2(x)$ *and* $G_1(y) = G_2(y)$ *for all* x, y.

(7) $H_2 \overset{c}{\geq} H_1 \Rightarrow H_2 \overset{w-r}{\geq} H_1$ *and* $H_2 \overset{w-c}{\geq} H_1$.

PROOF. For the proof of (1) to (4) see Schriever (1985, Proposition 4.1.2). The proof of (5) is obvious. For the proof of (6) see Schriever (1987a, Proposition 2.1). The proof of (7) is also obvious.

We adopt the following definition and notation of rank order for n bivariate observations $(X_1, Y_1), \ldots, (X_n, Y_n)$ where we assume that there are no ties among the x-values and among the y-values.

DEFINITION 4.6. Let $(j(1), \ldots, j(n))$ be the permutation such that $y_{j(1)} < \ldots < y_{j(n)}$. Then, $\mathbf{r}$ is said to be the row rank order of $(x_1, y_1), \ldots, (x_n, y_n)$ if $\mathbf{r}$ is the rank order of $x_{j(1)}, \ldots, x_{j(n)}$.

Appropriately interchanging the roles of x's and y's in Definition 4.6 one could define the column rank order, denoted by $\mathbf{s}$, of $(x_1, y_1), \ldots, (x_n, y_n)$. Note that $\mathbf{s} = \mathbf{r}^{-1}$.

Let $\hat{H}$ denote the empirical rank distribution based on $(X_1, Y_1), \ldots, (X_n, Y_n)$, i.e., $\hat{H}$ is the c.d.f. of a bivariate discrete random vector (R,S) putting mass n^{-1} at the points $(r(1), 1), \ldots, (r(n), n)$ (or equivalently $(1, s(1)), \ldots, (n, s(n))$) where $\mathbf{r}$ and $\mathbf{s}$ are the row and column rank orders of $(X_1, Y_1), \ldots, (X_n, Y_n)$. To compare, according to a positive dependence ordering, two bivariate c.d.f.'s H_1 and H_2 based upon samples of size n from each, it is natural to compare $\hat{H}_1$ and $\hat{H}_2$, where $\hat{H}_1$ and $\hat{H}_2$ are empirical rank distributions based on samples of size n from H_1 and H_2, respectively. For $k = 1, 2$, let $\mathbf{s}^{(k)}$ and $\mathbf{r}^{(k)}$ correspondingly denote the column and row rank order of the sample from H_k. With this notation, BCFS(a) proved essentially the following theorem.

THEOREM 4.7.

(1) $\hat{H}_1 \overset{c}{\geq} \hat{H}_1 \Leftrightarrow \mathbf{s}^{(2)} \overset{b_1}{\geq} \mathbf{s}^{(1)} \Leftrightarrow \mathbf{r}^{(2)} \overset{b_1}{\geq} \mathbf{r}^{(1)}$.

(2) $\hat{H}_2 \overset{rr}{\geq} \hat{H}_1 \Leftrightarrow \mathbf{s}^{(2)} \overset{b_2}{\geq} \mathbf{s}^{(1)} \Leftrightarrow \mathbf{r}^{(2)} \overset{b_3}{\geq} \mathbf{r}^{(1)}$.

(3) $\hat{H}_2 \overset{cr}{\geq} \hat{H}_1 \Leftrightarrow \mathbf{s}^{(2)} \overset{b_3}{\geq} \mathbf{s}^{(1)} \Leftrightarrow \mathbf{r}^{(2)} \overset{b_2}{\geq} \mathbf{r}^{(1)}$.

(4) $\hat{H}_2 \overset{a}{\geq} \hat{H}_1 \Leftrightarrow \mathbf{s}^{(2)} \overset{b_4}{\geq} \mathbf{s}^{(1)} \Leftrightarrow \mathbf{r}^{(2)} \overset{b_4}{\geq} \mathbf{r}^{(1)}$.

(5) $\hat{H}_2 \overset{w-c}{\geq} \hat{H}_1 \Leftrightarrow \mathbf{s}^{(2)} \overset{b_0}{\geq} \mathbf{s}^{(1)} \Leftrightarrow (\mathbf{r}^{(2)})^{-1} \overset{b_0}{\geq} (\mathbf{r}^{(1)})^{-1}$.

(6) $\hat{H}_2 \overset{w-r}{\geq} \hat{H}_1 \Leftrightarrow \mathbf{r}^{(2)} \overset{b_0}{\geq} \mathbf{r}^{(1)} \Leftrightarrow (\mathbf{s}^{(2)})^{-1} \overset{b_0}{\geq} (\mathbf{s}^{(1)})^{-1}$.

PROOF. See Theorems 3.6, 3.7, and 3.8 of BCFS(a) for the proof of parts (1) through (4). Parts (5) and (6) are obvious proofs.

BCFS(a) also proved that $\hat{H}_2 \overset{a}{\geq} \hat{H}_1$ if and only if there exist an empirical rank distribution $\hat{H}^*$ such that $\hat{H}_2 \overset{rr}{\geq} \hat{H}^* \overset{cr}{\geq} \hat{H}_1$ or there exists an empirical rank distribution $\hat{H}_*$ such that $\hat{H}_2 \overset{cr}{\geq} \hat{H}_* \overset{rr}{\geq} \hat{H}_1$.

We close this section by mentioning an important result of Yanagimoto and Okamoto (1969). For this we need the following definition of Yanagimoto and Okamoto.

DEFINITION 4.8. Let $\mathbf{r}$ and $\mathbf{s}$ correspondingly denote the row and column rank orders of $(X_1, Y_1), \ldots, (X_n, Y_n)$. A statistic T is said to be a rank statistic if there exists a function f (or equivalently g) from S_n to R^1 such that

$$T((X_1, Y_1), \ldots, (X_n, Y_n)) = f(\mathbf{r}) = g(\mathbf{s})$$

Moreover, T is said to be AI with respect to the ordering b_3 if f preserves the b_3 ordering.

THEOREM 4.9. *Assume that H_1 and H_2 have continuous marginals with $F_1(x) = F_2(x)$ and $G_1(y) = G_2(y)$. If (a) $H_2 \overset{rr}{\geq} H_1$ and (b) the rank statistic T is AI with respect to the ordering b_3, then T is stochastically not smaller under H_2 than under H_1, i.e.,*

$$Prob_{H_2}(T \geq c) \geq Prob_{H_1}(T \geq c) \ \textit{for all } c.$$

PROOF. See Yanagimoto and Okamoto (1969, Theorem 6.1).

Lehmann (1966) considered the rank statistic $T((X_1, Y_1, \ldots, X_n, Y_n)) = \sum_{k=1}^{n} A(k)B(\mathrm{s}(k)) \equiv g(\mathrm{s})$, where A and B are nondecreasing. It follows that $g \in \mathrm{AI}(b_3)$. Hence, T is AI with respect to the ordering b_3. Theorem 4 of Lehmann (1966), therefore, follows from Theorem 4.9.

REFERENCES

AHMED, A.N., LANGBERG, N.A., LÉON, R., and PROSCHAN, F. (1979). Partial orderings of positive quadrant dependence, with applications. Technical Report No. 78–3. Department of Statistics, Florida State University.

BLOCK, H.W., CHHETRY, D., FANG, Z., and SAMPSON, A.R. (1987). Metrics on permutations based on partial orderings. Technical Report No. 87–11. University of Pittsburgh.

BLOCK, H.W., CHHETRY, D., FANG, Z., and SAMPSON, A.R. (1990). Partial orders on permutations and dependence orderings on bivariate empirical distributions. *Ann. Statist.*, to appear.

BLOCK, H.W. and SAMPSON, A.R. (1988). Conditionally ordered distributions. *J. Multi. Anal.* **27** 91–104.

BOLAND, P.J., EL-NEWEIHI, E., and PROSCHAN, F. (1989). Stochastic order for inspection and repair policies. Technical Report. Florida State University.

CAMBANIS, S., SIMONS, G., and STOUT, W. (1976). Inequalities for EK(x, y) when marginals are fixed. Z. WAHR. **36** 285–294.

DIACONIS, P. and GRAHAM, R.L. (1977). Spearman's foot rule as a measure of disarray. *J. Roy. Statist. Soc. B* **39** 262–268.

FLIGNER, M.A. and VERDUCCI, J.S. (1986). Distance based ranking models. *J. Roy. Statist. Soc. B* **48** 359–369.

HĂJEK, J. (1969). Miscellaneous problems of rank testing. *Proc. Symp. Nonparametric Tech.*, Bloomington, IN.

HOLLANDER, M., PROSCHAN, F., and SETHURAMAN, J. (1977). Functions decreasing in transposition and their applications in ranking problems. *Ann. Statist.* **5** 722–733.

KARLIN, S. and RINOTT, Y. (1980). Classes of orderings of measures and related correlation inequalities I. Multivariate totally positive distributions. *J. Multi. Anal.* **10** 476–498.

KENDALL, M.G. (1970). *Rank Correlation Methods, 4th Ed.* Griffin, London.

KIMELDORF, G. and SAMPSON, A.R. (1987). Positive dependence orderings. *Ann. Inst. Statist. Math. A* **39** 113–128.

KNUTH, D. (1973). *The Art of Computer Programming, Vol. 2.* Addison-Wesley, Reading, MA.

LEHMAN, E.L. (1966). Some concepts of dependence. *Ann. Math. Statist.* **37** 1137–1153.

METRY, M. and SAMPSON, A.R. (1988a). Characterizing and generating bivariate empirical rank distributions satisfying certain positive dependence concepts. Technical Report No. 88–04. University of Pittsburgh.

METRY, M. and SAMPSON, A.R. (1988b). Positive dependence concepts for multivariate empirical rank distributions. Technical Report No. 88–09. University of Pittsburgh.

SAVAGE, I.R. (1957). Contributions to the theory of rank order statistics—the 'trend' case. *Ann. Math. Statist.* **28** 968–977.

SCHRIEVER, B.F. (1985). *Order Dependence.* Ph.D. Dissertation, Free University of Amsterdam.

SCHRIEVER, B.F. (1987a). An ordering for positive dependence. *Ann. Statist.* **15** 1208–1214.

SCHRIEVER, B.F. (1987b). Monotonity of rank statistics in some non-parametric testing problems. *Statistica Neerlandica* **41** 91–109.

SOBEL, M. (1955). On a generalization of an inequality of Hardy, Littlewood, and Polya. *Proc. Amer. Math. Soc.* **5** 596–602.

TCHEN, A.H.T. (1980). Inequalities for distributions with given marginals. *Ann. Prob.* **8** 814–927.

WHITT, W. (1982). Multivariate monotone likelihood ratio and uniform conditional stochastic order. *J. Appl. Prob.* **19** 695–701.

YANAGOMOTO, T. (1990). Dependence ordering in statistical models and other notions. To appear in *Topics in Statistical Dependence*, Institute of Mathematical Statistics Lectures Notes—Monograph Series.

YANAGIMOTO, T. and OKAMOTO, M. (1969). Partial orderings of permutations and monotonicity of a rank correlation statistics. *Ann. Inst. Statist. Math.* **21** 489–506.

DEPARTMENT OF MATHEMATICS AND STATISTICS
UNIVERSITY OF PITTSBURGH
PITTSBURGH, PA 15260

DEPARTMENT OF MATHEMATICS
KIRTIPUR CAMPUS
TRIBHUVAN UNIVERSITY
KATHMANDU, NEPAL

DEPARTMENT OF MATHEMATICS
UNIVERSITY OF SCIENCE
AND TECHNOLOGY OF CHINA
HEFEI, ANHUI 230029
PEOPLES REPUBLIC OF CHINA

DEPARTMENT OF MATHEMATICS AND STATISTICS
UNIVERSITY OF PITTSBURGH
PITTSBURGH, PA 15260

COMPARISONS FOR MAINTENANCE POLICIES INVOLVING COMPLETE AND MINIMAL REPAIR

BY HENRY W. BLOCK,[1], NAFTALI A. LANGBERG,[2] AND THOMAS H. SAVITS[1]

University of Pittsburgh, Haifa University, and University of Pittsburgh

Maintenance policies are compared under various types of aging. Formerly, a standard assumption was that when a component or system failed, it was replaced by a new one. Preventive maintenance usually took the form of replacement according to an age or block policy. Under the assumption that a component can be minimally repaired, new results involving block replacement policies can be obtained. An analog of age replacement, called repair replacement, is also discussed and compared with other policies.

1. Introduction. The study of operating characteristics of maintenance policies in reliability has a long history. For a survey of the very early developments see Barlow and Proschan (1965). In this article we shall review one aspect of this area, the comparison of maintenance policies.

A maintenance policy involves repairing or replacing a system or component when it fails. This cycle is continued indefinitely. We shall not consider the time taken to repair the component in this paper. An assumption equivalent to not considering these repair times, which we shall make, is that repairs or replacements are instantaneous. Rather than waiting for components to fail, intervention is possible in the sense that replacements may be planned. That is, working components can be replaced (or overhauled, but we shall consider only replacements for planned intervention in this paper) before failure. Two standard forms of intervention are block and age policies. A block policy is said to be in effect if components are replaced on a fixed schedule determined a priori and not depending on unplanned failures which may occur. Unplanned failures are handled as if there were no block policy. An age policy mandates replacement on a fixed schedule starting at time zero and continuing until an unplanned failure occurs at which time a new schedule starts.

If components are used which wear out it would seem that intervention of the age or block type would result in fewer unplanned failures. It has been shown

[1]Supported by AFOSR Grants No. AFOSR–84–0113 and AFOSR–89–0370.

[2]Partially supported by AFOSR Grant No. AFOSR–84–0113.

AMS 1980 subject classification. 62N05.

Key words and phrases. Minimal repair, complete repair, block replacement, age replacement, repair replacement.

that under certain types of wearout that this is in fact the case. Also comparisons between age and block policies have been made in order to determine which of these two policies is preferable under various types of wearout. Most of these results have been obtained under the assumption that unplanned replacements are complete (i.e., replacement is with new components). Equivalent to saying that replacements are complete is that components are repaired to a good as new state. This type of assumption is unrealistic in the situation where components can be repaired so they are only as good as they were immediately prior to failure. This is called the "bad as old" or "minimal repair" assumption and although the concept is not a new one, recent research has shown that many results holding in the good as new case also hold in the bad as old situation. We shall discuss some of these results for the comparison of maintenance policies.

In Section 3 we give a review of the comparison of policies where failed components are completely repaired. The comparison of policies where components are minimally repaired and block replacements are undertaken is discussed in Section 4. Section 5 is similar to Section 4 where instead of block replacement, a generalization of age replacement which we call repair replacement is considered. Section 6 considers comparisons for the totality of all planned replacements and unplanned repairs. An appendix gives mathematical details for the type of stochastic comparisons considered.

2. Preliminaries. We use a counting process $\{N(t),\ t \geq 0\}$ to model the number of failures of a system or a component, where the times between failure are given by $\{X_n,\ n \geq 1\}$. If the failed system or component is repaired and if there is no block or age policy, N will count only these unplanned repairs. If the repairs are complete then $\{X_n, n \geq 1\}$ will consist of independent failure times. If, in addition, the components used for replacement are alike the process N can be assumed to be a renewal process. The assumption that the repairs are minimal is modeled by assuming that the process is nonhomogeneous Poisson and in this case we use the notation $\{N_m(t),\ t \geq 0\}$. For background concerning minimal repair see Asher and Feingold (1984).

For a block replacement policy where replacements are planned at times T, $2T$, $3T, \ldots$ we use the notation $N^B(T,t)$ to indicate the number of unplanned complete repairs up to time t. We designate the process by $N^B(T)$. The process of all unplanned repairs and planned replacements is given by $R^B(T)$. If the underlying repair process is minimal we write $N_m^B(T)$ and $R_m^B(T)$ for the corresponding block replacement processes. Formerly R processes were called removal processes (see Barlow and Proschan (1981), p. 181). We shall call these *renovation processes.*

For an age replacement policy where the replacements occur every T units after a complete repair we let $N^A(T,t)$ be the number of unplanned repairs up to time t and $R^A(T,t)$ be the total number of unplanned repairs and planned replacements up to time t where $N^A(T)$ and $R^A(T)$ designate whole processes. If we want to emphasize the distribution of the underlying process, N or N_m, we will write N_F or $N_{m,F}$ where F is the distribution time until the first event of the process.

The basic results we will discuss will involve counting processes N_1 and N_2 related to two different maintenance policies. Some of the earliest results (see Barlow and Proschan, 1965) were of the type

(1) $$E[N_1(t)] \geq E[N_2(t)] \quad \text{for all } t.$$

Other results involve limiting relationships which can be obtained from the above by letting $t \to \infty$. Another standard result is of the form

(2) $$P\{N_1(t) \geq n\} \geq P\{N_2(t) \geq n\} \quad \text{for } n = 1, 2, \ldots, \text{ all } t,$$

which is equivalent to assuming that $E[f(N_1(t))] \geq E[f(N_2(t))]$ holds for all increasing functions f and all t. In this case we write

$$N_1(t) \overset{\text{st}}{\geq} N_2(t) \quad \text{for all } t.$$

We call this type of result a marginal stochastic comparison since it only involves the one dimensional marginals of the processes N_1 and N_2. A more general type of result would be

(3) $$(N_1(t_1), N_1(t_2)) \overset{\text{st}}{\geq} (N_2(t_1), N_2(t_2)) \quad \text{for all } t_1 < t_2,$$

which means

$$E[f(N_1(t_1), N_1(t_2))] \geq E[f(N_2(t_1), N_2(t_2))] \quad \text{for all } t_1 < t_2$$

and all f increasing in both variables.

The stochastic comparisons mentioned above are subsumed under a more general stochastic comparison denoted by

(4) $$N_1 \overset{\text{st}}{\geq} N_2.$$

This stochastic comparison is defined in the Appendix following (A.3). The definition in essence gives that the measure induced by N_1 on the space of counting processes stochastically dominates the measure induced by N_2. It is shown that this latter comparison is equivalent to the stochastic ordering of all finite dimensional distributions. Consequently this implies the two dimensional version given by (3).

3. Comparison of Policies with Complete Repair. Various results have been given which compare $N(t)$, $N^A(T,t)$, $N^B(T,t)$, $R^A(T,t)$ and $R^B(T,t)$. One elementary result is that without any assumptions on the lifetime

(5) $$R^A(T,t) \overset{\text{st}}{\geq} R^B(T,t) \quad \text{for all } t.$$

This is Theorem 4.1 (in different notation) of Barlow and Proschan (1965, p. 67). A second more interesting, but still intuitive, result is that if the lifetimes are IFR, then

$$N(t) \overset{\text{st}}{\geq} N^A(T,t) \overset{\text{st}}{\geq} N^B(T,t) \quad \text{for all } t. \tag{6}$$

This is also proven in Barlow and Proschan (1965, Theorem 4.4, p. 69). As an immediate consequence, using the second stochastic inequality and asymptotic results, these authors conclude (Corollary 4.5, p. 71) that if F is an IFR lifetime, then

$$E[N(t)] \leq \frac{tF(t)}{\int_0^t \bar{F}(x)dx}. \tag{7}$$

In Marshall and Proschan (1972) various improvements of these results were obtained where the weaker property of NBU is assumed rather than IFR. These results appear in Barlow and Proschan (1981) and we mention the most important of these:

$$N(t) \overset{\text{st}}{\geq} N^A(T,t) \quad \text{for all } t,T \quad \text{if and only if } F \text{ is NBU;} \tag{8}$$

$$N(t) \overset{\text{st}}{\geq} N^B(T,t) \quad \text{for all } t,T \quad \text{if and only if } F \text{ is NBU;} \tag{9}$$

$$N^B(T,t) \overset{\text{st}}{\leq} N^B(kT,t) \quad \text{for all } t,T,k \quad \text{if and only if } F \text{ is NBU;} \tag{10}$$

$$N^A(T,t) \overset{\text{st}}{\leq} N^A(kT,t) \quad \text{for all } t,T,k \quad \text{if and only if } F \text{ is NBU;} \tag{11}$$

$$N^A(T_1,t) \overset{\text{st}}{\leq} N^A(T_2,t) \quad \text{for all } T_1 \leq T_2 \text{ for all } t \text{ if and only if } F \text{ is IFR} \tag{12}$$

where $kT =< kT, 2kT, 3kT, \cdots >$ and k is a positive integer. These are all proven in Chapter 6, Section 4 of Barlow and Proschan (1981).

One other result which is of the above type, but is given in a disguised form as Theorem 3.2 of Chapter 6 of Barlow and Proschan (1981) is that if F is the distribution function of X_1 from the nonhomogeneous Poisson process N_m and also the distribution function associated with the renewal process N then

$$F \text{ is NBU implies } N_m(t) \overset{\text{st}}{\geq} N(t) \quad \text{for all } t. \tag{13}$$

This was also proven by Blumenthal, Greenwood, and Herbach (1976).

4. Block Replacement Policies. Block, Langberg, and Savits (1990) have generalized the previous results in several ways. First, they considered more general policies Z. By a policy Z we mean that replacements occur at times $z_1 < z_2 < z_3 \cdots$ and we use the notation $Z =< z_1, z_2, \cdots >$. In the special case that the z's occur every T units (i.e., $z_i = iT$) we write $Z = T$. Consequently, we use the notation $N^B(Z)(N_m^B(Z))$ to designate the process giving the number of unplanned replacements in a complete (minimal) repair block process. We continue to use the term block here in the sense that the replacement schedule is made a priori (in a block) even though the z_i need not be equal. Similarly $R^B(Z)$ and $R_m^B(Z)$ will designate the renovation process. These authors obtained comparisons of the whole processes of type (4) given in Section 2. Also, these authors considered the comparison of minimal repair policies involving block replacement. A basic result used to obtain some of these generalizations involves comparison of minimal repair processes with stochastically ordered lifetimes. That is, let F and G be the distribution functions of the lifetimes associated with the nonhomogeneous Poisson processes $\{N_{m,F}(t), t \geq 0\}$ and $\{N_{m,G}(t), t \geq 0\}$. The result is

$$N_{m,F} \overset{\text{st}}{\geq} N_{m,G} \quad \text{if and only if} \quad F \overset{\text{st}}{\leq} G \quad (\text{i.e., } F(t) \geq G(t) \quad \text{for all } t).$$

See, for example, Theorem 3.1.1(a) of the paper cited above. The following comparisons are also obtained:

$$N_m^B(Z) \overset{\text{st}}{\geq} N^B(Z) \quad \text{for all } Z \text{ if } F \text{ is NBU;} \tag{14}$$

$$N_m \overset{\text{st}}{\geq} N_m^B(Z) \quad \text{for all } Z \text{ if and only if } F \text{ is NBU;} \tag{15}$$

$$N_m \overset{\text{st}}{\geq} N^B(Z) \quad \text{for all } Z \text{ if and only if } F \text{ is NBU.} \tag{16}$$

Notice that (14) above generalizes the result (13) and gives that for any block replacement policy Z, a minimal repair policy for unplanned failures produces stochastically more failures than a complete repair process. The result (15) gives a result which generalizes (9) and the result (16) is a hybrid of (14) and (15).

In Theorem 4.1 of Block, Langberg, and Savits (1990), several other results which generalize (5) are also given.

5. Repair Replacement Policies. Block, Langberg, and Savits (1989) consider repair replacement policies. In these policies items are either minimally or completely repaired at unplanned failures or they are replaced if they survive a certain fixed time from the last repair without suffering an unplanned failure. If at failure only complete repairs are allowed, then the repair replacement policy reduces to an age replacement policy. Consequently, the concept of a repair replacement policy is a more general type of replacement policy than an age replacement policy. We consider repair replacement policies for two reasons. First, upon repair it may be that a replacement is scheduled within a short period of time; however, when the repair was made it may be clear that the component was in good shape and did not require an immediate replacement. Consequently, the replacement should be deferred. One way of deferring it is to start a new replacement schedule from the time of the repair. Secondly, as we will see in (18), a repair replacement policy has fewer unscheduled repairs than a minimal repair policy under IFR lifetimes. This is often a desirable outcome if cost is not a consideration.

As before we let N and N_m denote the processes with only complete and only minimal repair respectively with no intervention. We now define the repair replacement processes with scheduled planned replacement determined by $Z = < z_k >$. Planned replacements occur at times $z_1, z_1 + z_2, \ldots$ until an unplanned repair occurs. Assume this occurs between times $\Sigma_{i=1}^{n_1 - 1} z_i$ and $\Sigma_{i=1}^{n_1} z_i$. The planned replacement schedule is then restarted from the time of the unplanned repair and the schedule of planned replacements is given by $z_{n_1+1}, z_{n_1+1} + z_{n_1+2}, \ldots$ units of time *after the unplanned repair.* This process continues. Notice that Z here is different than in the previous section in that the z_i here give times between planned replacements. See Block, Langberg, and Savits (1989) for more details.

If only complete repairs are allowed, i.e., the N process is used, and $z_k = T$ for all k we have that at each unplanned complete repair, the schedule of planned replacement times *from that repair* is $T, 2T, 3T, \ldots$. This yields the usual age replacement policy.

If the repairs are minimal, i.e., the process N_m is used, the general repair replacement policy results.

If only complete repairs are permitted we let $N^A(Z)$ be the process counting the number of unplanned (complete) repairs of the repair replacement policy (i.e., the age replacement policy). For $Z = T$, $N^A(Z)$ reduces to the process $N^A(T)$ discussed in (6).

For minimal repairs, we use $N_m^R(Z)$ for the process counting the number of unplanned (minimal) repairs of the repair replacement.

We are now able to state some results of Block, Langberg, and Savits (1989). If we let F be the underlying distribution of the renewal process N we have

$$(17) \qquad N \overset{\text{st}}{\geq} N^A(Z) \quad \text{for all } Z \text{ if and only if } F \text{ is NBU.}$$

This is given in Theorem 3.2(b) of the aforementioned paper. This result extends the first inequality of (6) in two directions. First, instead of T, a general time schedule Z is used and, second, the above result is a comparison of the type (4).

A second result, given in Theorem 3.2(d) of Block, Langberg, and Savits (1989), is that

$$(18) \qquad N_m \overset{\text{st}}{\geq} N_m^R(Z) \quad \text{for all } Z \text{ if } F \text{ is IFR.}$$

This is a companion result to (15) and gives a generalization of (17) from complete to minimal repairs. Various other comparisons can be obtained among the processes N_m, $N^A(Z)$, and $N_m^R(Z)$. An example which is an analog to (14) for block policies is that

$$N_m^R(Z) \overset{\text{st}}{\geq} N^A(Z) \quad \text{for all } Z \text{ if } F \text{ is NBU.} \tag{19}$$

This is given in Theorem 3.2(c) of Block, Langberg, and Savits (1989).

Finally, the second inequality of (6) holds for processes as well as marginally and is given by Theorem 5.2 of Block, Langberg, and Savits (1989).

6. Comparisons for Renovations. We now consider the total number of renovations, i.e., the total number of planned replacements and unplanned repairs for both the block and repair replacement policies. We recall that $R^B(Z)$ and $R_m^B(Z)$ are the total renovations for the renewal process and minimal repair process with block policy Z. Similarly for the renewal process we use $R^A(Z)$ and $R_m^R(Z)$.

The following results have been shown in Block, Langberg, and Savits (1990):

$$N \overset{\text{st}}{\leq} R^B(Z) \quad \text{for all } Z; \tag{20}$$

$$R^B(Z) \overset{\text{st}}{\leq} R_m^B(Z) \quad \text{for all } Z \text{ if } F \text{ is NBU;} \tag{21}$$

$$N_m \overset{\text{st}}{\leq} R^B(Z) \quad \text{for all } Z \text{ if } F \text{ is NWU.} \tag{22}$$

Result (22) says that if a lifetime undergoes beneficial aging, there are more renovations (i.e., removals) for a renewal process with block replacement, then there are repairs for a minimal repair process.

For repair replacement policies Block, Langberg, and Savits (1989) have shown, among other results, that

$$N \overset{\text{st}}{\leq} R^A(Z) \quad \text{for all } Z, \tag{23}$$

$$R^A(Z) \overset{\text{st}}{\leq} R_m^R(Z) \quad \text{for all } Z \text{ if } F \text{ is NBU} \tag{24}$$

and

$$N_m \overset{\text{st}}{\leq} R^A(Z) \quad \text{for all } Z \text{ if } F \text{ is NWU.} \tag{25}$$

Appendix. In order to define the stochastic ordering of two processes N_1 and N_2 in (4), it would be enough to restrict ourselves to the class of all counting processes by which we mean the class of stochastic processes whose sample paths are nonnegative right-continuous step functions, starting at 0, only increasing by jumps of size one, and endowed with the Skorohod topology. We denote the set of all such sample paths by $S([0,\infty))$. However, it is convenient to enlarge our viewpoint somewhat. The framework which we shall follow is that delineated in Kamae, Krengel, and O'Brien (1977).

Let E be a Polish space (i.e., a complete separable metric space) equipped with a closed partial ordering $\leq$. A partial ordering $\leq$ is said to be closed if its graph $\{(x,y) : x \leq y\}$ is a closed subset of $E \times E$ in the product topology. The Borel σ-algebra on E is denoted by $\mathcal{E}$.

The principal examples considered in this paper are listed below. First, however, we state two useful facts.

(A.1) A countable product of partially ordered Polish spaces is also a partially ordered Polish space under the product topology and the coordinatewise partial ordering.

(A.2) A closed subset of a partially ordered Polish space is itself a partially ordered Polish space with the induced topology and partial ordering.

Examples:

(1) If $E_i = [0,\infty)$ with the usual topology and ordering, then we denote the partially ordered Polish space $\prod_{i=1}^{\infty} E_i$ by $\mathbf{R}_+^\infty$.

(2) Let I be the interval $[a,b]$ and $D(I)$ the set of all functions $x : I \to R$ which are right-continuous with left-hand limits. Then it is well-known that $D(I)$ equipped with the Skorohod metric and pointwise partial ordering (i.e., $x \leq y$ if and only if $x(t) \leq y(t)$ for all $t \in I$) is a partially ordered Polish space. Sometimes we also want to consider the case $I = [a,b)$ with b possibly infinite. In this case, the metric can be modified so that $D(I)$ is again a partially ordered Polish space (see Stone (1963)).

(3) For $D(I)$ as above, let $S(I)$ denote the subset of all nondecreasing step-functions $s : I \to \{0,1,2,\ldots\}$ having only jumps of size one and satisfying $s(a) = 0$. It is not hard to show that $S(I)$ is closed and hence itself a partially ordered Polish space.

We now consider the notion of stochastic order on a partially ordered Polish space E. A Borel set $U \in \mathcal{E}$ is said to be an upper set if $x \in U$ and $x \leq y$ implies $y \in U$. If λ_1 and λ_2 are two probability measures on $(E,\mathcal{E})$, we say that λ_1 is stochastically smaller than λ_2, written as $\lambda_1 \overset{\text{st}}{\leq} \lambda_2$, if $\lambda_1(U) \leq \lambda_2(U)$ for all upper sets U. This is equivalent to the condition that $\int f d\lambda_1 \leq \int f d\lambda_2$ for all nonnegative nondecreasing Borel measurable functions f. (A function $f : E \to \mathbf{R}$ is said to be nondecreasing if $x \leq y$ implies that $f(x) \leq f(y)$).

Another preservation property which is useful is the following:

(A.3) If $\phi : E \to F$ is a nondecreasing measurable mapping between two partially ordered Polish spaces and λ_1, λ_2 are two probability measures on $(E,\mathcal{E})$ with $\lambda_1 \overset{\text{st}}{\leq} \lambda_2$, then the induced measures $\mu_i = \lambda_i \circ \phi^{-1}$ on $(F,\mathcal{F})$ satisfy $\mu_1 \overset{\text{st}}{\leq} \mu_2$.

Let E be a partially ordered Polish space. By an E-valued stochastic process X we mean a measurable mapping from a probability space $(\Omega, \mathcal{F}, P)$ into $(E, \mathcal{E})$. We denote the induced probability measure $P \circ X^{-1}$ on $(E, \mathcal{E})$ by λ_X. If X and Y are two E-valued stochastic processes, we say that X is stochastically smaller than Y, denoted by $X \overset{\text{st}}{\leq} Y$, if $\lambda_X \overset{\text{st}}{\leq} \lambda_Y$. We say that X and Y are stochastically equivalent if $\lambda_X \overset{\text{st}}{\leq} \lambda_Y$ and $\lambda_Y \overset{\text{st}}{\leq} \lambda_X$; i.e., $\lambda_X = \lambda_Y$.

Several of the results in Block, Langberg, and Savits (1990) are a consequence of the following theorem. Let E_i, $i = 1, 2, \ldots$ be a sequence of partially ordered Polish spaces and set $E = \prod_{i=1}^{\infty} E_i$. We define the projection map $\pi_i : E \to E_i$ as usual. If X is an E-valued stochastic process, denote the ith coordinate of X by X_i, i.e., $X_i = \pi_i X$. For $n = 1$, define the probability measure p_1 on E_1 by $p_1(A) = P(X_1 \in A)$. For $n \geq 2$, let $p_n(A|x_1, \ldots, x_{n-1})$ be a regular conditional probability of $P(X_n \in A | X_1 = x_1, \ldots, X_{n-1} = x_{n-1})$. Such exists because $E_1 \times \cdots \times E_{n-1}$ is Polish. (See, e.g., Breiman (1968)). We shall call the collection $< p_n >$ a system of transition probabilities (for X). Note that such a system completely determines the induced probability measure λ_X on $E = \prod_{i=1}^{\infty} E_i$.

Now suppose X and Y are two $E = \prod_{i=1}^{\infty} E_i$-valued stochastic processes with corresponding systems $< p_n >$ and $< q_n >$ of transition probabilities. We then write $< p_n > \overset{\text{st}}{\leq} < q_n >$ if

(i) $p_1 \overset{\text{st}}{\leq} q_1$ and

(ii) $p_n(\cdot|x_1, \ldots, x_{n-1}) \overset{\text{st}}{\leq} q_n(\cdot|y_1, \ldots, y_{n-1})$

whenever $x_i \leq y_i$, $i = 1, \ldots, n-1$ and $n = 2, 3, \ldots$.

We are now ready to state the following result. It is essentially a reformulation of Theorem 2 in Kamae, Krengel, and O'Brien (1977).

THEOREM A.4. *Let $E = \prod_{i=1}^{\infty} E_i$ be the product of partially ordered Polish spaces and X, Y two E-valued stochastic processes with corresponding systems $< p_n >$ and $< q_n >$ of transition probabilities. If $< p_n > \overset{st}{\leq} < q_n >$, then $X \overset{st}{\leq} Y$.*

COROLLARY 1. *Let $\{K(t); t \geq 0\}$ and $\{L(t); t \geq 0\}$ be two counting processes with corresponding interarrival times $< X_n >$ and $< Y_n >$. Assume that*

(i) $X_1 \overset{st}{\leq} Y_1$

and

(ii) $(X_n | X_1 = x_1, \ldots, X_{n-1} = x_{n-1}) \overset{st}{\leq} (Y_n | Y_1 = y_1, \ldots, Y_{n-1} = y_{n-1})$

for $n \geq 2$ and whenever $x_1 \leq y_1, \ldots, x_{n-1} \leq y_{n-1}$. Then $K \overset{st}{\geq} L$.

PROOF. Let $E_i = [0,\infty)$. Then $\mathbf{X} =< X_1, X_2, \ldots >$ is an $E = \prod_{i=1}^{\infty} E_i$-valued stochastic process with a system $< p_n >$ of transition probability given by

$$p_n(A|x_1,\ldots,x_{n-1}) = P\{X_n \in A | X_1 = x_1, \ldots, X_{n-1} = x_{n-1}\}.$$

Similarly $\mathbf{Y} =< Y_1, Y_2, \ldots >$ is an E-valued stochastic process with system $< q_n >$ where

$$q_n(A|y_1,\ldots,y_{n-1}) = P\{Y_n \in A | Y_1 = y_1, \ldots, Y_{n-1} = y_{n-1}\}.$$

By assumptions (i) and (ii), we have $< p_n > \overset{\text{st}}{\leq} < q_n >$. Hence $\mathbf{X} \overset{\text{st}}{\leq} \mathbf{Y}$. Since $K = \Phi(< \mathbf{X}_i >)$ and $L = \Phi(< \mathbf{Y}_i >)$ with Φ a nonincreasing function on E, we deduce that $K \overset{\text{st}}{\geq} L$ by (A.3). Here Φ is defined as the mapping $\Phi : \mathbf{R}_+^\infty \to S([0,\infty))$ given by

$$\Phi(< x_i >)(t) = \sum_{i=1}^{\infty} I_{[0,t]}(x_1 + \cdots + x_i).$$

For the next result we need to introduce another mapping. Let $E_i = S([0, w_i])$ for $i = 1, 2, \ldots$ and set $w = \sum_{i=1}^{\infty} w_i (0 < w \leq \infty)$. We define the mapping

$$\Psi : \prod_{i=1}^{\infty} E_i \to E = S([0,w))$$

by

$$\Psi(< s_i >)(t) = \begin{cases} s_1(t) & \text{if } 0 \leq t < w_1, \\ \sum_{i=1}^{k-1} s_i(w_i) + s_k(t - \sum_{i=1}^{k-1} w_i) & \text{if } \sum_{i=1}^{k-1} w_i \leq t < \sum_{i=1}^{k} w_i. \end{cases}$$

COROLLARY 2. *Let $K_i = \{K_i(t);\ 0 \leq t \leq w_i\}$ and $L_i = \{L_i(t);\ 0 \leq t \leq w_i\}$ be counting processes such that $K_i \overset{st}{\leq} L_i$ for each $i = 1, 2, \ldots$. Suppose that each sequence $< K_i >$ and $< L_i >$ is independent. Then if $K = \Psi(< K_i >)$ and $L = \Psi(< L_i >)$ we have $K \overset{st}{\leq} L$.*

PROOF. Let $E_i = S([0, w_i])$ and set $X_i = K_i$ and $Y_i = L_i$. Then $\mathbf{X} = < X_1, X_2, \ldots >$ and $\mathbf{Y} =< Y_1, Y_2, \ldots >$ are $E = \prod_{i=1}^{\infty} E_i$-valued stochastic processes. By independence, the corresponding system of transition probabilities are given by

$$p_1(A) = P\{X_1 \in A\} = P\{K_1 \in A\},$$

$$q_1(A) = P\{Y_1 \in A\} = P\{L_1 \in A\},$$

and, for $n \geq 2$,

$$p_n(A|x_1, \ldots, x_{n-1}) = P\{X_n \in A|X_1 = x_1, \ldots, X_{n-1} = x_{n-1}\} = P\{K_n \in A\},$$

$$q_n(A|y_1, \ldots, y_{n-1}) = P\{Y_n \in A|Y_1 = y_1, \ldots, Y_{n-1} = y_{n-1}\} = P\{L_n \in A\}.$$

Thus $< p_n > \stackrel{\text{st}}{\leq} < q_n >$ since $K_n \stackrel{\text{st}}{\leq} L_n$ for all n by assumption. Consequently, $\mathbf{X} \stackrel{\text{st}}{\leq} \mathbf{Y}$. Since Ψ is a nondecreasing function on E, the result follows from (A.3).

Finally, in Section 2 following (4), we stated that the notion of stochastic ordering for counting processes could be equivalently expressed in terms of their finite dimensional distributions. We now give a proof of this obvious result.

Let X be a $D(I)$-valued stochastic process and $T = \{t_1, \ldots, t_n\} \subset I$. We assume without loss of generality that $t_1 < \cdots < t_n$. The induced probability measure $P\{(X(t_1), \ldots, X(t_n)) \in A\}$ on $(\mathbf{R}^n, \mathcal{B}^n)$ is called a *finite dimensional distribution of* X and is denoted by $\lambda_X^T(A)$.

The next lemma will be used in the proof of Theorem A.6 and may be of some independent interest.

LEMMA A.5. *Let $\lambda \stackrel{st}{\leq} \mu$ on a partially ordered Polish space E and suppose F is a closed subset of E. Then if $\lambda(F) = \mu(F) = 1$, it follows that $\lambda|F \stackrel{st}{\leq} \mu|F$.*

PROOF. Let U be an upper set in F and $\epsilon > 0$. Then there exists a compact set $K \subset U$ such that $\lambda(K) \geq \lambda(U) - \epsilon$. If $V = \{y \in E : y \geq x \text{ for some } x \in K\}$, then V is a closed upper set in E which contains K (see, e.g., Nachbin (1965)). Hence $\lambda(V) \leq \mu(V)$ and consequently $\lambda(V \cap F) = \lambda(V) \leq \mu(V) = \mu(V \cap F)$. But $V \cap F \subset U$ and so

$$\lambda(U) - \epsilon \leq \lambda(K) \leq \lambda(V) \leq \mu(V \cap F) \leq \mu(U).$$

THEOREM A.6. *Let X and Y be two $S(I)$-valued stochastic processes. Then $X \stackrel{st}{\leq} Y$ if and only if $\lambda_X^T \stackrel{st}{\leq} \lambda_Y^T$ for all finite sets $T \subset I$.*

PROOF. The necessity is clear since the mapping $\pi^T : S(I) \to \mathbf{R}^n$ given by $\pi^T(s) = (s(t_1), \ldots, s(t_n))$ is nondecreasing. Thus the result follows from (A.3).

Now we suppose that $\lambda_X^T \stackrel{\text{st}}{\leq} \lambda_Y^T$ for all finite sets $T \subset I$. Since $S(I) \subset D(I)$ we may consider X and Y as stochastic processes on $D(I)$. According to the proof of Theorem 4 in Kamae, Krengel, and O'Brien (1977) we can assert that $X \stackrel{\text{st}}{\leq} Y$ as $D(I)$-valued stochastic processes. The conclusion now follows from Lemma A.5.

REMARK A.7. Although Theorem 4 of Kamae, Krengel, and O'Brien (1977) is stated in terms of stochastic ordering between conditional probabilities, it is easy to reformulate it in terms of stochastic ordering between finite dimensional distributions. Note the equivalent version of Theorem A.6 for the space $D(I)$ is a consequence of their Theorem 4. In fact, if their Lemma 1 was adapted to the space $S(I)$, then Theorem A.6 would be a direct consequence of a suitably revised version of their Theorem 4.

REFERENCES

ASCHER, H. and FEINGOLD, H. (1984). *Repairable Systems Reliabililty: Modeling, Inference.* Marcel Dekker, New York.

BARLOW, R.E. and PROSCHAN, F. (1965). *Mathematical Theory of Reliability.* John Wiley and Sons, New York.

BARLOW, R.E. and PROSCHAN, F. (1981). *Statistical Theory of Reliability and Life Testing: Probability Models.* To Begin With, Silver Spring, MD.

BLOCK, H.W., LANGBERG, N., and SAVITS, T.H. (1990). Maintenance comparisons: Block policies. *J. Appl. Prob.*, to appear.

BLOCK, H.W., LANGBERG, N., and SAVITS, T.H. (1989). Repair replacement policies. University of Pittsburgh, Technical Report 89–07.

BLUMENTHAL, S.J., GREENWOOD, J.A., and HERBACH, L.H. (1976). A comparison of the bad as old and superimposed renewal models. *Management Science* **23** 280–285.

BREIMAN, L. (1968). *Probability.* Addison-Wesley Publishing Company, Reading, MA.

KAMAE, T., KRENGEL, U., and O'BRIEN, G.L. (1977). Stochastic inequalities on partially ordered spaces. *Ann. Prob.* **5** 899–912.

LANGBERG, N.A. (1988). Comparisons of replacement policies. *J. App. Prob.* **25** 780–788.

MARSHALL, A.W. and PROSCHAN, F. (1972). Classes of distribution applied in replacement, with renewal theory implications. *Proceedings of the 6th Berkeley Symposium on Mathematical Statistics and Probability* **Vol. 1** 395–415.

NACHBIN, L. (1965). *Topology and Order.* Van Nostrand, Princeton, NJ.

STONE, C. (1963). Weak convergence of stochastic processes defined on semi-infinite time intervals. *Proc. Amer. Math. Soc.* **14** 694–696.

DEPARTMENT OF MATHEMATICS AND STATISTICS
UNIVERSITY OF PITTSBURGH
PITTSBURGH, PA 15260

DEPARTMENT OF STATISTICS
HAIFA UNIVERSITY
MOUNT CARMEL, HAIFA, ISRAEL

DEPARTMENT OF MATHEMATICS AND STATISTICS
UNIVERSITY OF PITTSBURGH
PITTSBURGH, PA 15260

TIME SERIES MODELS FOR NON-GAUSSIAN PROCESSES

BY H.W. BLOCK[1], N.A. LANGBERG[1], D.S. STOFFER[1,2]

University of Pittsburgh, Haifa University, Israel, and University of Pittsburgh

In this paper we present univariate and multivariate time series models for processes with non-Gaussian marginal distributions. These include bivariate autoregressive-type models for processes with bivariate exponential marginals, nonlinear autoregressive-type models for processes with Dirichlet marginals, and nonlinear models for univariate time series with arbitrary marginal distributions. Examples of applications to real data sets are given for some of the models discussed. When applicable, the theory of positive dependence is used to establish the association of the processes.

1. Introduction. The classical model in multivariate time series analysis is the $m \times 1$ vector linear process given by

$$\mathbf{X}(n) = \Sigma_{j=-\infty}^{\infty} A(j)\epsilon(n-j), \quad n \in \mathcal{Z} \tag{1}$$

where $\mathcal{Z} = \{0, \pm 1, \pm 2, \ldots\}$, $\{\epsilon(n), n \in \mathcal{Z}\}$ is a sequence of iid $m \times 1$ random vectors with mean zero and unknown covariance matrix, and $\{A(n), n \in \mathcal{Z}\}$ is a sequence of unknown $m \times m$ matrices such that $\Sigma_{j=-\infty}^{\infty} \parallel A(j) \parallel < \infty$ where $\parallel \cdot \parallel$ denotes the usual eigenvalue norm. Note that autoregressive (AR), moving average (MA), and mixed autoregressive-moving average (ARMA) models are important particular cases of the classical linear process (1).

If the $\epsilon(n)$'s are Gaussian then clearly so are the $\mathbf{X}(n)$'s in (1). Furthermore, if the $\mathbf{X}(n)$'s are Gaussian with mean zero and absolutely continuous spectrum, then there is a sequence of iid normal mean-zero random vectors $\epsilon(n)$, $n \in \mathcal{Z}$, and a sequence of matrices $A(n), n \in \mathcal{Z}$, such that the two processes $\mathbf{X}(n)$ and

[1] Supported in part by the Air Force Office of Scientific Research grant AFOSR–84–0113.

[2] Supported in part by National Science Foundation Grant DMS–9000522 and by a grant from the Centers for Disease Control through a cooperative agreement with the Association of Schools of Public Health.

AMS 1980 subject classifications. Primary 62M10; secondary 62E99, 62H20.

Key words and phrases. Bivariate exponential distributions, Dirichlet distribution, nonlinear time series models, positive dependence, time series with arbitrary marginals.

The authors are grateful to Robert H. Shumway, University of California, Davis for supplying the data sets used in the examples. The authors would also like to thank a referee and the editors for comments and suggestions that improved the presentation.

$\Sigma_{j=-\infty}^{\infty} A(n)\epsilon(n-j), n \in \mathcal{Z}$, are stochastically equal (see Hannan, 1970, p. 221). If the $\mathbf{X}(n)$ process is non-Gaussian, then a decomposition (1) may not exist, and the statistical inference procedures developed for processes satisfying (1) do not apply.

There are, however, many physical situations in which time series are patently non-normal. It has been suggested that time series that depart slightly from normality be handled by data transformations. For other cases, where the departure from normality is more substantial, it has been suggested that new time series models be developed (see, for example, Lewis, 1980).

Over the past decade, there has been a considerable amount of research on modeling time series with exponential, gamma, geometric, or general discrete marginal distributions. For example, Lawrance and Lewis (1980, 1981, 1985) and Jacobs and Lewis (1977) present univariate ARMA-type models with exponential marginals. Raftery (1982) generalizes their models to include multivariate time series with exponential marginals. Gaver and Lewis (1980) present univariate AR-type models with gamma marginals, and Jacobs and Lewis (1978, 1983) present ARMA-type models for univariate discrete-valued time series. Block, Langberg, and Stoffer (1988) present bivariate ARMA-type models with bivariate exponential and geometric marginals, and Langberg and Stoffer (1987) develop bivariate MA-type processes with exponential and geometric marginals. Models and statistical methodologies for the analysis of categorical time series have recently been developed by Fahrmeir and Kauffmann (1987), Kauffmann (1987), and Stoffer (1987).

In this paper we focus on three different problems. In Section 2 we present a model and corresponding statistical methodology for analyzing univariate time series with arbitrary continuous marginal distributions. The model is then used to analyze wind speed data. In Section 3 we present the bivariate exponential autoregressive (BEAR) model which can be used to analyze bivariate time series in which the process of interest has bivariate exponential marginals. We show that a BEAR process can have well known bivariate exponential marginal distribution such as the Marshall and Olkin (1967) bivariate exponential distribution. The theory of positive dependence is used to show that the BEAR model can consist of associated random variables. Finally, in Section 4 we present a multivariate AR-type model with Dirichlet marginal distributions. This model is useful for modeling and forecasting vector processes in which the distribution of the random vector at each point in time is a Dirichlet distribution. Estimation and prediction methods are given for the model and the techniques are illustrated on a data set from soil science.

2. Univariate Processes With Arbitrary Continuous Marginals. In this section we present models and corresponding statistical methodology that will allow an investigator to model and analyze time series with any continuous marginal distribution, whether or not the investigator is willing to specify the family of distributions. The method used here is essentially the translation method (see Mardia, 1970) via the probability integral transform. That is, the processes are constructed by a monotone nondecreasing transform of a Gaussian linear process and the data are assumed to be obtained by a simple instantaneous nonlinear filter acting on a Gaussian process (see Hannan, 1970, section 2.7, for a discussion of nonlinear filters applied to Gaussian processes). Both a nonparametric approach (when the family of distributions is not specified) and a parametric approach (when the family of distributions is specified) are taken. The techniques of this section are then illustrated by analyzing a real data set.

Underlying both the nonparametric models and the parametric models is a mean zero Gaussian process $\{Y_n, n \in \mathcal{Z}\}$, such that $E\{Y_n^2\} = 1$. Thus, for each n, $Y_n \sim \Phi$ where Φ represents the Gaussian cdf. As an example of such a process, consider a linear process where $\{\epsilon_n, n \in \mathcal{Z}\}$ is a sequence of iid $N(0, \sigma^2)$ random variables, $\{\psi_0, \psi_1, \ldots\}$ is a sequence of parameters such that $\Sigma_{q=0}^{\infty} \mid \psi_q \mid < \infty$ and $\sigma^{-2} = \Sigma_{q=0}^{\infty} \psi_q^2$, and

$$(2) \qquad Y_n = \Sigma_{q=0}^{\infty} \psi_q \epsilon_{n-q}, \quad n \in \mathcal{Z}.$$

First we present the nonparametric approach. Let $\{X_n, n \in \mathcal{Z}\}$ be a process of interest such that the cdf of X_n is H (that is, $X_n \sim H, n \in \mathcal{Z}$) where H is a continuous, unspecified cdf. For H, let H^{-1} be a right continuous inverse of H given by:

$$(3) \qquad H^{-1}(p) = \begin{cases} \inf\,\{x : H(x) > p\} & 0 \le p < 1 \\ \sup\,\{x : H(x) < 1\} & p = 1. \end{cases}$$

We define the process $\{X_n, n \in \mathcal{Z}\}$ as follows:

$$(4) \qquad X_n = H^{-1}[\Phi(Y_n)], \quad n \in \mathcal{Z}$$

where $\{Y_n, n \in \mathcal{Z}\}$ is the Gaussian process. Noting that for all n, $\Phi(Y_n)$ has a uniform cdf on (0,1) it follows that $X_n \sim H, n \in \mathcal{Z}$.

Next we present the parametric approach. Let Ω be a subset of the kth Euclidean space and let $\{H_\theta : \boldsymbol{\theta} \in \Omega\}$ be a specified family of absolutely continuous cdf's which depends on the parameter $\boldsymbol{\theta}$. We define the process $\{X_n(\boldsymbol{\theta}); n \in \mathcal{Z}, \boldsymbol{\theta} \in \Omega\}$ as follows:

$$(5) \qquad X_n(\boldsymbol{\theta}) = H_\theta^{-1}[\Phi(Y_n)], \quad n \in \mathcal{Z}.$$

By a preceding argument, it follows that $X_n(\boldsymbol{\theta}) \sim H_\theta; n \in \mathcal{Z}, \boldsymbol{\theta} \in \Omega$.

For an example of some parametric models, let $\Omega = (0, \infty) \times (-\infty, \infty)$, let Z be a random variable with cdf F, and let H_θ be the cdf of $\theta_1 Z + \theta_2, \boldsymbol{\theta} \in \Omega$. Then the class of processes $\{X_n(\boldsymbol{\theta}) = \theta_1 F^{-1}[\Phi(Y_n)] + \theta_2; n \in \mathcal{Z}, \boldsymbol{\theta} \in \Omega\}$ has H_θ marginals. So, for example, the class of processes could have Cauchy marginals: $F^{-1}(p) = \tan \pi p$, logistic marginals: $F^{-1}(p) = -\ell n(p^{-1} - 1)$, or shifted exponential marginals: $F^{-1}(p) = -\ell n(1-p)$, to mention a few.

Estimation and prediction for each of the models is relatively simple if the underlying Gaussian process $\{Y_n, n \in \mathcal{Z}\}$ is an ARMA(p,q) process. Henceforce we assume that

$$(6) \qquad \Sigma_{j=0}^{p}\alpha_j Y_{n-j} = \Sigma_{k=0}^{q}\beta_k \epsilon_{n-k}, \quad n \in \mathcal{Z}, \quad \alpha_0 = \beta_0 = 1$$

where $\{\epsilon_n, n \in \mathcal{Z}\}$ is the process described in (2), and with appropriate conditions on $\{\alpha_1, \ldots, \alpha_p, \beta_1, \ldots, \beta_q\}$ for (6) to be a stationary, causal and invertible process (see for example, Brockwell and Davis, 1987). If $x_1, \ldots, x_N$ are N observations from a process given by (4) or (5), let $x_{(1)}, \ldots, x_{(N)}$ be the corresponding order statistics and let $H_N(t), t \in (-\infty, \infty)$, be the empirical cdf of the observations given by:

$$(7) \qquad H_N(t) = N^{-1}\Sigma_{n=1}^{N-1} I(t)_{[x_{(n)},\infty)} + N^{-1} I(t)_{(x_{(N)},\infty)}$$

where I_A denotes the indicator function of A. Note that in (7), $H_N(t) < 1$ for $t \leq x_{(N)}$; the necessity of this definition will be apparent in (8). The properties of the empirical cdf for dependent processes (that are mixing) are given in Gastwirth and Rubin (1975). A Kolmogorov-Smirnov statistic for (correlated) data sampled from an autoregressive process is given in Weiss (1978).

For the nonparametric model, estimation and prediction are accomplished by first transforming the data as follows:

$$(8) \qquad \hat{y}_n = \Phi^{-1}[H_N(x_n)], \quad n = 1, \ldots, N$$

where H_N is given in (7), and then by treating the $\hat{y}_n$ as a sample of length N from the Gaussian process (6). Parameter estimates and forecasts based on the $\hat{y}_n$ are obtained via standard methods and the forecasts of the original model are approximated by setting $x_{N+j}^N = H_N^{-1}[\Phi(\hat{y}_{N+j}^N)]$, $j = 1, 2, \ldots$, where $\hat{y}_{N+j}^N$ are the estimated j-step-ahead forecasts of the Y_n process based on the data $\hat{y}_1, \ldots, \hat{y}_N$, and H_N^{-1} is given in (3).

For the parametric model, parameter estimation can be obtained via maximum likelihood. Let $\boldsymbol{\alpha} = (\alpha_1, \ldots, \alpha_p), \boldsymbol{\beta} = (\beta_1, \ldots, \beta_q)$, and for $y_1, \ldots, y_N$, N observations from the Gaussian ARMA(p,q) process given by (6), let $L_Y(\boldsymbol{\alpha}, \boldsymbol{\beta}, y_1, \ldots, y_N)$ be the likelihood function of $y_1, \ldots, y_N$. Recall that σ^2 is a function of $\boldsymbol{\alpha}$ and $\boldsymbol{\beta}$ alone. The likelihood function of the data $x_1, \ldots, x_N$, is given by

$$(9) \quad L_X(\boldsymbol{\theta}, \boldsymbol{\alpha}, \boldsymbol{\beta}, x_1, \ldots, x_N) = L_Y(\boldsymbol{\alpha}, \boldsymbol{\beta}, \Phi^{-1}[H_\theta(x_1)], \ldots, \Phi^{-1}[H_\theta(x_N)]) \times \Pi_{n=1}^{N}\{\dot{\Phi}^{-1}[H_\theta(x_n)]\}\dot{H}_\theta(x_n)$$

where $\cdot$ denotes differentiation. We note that in (9) we may write

$$\dot{\Phi}^{-1}[H_\theta(x_n)] = 1/\dot{\Phi}\{\Phi^{-1}[H_\theta(x_n)]\}.$$

Let $\hat{\boldsymbol{\theta}}_N$ denote the MLE of $\boldsymbol{\theta}$, and let

$$(10) \qquad \tilde{y}_n = \Phi^{-1}[H_{\hat{\theta}_N}(x_n)], \quad n = 1, 2, \ldots, N,$$

then the forecasts for the parametric process (5) may be approximated by setting $x^N_{N+j}(\hat{\boldsymbol{\theta}}_N) = H^{-1}_{\hat{\theta}_N}[\Phi(\tilde{y}^N_{N+j})], j = 1, 2, \ldots,$ where $\tilde{y}^N_{N+j}$ denotes the forecast obtained from the $\tilde{y}_n$, $n = 1, \ldots, N$, based on the MLE's of $\boldsymbol{\alpha}$ and $\boldsymbol{\beta}$, and the model (6).

As an example of the kind of data that can be handled by the models, we fit both a parametric and a nonparametric model to wind speed (mph) measurements in Washington, D.C., May – September, 1977, 133 observations (which we denote $x_0, \ldots, x_{132}$) made daily at noon. Since wind speeds have been modeled using the Weibull distribution (see Lawrance and Lewis, 1985), our parametric model assumes that the data comes from a process with shifted Weibull marginals, that is,

$$(11) \qquad H_\theta(x) = 1 - \exp\{-\mu(x - \xi)^\gamma\}, \quad x \geq \xi$$

where $\boldsymbol{\theta} = (\gamma, \mu, \xi)$ with $\gamma > 0$, $\mu > 0$, and $-\infty < \xi < \infty$. Also, the data indicated an AR(1) model for the $\{Y_n\}$ process given by (6), that is,

$$(12) \qquad Y_n = \alpha Y_{n-1} + \epsilon_n, \quad | \alpha | < 1$$

where $\{\epsilon_n\}$ is white Gaussian noise, $\epsilon_n \sim N(0, \sigma^2)$, with $\sigma^2 = 1 - \alpha^2$.

Figure 2.1 shows a plot of the data as well as the one-step-ahead forecasts using the parametric approach. The actual data is shown by a solid line in Figure 2.1, with the data points represented by circles; the extreme observations are 5.0 mph and 18.8 mph. Maximum likelihood estimation (cf. 9) under model assumptions (5), (11) and (12) yielded the following MLE's:

$$\hat{\gamma}_N = 1.732, \ \hat{\mu}_N = .069, \ \hat{\xi}_N = 4.920, \ \hat{\alpha}_N = .388.$$

The one-step-ahead forecasts (shown in Figure 2.1 by a dashed line) were calculated by first computing the transformed data (10), then forecasting $\tilde{y}_n$ as $\tilde{y}^{n-1}_n = .388\tilde{y}_{n-1}$, and then setting $x^{n-1}_n(\hat{\boldsymbol{\theta}}_N) = H^{-1}_{\hat{\theta}_N}[\Phi(\tilde{y}^{n-1}_n)]$, $n = 1, \ldots, 132$, as the one-step-ahead wind speed forecasts. The analysis of the $\tilde{y}_n$ values $(n = 0, 1, \ldots, 132)$ using the parametric model verified the model assumption (12).

Figure 2.2 compares the empirical distribution function (solid line) of the wind speed data with $H_{\hat{\theta}_N}$ (dashed line), showing satisfactory results. It is clear from Figure 2.2 that the nonparametric approach for this data set leads to relatively the same results as the parametric approach. The only parameter to be estimated in the nonparametric case is α in (12). For this example we obtained a value of .386 as an estimate of α; this compares well with the corresponding estimate $\hat{\alpha}_N$ in the parametric case. As the one-step-ahead forecasts of the wind speed data do not differ visually from those obtained from the parametric model, we do not show these forecasts. The analysis of the $\hat{y}_n(n = 0, \ldots, 132)$ using the nonparametric approach (cf. 8) verified the model assumption (12).

Figure 2.1: Observed (solid line) and predicted (dashed line) wind speed data (mph) using a parametric model.

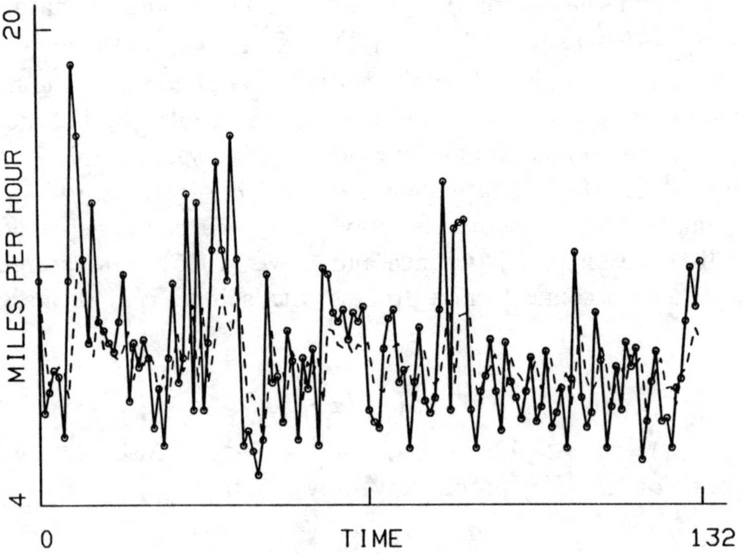

Figure 2.2: Empirical distribution (solid line) of the wind speed measurements and the fitted Weibull c.d.f. (dashed line).

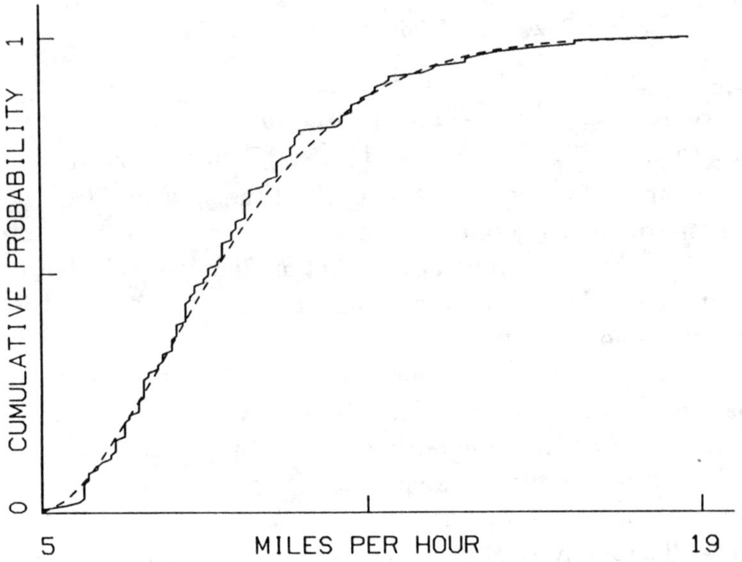

3. Bivariate Exponential Autoregressive Processes. In this section we present the BEAR(1) model and discuss some of its properties. A detailed account of the BEAR(1) process as well as some related models, statistical methodology, bivariate dependence mechanisms, and a data example may be found in Block, Langberg, and Stoffer (1988). Throughout this section we will say that a random vector $\mathbf{E} = (E_1, E_2)$ has a bivariate exponential distribution if the marginal distributions of E_1 and E_2 are exponential. First we introduce the concepts of random mixing (Lemma 3.1) and random summation (Lemma 3.2). We shall write $U \stackrel{S}{=} V$ to mean that random variables (or vectors) U and V are stochastically equal.

LEMMA 3.1. *(Random Mixing). Let X and Z be independent random variables with exponential distributions where $X \stackrel{S}{=} Z$. Let I be a Bernoulli random variable independent of X and Z, and such that* $\Pr\{I = 1\} = 1 - \pi$. *Then the random variable given by*

$$Y \stackrel{S}{=} IZ + \pi X$$

has the same exponential distribution as X and Z.

LEMMA 3.2. *(Random Summation). Let X_j, $j = 1, 2, \ldots$, be iid exponential random variables with mean $\pi/\lambda, 0 < \pi < 1$. Let N have a geometric distribution $(N \geq 1)$ with mean π^{-1} and be independent of the X_j. Then the random variable given by*

$$Y \stackrel{S}{=} \Sigma_{j=1}^{N} X_j$$

has an exponential distribution with mean λ^{-1}.

The proofs of Lemmas 3.1 and 3.2 follow easily by computing characteristic functions. In the following theorem we connect the concepts of these two lemmas and show how these results may be used to extend the concepts of random mixing and random summation to bivariate exponential random vectors. The proof of this theorem may be found in Block, Langberg, and Stoffer (1988, Lemma 2.17).

THEOREM 3.1. *Let (X_{1j}, X_{2j}), $j = 1, 2, \ldots$, be iid random vectors with a fixed bivariate exponential distribution. Let (N_1, N_2) be independent of the (X_{1j}, X_{2j}) and have the bivariate geometric distribution given by*

$$\Pr\{N_1 > n_1, N_2 > n_2\} = \begin{cases} p_{11}^{n_1}[p_{01} + p_{11}]^{n_2 - n_1} & 0 \leq n_1 \leq n_2 \\ p_{11}^{n_2}[p_{10} + p_{11}]^{n_1 - n_2} & 0 \leq n_2 \leq n_1 \end{cases} \tag{13}$$

where $p_{ij} \geq 0$ $(i, j = 0, 1)$, $p_{00} + p_{01} + p_{10} + p_{11} = 1$, $p_{10} + p_{11} < 1$, and $p_{01} + p_{11} < 1$. Let (I_1, I_2) be a bivariate Bernoulli random vector where $\Pr\{I_1 = i, I_2 = j\} = p_{ij}$, *$i, j = 0, 1$, such that $1 - \pi_1 = p_{10} + p_{11} < 1$, and $1 - \pi_2 = p_{01} + p_{11} < 1$. Then*

$$\left(\Sigma_{j=1}^{N_1} X_{1j}, \Sigma_{j=1}^{N_2} X_{2j}\right) \stackrel{S}{=} (I_1 Z_1, I_2 Z_2) + (\pi_1 X_1, \pi_2 X_2)$$

where $(Z_1, Z_2) \stackrel{S}{=} \left(\Sigma_{j=1}^{N_1} X_{1j}, \Sigma_{j=1}^{N_2} X_{2j}\right), (X_1, X_2) \stackrel{S}{=} (\pi_1^{-1} X_{1j}, \pi_2^{-1} X_{2j})$, *and* (I_1, I_2), (X_1, X_2), *and* (Z_1, Z_2) *are mutually independent.*

The bivariate geometric distribution given in (13) is discussed in detail in Block (1977). In order to provide insight into the connection between the concepts of random summation and random mixing and the BEAR(1) process, we present the following theorem.

THEOREM 3.2. *Let* (Y_1, Y_2) *and* (Z_1, Z_2) *have the same unknown bivariate exponential distribution. Let* (X_1, X_2) *have a known bivariate exponential distribution. Let* (I_1, I_2) *be as defined in Theorem 3.1, and let all the aforementioned random vectors be mutually independent. If*

$$(Y_1, Y_2) \stackrel{S}{=} (I_1 Z_1, I_2 Z_2) + (\pi_1 X_1, \pi_2 X_2)$$

then (Y_1, Y_2) *and* (Z_1, Z_2) *have the same bivariate exponential distribution as* (X_1, X_2).

PROOF. Let Ψ_Y, Ψ_X, and Ψ_Z be the characteristic functions of the random vectors (Y_1, Y_2), (X_1, X_2), and (Z_1, Z_2), respectively, and note that $\Psi_Y = \Psi_Z$. Then for real numbers u, v, we have the characteristic function equation:

$$\Psi_Y(u, v) = \Psi_X(\pi_1 u, \pi_2 v)[p_{00} + p_{10}\Psi_Z(u, 0) + p_{01}\Psi_Z(0, v) + p_{11}\Psi_Z(u, v)].$$

Letting $u = 0$, we may solve for $\Psi_Y(0, v)$ and $\Psi_Z(0, v)$. Similarly, letting $v = 0$, we may solve for $\Psi_Y(u, 0)$ and $\Psi_Z(u, 0)$. Inserting these results into the characteristic function equation and noting that Ψ_X is specified, the result then follows easily by solving for $\Psi_Y(u, v)$ and $\Psi_Z(u, v)$. ||

We are now ready to present the BEAR(1) model. First, we shall need some notation. Let $(I_1(n), I_2(n))$ and (N_1, N_2) be as in Theorem 3.1 and let $\{\mathbf{E}(n), n = \pm 1, \pm 2, \ldots\}$ be an iid sequence of bivariate exponential random vectors with mean vector $(\lambda_1^{-1}, \lambda_2^{-1})$ which is independent of (N_1, N_2) and $(I_1(n), I_2(n))$. We set

$$\mathbf{E}(0) = \left(\Sigma_{j=1}^{N_1} \pi_1 E_1(-j),\ \Sigma_{j=1}^{N_2} \pi_2 E_2(-j)\right)' \tag{14}$$

where we have written $\mathbf{E}(n) = (E_1(n), E_2(n))'$. Note that by Lemma 3.2, $\mathbf{E}(0)$ has a bivariate exponential distribution with mean vector $(\lambda_1^{-1}, \lambda_2^{-1})$, but the joint distribution of $\mathbf{E}(0)$ is not completely specified unless the distribution of $\mathbf{E}(n)$ is given. Define $A(n)$ to be the 2×2 diagonal random matrix $A(n) = \text{diag}\{I_1(n), I_2(n)\}$ and B to be the 2×2 diagonal matrix $B = \text{diag}\{\pi_1, \pi_2\}$.

The BEAR(1) process is defined as follows:

$$\mathbf{X}(n) = \begin{cases} \mathbf{E}(0) & n = 0 \\ A(n)\mathbf{X}(n-1) + B\mathbf{E}(n) & n = 1, 2, \ldots . \end{cases}$$

The following useful characterization of the BEAR(1) process is established in Block, Langberg, and Stoffer (1988). This characterization is primarily based on the preceding lemmas and theorems.

LEMMA 3.3. *Let* $\{\mathbf{X}(n), n = 0,1,2,\ldots\}$ *be a BEAR(1) process. Then for* $n = 0,1,2,\ldots,$

$$\mathbf{X}(n) \stackrel{S}{=} \mathbf{E}(0)$$

where $\mathbf{E}(0)$ *is given in (14).*

From Lemma 3.3 and results given in Block (1977), we may show that a BEAR(1) process can have well known bivariate exponential distributions. These include the Marshall-Olkin (1967), Downton (1970), Hawkes (1972), and Paulson (1973) bivariate exponential distributions. Details may be found in Block, Langberg, and Stoffer (1988, Remark 4.5). As an example we specify a particular $\mathbf{E}(0)$. Let E be an exponential random variable with mean θ, $0 < \theta < (\lambda_1 + \lambda_2)^{-1}$, $\pi_1 = \lambda_1\theta$, $\pi_2 = \lambda_2\theta$, and let $\mathbf{E}(1) = (\pi_1^{-1}E, \pi_2^{-1}E)'$. Let b_1, b_2, b_{12} be nonnegative real numbers such that $\lambda_1 = b_1 + b_{12}$, $\lambda_2 = b_2 + b_{12}$, and let $p_{00} = \theta b_{12}$, $p_{10} = \theta b_2$, $p_{01} = \theta b_1$, and $p_{11} = 1 - \theta(b_1 + b_2 + b_{12})$, then the resulting $\mathbf{X}(n)$ has a Marshall-Olkin (1967) bivariate exponential distribution.

In view of Lemma 3.3, we also have the following result on the positive dependence of the random variables of the BEAR(1) process. If $\mathbf{Z} = (Z_1, \ldots, Z_q)$, $q = 1,2,\ldots$, is a random vector we say that the random variables $Z_1, \ldots, Z_q$ are *associated* if $\operatorname{cov}\{f(Z), g(Z)\} \geq 0$ for all f and g monotonically nondecreasing in each argument, such that the expectations exist.

LEMMA 3.4. *Let* $\{\mathbf{X}(n)\}$ *be a BEAR(1) process and suppose* $E_1(1)$ *and* $E_2(1)$ *are associated, then* $\{X_i(n_j); i = 1,2, j = 1,2,\ldots,k\}$, $k > 0$ *integer, are associated.*

The proof of Lemma 3.4 follows by using the result of Lemma 3.3 in conjunction with a theorem on the association of a sequence of random variables that is conditionally increasing in sequence, see Barlow and Proschan (1981, Theorem 4.7).

4. Multivariate Dirichlet Processes. In this section we are concerned with modeling $(k+1)$-dimensional series, say $\mathbf{P}(n) = (P_1(n), \ldots, P_{k+1}(n))$; $n = 0,1,2,\ldots$, where at each time point n, $\mathbf{P}(n)$ has a Dirichlet distribution with parameter vector $(\alpha_1, \ldots, \alpha_{k+1})$. That is, we are interested in modeling and forecasting multivariate time series in which the data are proportions and are constrained so that $\Sigma_{j=1}^{k+1} P_j(n) = 1$ at each point in time.

Before presenting the model we establish some results. Throughout this section let $\alpha_1, \ldots, \alpha_{k+1} > 0$, let Y_j be independent gamma$(1, \alpha_j)$ random variables, and define $Z_j = Y_j / \Sigma_{\ell=1}^{j} Y_\ell$, $j = 1, \ldots, k+1$. Note that $Z_1 \equiv 1$ and that Z_j, $j = 2, \ldots, k+1$ are independent beta$(\alpha_j, \Sigma_{\ell=1}^{j-1}\alpha_\ell)$ random variables. Finally let

$$U_j = \begin{cases} Z_j \Pi_{\ell=j+1}^{k+1}(1 - Z_\ell) & j = 1, \ldots, k \\ Z_{k+1} & j = k+1 \end{cases}$$

and note that $U_j = Y_j / \Sigma_{\ell=1}^{k+1} Y_\ell$, $j = 1, \ldots, k+1$. It then follows that

$$(U_1, \ldots, U_{k+1}) \sim \text{Dirichlet } (\alpha_1, \ldots, \alpha_{k+1}).$$

We now define a univariate autoregressive-type process with beta marginals. This process will then be used to build the multivariate Dirichlet process. To ease the notation, we define $A_q = \Sigma_{\ell=1}^{q} \alpha_\ell$.

LEMMA 4.1. *(Beta Processes). For each j, $j = 2, \ldots, k+1$, let $\{B_{nj}, n = 1, 2, \ldots\}$ be a sequence of iid Bernoulli random variables such that* $\Pr\{B_{nj} = 1\} = \alpha_j A_j^{-1}$, *let $\{Q_{nj}, n = 1, 2, \ldots\}$ be a sequence of iid beta$(1, A_j)$ random variables, and let Z_{0j} be a beta(α_j, A_{j-1}) random variable. Assume that $\{B_{nj}\}$, $\{Q_{nj}\}$, and $\{Z_{0j}\}$ are mutually independent, $n = 1, 2, \ldots; j = 2, \ldots, k+1$. Then the processes $\{Z_{nj}, n = 1, 2, \ldots; j = 2, \ldots, k+1\}$ defined by*

$$Z_{nj} = Q_{nj} B_{nj} + (1 - Q_{nj}) Z_{n-1,j} \tag{15}$$

have beta(α_j, A_{j-1}) marginals, that is,

$$Z_{nj} \stackrel{S}{=} Z_{0j}, \quad n = 1, 2, \ldots; \; j = 2, \ldots, k+1. \tag{16}$$

PROOF. Let Y_0 be a gamma(1,1) random variable which is independent of $Y_1, \ldots, Y_{k+1}$. Then we have the following results:

(i) $Q_{nj} \stackrel{S}{=} Y_0 (\Sigma_{\ell=0}^{j} Y_\ell)^{-1}$, and

(ii) $Z_{nj} \stackrel{S}{=} Y_j (\Sigma_{\ell=1}^{j} Y_\ell)^{-1}$ for $n = 1, 2, \ldots; j = 2, \ldots, k+1$,

and by Basu's Theorem the random variables

(iii) $Y_0 (\Sigma_{\ell=0}^{j} Y_\ell)^{-1}$ and $Y_j (\Sigma_{\ell=0}^{j} Y_\ell)^{-1}$ are independent, $j = 2, \ldots, k+1$.

We now establish (16) by an induction argument on n. By (i), (ii), (iii), and (15) we have that

$$\begin{aligned} Z_{1j} &\stackrel{S}{=} Y_0 (\Sigma_{\ell=0}^{j} Y_\ell)^{-1} B_{1j} \\ &\quad + [1 - Y_0 (\Sigma_{\ell=0}^{j} Y_\ell)^{-1}] Y_j (\Sigma_{\ell=1}^{j} Y_\ell)^{-1} \\ &= [Y_0 B_{1j} + Y_j] (\Sigma_{\ell=0}^{j} Y_\ell)^{-1}, \quad j = 2, \ldots, k+1. \end{aligned} \tag{17}$$

Let f_j, g_j be the density function of Z_{1j}, Z_{0j}, respectively, $j = 2, \ldots, k+1$, and let $y \in (0,1)$. Then by (17), for $j = 2, \ldots, k+1$,

$$\begin{aligned} f_j(y) &= \alpha_j A_j^{-1} \alpha_j^{-1} A_j y g_j(y) + A_{j-1} A_j^{-1} A_j A_{j-1}^{-1} (1-y) g_j(y) \\ &= g_j(y). \end{aligned}$$

Hence (16) holds for $n = 1$. The induction proof now proceeds by assuming that (16) holds for a given n, $n > 1$, then using a similar argument, one can show that (16) holds for $n+1$. Hence (16) holds for all n. ||

We are now ready to define the stationary autoregressive-type sequence of Dirichlet random vectors, $\{\mathbf{P}(n), n = 0, 1, 2, \ldots\}$. Let $Z_{n1} \equiv 1, n = 0, 1, 2, \ldots$, and let $\{Z_{nj}, n = 0, 1, 2, \ldots; j = 2, \ldots k+1\}$ be as defined as (15). Let

$$P_j(n) = \begin{cases} Z_{nj} \Pi_{\ell=j+1}^{k+1} (1 - Z_{n\ell}) & j = 1, \ldots, k \\ Z_{n,k+1} & j = k+1 \end{cases} \tag{18}$$

and let

$$\mathbf{P}(n) = (P_1(n), \ldots, P_{k+1}(n)), \quad n = 0, 1, 2, \ldots. \tag{19}$$

It follows by Lemma 4.1 and the preceding results on the joint distribution of $(U_1, \ldots, U_{k+1})$ that for each n, $\mathbf{P}(n)$ has a Dirichlet distribution with parameter vector $(\alpha_1, \ldots, \alpha_{k+1}), n = 0, 1, 2, \ldots$.

Estimation of the parameter vector $\boldsymbol{\theta} = (\alpha_1, \ldots, \alpha_{k+1})$ of the Dirichlet process (19) may be carried out by maximum likelihood. Let $\mathbf{P}(1), \ldots, \mathbf{P}(N)$, be a realization of length N of the process (19). Maximum likelihood estimation is accomplished as follows:

(1) Transform the data to the beta sequences by setting

$$Z_{nj} = P_j(n)[\Sigma_{\ell=1}^{j} P_\ell]^{-1}, \quad j = 2, \ldots, k+1; n = 1, \ldots, N.$$

(2) From the Z_{nj}, calculate the Q_{nj} and B_{nj} defined in Lemma 4.1:

(a) If $Z_{nj} > Z_{n-1,j}$ then $B_{nj} = 1$ and

$$Q_{nj} = (Z_{nj} - Z_{n-1,j})/(1 - Z_{n-1,j}),$$

(b) If $Z_{nj} \leq Z_{n-1,j}$ then $B_{nj} = 0$ and

$$Q_{nj} = (Z_{nj} - Z_{n-1,j})/(-Z_{n-1,j}),$$

for $j = 2, \ldots, k+1$, and $n = 1, \ldots, N$. These calculations are direct consequences of (15).

(3) Maximize the log-likelihood function of the observations $\{Q_{nj}, B_{nj}; n = 1, \ldots, N, j = 2, \ldots, k+1\}$, say $L(\theta)$, given by

$$\begin{aligned} L(\theta) &= \Sigma_{\ell=2}^{k+1}\{A_\ell \ell n \Delta_\ell + S_\ell \ell n(A_\ell A_{\ell-1}^{-1} - 1)\} \\ &+ N\Sigma_{\ell=1}^{k} \ell n A_\ell - \Sigma_{\ell=2}^{k+1} \ell n \Delta_\ell, \end{aligned}$$

where $\Delta_\ell = \Pi_{n=1}^{N}(1 - Q_{n\ell})$, and $S_\ell = \Sigma_{n=1}^{N} B_{n\ell}$, for $\ell = 2, \ldots, k+1$.

For the cases $k = 1, 2$ an explicit solution to $\partial L(\theta)/\partial A_\ell = 0$, $\ell = 1, \ldots, k+1$, exists so that for $k = 1$ (that is, the process of interest is the univariate beta process given in (15) with $j = 2$)

$$\hat{\alpha}_1 = (N - S_2)/(-\ell n \Delta_2)$$

$$\hat{\alpha}_2 = S_2/(-\ell n \Delta_2)$$

and for $k = 2$,

$$\hat{\alpha}_1 = (2N - S_3)(N - S_2)/[-N\ell n(\Delta_2\Delta_3)]$$

$$\hat{\alpha}_2 = (2N - S_3)S_2/[-N\ell n(\Delta_2\Delta_3)]$$

$$\hat{\alpha}_3 = S_3/(-\ell n(\Delta_3)$$

are the MLE's. For arbitrary k, the MLE's can be obtained by a numerical method such as a Newton-Raphson or scoring procedure.

As an example of the kind of data that can be handled by the model, we consider spatial data presented in Mechergui (1984). As part of a study of the water table in the vicinity of drainage tiles, Mechergui analyzes the content of 72 equally (and linearly) spaced auger holes each dug to a depth of 2.4 meters. The sampling rate was 10 auger holes per 52 meters and the proportions of silt, clay, and sand are obtained for each sample with the constraint that the percentages of silt, clay, and sand in each auger hole sample sum to 100%.

We fit a Dirichlet model (19) to the proportions of silt, clay, and sand, $P_1(n)$, $P_2(n)$, and $P_3(n)$, respectively, based on the 72 observation vectors $\mathbf{P}(n) = (P_1(n), P_2(n), P_3(n))$, $n = 0, 1, 2, \ldots, 71$, and then used the estimated model to obtain one-step-ahead forecasts for the time points $n = 1, \ldots, 71$. Figure 4.1 shows the silt series, $P_1(n)$, as well as the one-step-ahead forecasts based on the estimated model; Figure 4.2 shows similar plots for the clay series, $P_2(n)$.

Figure 4.1: Observed (solid line) and predicted (dashed line) silt content series.

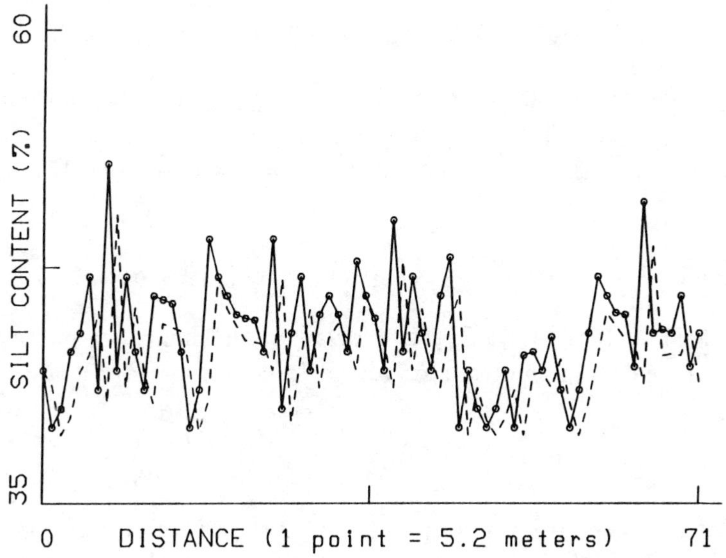

Figure 4.2: Observed (solid line) and predicted (dashed line) clay content series.

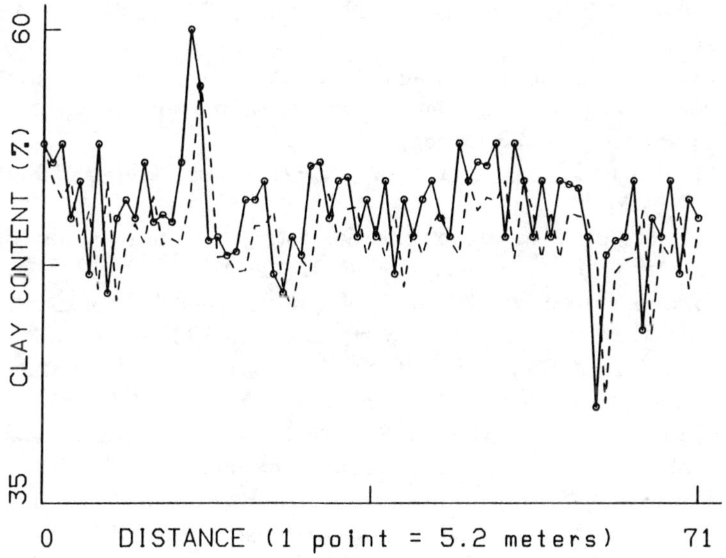

Maximum likelihood estimation was carried out as previously discussed. In this case we found the MLE's to be:

$$\hat{\alpha}_1 = 2.573, \quad \hat{\alpha}_2 = 2.963, \quad \hat{\alpha}_3 = 2.137.$$

Forecasting was accomplished by first forecasting Z_{n2} and Z_{n3} obtained in estimation step (1), and then using these forecasts to predict $\mathbf{P}(n)$ via the transformation (18). Then one-step-ahead forecasts of Z_{nj}, $j = 2,3$ based on (15) is given as follows:

$$\hat{Z}_{nj} = (1 + \Sigma_{\ell=1}^{j}\hat{\alpha}_\ell)^{-1}[\hat{\alpha}_j(\Sigma_{\ell=1}^{j}\hat{\alpha}_\ell)^{-1} + (\Sigma_{\ell=1}^{j}\hat{\alpha}_\ell)Z_{n-1,j}], \quad j = 2,3$$

for $n = 1,\ldots,71$. Recall that $Z_{n1} \equiv 1$ for all n, and hence we put $\hat{Z}_{n1} \equiv 1$ for all n. From (18) we obtain the one-step-ahead forecasts for $P_j(n)$, $j = 1,2,3$:

$$\hat{P}_j(n) = \begin{cases} \hat{Z}_{nj}\Pi_{\ell=j+1}^{3}(1 - \hat{Z}_{n\ell}) & j = 1,2 \\ \hat{Z}_{n3}, & j = 3. \end{cases} \tag{20}$$

The root mean square prediction errors for the silt, clay, and sand series based on the preceding prediction equations are .040, .036, and .C37 respectively.

REFERENCES

BARLOW, R.E. and PROSCHAN, F. (1981). *Statistical Theory of Reliability and Life Testing: Probability Models.* To Begin With, Silver Spring, MD.

BLOCK, H.W. (1977). A family of bivariate life distributions. In *The Theory and Applications of Reliability, Vol. 1*, C.P. Tsokos and I. Shimi, Eds. Academic Press, New York.

BLOCK, H.W., LANGBERG, N.A., and STOFFER, D.S. (1988). Bivariate exponential and geometric autoregressive and autoregressive-moving average models. *Advances in Applied Probability* **20** 798–821.

BROCKWELL, P.J. and DAVIS, R.A. (1987). *Time Series: Theory and Methods*, Springer-Verlag, New York.

DOWNTON, F. (1970). Bivariate exponential distributions in reliability theory. *Journal of the Royal Statistical Society, Ser. B.* **32** 408–417.

FAHMEIR, L. and KAUFMANN, H. (1987). Regression models for nonstationary categorical time series. *Journal of Time Series Analysis* **8** 147–160.

GASTWIRTH, J.L. and RUBIN, H. (1975). The asymptotic distribution theory of the empiric c.d.f. for mixing processes. *Annals of Statistics* **3** 809–824.

GAVER, D.P. and LEWIS, P.A.W. (1980). First-order autoregressive gamma sequences and point processes. *Advances in Applied Probability* **12** 727–745.

HANNAN, E.J. (1970). *Multiple Time Series.* Wiley and Sons, New York.

HAWKES, A.G. (1972). A bivariate exponential distribution with applications in reliability. *Journal of the Royal Statistical Society, Ser. B.* **34** 129–131.

JACOBS, P.A. and LEWIS, P.A.W. (1977). A mixed autoregressive-moving average exponential sequence and point process (EARMA (1,1)). *Advances in Applied Probability* **9** 87–104.

JACOBS, P.A. and LEWIS, P.A.W. (1978). Discrete time series generated by mixtures I: Correlational and runs properties. *Journal of the Royal Statistical Society, Ser. B* **40** 94–105.

JACOBS, P.A. and LEWIS, P.A.W. (1983). Stationary discrete autoregressive moving average time series generated by mixtures. *Journal of Time Series Analysis* **4** 18–36.

KAUFMANN, H. (1987). Regression models for nonstationary categorical time series: Asymptotic estimation theory. *Annals of Statistics* **15** 79–98.

LANGBERG, N.A. and STOFFER, D.S. (1987). Moving average models with bivariate exponential and geometric distributions. *Journal of Applied Probability* **24** 48–61.

LAWRANCE, A.J. and LEWIS, P.A.W. (1980). The exponential autoregressive-moving average process EARMA (p, q). *Journal of the Royal Statistical Society, Ser. B* **42** 150–161.

LAWRANCE, A.J. and LEWIS, P.A.W. (1981). A new autoregressive time series model in exponential variables (NEAR(1)). *Advances in Applied Probability* **13** 826–845.

LAWRANCE, A.J. and LEWIS, P.A.W. (1985). Modeling and residual analysis of non-linear autoregressive time series in exponential variables (with Discussion). *Journal of the Royal Statistical Society, Ser. B* **47** 165–202.

LEWIS, P.A.W. (1980) Simple models for positive-valued and discrete-valued time series with ARMA correlation structure. In *Multivariate Analysis* V, P.R. Krishnaiah, ed., North Holland, 151–166.

MARDIA, K.V. (1970). *Families of Bivariate Distributions.* Hafner Publishing Co., Darien, Conn.

MARSHALL, A.W. and OLKIN, I. (1967). A multivariate exponential distribution. *Journal of the American Statistical Association* **62** 30–44.

MECHERGUI, M. (1984). Stochastic modeling of the water table in the vicinity of drainage tiles. Unpublished dissertation in Soil Science, Graduate Division, University of California, Davis.

PAULSON, A.S. (1973). A characterization of the exponential distribution and a bivariate exponential distribution. *Sankhȳa, Ser. A* **35** 69–78.

RAFTERY, A.E. (1982). Generalized non-normal time series. In *Time Series Analysis: Theory and Practice I*, O.D. Anderson, ed., North Holland, Amsterdam, 621–640.

STOFFER, D.S. (1987). Walsh-Fourier analysis of discrete-valued time series. *Journal of Time Series Analysis* **8** 449–467.

WEISS, M.S. (1978). Modification of the Kolmogorov-Smirnov statistic for use with correlated data. *Journal of the American Statistical Association* **73** 872–875.

DEPARTMENT OF MATHEMATICS AND STATISTICS
UNIVERSITY OF PITTSBURGH
PITTSBURGH, PA 15260

DEPARTMENT OF STATISTICS
HAIFA UNIVERSITY
MOUNT CARMEL, HAIFA, ISRAEL

DEPARTMENT OF MATHEMATICS AND STATISTICS
UNIVERSITY OF PITTSBURGH
PITTSBURGH, PA 15260

SOME MAJORIZATION INEQUALITIES FOR FUNCTIONS OF EXCHANGEABLE RANDOM VARIABLES

By Philip J. Boland,[1] Frank Proschan,[2] and Y.L. Tong[3]

University College, Dublin, Florida State University, and Georgia Institute of Technology

This paper contains inequalities for the expectations of permutation-invariant concave functions and Schur-concave functions of the partial sums of nonnegative exchangeable random variables. Two majorization inequalities are derived, and an application in reliability theory is presented.

1. Introduction and Summary. For fixed $n > 1$ let $\mathbf{X} = (X_1, \ldots, X_n)$ denote an n-dimensional random vector with density function $f(\mathbf{x})$ that is absolutely continuous w.r.t. the Lebesgue measure or the product measure of counting measures. $X_1, \ldots, X_n$ are said to be exchangeable[†] if f is invariant under permutations of its arguments. This paper develops inequalities for the expectations of functions of partial sums of $X_1, \ldots, X_n$.

The notion of majorization defines a partial ordering of the diversity of the components of vectors. Let $\mathbf{a} = (a_1, \ldots, a_n)$, $\mathbf{b} = (b_1, \ldots, b_n)$ be two n-dimensional vectors and let $a_{[1]} \geq \cdots \geq a_{[n]}$, $b_{[1]} \geq \cdots \geq b_{[n]}$ denote their ordered components. $\mathbf{a}$ is said to *majorize* $\mathbf{b}$ (in symbols $\mathbf{a} \succ \mathbf{b}$) if

$$\Sigma_1^h a_{[i]} \geq \Sigma_1^h b_{[i]} \quad \text{for } h = 1, \ldots, n-1$$

and $\Sigma_1^n a_i = \Sigma_1^n b_i$. It is known that $\mathbf{a} \succ \mathbf{b}$ iff there exists a doubly stochastic matrix Q such that $\mathbf{b} = \mathbf{a}Q$, i.e., $\mathbf{b}$ is an "average" of $\mathbf{a}$. A function $\psi : R^n \to R$ is said to be a Schur-concave function if $\mathbf{a} \succ \mathbf{b}$ implies $\psi(\mathbf{a}) \leq \psi(\mathbf{b})$. For a comprehensive treatment of majorization and Schur functions, see Marshall and Olkin (1979).

[1]Research partially supported by the Air Force Office of Scientific Research, AFSC, USAF, under Grant AFOSR 88–0040.

[2]Research supported by the Air Force Office of Scientific Research, AFSC, USAF, under Grant AFOSR 88–0040.

[3]Research partially supported by NSF Grants DMS–8502346 and DMS–8801327.

AMS 1980 subject classifications. 60E15, 62H99.

Key words and phrases. Majorization inequalities, exchangeable random variables, concave and Schur-concave functions, moment inequalities.

[†] More precisely, $X_1, \ldots, X_n$ are *finitely* exchangeable instead of exchangeable. For the minor distinction between finite exchangeability and exchangeability see e.g., Tong ((1980), p. 96).

In an earlier paper Marshall and Proschan (1965) proved the following inequality: Let $X_1, \ldots, X_n$ be exchangeable and let $\phi : R^n \to R$ be Borel measurable, permutation invariant, and concave. If $a_1, \ldots, a_n$ are more diverse than $b_1, \ldots, b_n$ in the sense of majorization, i.e., if $\mathbf{a} \succ \mathbf{b}$, then $E\phi(b_1X_1, \ldots, b_nX_n) \geq E\phi(a_1X_1, \ldots, a_nX_n)$ provided the expectation exists. This inequality yields a number of useful results and implies many previously-known results as special cases (see, e.g., Corollaries 1–3 in their paper). In this paper, we prove some related results and discuss an application in reliability theory. The results (Theorems 1 and 2) involve the expectations of functions of partial sums of exchangeable random variables, and depend on the notion of majorization in a different fashion. For fixed $k < n$, let $\mathbf{r} = (r_1, \ldots, r_k)$ be a vector of positive integers such that $\Sigma_1^k r_i = n$. Let $X_1, \ldots, X_n$ be exchangeable random variables and let $\mathbf{Y_r} = (Y_1^{(\mathbf{r})}, \ldots, Y_k^{(\mathbf{r})})$ denote a k-dimensional random vector such that

$$Y_1^{(\mathbf{r})} = \Sigma_1^{r_1} X_i, \quad Y_2^{(\mathbf{r})} = \Sigma_{r_1+1}^{r_1+r_2} X_i, \ldots, \quad Y_k^{(\mathbf{r})} = \Sigma_{r_1+\cdots+r_{k-1}+1}^{n} X_i;$$

that is, $Y_j^{(\mathbf{r})}$ is the sum of r_j such X_i's and $Y_1^{(\mathbf{r})}, \ldots, Y_k^{(\mathbf{r})}$ do not contain any common elements. Let $\mathbf{s} = (s_1, \ldots, s_k)$ denote another such vector and $\mathbf{Y}^{(s)}$ be defined similarly. Let $\phi(\mathbf{y}) = \phi(y_1, \ldots, y_k)$ denote a real-valued function that is permutation invariant and concave. We show that (Theorem 1) if $\mathbf{s} \succ \mathbf{r}$ and if the X_i's are nonnegative exchangeable random variables, then $E\phi(\mathbf{Y}^{(\mathbf{r})}) \geq E\phi(\mathbf{Y}^{(\mathbf{s})})$. The reasons for considering such a random vector $\mathbf{Y}$ and for studying inequalities of this type arise from certain applications. One such application concerns the optimal arrangement policy for parallel and series systems in reliability theory, and is given in Section 4. In Theorem 2 we show that, by imposing an additional condition on the joint density f, the same inequality holds for all Schur-concave functions ϕ.

Since the theorems apply to nonnegative random variables only, a natural question is whether the same statements hold for random variables which may take negative values. We show in Section 3 that the answer is negative even for i.i.d. normal variables.

2. The Main Results. For the theorems stated in this section, the density function f of $\mathbf{X} = (X_1, \ldots, X_n)$ is assumed to be absolutely continuous w.r.t. the Lebesgue measure or the product measure of the counting measures. The proofs will be given for the former. For the product of counting measures, simply change the integral signs to summation signs.

THEOREM 1. *If (i) f is permutation invariant and $f = 0$ for any $x_i < 0$ $(i = 1, \ldots, n)$, (ii) $\phi(y_1, \ldots, y_k)$ is a permutation invariant concave function, and (iii) $\mathbf{s} \succ \mathbf{r}$, then*

$$E\phi(Y_1^{(\mathbf{r})}, \ldots, Y_k^{(\mathbf{r})}) \geq E\phi(Y_1^{(\mathbf{s})}, \ldots, Y_k^{(\mathbf{s})}) \tag{1}$$

holds provided that the expectations exist.

PROOF. It is well-known (Marshall and Olkin (1979), Chapter 2) that it suffices to assume that

$$s_1 > r_1 \geq r_2 > s_2 \equiv t, \quad r_1 + r_2 = s_1 + s_2 \equiv d$$

and $r_j = s_j$ for $j = 3, \ldots, k$. Let us define

$$Z_1 = \Sigma_1^t X_i, \quad Z_2 = \Sigma_{s_1+1}^d X_i$$

and $Y_j = Y_j^{(\mathbf{r})} = Y_j^{(\mathbf{s})}$ for $j = 3, \ldots, k$. Let

$$g(z_1, z_2) = g(z_1, z_2 \mid \mathbf{x}_0, y_3, \ldots, y_k) \tag{2}$$

denote the conditional density of (Z_1, Z_2) given $\mathbf{X}_0 \equiv (X_{t+1}, \ldots, X_{s_1}) = \mathbf{x}_0$ and $Y_j = y_j$ $(j = 3, \ldots, k)$. Then it is easy to check that $g(z_1, z_2) = g(z_2, z_1)$ and that

$$\begin{aligned}
& E[\phi(Y_1^{(\mathbf{r})}, \ldots, Y_k^{(\mathbf{r})}) \mid (\mathbf{X}_0, Y_3, \ldots, Y_k) = (\mathbf{x}_0, y_3, \ldots, y_k)] \\
= & \int\int \phi(z_1 + u_1, z_2 + u_2, y_3, \ldots, y_k) g(z_1, z_2) dz_1 dz_2 \\
= & \iint_{z_1 \geq z_2} \phi(z_1 + u_1, z_2 + u_2, y_3, \ldots, y_k) g(z_1, z_2) dz_1 dz_2 \\
& + \iint_{z_1 < z_2} \phi(z_1 + u_1, z_2 + u_2, y_3, \ldots, y_k) g(z_1, z_2) dz_1 dz_2 \\
= & \iint_{z_1 \geq z_2} \{\phi(z_1 + u_1, z_2 + u_2, y_3, \ldots, y_k) \\
& \qquad + \phi(z_1 + u_2, z_2 + u_1, y_3, \ldots, y_k)\} g(z_1, z_2) dz_1 dz_2
\end{aligned}$$

where $(u_1, u_2) = \left(\Sigma_{i=t+1}^{r_1} x_i, \Sigma_{i=r_1+1}^{s_1} x_i\right)$. Now let $(v_1, v_2) = \left(\Sigma_{i=t+1}^{s_1} x_i, 0\right)$. Since $x_i \geq 0$, there exists an $\alpha = \frac{u_1}{v_1} \in [0, 1]$ which satisfies

$$\begin{aligned}
(z_1 + u_1, z_2 + u_2) &= \alpha(z_1 + v_1, z_2 + v_2) + (1 - \alpha)(z_1 + v_2, z_2 + v_1), \\
(z_1 + u_2, z_2 + u_1) &= (1 - \alpha)(z_1 + v_1, z_2 + v_2) + \alpha(z_1 + v_2, z_2 + v_1)
\end{aligned}$$

for every point in $\{(z_1, z_2) : z_1 \geq z_2\}$. Thus for every fixed $(\mathbf{x}_0, y_3, \ldots, y_k)$ and every such (z_1, z_2),

$$\begin{aligned}
& \phi(z_1 + u_1, z_2 + u_2, y_3, \ldots, y_k) + \phi(z_1 + u_2, z_2 + u_1, y_3, \ldots, y_k) \\
\geq & \ \alpha\phi(z_1 + v_1, z_2 + v_2, y_3, \ldots, y_k) + (1 - \alpha)\phi(z_1 + v_2, z_2 + v_1, y_3, \ldots, y_k) \\
& + (1 - \alpha)\phi(z_1 + v_1, z_2 + v_2, y_3, \ldots, y_k) + \alpha\phi(z_1 + v_2, z_2 + v_1, y_3, \ldots, y_k).
\end{aligned}$$

Consequently, we have

$$
\begin{aligned}
& E\left[\phi(Y_1^{(\mathbf{r})},\ldots,Y_k^{(\mathbf{r})}) \mid (\mathbf{X}_0,Y_3,\ldots,Y_k)=(\mathbf{x}_0,y_3,\ldots,y_k)\right] \\
\geq & \iint_{z_1\geq z_2} \{\phi(z_1+v_1,z_2+v_2,y_3,\ldots,y_k)+ \\
& \qquad \phi(z_1+v_2,z_2+v_1,y_3,\ldots,y_k)\}g(z_1,z_2)dz_1dz_2 \\
= & E\left[\phi(Y_1^{(\mathbf{s})},\ldots,Y_k^{(\mathbf{s})}) \mid (\mathbf{X}_0,Y_3,\ldots,Y_k)=(\mathbf{x}_0,y_3,\ldots,y_k)\right],
\end{aligned}
$$

and the conclusion follows by unconditioning. ||

In the next theorem we change the condition on ϕ to be any measurable Schur-concave function, and impose a stronger condition on the conditional density g.

THEOREM 2. *If (i) f is permutation invariant, $f=0$ for any $x_i<0$ ($i=1,\ldots,n$), and such that the conditional density $g(z_1,z_2)$ defined in (2) is a Schur-concave function of (z_1,z_2) for every fixed $(\mathbf{x}_0,y_3,\ldots,y_k)$ and every $t>0$, (ii) $\phi(y_1,\ldots,y_k)$ is a Borel-measurable Schur-concave function, and (iii) $\mathbf{s}\succ\mathbf{r}$, then (1) holds provided the expectations exist.*

PROOF. We shall follow the notation developed in the proof of Theorem 1 and compare $E\phi(\mathbf{Y}^{(\mathbf{r})})$ with $E\phi(\mathbf{Y}^{(\mathbf{s})})$ for $\mathbf{s}\succ\mathbf{r}$. Again it suffices to assume that $s_1>r_1\geq r_2>s_2$ and $r_j=s_j$ for $j>2$. Then the conditional expectation of

$$\phi(Y_1^{(\mathbf{r})},Y_2^{(\mathbf{r})},Y_3,\ldots,Y_k)-\phi(Y_1^{(\mathbf{s})},Y_2^{(\mathbf{s})},Y_3,\ldots,Y_k)$$

given $(\mathbf{X}_0,Y_3,\ldots,Y_k)=(\mathbf{x}_0,y_3,\ldots,y_k)$ is

$$\Delta\equiv\int\int\{\phi^*(z_1+u_1,z_2+u_2)-\phi^*(z_1+u_1+u_2,z_2)\}g(z_1,z_2)dz_1dz_2$$

where

$$\phi^*(y_1,y_2)=\phi(y_1,y_2,y_3,\ldots,y_k)$$

and g is the conditional density of (Z_1,Z_2). It is straightforward to verify that, after following the same steps as in the proof of Theorem J.1 in Marshall and Olkin (1979, p. 100), we have

$$\Delta=\iint_{z_1\geq z_2}\{\phi^*(z_1,z_2+u_2)-\phi^*(z_1+u_2,z_2)\}\{g(z_1-u_1,z_2)-g(z_1,z_2-u_1)\}dz_1dz_2.$$

Since ϕ^* and g are Schur-concave functions and $u_i>0$ ($i=1,2$), we have

$$
\begin{aligned}
(z_1+u_2,z_2) &\succ (z_1,z_2+u_2),\\
(z_1,z_2-u_1) &\succ (z_1-u_1,z_2)
\end{aligned}
$$

and $\Delta\geq 0$. Thus the conclusion follows by unconditioning. ||

REMARK. Proschan and Sethuraman (1977) previously proved that if $X_1, \ldots, X_n$ are i.i.d. nonnegative random variables with a common density that is log-concave, then the conclusion in Theorem 2 holds. Their proof depends on an application of the main theorem in their paper and on a TP_2 property of the convolution of log-concave densities given in Karlin and Proschan (1960). It is noted here that their result now follows immediately from Theorem 2. This is so because if $X_1, \ldots, X_n$ are i.i.d. random variables with a common density that is log-concave, then $\Sigma_1^{s_2} X_i$ and $\Sigma_{s_1+1}^{s_1+s_2} X_i$ are independent random variables with a common density that is also log-concave (see e.g., Das Gupta (1973, Theorem 4.2)). Consequently, the joint density of $\left(\Sigma_1^{s_2} X_i, \Sigma_{s_1+1}^{s_1+s_2} X_i\right)$ is a Schur-concave function and Theorem 2 applies.

In most applications, the assumption on the Schur-concavity of the conditional density $g(z_1, z_2)$ is not easy to verify. It is clear that if the following conjecture concerning the convolution of Schur-concave random variables is true, then the assumption holds when f (the joint density of $\mathbf{X}$) is a Schur-concave function. We state the conjecture in a more general form without assuming that the random variables are nonnegative.

CONJECTURE. For $n = mk$ and $\mathbf{X} = (X_1, \ldots, X_n)$ let

$$Z_j = \Sigma_{(j-1)m+1}^{jm} X_i, \quad j = 1, 2, \ldots, k.$$

If the joint density of $\mathbf{X}$ is a Schur-concave function of $\mathbf{x}$ for $\mathbf{x} \in \Re^n$, then the joint density of $\mathbf{Z} = (Z_1, \ldots, Z_k)$ is a Schur-concave function of $\mathbf{z}$ for $\mathbf{z} \in \Re^k$ for all positive integers k and m.

It is not yet known to us whether this conjecture is true for continuous random variables. However, the following counterexample shows that at least for the discrete case, it is not true.

EXAMPLE. Consider $k = m = 2$, and assume that (X_1, X_2, X_3, X_4) takes only integer values $0, 1, 2, 3$. Let $Z_1 = X_1 + X_2$, $Z_2 = X_3 + X_4$. Then $P[Z_1 = 4, Z_2 = 2]$ is the probability of the set of the following points:

(3,1,1,1), (1,3,1,1), (2,2,1,1), (2,2,2,0) (2,2,0,2)
(3,1,2,0), (3,1,0,2), (1,3,2,0), (1,3,0,2)

Similarly $P[Z_1 = Z_2 = 3]$ is the probability of the set consisting of

(2,1,2,1), (2,1,1,2), (1,2,2,1), (1,2,1,2),
(2,1,3,0), (2,1,0,3), (1,2,3,0), (1,2,0,3),
(3,0,2,1), (3,0,1,2), (0,3,2,1), (0,3,1,2),
(3,0,3,0), (3,0,0,3), (0,3,3,0), (0,3,0,3).

If the joint density of (X_1, X_2, X_3, X_4) takes values of $1/14$ for each of the points (3,1,1,1), (2,2,2,0), (2,2,1,1) and each of their permutations, and zero otherwise, then it is a Schur-concave function on the product of integer space, and we have

$$P[Z_1 = 4, Z_2 = 2] = \frac{5}{14} > \frac{4}{14} = P[Z_1 = 3, Z_2 = 3]. \quad \|$$

3. An Example For Random Variables Which Are Not Nonnegative. It might be tempting to think that results similar to our Theorems 1 and 2 also hold when the condition that $X_i \geq 0$ a.s. is removed. This is not true even for i.i.d. normal variables. In the following, we give an example to show that the conclusion of Theorem 2 does not hold without this condition. An example for Theorem 1 can be obtained similarly.

Consider, for $n = 2m$, independent normal variables $X_1, \ldots, X_n$ with mean zero and variance one. For $t \leq m$ consider

$$Y_1 = \Sigma_{i=1}^{t} X_i, \quad Y_2 = \Sigma_{t+1}^{n} X_i,$$

and denote $W = Y_1 - Y_2$, $V = Y_1 + Y_2 = \Sigma_1^n X_i$. Then (W, V) has a bivariate normal distribution with means zero, variances n, and correlation coefficient $(2t/n) - 1$. Thus the conditional distribution of W given $V = v$ is normal with mean $\frac{1}{n}(2t - n)v$ and variance $\sigma^2_{W|V=v} = 4t(n-t)/n$. Now choose $n = 4$, $\mathbf{s} = (3, 1)$, $\mathbf{r} = (2, 2)$. Clearly the conditional density function $g(z_1, z_2)$ of (X_1, X_4) given $\mathbf{X}_0 = (X_2, X_3)$ is Schur-concave. For an arbitrary but fixed $\epsilon > 0$ let us define

$$\phi(y_1, y_2) = \begin{cases} -(y_1 - y_2)^2 & \text{for } 0 \leq |\, y_1 + y_2 \,| \leq \epsilon \\ 0 & \text{otherwise,} \end{cases}$$

then ϕ is also Schur-concave. From the joint distribution of (W, V) clearly we have

$$\begin{aligned} & E\left[\{\phi(X_1, \Sigma_2^4 X_i) - \phi(\Sigma_1^2 X_i, \Sigma_3^4 X_i)\} \mid \Sigma_1^4 X_i = 0\right] \\ = \; & -\mathrm{Var}((X_1 - \Sigma_2^4 X_i) \mid V = \Sigma_1^4 X_i = 0) + \mathrm{Var}((\Sigma_1^2 X_i - \Sigma_3^4 X_i) \mid V = 0) \\ = \; & (-3 + 4) = 1 > 0. \end{aligned}$$

Thus by continuity there exists a small $\epsilon > 0$ such that

$$\begin{aligned} & E\left[\phi(X_1, \Sigma_2^4 X_i) - \phi(\Sigma_1^2 X_i, \Sigma_3^4 X_i)\right] \\ = & \int_{-\epsilon}^{\epsilon} E\left[\{- \mid X_1 - \Sigma_2^4 X_i \mid^2 + \mid \Sigma_1^2 X_i - \Sigma_3^4 X_i \mid^2\} \mid \Sigma_1^4 X_i = v\right] dP\left[\Sigma_1^4 X_i \leq v\right] > 0. \end{aligned}$$

4. An Application in Reliability Theory. In this section we state an application of Theorem 1 in reliability theory. Consider n exchangeable components with life lengths $X_1, \ldots, X_n$ which are obviously nonnegative. If the components are manufactured independently, then the joint density f of the X_i's is the product of the common marginal densities; otherwise if they are manufactured under the influence of some common factors or under a common environment, then it is well-known that f is a mixture and the random variables are conditionally i.i.d. In either case f is permutation invariant.

Suppose that a system consists of k subsystems, and that the j-th subsystem, consisting of $r_j \geq 1$ such components, is required to operate properly with one component in operation and the others in a standby capacity ($j = 1, \ldots, k$). Then the life length Y_j of the j-th subsystem is $\Sigma_{r_1+\cdots+r_{j-1}+1}^{r_1+\cdots+r_j} X_i$. Let $Y_{(1)} \leq Y_{(2)} \leq \cdots \leq Y_{(k)}$ denote the order statistics of $Y_1, \ldots, Y_k$ and $\mathbf{r} = (r_1, \ldots, r_k)$ be an allocation vector such that $r_j \geq 1$ and $\Sigma_1^k r_j = n$. When the subsystems are connected in series, then the life length of the system is $Y_{(1)}$. On the other hand if they are connected in parallel, then it is $Y_{(k)}$. Now for fixed $c_i \geq 0$, $\Sigma_1^k c_j y_{(j)}$ is a permutation invariant and concave (convex) function of $(y_1, \ldots, y_k)$ if $c_1 \geq \cdots \geq c_k$ (if $c_1 \leq \cdots \leq c_k$). Consequently, Theorem 1 provides a partial ordering for the expected life length of the system for series and parallel systems. In particular, for series systems the optimal allocation policy is such that $| r_j - r_{j'} | \leq 1$ for all $j \neq j'$, and for parallel systems an optimal policy is that $r_1 = n - k + 1$ and $r_2 = \cdots = r_k = 1$.

REFERENCES

DAS GUPTA, S. (1973). S-unimodal functions: Related inequalities and statistical applications. *Sankhyā Ser. B* **38** 301–314.

KARLIN, S. and PROSCHAN, F. (1960). Pólya type distributions of convolutions. *Ann. Math. Statist.* **31** 721–736.

MARSHALL, A.W. and OLKIN, I. (1979). *Inequalities: Theory of Majorization and Its Applications.* Academic Press, New York.

MARSHALL, A.W. and PROSCHAN, F. (1965). An inequality for convex functions involving majorization. *J. Math. Anal. Appl.* **12** 87–90.

PROSCHAN, F. and SETHURAMAN, J. (1977). Schur functions in statistics. I. The preservation theorem. *Ann. Math. Statist.* **5** 256–262.

TONG, Y.L. (1980). *Probability Inequalities in Multivariate Distributions.* Academic Press, New York.

DEPARTMENT OF STATISTICS
UNIVERSITY COLLEGE, DUBLIN
BELFIELD, DUBLIN 4, IRELAND

DEPARTMENT OF STATISTICS
FLORIDA STATE UNIVERSITY
TALLAHASSEE, FL 32306-3033

SCHOOL OF MATHEMATICS
GEORGIA INSTITUTE OF TECHNOLOGY
ATLANTA, GA 30332

STOCHASTICALLY MONOTONE DEPENDENCE

BY BENNETT BRADY AND NOZER D. SINGPURWALLA[1]

U.S. Nuclear Regulatory Commission and George Washington University

The notion of monotone dependence, which has played a key role in reliability theory, is generalized to that of "stochastically monotone dependence." The idea here is that since two lifelengths are dependent or independent based on the disposition of a conditioning variable, they are unconditionally stochastically dependent or independent. A measure of stochastic dependence is introduced and the measure used for comparing the correlations of pairs of random variables which can now be described as being "highly stochastically correlated" or "weakly stochastically correlated." Extensions to the multivariate case are possible and the ideas illustrated via examples. This paper is expository; its purpose is to propose a natural idea and to explore its ramifications.

1. Introduction and Motivation. An important, though little noticed, principle of probability theory is that the notions of dependence and independence are conditional, the conditioning being done on some observable or unobservable quantity, say Θ. It is common to think of Θ as a "parameter" and this is the point of view that we adopt. A consequence of the above is that unconditionally the notions of dependence and independence must be stochastic. That is, one should not make an unqualified judgment that lifelengths X_1 and X_2 are dependent or independent—rather one may talk in terms of the *probability that they are dependent or independent.* This is contrary to current thinking although the literature on artificial intelligence [cf. Pearl (1989)] appears to be taking cognizance of this fact. In this paper, we explore the ramifications of the above formulation, and in the sequel raise questions pertaining to the everyday used notions of covariance and correlation.

By way of some motivation, consider a system of two components with lifelengths X_1 and X_2, operating in an environment which is characterized by an

---

[1]Research supported by Grant DAAL 03–87–K–0056, the U.S. Army Research Office, Contract N00014–85–K–0202, Project NR 042–372 Office of Naval Research, and the Air Force Office of Scientific Research, Grant AFOSR–89–0381.

AMS 1980 subject classifications. Primary 60K10; secondary 62N05.

Key words and phrases. Biometry, multivariate lifelengths, reliability.

Discussions with Henry Block and Allan Sampson have helped clarify several issues; the operational scheme following Example 2.1 is a consequence of such discussions.

abstract (idealized and unobservable) parameter $\Theta \in \mathbf{R}$. Suppose that I_1, I_2, and I_3 partition the real line $\mathbf{R}$ such that $I_1 \cup I_2 \cup I_3 = \mathbf{R}$, and suppose that when $\Theta \in I_1$, the operating environment is classified as being "average" or "normal" whereas when $\Theta \in I_2$ of I_3 the operating environment is classified as being "mild" or "harsh," respectively. Now it is possible to conceive of a situation in which X_1 is independent of X_2—denoted henceforth as $X_1 \perp\!\!\!\perp X_2$—whenever $\Theta \in I_1$, and that X_1 and X_2 are positively or negatively dependent whenever $\Theta \in I_2$ or I_3 respectively. In any particular application, the exact disposition of Θ will be unknown or will change—from the point of view of an analyst—and so the nature of dependence between X_1 and X_2 is stochastic, depending on the probability that Θ belongs to I_1, I_2, or I_3.

Other scenarios which motivate the thesis of this paper arise from the biological sciences in which the nature of dependence between the lifelengths of two organs depends on the stochastic behavior of a conditioning covariate, such as the "life-style" of an individual.

Whereas the rationalization of positive dependence under common environmental conditions is relatively straightforward, see for example Lindley and Singpurwalla (1986), the rationalization of independence and negative dependence, particularly the latter, is more difficult. One possible argument is to suppose that under harsh conditions there may be a tendency to devote more resources and maintenance to the more important components of the system with the result that such components perform better than expected than those components which receive less attention. Such a policy would result in negative dependence.

Latent variable methods (cf. Holland and Rosenbaum (1986)) consider the concept of the distribution of a set of random variables given the latent variable. Such models consist of a set of "manifest variables" and the "latent or parametric variable." The manifest variables which are real or integer valued can be observed directly while the latent variable is unobservable. A basic assumption of the model is that the manifest variables are conditionally independent given the latent variable. Certain classes of latent variable models imply that the manifest variables exhibit stronger forms of positive dependence with the latent variable. However, these models do not incorporate the notion that the dependence of the manifest variables may change with a change in the latent variable.

In view of the preceding arguments it is necessary to reconsider the various notions of monotone dependence and their resulting bounds and inequalities. In this paper we define a new concept of dependence between random variables. Two variables are not unconditionally dependent or independent but are probably dependent or independent, depending on the disposition of the conditioning variable.

2. Stochastic Monotone Dependence. Let A be a σ-field of events generated by a sample space $\mathcal{X}$ and $\mathbf{P}$ be a family of probability measures defined on A_i, $i = 1, 2, \ldots$, the elements of A. Let $\mathbf{X}$ and $\mathbf{Y}$ be two vector valued random variables, of dimension p and q respectively, defined on $\mathcal{X}$; assume for now that $p = q \geq 1$.

The notation $((\mathbf{X} \perp\!\!\!\perp \mathbf{Y})|\boldsymbol{\theta})$ means that $\mathbf{X}$ is independent of $\mathbf{Y}$ given $\boldsymbol{\theta}$, where $\boldsymbol{\theta}$ is an s-dimensional vector of parameters. Without any loss of generality assume that $s = 1$.

2.1 Stochastic Dependence and Independence. Suppose we have a partial ordering on $\mathbf{R}^p$ such that for vectors $\mathbf{a} = (a_1, a_2, \ldots, a_p)$ and $\mathbf{b} = (b_1, b_2, \ldots, b_p)$ in $\mathbf{R}^p$

$$\mathbf{a} \leq \mathbf{b} \text{ means } a_i \leq b_i, \quad i = 1, 2, \ldots, p$$

and suppose that the elements A_i are open upper sets, i.e., A_i is an upper set if $\mathbf{a} \in A_i$, and $\mathbf{a} \leq \mathbf{b}$ implies $\mathbf{b} \in A_i$ (Shaked (1982)).

DEFINITION 2.1. The random vectors $\mathbf{X}$ and $\mathbf{Y}$ are independent given $\boldsymbol{\theta}$, denoted by $\{(\mathbf{X} \perp\!\!\!\perp \mathbf{Y})|\boldsymbol{\theta}\}$ if

$$P\{\mathbf{X} \in A_i | \mathbf{Y} \in A_j, \ \boldsymbol{\theta}\} = P\{\mathbf{X} \in A_i | \boldsymbol{\theta}\}, \ \ \forall A_i, \ A_j, \ \boldsymbol{\theta}, \ \text{ and any } P \in \mathbf{P}.$$

Suppose that $\boldsymbol{\theta}$ takes values in $\mathbf{R}$, and suppose that the Borel σ-field generated by $\mathbf{R}$ is endowed with a family of probability measures $\tilde{\mathbf{P}}$. Let I_1, I_2, and I_3 be members of the Borel σ-field generated by $\mathbf{R}$ and let $\tilde{P} \in \tilde{\mathbf{P}}$.

DEFINITION 2.2. The random vector $\mathbf{X}$ is $\boldsymbol{\theta} \in I_1$ *conditionally independent of* $\mathbf{Y}$ and $\boldsymbol{\theta} \notin I_1$ *conditionally dependent on* Y, denoted by $\{(\mathbf{X} \perp\!\!\!\perp \mathbf{Y})|\boldsymbol{\theta} \in I_1, \neq\}$, if

i) $P\{\mathbf{X} \in A_i | \mathbf{Y} \in A_j, \ \boldsymbol{\theta} \in I_1\} = P\{\mathbf{X} \in A_i | \boldsymbol{\theta} \in I_1\}$,

ii) $P\{\mathbf{X} \in A_i | \mathbf{Y} \in A_j, \ \boldsymbol{\theta} \notin I_1\} \neq P\{\mathbf{X} \in A_i | \boldsymbol{\theta} \notin I_1\}$, and $\forall A_i, \ A_j, \ \boldsymbol{\theta}$.

DEFINITION 2.3. The random vectors $\mathbf{X}$ and $\mathbf{Y}$ are unconditionally independent, denoted by $(\mathbf{X} \perp\!\!\!\perp \mathbf{Y})$ if

$$\nexists \boldsymbol{\theta} \ni P\{\mathbf{X} \in A_i | \mathbf{Y} \in A_j, \ \boldsymbol{\theta}\} \neq P\{\mathbf{X} \in A_i | \boldsymbol{\theta}\}, \ \ \forall A_i, \ A_j.$$

The subjective nature of the notion of independence is revealed in Definition 2.3 if one interprets $\nexists \boldsymbol{\theta}$ as being the nonexistence—to a probability assessor—of a $\boldsymbol{\theta}$.

LEMMA 2.1. *If* $\{\mathbf{X} \perp\!\!\!\perp \mathbf{Y} | \boldsymbol{\theta} \in I_1, \neq\}$, *and if* $\{\boldsymbol{\theta} \in I_1\}$ *is the only event for which* $\mathbf{X}$ *and* $\mathbf{Y}$ *are conditionally independent, then*

$$P\{\mathbf{X} \perp\!\!\!\perp \mathbf{Y}\} = \tilde{P}\{\boldsymbol{\theta} \in I_1\}.$$

PROOF. From Definition 2.3 we see that

$$\begin{aligned} P\{\mathbf{X} \perp\!\!\perp \mathbf{Y}\} &= P\{\nexists \boldsymbol{\theta} \ni P\{\mathbf{X} \in A_i | \mathbf{Y} \in A_j,\ \boldsymbol{\theta}\} \neq P\{\mathbf{X} \in A_i | \boldsymbol{\theta}\}\} \\ &= 1 - \{P\{\nexists \boldsymbol{\theta} \ni P\{\mathbf{X} \in A_i | \mathbf{Y} \in A_j,\ \boldsymbol{\theta}\} \neq P\{\mathbf{X} \in A_i | \boldsymbol{\theta}\}\} \\ &= 1 - (1 - \Pi(\boldsymbol{\theta})) = \Pi(\boldsymbol{\theta}), \end{aligned}$$

where $\Pi(\boldsymbol{\theta}) = \tilde{P}(\boldsymbol{\theta} \in I_1)$.

A strengthening of Definition 2.2 is given next.

DEFINITION 2.4. The random vector $\mathbf{X}$ is $\boldsymbol{\theta} \in I_1$ *conditionally independent of* $\mathbf{Y}$, and is $\boldsymbol{\theta} \notin I_1$ *conditionally positively (negatively) dependent on* $\mathbf{Y}$, denoted $\{(\mathbf{X} \perp\!\!\perp \mathbf{Y}) | (\boldsymbol{\theta} \in I_1,\ > (<))\}$, if

i) $P\{\mathbf{X} \in A_i | \mathbf{Y} \in A_j,\ \boldsymbol{\theta} \in I_1\} = P\{\mathbf{X} \in A_i | \boldsymbol{\theta} \in I_1\}$, and $\forall A_i,\ A_j,\ \boldsymbol{\theta}$,

ii) $P\{\mathbf{X} \in A_i | \mathbf{Y} \in A_j,\ \boldsymbol{\theta} \notin I_1\} \geq (\leq) P\{\mathbf{X} \in A_i | \boldsymbol{\theta} \notin I_1\}$.

For convenience we denote $P\{\mathbf{X} \in A_i | \mathbf{Y} \in A_j,\ \boldsymbol{\theta} \notin I_1\} \geq P\{\mathbf{X} \in A_i | \boldsymbol{\theta} \notin I_1\}$ by $\mathbf{X} \perp\!\!\perp^{(+)} \mathbf{Y}$, and by $\mathbf{X} \perp\!\!\perp^{(-)} \mathbf{Y}$ when the above inequality is reversed.

LEMMA 2.2. *If* $\{(\mathbf{X} \perp\!\!\perp \mathbf{Y}) | \boldsymbol{\theta} \in I_1,\ > (<)\}$, *and* $\boldsymbol{\theta}$ *is unique, then*

$$\begin{aligned} &P\{\mathbf{X} \perp\!\!\perp \mathbf{Y}\} = \Pi(\boldsymbol{\theta}) \ \ \textit{and} \\ &P\{\mathbf{X} \perp\!\!\perp^{(+)} \mathbf{Y}\} \ \textit{or}\ P\{\mathbf{X} \perp\!\!\perp^{(-)} \mathbf{Y}\} = 1 - \Pi(\boldsymbol{\theta}). \end{aligned}$$

PROOF. Follows from Lemma 2.1 and the fact that it is not possible to have both $P\{\mathbf{X} \perp\!\!\perp^{(+)} \mathbf{Y}\}$ and $P\{\mathbf{X} \perp\!\!\perp^{(-)} \mathbf{Y}\}$.

A further strengthening of Definition 2.4 is given next.

DEFINITION 2.5. The random vector $\mathbf{X}$ is $\boldsymbol{\theta} \in I_1$ *conditionally independent of* $\mathbf{Y}$, and is $\boldsymbol{\theta} \in I_2(I_3)$ *conditionally positively (negatively) dependent on* $\mathbf{Y}$, denoted by $\{(\mathbf{X} \perp\!\!\perp \mathbf{Y}) | (\boldsymbol{\theta} \in I_1,\ > \boldsymbol{\theta} \in I_2,\ < \boldsymbol{\theta} \in I_3)\}$, if

i) $P\{\mathbf{X} \in A_i | \mathbf{Y} \in A_j,\ \boldsymbol{\theta} \in I_1\} = P\{\mathbf{X} \in A_i | \boldsymbol{\theta} \in I_1\}$,

ii) $P\{\mathbf{X} \in A_i | \mathbf{Y} \in A_j,\ \boldsymbol{\theta} \in I_2\} \geq P\{\mathbf{X} \in A_i | \boldsymbol{\theta} \in I_2\}$, and

iii) $P\{\mathbf{X} \in A_i | \mathbf{Y} \in A_j,\ \boldsymbol{\theta} \in I_3\} \leq P\{\mathbf{X} \in A_i | \boldsymbol{\theta} \in I_3\}$, $\quad \forall A_i,\ A_j,\ \boldsymbol{\theta}$.

We may now state

LEMMA 2.3. *If* $\Pi_i(\boldsymbol{\theta}) = \tilde{P}\{\boldsymbol{\theta} \in I_i\}$, *then*

$$\begin{aligned}
P\{\mathbf{X}\perp\!\!\!\perp\mathbf{Y}\} &= \Pi_1(\boldsymbol{\theta}) \\
P\{\mathbf{X}\perp\!\!\!\perp^{(+)}\mathbf{Y}\} &= \Pi_2(\boldsymbol{\theta}), \textit{ and} \\
P\{\mathbf{X}\perp\!\!\!\perp^{(-)}\mathbf{Y}\} &= \Pi_3(\boldsymbol{\theta}).
\end{aligned}$$

2.2. Equivalent Conditions and Definitions. Assume that $p = q = 1$. Definition 2.5 can also be stated in terms of the joint and the marginal distribution functions of X and Y. Let

$$\begin{aligned}
F(x, y|\boldsymbol{\theta}) &= P\{X \leq x,\ Y \leq y|\boldsymbol{\theta}\} \\
G(x|\boldsymbol{\theta}) &= P\{X \leq x|\boldsymbol{\theta}\}, \text{ and} \\
H(y|\boldsymbol{\theta}) &= P\{Y \leq y|\boldsymbol{\theta}\}.
\end{aligned}$$

Then Definition 2.5 is equivalent to:

DEFINITION 2.6. The random vector $\mathbf{X}$ is $\boldsymbol{\theta} \in I_1$ *conditionally independent of* $\mathbf{Y}$ and is $\boldsymbol{\theta} \in I_2(I_3)$ *conditionally positively (negatively) quadrant dependent on* $\mathbf{Y}$, denoted by $\{(\mathbf{X}\perp\!\!\!\perp\mathbf{Y})|\boldsymbol{\theta} \in I_1,\ > \boldsymbol{\theta} \in I_2,\ < \boldsymbol{\theta} \in I_3)\}$, if

i) $F(x, y|\boldsymbol{\theta} \in I_1) = G(x|\boldsymbol{\theta} \in I_1)H(y|\boldsymbol{\theta} \in I_1)$,

ii) $F(x, y|\boldsymbol{\theta} \in I_2) \geq G(x|\boldsymbol{\theta} \in I_2)H(y|\boldsymbol{\theta} \in I_2)$, and

iii) $F(x, y|\boldsymbol{\theta} \in I_3) \leq G(x|\boldsymbol{\theta} \in I_3)H(y|\boldsymbol{\theta} \in I_3)$.

Using a lemma by Hoeffding (1940), we are able to state Lemma 2.5.

LEMMA 2.4. *(Hoeffding) If F denotes the joint, and G and H the marginal distribution functions of X and Y respectively, then*

$$E(XY) - E(X)E(Y) = \int_{-\infty}^{+\infty}\int_{-\infty}^{+\infty}[F(x, y) - G(x)H(y)]dx\, dy,$$

provided the expectations exist.

LEMMA 2.5. *If conditions i), ii), and iii) of Definition 2.6 hold and if the conditional expectations $E(XY|\boldsymbol{\theta})$, $E(X|0)$ and $E(Y|\boldsymbol{\theta})$ exist, then Definition 2.6 implies that*

i) $E(XY|\boldsymbol{\theta} \in I_1) = E(X|\boldsymbol{\theta} \in I_1)E(Y|\boldsymbol{\theta} \in I_1)$,

ii) $E(XY|\boldsymbol{\theta} \in I_2) \geq E(X|\boldsymbol{\theta} \in I_2)E(Y|\boldsymbol{\theta} \in I_2)$, *and*

iii) $E(XY|\boldsymbol{\theta} \in I_3) \leq E(X|\boldsymbol{\theta} \in I_3)E(Y|\boldsymbol{\theta} \in I_3)$.

It follows then that

LEMMA 2.6. $\{(X \perp\!\!\!\perp Y) | \boldsymbol{\theta} \in I_1, > \boldsymbol{\theta} \in I_2, < \boldsymbol{\theta} \in I_3\} \Longrightarrow$

i) $COV(X, Y | \boldsymbol{\theta} \in I_1) = 0$,

ii) $COV(X, Y | \boldsymbol{\theta} \in I_2) \geq 0$, *and*

iii) $COV(X, Y | \boldsymbol{\theta} \in I_3) \leq 0$.

EXAMPLE 2.1. As an example illustrating the intent of Lemma 2.6, suppose that X and Y have a bivariate normal distribution with mean $\boldsymbol{\mu} = (\mu_1, \mu_2)$ and covariance

$$\sum = \begin{bmatrix} \sigma_1^2 & \rho\sigma_1\sigma_2 \\ \rho\sigma_1\sigma_2 & \sigma_2^2 \end{bmatrix}$$

where ρ is the coefficient of correlation. Then ρ is our conditioning variable and as is well known $\{(X \perp\!\!\!\perp Y) | \rho = 0\}$, and X and Y have positive (negative) dependence when $\rho > (<) 0$.

To describe the operational implication of the above, suppose that we are asked to make a prediction of X when $Y = y$ has been observed. To simplify matters suppose that $\boldsymbol{\mu} = (0, 0)$ and that $P(\rho \neq 0) = \Pi_1$. Assuming that the penalty of poor prediction is described by the squared error loss, we would specify $E(X|y, \rho = 0) = 0$, or $E(X|y, \rho \neq 0) = \rho\sigma_1 y/\sigma_2$. Operationally, we would toss a coin whose probability of heads is Π_1, and bet on $\rho\sigma_1 y/\sigma_2$ if the coin lands heads, and on 0 if it lands tails.

A strengthening of Lemma 2.6 is

LEMMA 2.7. *Let f and g be nondecreasing functions of X and Y, respectively. If X and Y satisfy Definition 2.6, then*

i) $COV(f(X), g(Y) | \boldsymbol{\theta} \in I_1) = 0$,

ii) $COV(f(X), g(Y) | \boldsymbol{\theta} \in I_2) \geq 0$, *and*

iii) $COV(f(X), g(Y) | \boldsymbol{\theta} \in I_3) \leq 0$.

PROOF. This follows by an extension of a proof due to Lehmann (1966).

When ii) holds, we shall say that X and Y are *conditionally $\theta \in I_2$ positively associated*; and when iii) holds, we shall say that X and Y are *conditionally $\theta \in I_3$ negatively associated.*

Our motivation for a consideration of the above material is the introduction of the notion of stochastic covariance and correlation. This is discussed next.

2.3. Stochastic Linear Dependence. The notions of covariance and correlation appear in everyday use of probability and statistics. In sample theory statistics, the covariance and correlation have been viewed as fixed but unknown quantities. In Bayesian statistics, all unknown quantities are to be assigned a prior distribution to reflect one's uncertainty about them. To see this, note that the conditional nature of Definition 2.6, and its implications prompt us to generalize the notion of covariance and make it stochastic. The situation here is not unlike that of hierarchical modelling [c.f. Good (1983)].

Suppose that $p = q = s = 1$, so that $\mathbf{X} = X$, $\mathbf{Y} = Y$, and $\boldsymbol{\theta} = \theta$. Then from Definition 2.6, it follows that COV $(X, Y|\theta \in (I_1 \cup I_2)) \geq 0$, where

$$\begin{aligned}\text{COV}\,(X,Y|\theta \in (I_1 \cup I_2)) &= E\,(XY|\theta \in (I_1 \cup I_2)) \\ &-E\,(X|\theta \in (I_1 \cup I_2))\,E\,(Y|\theta \in (I_1 \cup I_2)).\end{aligned}$$

Using an argument analogous to that of Lemma 2.1, we see that unconditionally $P\{\text{COV}(X,Y) \geq 0\} = P\{\theta \in (I_1 \cup I_2)\}$. Thus a prior distribution on the covariance would depend on the nature of the parameterization of the probability model for X and Y, and our uncertainty about the disposition of the parameter. If $\theta \in I_1$, or if $\theta \in I_2$, that is, if we judge $(X \perp\!\!\!\perp Y)$ or $(X \perp\!\!\!\perp^{(+)} Y)$, and if θ is unique, then $P\{\text{COV}(X,Y) \geq 0\} = 1$. Thus $P\{\text{COV}(X,Y) \geq 0\}$ gives us a measure of the strength of the linear relationship between X and Y.

The above motivates us to consider the quantity $\Pi(\alpha) = P\{|\text{COV}(X,Y)| \geq \alpha\}$, $\alpha \geq 0$, for characterizing the strength of linear dependence between X and Y.

DEFINITION 2.7. To simplify matters, suppose that $E(X|\theta) = E(Y|\theta) = 0$ and that VAR $(X|\theta) =$ VAR $(Y|\theta) = 1$, for all values of θ: then COV $(X,Y) = \rho(X,Y)$, the correlation coefficient between X and Y. Let $\Pi(\alpha) = P\{|\text{COV}(X,Y)| \geq \alpha\} = P\{|\rho(X,Y)| \geq \alpha\}$; we refer to $\Pi(\alpha)$ as the *correlation survival function.*

It is clear that $\Pi(\alpha) \downarrow \alpha$, $\Pi(0) = 1$, and $\Pi(1^+) = 0$. A plot of $\Pi(\alpha)$ versus α, $0 \leq \alpha \leq 1$, for $\Pi(\alpha) = 1 - \alpha$, is given on the following page.

EXAMPLE 2.2. As an example of the above, let X and Y be binary with $P\{X = 1\} = p_x$, $P\{Y = 1\} = p_y$, and $P\{X = 1,\ Y = 1\} = p_{xy}$. Clearly, $P\{X = 1,\ Y = 0\} = p_x - p_{xy}$, $P\{X = 0,\ Y = 1\} = p_y - p_{xy}$, $P\{X = 0,\ Y = 0\} = 1 - p_x - p_y + p_{xy}$, and thus $0 \leq p_{xy} \leq \min(p_x, p_y)$.

Suppose that p_x and p_y are specified, but the disposition of p_{xy} is unknown. Then it can be verified that

$$(\rho(X,Y)|p_{xy}) = \frac{(p_{xy} - p_x p_y)}{\sqrt{p_x p_y (1 - p_x)(1 - p_y)}}$$

from which it follows that

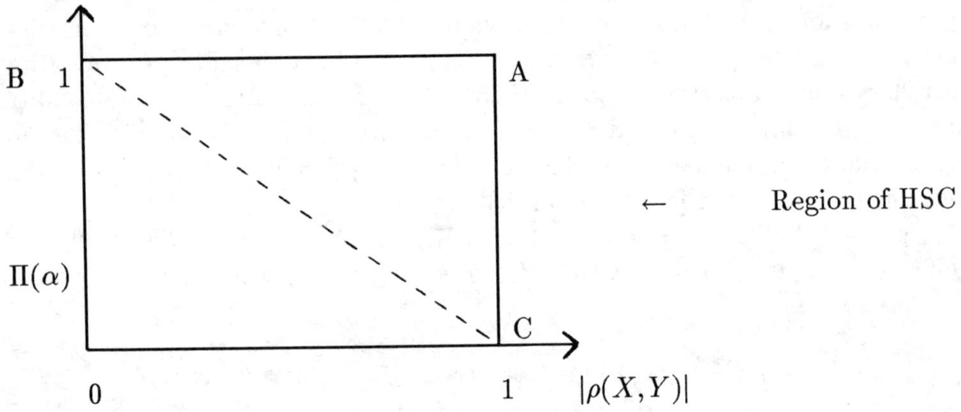

Figure 2.1. The correlation survival function when $\Pi(\alpha) = 1 - \alpha$

$$\Pi(\alpha) = 1 - \{\Delta(p_x p_y + \alpha\sqrt{p_x p_y(1-p_x)(1-p_y)})$$
$$-\Delta(p_x p_y - \alpha\sqrt{p_{xy}(1-p_x)(1-p_y)})\}$$

where $\Delta(\cdot)$ is the cumulative distribution function of p_{xy}, for $0 \le x \le \min(p_x, p_y)$.

To characterize the strength of linear dependence between X and Y via $\Pi(\alpha)$, we first note that the strongest case for linear dependence is when $P\{\rho(X,Y) = 1\} = 1$; that is, when we are absolutely sure that the values of X and Y match perfectly. When this happens $P\{|\rho(X,Y)| \ge \alpha\} = \Pi(\alpha) = 1$, for all $0 \le \alpha \le 1$; that is, the correlation survival curve is the locus BAC in Figure 2.1. The worst case for linear dependence is when $P\{|\rho(X,Y)| \ge \alpha\} = 0$ for all $\alpha > 0$. For this case the correlation survival rate is the locus BOC. Thus, any correlation survival function which is closer to the locus BAC is to be preferred, in the sense that the variables are more highly correlated, to the one which is closer to the locus BOC. The above considerations prompt us to suggest the following as a plausible criteria for describing a high stochastic correlation.

DEFINITION 2.8. Random variables X and Y are said to be *highly stochastically correlated* (HSC) if

$$P\{|\mathrm{COV}(X,Y)| \ge \alpha\} = P\{|\rho(X,Y)| \ge \alpha\} \ge 1 - \alpha,$$

$0 \le \alpha \le 1$; otherwise they are *weakly stochastically correlated* (WSC).

EXAMPLE 2.3. Suppose that in Example 2.2, that p_{xy} has a uniform distribution on $[0, \min(p_x, p_y)]$. Then,

$$\Pi(\alpha) = 1 - \frac{(2\alpha\sqrt{p_x p_y(1-p_x)(1-p_y)})}{\min(p_x, p_y)}.$$

The binary variables X and Y are HSC if $\Pi(\alpha) \geq 1 - \alpha$; i.e., if

$$\frac{1}{2} \geq \frac{\sqrt{p_x p_y(1-p_x)(1-p_y)}}{\min(p_x, p_y)}.$$

In order to compare the strength of linear dependence between two pairs of random variables, we introduce the following definition.

DEFINITION 2.9. Random variables (X, Y) are *stochastically more (less) correlated* than (X^1, Y^1) if

$$P\{|\rho(X,Y)| \geq \alpha\} \geq (\leq) P\{|\rho(X^1, Y^1)| \geq \alpha\}, \quad 0 \leq \alpha \leq 1.$$

We shall denote the above writing $\rho(X,Y) \overset{\text{st}}{\geq} (\overset{\text{st}}{\leq}) \rho(X^1, Y^1)$.

Definition 2.9 provides a basis for comparing the strength of the linear dependence between (X, Y) and (X^1, Y^1) when their respective correlation survival functions do not cross. To characterize linear dependence when the correlation survival functions do cross, we need the following.

DEFINITION 2.10. Random variables (X, Y) are *more (less) correlated in expectation* than (X^1, Y^1) if

$$\int_0^1 \Pi_{X,Y}(\alpha)\, d\alpha \geq (\leq) \int_0^1 \Pi_{X^1,Y^1}(\alpha)\, d\alpha$$

where

$$\Pi_{X,Y}(\alpha) = P\{|\rho(X,Y)| \geq \alpha\} \text{ and } \Pi_{X^1,Y^1}(\alpha) = P\{|\rho(X^1,Y^1)| \geq \alpha\}.$$

We shall denote the above by writing $\rho(X,Y) \overset{E}{\geq} (\overset{E}{\leq}) \rho(X^1, Y^1)$. It is obvious from the above that

PROPOSITION 2.1. $\rho(X,Y) \overset{\text{st}}{\geq} (\overset{\text{st}}{\leq}) \rho(X^1,Y^1) \Longrightarrow \rho(X,Y) \overset{E}{\geq} (\overset{E}{\leq}) \rho(X^1,Y^1)$.

EXAMPLE 2.4. The survival function of Marshall and Olkin's bivariate exponential distribution is given as

$$\bar{F}(x,y) = e^{-\lambda_1 x - \lambda_2 y - \lambda \max(x,y)},$$

where we suppose λ_1 and λ_2 known but λ is unknown. Then $\{X \perp\!\!\!\perp Y | \lambda = 0, > \lambda \in (0, \infty)\}$; i.e., X and Y are independent if $\lambda = 0$, and otherwise always positive dependent.

The coefficient of correlation is

$$\rho(X, Y|\lambda) = \lambda/(\lambda_1 + \lambda_2 + \lambda),$$

from which it follows that the correlation survival function is

$$\Pi(\alpha) = P\{\rho(X, Y) \geq \alpha\} = 1 - \Delta(\frac{\alpha}{1-\alpha}[\lambda_1 + \lambda_2])$$

where $\Delta(\cdot)$ is the cumulative distribution function of λ, for $\lambda \geq 0$.

If λ has an exponential distribution with parameter λ', then the correlation survival function is

$$\Pi(\alpha) = \exp\{-\frac{\alpha\lambda'(\lambda_1 + \lambda_2)}{1-\alpha}\}.$$

So X and Y are HSC if

$$\exp\{-\frac{\alpha\lambda'(\lambda_1 + \lambda_2)}{1-\alpha}\} \geq 1 - \alpha,$$

that is, if

$$\lambda' \leq -\frac{1-\alpha}{\alpha(\lambda_1 + \lambda_2)} \log(1-\alpha), \quad \text{for all } \alpha, \quad 0 \leq \alpha \leq 1.$$

An extension of the above ideas to the multivariate case together with notions of higher order stochastic dependence are discussed in Brady (1988).

REFERENCES

BRADY, B.M. (1988). Stochastic positive and negative dependence. Doctoral dissertation. Department of Operations Research, The George Washington University, Washington, DC 20052.

GOOD, I.J. (1983). *Good Thinking.* University of Minnesota Press, Minneapolis, MN.

HOEFFDING, W. (1940). A class of statistics with asymptotically normal distributions. *Ann. Math. Statist.* **19** 293–375.

HOLLAND, P.W. and ROSENBAUM, P.R. (1986). Conditional association and unidimensionality in monotone latent variable models. *Ann. Statist.* **14**, No. 4, 1523–1543.

LEHMANN, E.L. (1966). Some concepts of dependence. *Ann. Math. Statist.* **37** 1137–1152.

LINDLEY, D.V. and SINGPURWALLA, N.D. (1986). Multivariate distributions for life-lengths of components of a system sharing a common environment. *J. Appl. Prob.* **23** No. 2.

PEARL, J. (1989). *Probabilistic Reasoning in Intelligent System: Networks of Plausible Inference.* Morgan Kauffman Publishers, Inc., San Mateo, CA.

SHAKED, M. (1982). A general theory of some positive dependence notions. *J. Multi. Anal.* **12** 199–218.

U.S. NUCLEAR REGULATORY COMMISSION
WASHINGTON, D.C. 20555

DEPARTMENT OF OPERATIONS RESEARCH
GEORGE WASHINGTON UNIVERSITY
WASHINGTON, DC 20052

A DECISION APPROACH TO ORDERING STOCHASTIC DEPENDENCE

By T. Bromek and T. Kowalczyk

Polish Academy of Sciences

Editors' Note: This paper is being published posthumously and in dedicated remembrance of Tadeusz Bromek, who died in an automobile accident on August 23, 1988 in Warsaw. He was very active in the Polish statistical community as well as the international statistical community.

An ordering of global dependence is defined on the basis of a natural ordering of pairs of distributions describing two classes of objects. Its properties are investigated; the links with orderings of multinormal and 2×2 distributions are shown.

1. Introduction. Traditionally, two types of stochastic dependence of components of a vector $\mathbf{X}$ have been distinguished in statistical literature, namely monotone and global dependence. Orderings for monotone dependence were considered by many authors; an overview was given by Yanagimoto (1990). Kimeldorf and Sampson (1987) introduced an axiomatic approach to the matter of orders of monotone dependence. The abundance of formalizations for orderings of monotone dependence contrasts with the silence concerning orderings of global dependence (see Dabrowska (1985)). It seems that a good starting point could be two-class discriminant analysis, with one class reserved for the distribution of the vector $\mathbf{X}$ and the other class reserved for the respective product of marginal distributions of $\mathbf{X}$. Thus, a natural ordering of pairs of distributions describing two classes of objects (Niewiadomska-Bugaj (1987)), called prognostic ordering and denoted $\leq_p$, may be a base to define an ordering of global dependence, called global ordering and denoted $\leq_g$.

In Section 2 we recall the definition of $\leq_p$ and prove some of its new properties. Section 3 contains the definition and properties of $\leq_g$.

2. Prognostic Order $\leq_p$. Consider a two-class discriminant problem corresponding to a pair $(\mathbf{Z}_1, \mathbf{Z}_2)$ where $\mathbf{Z}_1$ and $\mathbf{Z}_2$ are random vectors supported on $\mathcal{Z} \subset R^k$. Distribution P_i of $\mathbf{Z}_i$ describes the ith class of the considered population $(i = 1, 2)$. Let a classification rule for $(\mathbf{Z}_1, \mathbf{Z}_2)$ be a Borel measurable function

AMS 1980 subject classifications. 62H20.

Key words and phrases. Global dependence, monotone dependence, likelihood ratio, two-class discriminant analysis.

$\partial : \mathcal{Z} \to [0,1]$, where $\partial(\mathbf{z})$ is the probability of classifying an object with features vector $\mathbf{z}$ to the 1st class. Performance of ∂ is measured by the error rates $a_{12}(\partial)$ and $a_{21}(\partial)$:

$$a_{12}(\partial) = \int_{\mathcal{Z}} (1 - \partial(\mathbf{z}))dP_1, \quad a_{21}(\partial) = \int_{\mathcal{Z}} \partial(\mathbf{z})dP_2.$$

For two pairs of random vectors $(\mathbf{Z}_1, \mathbf{Z}_2)$, $(\tilde{\mathbf{Z}}_1, \tilde{\mathbf{Z}}_2)$, there is defined an ordering with respect to their discriminant powers:

> $(\mathbf{Z}_1, \mathbf{Z}_2) \leq_p (\tilde{\mathbf{Z}}_1, \tilde{\mathbf{Z}}_2)$ iff for every decision rule ∂ for $(\mathbf{Z}_1, \mathbf{Z}_2)$ there exists a decision rule $\tilde{\partial}$ for $(\tilde{\mathbf{Z}}_1, \tilde{\mathbf{Z}}_2)$ such that $a_{12}(\tilde{\partial}) \leq a_{12}(\partial)$ and $a_{21}(\tilde{\partial}) \leq a_{21}(\partial)$.

It is clear that $\leq_p$ is a preorder. Thus, a relation $\underset{p}{\approx}$ defined by:

$$(\mathbf{Z}_1, \mathbf{Z}_2) \underset{p}{\approx} (\tilde{\mathbf{Z}}_1, \tilde{\mathbf{Z}}_2) \text{ iff } (\mathbf{Z}_1, \mathbf{Z}_2) \leq_p (\tilde{\mathbf{Z}}_1, \tilde{\mathbf{Z}}_2) \text{ and } (\tilde{\mathbf{Z}}_1, \tilde{\mathbf{Z}}_2) \leq_p (\mathbf{Z}_1, \mathbf{Z}_2)$$

is an equivalence.

THEOREM 1.

(i). For any Borel measurable function $f : \mathcal{Z} \to R^k$

$$1^\circ \ (f(\mathbf{Z}_1), f(\mathbf{Z}_2)) \leq_p (\mathbf{Z}_1, \mathbf{Z}_2);$$

if f is an injection, then

$$2^\circ \ (f(\mathbf{Z}_1), f(\mathbf{Z}_2)) \underset{p}{\approx} (\mathbf{Z}_1, \mathbf{Z}_2);$$

(ii). $(\mathbf{Z}_1, \mathbf{Z}_2)$ is a minimal element of $\leq_p$ iff $\mathbf{Z}_1$ and $\mathbf{Z}_2$ are distributed identically;

(iii). $(\mathbf{Z}_1, \mathbf{Z}_2)$ is a maximal element of $\leq_p$ iff there exist a set $A \subset \mathcal{Z}$ such that $P_1(A) = 1$ and $P_2(A) = 0$.

PROOF.

(i). To any classification rule $\bar{\partial}$ for $(f(\mathbf{Z}_1), f(\mathbf{Z}_2))$ there corresponds a classification rule $\partial = \bar{\partial} \circ f$ for $(\mathbf{Z}_1, \mathbf{Z}_2)$ with error rates a_{12} and a_{21}, respectively equal to those of $\bar{\partial}$. Thus 1° holds. Then, applying to $(f(\mathbf{Z}_1), f(\mathbf{Z}_2))$ the inverse function $f^{-1} : f(\mathcal{Z}) \to \mathcal{Z}$ (which is Borel measurable since f is the injection), we get $(\mathbf{Z}_1, \mathbf{Z}_2) \leq_p (f(\mathbf{Z}_1), f(\mathbf{Z}_2))$ which implies equivalence 2°.

(ii). Obviously, for any pair $(\mathbf{Z}_1, \mathbf{Z}_2)$ and any classification rule ∂ for that pair,

$$a_{12}(\partial) + a_{21}(\partial) = 1 \text{ iff } \int_{\mathcal{Z}} \partial(\mathbf{z}) dP_1 = \int_{\mathcal{Z}} \partial(\mathbf{z}) dP_2.$$

Therefore $\mathbf{Z}_1$ and $\mathbf{Z}_2$ are distributed identically iff, for any classification rule ∂ applied to $(\mathbf{Z}_1, \mathbf{Z}_2)$, $a_{12}(\partial) + a_{21}(\partial) = 1$. On the other hand, for any $(\tilde{\mathbf{Z}}_1, \tilde{\mathbf{Z}}_2)$ and any constant rule $\partial(\tilde{\mathbf{z}}) = \mathcal{L}$, $0 \leq \mathcal{L} \leq 1$, $a_{12}(\partial) = \mathcal{L}$ and $a_{21}(\partial) = 1 - \mathcal{L}$; hence $(\mathbf{Z}_1, \mathbf{Z}_2) \leq_p (\tilde{\mathbf{Z}}_1, \tilde{\mathbf{Z}}_2)$ iff $\mathbf{Z}_1$ and $\mathbf{Z}_2$ are distributed identically.

(iii). Let $A \subset \mathcal{Z}$ satisfy $P_1(A) = 1$, $P_2(A) = 0$, and let ∂ be a rule such that $\partial(\mathbf{z}) = 1$ if $\mathbf{z} \in A$ and $\partial(\mathbf{z}) = 0$ if $z \in \mathcal{Z} \setminus A$. Then $a_{12}(\partial) = a_{21}(\partial) = 0$, and hence $(\mathbf{Z}_1, \mathbf{Z}_2)$ is a maximal element of $\leq_p$.

Conversely, if for $(\mathbf{Z}_1, \mathbf{Z}_2)$ there exists a rule ∂ such that $a_{12}(\partial) = a_{21}(\partial) = 0$, then for a set $A = \{z \in \mathcal{Z}; \partial(\mathbf{z}) > 0\}$ we have $P_2(A) = 0$, $P_1(A) \geq \int_A \partial(\mathbf{z}) dP_1 = \int_{\mathcal{Z}} \partial(\mathbf{z}) dP_1 = 1 - a_{21}(\partial) = 1$. ||

It follows from the Neyman-Pearson Lemma that this set consists of threshold rules based on the likelihood ratio $h = f_2/f_1$, where f_i is a density function of $\mathbf{Z}_i$ with respect to some measure ν (we set $h(\mathbf{z}) = \infty$ if $f_1(\mathbf{z}) = 0$). These rules are defined by

$$\partial(\mathbf{z}) = \begin{cases} 1 & \text{if } h(\mathbf{z}) < \gamma \\ s & \text{if } h(\mathbf{z}) = \gamma \\ 0 & \text{if } h(\mathbf{z}) > \gamma, \end{cases}$$

for $\gamma > 0$ and $s \in [0, 1]$.

Now, let us extend this set of rules admitting $\gamma = 0$ and $\gamma = +\infty$, and let

$$\begin{aligned} C_{(\mathbf{Z}_1,\mathbf{Z}_2)} = & \{P_1(h(\mathbf{z}) > \gamma) + (1 - s)P_1(h(\mathbf{z}) = \gamma), P_2(h(\mathbf{z}) < \gamma) + sP_2(h(\mathbf{z}) = \gamma); \\ & 0 \leq \gamma \leq \infty,\ 0 \leq s \leq 1\}. \end{aligned}$$

It is easy to see that $C_{(\mathbf{Z}_1,\mathbf{Z}_2)}$ is a curve joining points (0,1) and (1,0) which is continuous, convex, and nonincreasing. It will be called the divergence curve for $(\mathbf{Z}_1, \mathbf{Z}_2)$. Obviously, $C_{(\mathbf{Z}_1,\mathbf{Z}_2)}$ is the set of errors $(a_{12}(\partial), a_{21}(\partial))$ for threshold rules from the extended set of rules with minimal error rates.

We will say that $C_{(\tilde{\mathbf{Z}}_1,\tilde{\mathbf{Z}}_2)} \leq C_{(\mathbf{Z}_1,\mathbf{Z}_2)}$ iff for any $(x_1, x_2) \in C_{(\mathbf{Z}_1,\mathbf{Z}_2)}$ there exists $(x_1, \tilde{x}_2) \in C_{(\tilde{\mathbf{Z}}_1,\tilde{\mathbf{Z}}_2)}$ such that $x_2 \geq \tilde{x}_2$.

The following is an equivalent definition of $\leq_g$:

$$(\mathbf{Z}_1, \mathbf{Z}_2) \leq_p (\tilde{\mathbf{Z}}_1, \tilde{\mathbf{Z}}_2) \text{ iff } C_{(\tilde{\mathbf{Z}}_1,\tilde{\mathbf{Z}}_2)} \leq C_{(\mathbf{Z}_1,\mathbf{Z}_2)}.$$

3. Global Dependence Order $\leq_g$. Given a random vector X, consider a discriminant problem corresponding to a pair $({}^{\perp}\mathbf{X}, \mathbf{X})$, where ${}^{\perp}\mathbf{X}$ is a random vector distributed according to the product of marginal distributions of $\mathbf{X}$. For any pair of random vectors $\mathbf{X}$ and $\mathbf{Y}$, we define the following ordering $\leq_g$ of global dependence:

$$\mathbf{X} \leq_g \mathbf{Y} \text{ iff } ({}^{\perp}\mathbf{X}, \mathbf{X}) \leq_p ({}^{\perp}\mathbf{Y}, \mathbf{Y}).$$

THEOREM 2.

(i). $\leq_g$ *is a preorder.*

(ii). For any random vectors $\mathbf{X}(n - \dim)$ *and* $\mathbf{Y}(k - \dim)$ *supported on* $\mathcal{X}$ *and* $\mathcal{Y}$, *respectively,* 1° *for Borel measurable functions* $f : \mathcal{X} \to R^n$, $g : \mathcal{Y} \to R^k$ *such that* $f(x_1, \ldots, x_n) = (f_1(x_1), \ldots, f_n(x_n))$, $g(y_1, \ldots, y_k) = (g_1(y_1), \ldots, g_k(y_k))$, *where* f_i, g_j *are injections,*

$$\mathbf{X} \leq_g \mathbf{Y} \textit{ iff } f(\mathbf{X}) \leq_g g(\mathbf{Y}), \tag{1}$$

2° *for any n-elements and k-elements permutations* $\Pi^{(n)}$ *and* $\Pi^{(k)}$

$$\mathbf{X} \leq_g \mathbf{Y} \textit{ iff } \Pi^{(n)}(\mathbf{X}) \leq_g \Pi^{(k)}(\mathbf{Y}). \tag{2}$$

(iii). $\mathbf{X}$ *is a minimal element of* $\leq_g$ *iff* $\mathbf{X} =^{\perp} \mathbf{X}$.

(iv). For $\mathbf{X}$ *with continuous marginal distribution: if the distribution of* $\mathbf{X}$ *is degenerate, then* $\mathbf{X}$ *is a maximal element of the preorder* $\leq_g$.

(v). For normally distributed k-dimensional random vectors $\mathbf{X}$ *and* $\mathbf{Y}$ *with identical sets of orthogonal eigenvectors for the correlation matrices of* $\mathbf{X}$ *and* $\mathbf{Y}$, *and for each pair of eigenvalues* β_i, $\bar{\beta}_i$ *of correlation matrices of* $\mathbf{X}$ *and* $\mathbf{Y}$, *respectively:*

$$\textit{if } \bar{\beta}_i > \beta_i > 1 \textit{ or } 1 > \beta_i > \bar{\beta}_i \;\; i = 1, \ldots, k \textit{ then } \mathbf{X} \leq_g \mathbf{Y}.$$

PROOF. (ii). It follows from Th.1 (i) that for an injection $f : \mathcal{X} \to R^n$, we have $({}^{\perp}\mathbf{X}, \mathbf{X}) \underset{p}{\approx} (f({}^{\perp}\mathbf{X}), f(\mathbf{X}))$. On the other hand for a function $f : \mathcal{X} \to R^n$ independently transforming vector components, we have $f({}^{\perp}\mathbf{X}) \sim^{\perp} f(\mathbf{X})$. Therefore, $({}^{\perp}\mathbf{X}, \mathbf{X}) \underset{p}{\approx} ({}^{\perp}f(\mathbf{X}), f(\mathbf{X}))$. Analogously, $({}^{\perp}\mathbf{Y}, \mathbf{Y}) \underset{p}{\approx} ({}^{\perp}g(\mathbf{Y}), g(\mathbf{Y}))$. Thus (1) holds due to the definition and transitivity of $\leq_g$. The proof of (2) is analogous since for any n-element permutation Π, $\Pi({}^{\perp}\mathbf{X}) =^{\perp} \Pi(\mathbf{X})$.

Proofs of (iii) and (iv) follow immediately from Th.1 (ii) and (iii), respectively.

(v). By (ii), we may restrict consideration to the vectors with standardized marginals. We shall show that, under the assumptions of (v), the error rate $a_{21}(\partial)$ of any threshold rule ∂ for $({}^{\perp}\mathbf{X}, \mathbf{X})$ would diminish and $a_{12}(\partial)$ would not change when ∂ was applied for $({}^{\perp}\mathbf{Y}, \mathbf{Y})$.

Let Σ and $\bar{\Sigma}$ denote the correlation matrices for $\mathbf{X}$ and $\mathbf{Y}$, respectively. The likelihood ratio of $\mathbf{X}$ against ${}^{\perp}\mathbf{X}$ is

$$h(\mathbf{x}) = \mid \Sigma \mid^{-1/2} \exp((-1/2)\mathbf{x}'(\Sigma^{-1} - I)\mathbf{x}).$$

Let ∂ be a threshold rule such that $\partial(\mathbf{x}) = 1$ if $h(\mathbf{x}) < \gamma$ and $\partial(\mathbf{x}) = 0$ if $h(\mathbf{x}) > \gamma$. Let $a_{ij}(\partial)$ denote the error rates of ∂ for $({}^{\perp}\mathbf{X}, \mathbf{X})$ and $\bar{a}_{ij}(\partial)$ be the error rates of ∂ for $({}^{\perp}\mathbf{Y}, \mathbf{Y})$. Then

$$a_{12}(\partial) = 1 - \int_{D_\gamma} \cdots \int (2\pi)^{-n/2} \exp((-1/2)\mathbf{x}'\mathbf{x}) dx_1 \ldots dx_k,$$
$$a_{21}(\partial) = \int_{D_\gamma} \cdots \int (2\pi)^{-n/2} \mid \Sigma \mid^{-1/2} \exp((-1/2)\mathbf{x}'\Sigma^{-1}\mathbf{x}) dx_1 \ldots dx_k$$

and

$$\bar{a}_{12}(\partial) = 1 - \int_{D_\gamma} \cdots \int (2\pi)^{-n/2} \exp((-1/2)\mathbf{y}'\mathbf{y}) dy_1 \ldots dy_k$$
$$\bar{a}_{21}(\partial) = \int_{D_\gamma} \cdots \int (2\pi)^{-n/2} \mid \bar{\Sigma} \mid^{-1/2} \exp((-1/2)\mathbf{y}'\bar{\Sigma}^{-1}\mathbf{y}) dy_1 \ldots dy_k,$$

where

$$D_\gamma = \{\mathbf{x} : \mid \Sigma \mid^{-1/2} \exp((-1/2)\mathbf{x}'(\Sigma^{-1} - I)\mathbf{x} < \gamma\}.$$

Putting

$$\mathbf{y} = \bar{\Sigma}^{1/2}\Sigma^{-1/2}\mathbf{x}, \quad \text{we get}$$
$$\bar{a}_{21}(\partial) = \int_{\bar{D}_\gamma} \cdots \int (2\pi)^{-n/2} \mid \Sigma \mid^{-1/2} \exp((-1/2)\mathbf{x}'\Sigma^{-1}\mathbf{x}) dx_1 \ldots dx_k,$$

where

$$\bar{D}_\gamma = \{\mathbf{x} : \mid \Sigma \mid^{-1} \exp((-1/2)\mathbf{x}'\Sigma^{-1/2}\bar{\Sigma}^{1/2}(\Sigma^{-1} - I)\bar{\Sigma}^{1/2}\Sigma^{-1/2}\mathbf{x} < \gamma\}.$$

We shall show that $\bar{D}_\gamma \subset D_\gamma$. Let $\mathbf{x}^1, \ldots, \mathbf{x}^k$ be the orthonormal set of eigenvectors, common for Σ and $\bar{\Sigma}$. Substituting $\mathbf{x} = \Sigma_{i=1}^k \zeta_i \mathbf{x}^i$ to D_γ and $\bar{D}_\gamma$ we get:

$$\mathbf{x} \in D_\gamma \quad \text{iff} \quad (2\pi)^{-n/2} \mid \Sigma \mid^{-1} \exp((-1/2)\Sigma_{i=1}^k (\beta_i^{-1} - 1)\zeta_i^2 < \gamma,$$
$$\mathbf{x} \in \bar{D}_\gamma \quad \text{iff} \quad (2\pi)^{-n/2} \mid \Sigma \mid^{-1} \exp((-1/2)\Sigma_{i=1}^k (\bar{\beta}_i/\beta_i)(\beta_i^{-1} - 1)\zeta_i^2 < \gamma.$$

Under the assumptions of (v), $\bar{D}_\gamma \subset D_\gamma$ and $\bar{a}_{21}(\partial) \le a_{21}(\partial)$. Thus, $({}^{\perp}\mathbf{X}, \mathbf{X}) \le_p ({}^{\perp}\mathbf{Y}, \mathbf{Y})$ since $\bar{a}_{12}(\partial) = a_{12}(\partial)$. $\|$

COROLLARY 1. *Let $\mathbf{X}, \mathbf{Y}$ be $k-\dim$ normally distributed random vectors and let ζ_{ij} and $\bar{\zeta}_{ij}$ be the elements of the correlation matrices of $\mathbf{X}$ and $\mathbf{Y}$.*

(i). For $k = 2$, $\mathbf{X} \le_g \mathbf{Y}$ iff $\mid \zeta_{12} \mid \le \mid \bar{\zeta}_{12} \mid$.

(ii). For $k = 3$, if $\zeta_{12}\zeta_{23}\zeta_{31} = 0$ and $\bar{\zeta}_{ij} = \eta\zeta_{ij}$, for $1 \le \eta \le (\zeta_{12}^2 + \zeta_{23}^2 + \zeta_{13}^2)^{-1/2}$, $i \ne j, i, j = 1, 2, 3$, then $\mathbf{X} \le_g \mathbf{Y}$.

Now, we will consider pairs of binary random variables $\mathbf{X} = (X^{(1)}, X^{(2)})$. A 2×2 distribution of $\mathbf{X}$ is specified by $P = (p_{ij})$, p_{ij} being the probability that $X^{(1)} = i$, $X^{(2)} = j$ $(i, j = 1, 2)$. The set of 2×2 distributions will be denoted $\mathcal{P}_{2\times 2}$.

A distribution $P \in \mathcal{P}_{2\times 2}$ is either positive dependent $(p_{11}p_{22} \ge p_{12}p_{21})$ or negative dependent $(p_{11}p_{22} \le p_{12}p_{21})$. A natural monotone ordering $\le_m$ of $\mathcal{P}_{2\times 2}$ is given by:

$$P \leq_m P' \text{ if } p_{11} \leq p'_{11},\ p_{22} \leq p'_{22},\ p_{12} \geq p'_{12},\ p_{21} \geq p'_{21}$$

and P is positive dependent, or

$$p_{11} \geq p'_{11},\ p_{22} \geq p'_{22},\ p_{12} \leq p'_{12},\ p_{21} \leq p'_{21}$$

and P is negative dependent.
We shall show that $P \leq_m P'$ implies $P \leq_g P'$ (but the opposite implication is not true).

LEMMA 1. *Let $Q = (q_{ij})$, $R = (r_{ij})$ belong to $\mathcal{P}_{2\times 2}$ and let $\mathcal{E}_{ii} = r_{ii} - q_{ii}$ for $i = 1,2$, $\mathcal{E}_{ij} = q_{ij} - r_{ij}$ for $i \neq j, i,j = 1,2$.*
If $q_{11}q_{22} \geq q_{12}q_{21}$ ($q_{11}q_{22} \leq q_{12}q_{21}$); $\mathcal{E}_{ij} \geq 0$, $i,j = 1,2$ ($\mathcal{E}_{ij} \leq 0$, $i,j = 1,2$); and $\min(\mathcal{E}_{ij}; i,j = 1,2) = 0$ ($\max(\mathcal{E}_{ij}; i,j = 1,2) = 0$), then

$$\frac{r_{ij}}{r_{i.}r_{.j}} \leq \frac{q_{ij}}{q_{i.}q_{.j}},\ \frac{r_{ii}}{r_{i.}r_{.i}} \geq \frac{q_{ii}}{q_{i.}q_{.i}} \quad i \neq j, i,j = 1,2. \tag{3}$$

PROOF. Replacing in (3) r_{ij} by $q_{ij} - \mathcal{E}_{ij}$ for $i \neq j$, $i,j = 1,2$, and r_{ii} by $q_{ii} + \mathcal{E}_{ii}$ for $i = 1,2$, and taking into account that $\mathcal{E}_{11} + \mathcal{E}_{22} = \mathcal{E}_{12} + \mathcal{E}_{21}$, we obtain inequalities equivalent to (3), which obviously hold. ||

THEOREM 3. *If $P \leq_m P'$ then $P \leq_g P'$.*

PROOF. Let $\mathbf{X} \sim P \in \mathcal{P}_{2\times 2}$. Instead of $C_{(\perp\mathbf{X},\mathbf{X})}$ we will write $C_{(P)}$. Then the divergence curve $C_{(P)}$ is convex and piece-wise linear; it joins points (0,1) and (1,0) and consists of four segments with slopes equal to $(-p_{ij}/p_{i.}p_{.j})$, ordered nondecreasingly.

Let P, $P' \in \mathcal{P}_{2\times 2}$ and let a_k, a'_k, $k = 1,\ldots,4$, be the slopes of consecutive segments of $C_{(P)}$ and $C_{(P')}$. Clearly, $C_{(P')} \leq C_{(P)}$ if

$$a'_k \leq a_k \text{ for } k = 1,2, \quad a'_k \geq a_k \text{ for } k = 3,4 \tag{4}$$

If P is positive dependent then

$$\begin{aligned} -a_1 &= \max(p_{11}/p_{1.}p_{.1},\ p_{22}/p_{2.}p_{.2}), \\ -a_2 &= \min(p_{11}/p_{1.}p_{.1},\ p_{22}/p_{2.}p_{.2}), \\ -a_3 &= \max(p_{12}/p_{1.}p_{.2},\ p_{21}/p_{2.}p_{.1}), \\ -a_4 &= \min(p_{12}/p_{1.}p_{.2},\ p_{21}/p_{2.}p_{.1}), \end{aligned}$$

and a'_k are expressed analogously. To prove (4), it suffices to show that

$$\frac{p'_{ij}}{p'_{i.}p'_{.j}} \leq \frac{p_{ij}}{p_{i.}p_{.j}},\ \frac{p'_{ii}}{p'_{i.}p'_{.i}} \geq \frac{p_{ii}}{p_{i.}p_{.i}} \quad i \neq j, i,j = 1,2. \tag{5}$$

Denote $\mathcal{E}_{ij} = | p'_{ij} - p_{ij} |$, $i,j = 1,2$, $\mathcal{E}_\circ = \min(\mathcal{E}_{ij}; i,j = 1,2)$, $p^\circ_{ii} = p_{ii} + \mathcal{E}_\circ$, $p^\circ_{ij} = p_{ij} - \mathcal{E}_\circ$, $i \neq j$, $i,j = 1,2$, $P^\circ = (p^\circ_{ij})$. Then

$$P' = \begin{pmatrix} p_{11} + \mathcal{E}_{11}, & p_{12} - \mathcal{E}_{12} \\ p_{21} - \mathcal{E}_{21}, & p_{22} + \mathcal{E}_{22} \end{pmatrix} = \begin{pmatrix} p^\circ_{11} + (\mathcal{E}_{11} - \mathcal{E}_\circ), & p^\circ_{12} - (\mathcal{E}_{12} - \mathcal{E}_\circ) \\ p^\circ_{21} - (\mathcal{E}_{21} - \mathcal{E}_\circ), & p^\circ_{22} + (\mathcal{E}_{22} - \mathcal{E}_\circ) \end{pmatrix}.$$

Since $P \leq_m P^\circ$ and $p^\circ_{1.} = p_{1.}$, $p^\circ_{.1} = p_{.1}$, we see that for P and P° the inequalities in (5) hold with (P, P') replaced by (P, P°). By Lemma 1, since $\min\{\mathcal{E}_{ij} - \mathcal{E}_\circ; i,j = 1,2\} = 0$, inequalities (5) hold with (P, P') replaced by (P°, P'). But P° was constructed so that $P \leq_m P^\circ \leq_m P'$. Thus, inequalities (5) hold.

The proof for negative dependent distribution can be easily reduced to that for positive dependent one. ||

REFERENCES

DABROWSKA, D. (1985). Descriptive parameters of location, dispersion and stochastic dependence. *Math. Operationsforsch Stat.* **16** 63–88.

KIMELDORF, G. and SAMPSON, A.R. (1987). Positive dependence orderings. *Ann. Inst. Stat. Math. A* **39** 113–118.

NIEWIADOMSKA-BUGAJ, M. (1987). Standard families of two-class discriminant problems with deferred decision. *Proceedings of Conference on Discriminant Analysis, Cluster Analysis and Other Methods of Data Classification*, 223–229.

YANAGIMOTO, T. (1990). Dependence ordering in statistical models and other notions. In *Topics in Statistical Dependence* (H. Block, A.R. Sampson, and T. Savits, eds.). IMS Lecture Notes–Monograph Series, Hayward, CA.

INSTITUTE OF COMPUTER SCIENCE
POLISH ACADEMY OF SCIENCES
WARSAW, POLAND

ON A CORRELATION INEQUALITY AND ITS APPLICATIONS

By Mark Brown[1]

The City College, CUNY

Consider a continuous distribution on $[0, \infty)$ with cdf F, survival function $\bar{F} = 1 - F$ and cumulative hazard function $H = -Ln\bar{F}$. For F NBUE it is shown that the correlation coefficient between $X \sim F$ and $H(X)$ is bounded below by σ/μ, the coefficient of variation of F, while for F NWUE the correlation coefficient is bounded below by μ/σ. Several applications of this inequality and its generalizations are discussed, including Monte Carlo simulation of the renewal function, exponential approximation of DMRL distributions, moment inequalities for record values, and a variance inequality for random event epochs in a homogeneous Poisson process.

1. Introduction and Summary. Consider a continuous distribution on $[0, \infty)$, with cdf F, survival function $\bar{F} = 1 - F$ and cumulative hazard function $H = -Ln\bar{F}$. If $X \sim F$ then $H(X)$ is exponentially distributed with mean 1. The random variable $H(X)$ measures lifetime by total hazard overcome until death, while X measures lifetime in ordinary time units. Since H is an increasing function we know that $H(X)$ and X are positively correlated. The question of how positively correlated arose naturally in Brown, Solomon, and Stephens (1981) and Brown (1987) in different contexts. In the former paper the asymptotic relative savings in risk between two Monte Carlo estimators of the renewal function was given by the square of the correlation coefficient between X and $H(X)$. In Brown (1987), a quantity closely related to the correlation coefficient was needed to bound the distance between a DMRL (decreasing mean residual life) distribution and its stationary renewal distribution.

In this paper we show that for X NBUE (new better than used in expectation):

$$\rho(X, H(X)) \geq \frac{\sigma}{\mu} \tag{1}$$

while for X NWUE (new worse than used in expectation):

---

[1]Research supported by the Air Force Office of Scientific Research under Grant No. 84-0095.

AMS 1980 subject classifications. Primary 60E15; secondary 60G55.

Key words and phrases. Inequalities, correlation coefficient, record values, exponential approximation, Poisson processes, NBUE and NWUE distributions.

I wish to thank the editors and the referee for their helpful suggestions.

$$\rho(X, H(X)) \geq \frac{\mu}{\sigma}. \tag{2}$$

The lower bound for the correlation coefficient is also applicable to record value processes. Record value processes are of interest in reliability theory due to their connection with minimal repair (Crow, 1974). Consider an i.i.d. sequence $\{X_i, i \geq 1\}$ with $X_i \sim F$. Define $S_1 = X_1$, $N_2 = \min\{i : X_i > X_1\}$, $S_2 = X_{N_2}$, $N_k = \min\{i : X_i > X_{N_{k-1}}\}$, $S_k = X_{N_k}$, $k = 3, 4, \ldots$; S_k is referred to as the kth record value. Using (1) and (2) we show that for F NBUE:

$$\frac{\sigma^2}{\mu} \leq E(S_2 - S_1) \leq \sigma \tag{3}$$

and for F NWUE:

$$\mu \leq E(S_2 - S_1) \leq \sigma. \tag{4}$$

Using similar methodology, inequalities are derived for the moments of higher record values. For example, it is shown (Section 3.4) that for F IFRA:

$$\frac{\mu_{k+r-1}}{\mu_{r-1}} \leq ES_r^k \leq \binom{k+r-1}{k} \mu_k \tag{5}$$

where $\mu_j = \int x^j dF(x)$.

In Section 3.5 we show that if $\{T_i, i \geq 1\}$ are the arrival epochs for a homogeneous Poisson process with parameter λ, and N is a stopping time, then Var $T_N \geq \lambda^{-2}$. We further demonstrate among distributions with failure rate uniformly bounded above by λ, the exponential distribution with parameter λ has minimum variance for the kth record value, for $k \geq 1$. Equivalently, among non-homogeneous Poisson processes with intensity function uniformly bounded above by λ, the homogeneous Poisson process with parameter λ has minimum variance for S_k, the time of the kth event, for $k \geq 1$.

2. A Correlation Inequality. Consider a non-negative random variable X with continuous cdf F. Denote by X^* a random variable with cdf $G(x) = \mu^{-1}\int_0^x \bar{F}(t)dt$, the stationary renewal distribution corresponding to F. Let T denote a random variable with distribution $dF_T(t) = t\mu^{-1}dF(t)$. T is distributed as the length of the interval covering an arbitrary fixed point in a stationary renewal process with interarrival time distribution F (Feller, 1971, p. 371). For fixed $x \geq 0$ define $g(t) = \bar{F}(x \vee t)/\bar{F}(t)$, where $x \vee t = \max(x, t)$. Note that:

$$\begin{aligned}\Pr(T > x) &= \int_x^\infty t\mu^{-1}dF(t) = x\mu^{-1}\bar{F}(x) + \bar{G}(x) \\ &= \int \left(\frac{\bar{F}(x \vee t)}{\bar{F}(t)}\right)\left(\frac{\bar{F}(t)}{\mu}\right) dt = Eg(X^*).\end{aligned}$$

Thus:

$$\Pr(T > x) = Eg(X^*). \tag{6}$$

Next, consider the record value process corresponding to F, described in Section 1. The sequence of record values $\{S_i, i \geq 1\}$ generates a nonhomogeneous Poisson process with $EN(t) = -Ln\bar{F}(t) = H(t)$ (Shorrock, 1972). Now:

$$\begin{aligned} \Pr(S_2 > x) = \Pr(N(x) \leq 1) &= \bar{F}(x)[1 + H(x)] = \\ &\int \left(\frac{\bar{F}(x \vee t)}{\bar{F}(t)}\right) dF(t) = Eg(X). \end{aligned} \tag{7}$$

LEMMA 2.1. *If F is NBUE (NWUE) then S_2 is stochastically larger (smaller) than T.*

PROOF. F NBUE is equivalent to $X \overset{\text{st}}{\geq} X^*$. Since g is increasing in t, it follows from (6) and (7) that:

$$\Pr(T > x) = Eg(X^*) \leq Eg(X) = \Pr(S_2 > x).$$

The NWUE case similarly follows.

LEMMA 2.2. *If F is NBUE then $\rho(X, H(X)) \geq \sigma/\mu$. If F is NWUE then $\rho(X, H(X)) \geq \mu/\sigma$.*

PROOF. Note that $dF_{S_2}(t) = H(t)dF(t)$ while $dF_T(t) = t\mu^{-1}dF(t)$. By Lemma 2.1, if F is NBUE then:

$$ES_2 = E(XH(X)) \geq ET = \mu_2/\mu. \tag{8}$$

Now subtract μ and divide by σ on both sides of (8) and the NBUE result follows.

Next, assume that X is NWUE. It follows from Lemma 2.1 that for any increasing function ℓ (with the expectations existing):

$$E\ell(S_2) = \int \ell(x)H(x)dF(x) \leq \mu^{-1} \int x\ell(x)dF(x) = E\ell(T) \tag{9}$$

choose $\ell(x) = H(x)$, then:

$$EH^2(X) = 2 \leq \mu^{-1}E(XH(X)). \tag{10}$$

From (10) the NWUE result easily follows.

3. Applications.

3.1. Monte Carlo Estimation of the Renewal Function. Suppose we wish to estimate $M(t)$, the expected number of renewals in $[0, t]$ for a renewal process with interarrival distribution F, by Monte Carlo simulation. An obvious approach is to simulate $N(t)$, the number of renewals in $[0,t]$, K times $(N_1(t),\ldots,N_K(t))$, and to estimate $M(t)$ by the sample mean. In Brown, Solomon, and Stephens (1981) an unbiased estimator $M^*(t)$ was proposed and it was shown that as $t\to\infty$ the asymptotic relative savings in risk between $M^*(t)$ and the estimator based on $N(t)$ was given by $\rho^2(X,H(X))$. Lemma (2.2) gives a lower bound on ρ and thus a lower bound on the asymptotic relative savings in risk.

3.2. Exponential Approximation of DMRL Distributions. Consider a continuous DMRL (decreasing mean residual life) distribution F on $[0,\infty)$ with stationary renewal distribution G. In Brown (1987) it is shown that:

$$\mathcal{D}^*(F,G)=\sup\mid F(B)-G(B)\mid\leq 1-EH(X^*) \tag{11}$$

where $H=-Ln\bar{F}$, the cumulative hazard function, $X^*\sim G$, and the sup is taken over all Borel sets. Now:

$$ES_2=\int(H(t)+1)\bar{F}(t)dt=\mu[1+EH(X^*)]. \tag{12}$$

But F DMRL implies F NBUE, thus (12) and Lemma 2.1 give:

$$ES_2=\mu[1+EH(X^*)]\geq ET=\mu_2/\mu \tag{13}$$

thus:

$$EH(X^*)\geq\sigma^2/\mu^2. \tag{14}$$

From (11) and (14) we obtain:

$$\mathcal{D}^*(F,G)\leq 1-(\sigma^2/\mu^2). \tag{15}$$

The inequality (15) thus extends the result of Brown (1987) from F IFR to F DMRL. Moreover it follows from (15), employing the methodology of Brown (1987), that for F DMRL:

$$\sup\mid\bar{F}(t)-e^{-t/\mu}\mid\leq 1-(\sigma^2/\mu^2). \tag{16}$$

Thus if F is DMRL with coefficient of variation close to 1, then F is approximately exponential.

3.3. The Second Record Value. Consider S_2-S_1 the interarrival time between the first and second record values in a record value process corresponding to F (equivalently the interarrival time between the first and second events in a non-homogeneous Poisson process with $EN(t)=H(t)=-Ln\bar{F}(t)$). It follows from Lemma 2.1 that F NBUE implies:

$$E(S_2-S_1)\geq ET-\mu=\frac{\mu_2}{\mu}-\mu=\frac{\sigma^2}{\mu} \tag{17}$$

while F NWUE leads to $E(S_2 - S_1) \leq (\sigma^2/\mu)$.

The quantity $E(S_2 - S_1)$ is the expected residual life for an item which is minimally repaired at its first failure. It is of interest in the evaluation and planning of maintenance policies.

Lemma 3.3.1, below, presents an upper bound of σ for $E(S_2 - S_1)$, derived without aging assumptions of F. As is done throughout this paper we assume that F is a continuous distribution on $[0, \infty)$.

LEMMA 3.3.1. *Let $X \sim F$ and g a function on $[0, \infty)$ with $Eg^2(X) < \infty$. Then:*

$$| E(g(S_2) - g(S_1)) | \leq \sigma_g$$

where σ_g is the standard deviation of $g(X)$. In particular the choice $g(x) = x$ gives:

$$E(S_2 - S_1) \leq \sigma$$

where σ is the standard deviation of X.

PROOF. $Eg(S_2) = E(g(X)H(X)) = Eg(X)EH(X) + \sigma_g \sigma_{H(X)} \rho(g(X), H(X)) \leq Eg(X) + \sigma_g$. Thus $E(g(S_2) - g(S_1)) \leq \sigma_g$. Substituting $-g$ for g yields $E(g(S_1) - g(S_2)) \leq \sigma_g$ from which the result follows. ||

COROLLARY 3.3.1. *For F NBUE, $\sigma^2/\mu \leq E(S_2 - S_1) \leq \sigma$. For F NWUE, $\mu \leq E(S_2 - S_1) \leq \sigma$.*

PROOF. The NBUE case follows from expression (17) and Lemma 3.3.1. The NWUE case follows from Lemma 3.3.1 and the obvious NWUE inequality $E(S_2 - S_1) \geq \mu$. ||

A function $g(x)$ on $[0, \infty)$ is defined to be starshaped if $\frac{g(x)}{x}$ is increasing (meaning non-decreasing). If g is non-negative and starshaped then g is increasing.

Consider, now, a function g which is non-negative and starshaped on $[0, \infty)$, with $\mu_g = Eg(X) < \infty$. Define:

$$dF_g(t) = g(t)dF(t)/\mu_g.$$

Then: $dF_g(t)/dF_T(t) = \mu_g^{-1}\mu(g(t)/t)$ which is increasing. Thus F_g is larger than F_T under the partial ordering of monotone likelihood ratio (Lehmann [1959] p.73) and is thus stochastically larger. It follows that:

(18) $$E[Xg(X)] \geq \mu_g \mu_2 / \mu.$$

Now assume that F is NBUE. By Lemma 2.1 and (18):

(19) $$Eg(S_2) \geq Eg(T) = \mu^{-1}E(Xg(X)) \geq \mu_g \mu_2 / \mu^2.$$

Thus for F NBUE and g non-negative and starshaped it follows from Lemma 3.3.1 and (19) that:

$$\frac{\sigma^2}{\mu^2}\mu_g \leq E(g(S_2) - g(S_1)) \leq \sigma_g. \tag{20}$$

The choice $g(x) = x$ leads to the NBUE inequality of Corollary 3.3.1.

3.4. Higher Record Values. Let S_k denote the kth record value in a record value process corresponding to F continuous. Since S_k is the kth event epoch in a non-homogeneous Poisson process with $EN(t) = H(t)$ it follows that:

$$dF_{S_k}(t) = [(H(t))^{k-1}/(k-1)!]dF(t) \tag{21}$$

and also that:

$$dF_{S_k}(t) = [(H(t)/k-1)]dF_{S_{k-1}}(t), \;\; k \geq 2. \tag{22}$$

Consequently (from 22):

$$Eg(S_k) = (k-1)^{-1}E[g(S_{k-1})H(S_{k-1})]. \tag{23}$$

Now $H(S_{k-1})$ is gamma distributed with parameters $k-1$ and 1 (the sum of $k-1$ i.i.d. exponentials with parameter 1) thus $ES_{k-1} = \text{Var}S_{k-1} = k-1$.

Using the mean and variance of $H(S_{k-1})$, (23) and the upper bound for the product moment, EUV $\leq$ EUEV $+\sigma_U\sigma_V$ with $U = S_{k-1}, V = H(S_{k-1})$ we obtain:

$$Eg(S_k) \leq Eg(S_{k-1}) + (\sigma(g(S_{k-1}))/\sqrt{k-1}). \tag{24}$$

From (24) we obtain the following generalizations of Lemma (3.3.1):

$$\mid E[(g(S_k) - g(S_{k-1})] \mid \leq \sigma(g(S_{k-1}))/\sqrt{k-1}. \tag{25}$$

The case $k = 2$ corresponds to Lemma 3.3.1. However the more general inequality appears to be computationally useful only when $k = 2$. For general k, $\sigma(g(S_{k-1}))$ is no easier to compute than $E(g(S_k) - g(S_{k-1}))$.

We have no analogue of Lemma 2.1 for F NBUE or NWUE. However if we strengthen the restriction on F from NBUE (NWUE) to IFRA (DFRA) then we obtain the following:

LEMMA 3.4.2. *Let F be a continuous IFRA distribution, and T_r be a random variable with distribution $dF_{T_r}(t) = x^{r-1}dF(x)/\mu_{r-1}$, where μ_m is the mth moment of F. Then S_r is stochastically larger than T_r and:*

$$\frac{\mu_{k+r-1}}{\mu_{r-1}} \leq ES_r^k \leq \binom{k+r-1}{k}\mu_k.$$

If F is a continuous DFRA distribution with finite $(r-1)^{st}$ moment then S_r is stochastically smaller than T_r. If in addition F has finite $(k+r-1)^{st}$ moment then the above inequality reverses.

For $r = 2$ the above inequalities hold under the weaker condition that F is NBUE or NWUE.

PROOF. Note that $dF_{S_r}(t)/dF_{T_r}(t) = \frac{\mu_{r-1}}{(r-1)!}\left(\frac{H(t)}{t}\right)^{r-1}$ which is increasing, as F is IFRA. Thus S_r is larger than T_r under monotone likelihood ratio and is thus stochastically larger. Thus:

$$ES_r^k \geq ET_r^k = \mu_{k+r-1}/\mu_{r-1}. \tag{26}$$

Next:

$$\frac{H^k}{k!}dF \overset{\text{st}}{\geq} \frac{x^k}{\mu_k}dF. \tag{27}$$

Multiply both sides of (27) by $H^{r-1}/(r-1)!$ and integrate obtaining:

$$\binom{k+r-1}{k} \geq \frac{1}{\mu_k}ES_r^k. \tag{28}$$

Thus $ES_r^k \leq \mu_k\binom{k+r-1}{k}$ and this inequality and (26) yield the IFRA result. The DFRA case similarly follows. By Lemma 2.1, for F NBUE and $r=2$, $S_2 \overset{\text{st}}{\geq} T_2 = T$ which is sufficient, by our above derivation, for (26) and (28) to follow (with $r=2$). ||

Note that the various inequalities derived above for record value processes hold for non-homogeneous Poisson processes.

3.5. A Variance Inequality. Consider an absolutely continuous distribution F with failure rate function $h(t)$ bounded above by $\lambda(h(t) \leq \lambda$ for all $t \geq 0)$. Let S_1 and S_2 denote the first two record values in a record value process corresponding to F. The failure rate function of $S_2 - S_1$ evaluated at t is a mixture of the values $\{h(s), s \geq t\}$ and is thus bounded above by λ for all t. Consequently:

$$E(S_2 - S_1) \geq \lambda^{-1}. \tag{29}$$

By Lemma 3.3.1:

$$E(S_2 - S_1) \leq \sigma \tag{30}$$

where σ is the standard deviation corresponding to F. From (29) and (30) we obtain:

$$\sigma^2 \geq \lambda^{-2}. \tag{31}$$

Thus among distributions on $[0,\infty)$ with failure rate bounded above by λ, the exponential distribution with parameter λ has smallest variance.

Next, consider a homogeneous Poisson process on $[0,\infty)$ with intensity λ and event epochs $\{T_i, i \geq 1\}$. Let N be a stopping time and consider the random variable T_N, letting h^* denote its failure rate function. Since T_N can only occur at an event epoch for the Poisson process, and since the conditional intensity of an event at t (given the past) for the Poisson process is always λ, it follows that:

$$h^*(t) = \lambda \Pr(T_N = t \mid T_N \geq t, T_i = t \text{ for some } i) \leq \lambda. \tag{32}$$

Thus (31) and (32) imply:

$$\operatorname{Var}(T_N) \geq \lambda^{-2}. \tag{33}$$

Note that λ^{-2} is the variance of T_1 as well as the variance of $T_{N(t)+1}$, the time of the first event after time t. These event epochs have smallest variance among all random event epochs for the Poisson process.

The inequality (33) holds for a large variety of random variables arising in secondary processes generated by a Poisson process. These include counter models, queues with Poisson input and uniformizable Markov chains.

Also note that if the failure rate of F is uniformly bounded above by λ, then by (25) and the argument used to derive (29):

$$\lambda^{-1} \leq E(S_{k+1} - S_k) \leq k^{-1/2}\sigma(S_k). \tag{34}$$

Thus $\sigma(S_k) \geq k^{1/2}\lambda^{-1}$, the lower bound being achieved in the exponential case. Thus for $k \geq 1$, the exponential distribution with parameter λ minimizes the variance of the kth record value, among distributions with failure rate function uniformly bounded above by λ. Equivalently, consider a non-homogeneous Poisson process with intensity function $\lambda(t)$ bounded above by λ. Then Var $S_k \geq k\lambda^{-2}$ where S_k is the kth arrival epoch. Thus among all non-homogeneous Poisson processes with intensity functions uniformly bounded above by λ, the homogeneous Poisson process with intensity λ minimizes the variance of S_k, for all $k \geq 1$.

REFERENCES

BROWN, M. (1987). Inequalities for distributions with increasing failure rate. *Contributions to the Theory and Application of Statistics, A Volume in Honor of Herbert Solomon*, pp. 3–17. Academic Press, New York.

BROWN, M., SOLOMON, H., and STEPHENS, M.A. (1981). Monte Carlo simulation of the renewal function. *J. Appl. Prob.* **18** 426–434.

CROW, L.H. (1974). Reliabilitly analysis for complex repairable systems. *Reliability and Biometry, Statistical Analysis of Lifelength*, (F. Proschan and R.J. Serfling, eds.), SIAM, Philadelphia, 379–414.

FELLER, W. (1971). *An Introduction to Probability Theory and Its Applications, II, 2nd Edition.* John Wiley and Sons, New York.

LEHMANN, E.L. (1959). *Testing Statistical Hypotheses.* John Wiley and Sons, New York.

SHORROCK, R.W. (1972). A limit theorem for inter-record times. *J. Appl. Prob.* 9 219–223.

DEPARTMENT OF MATHEMATICS
CITY COLLEGE, CUNY
138 ST. & CONVENT AVENUE
NEW YORK, NY 10031

CONVEX-ORDERING AMONG FUNCTIONS, WITH APPLICATIONS TO RELIABILITY AND MATHEMATICAL STATISTICS

By Wai Chan,[1] Frank Proschan,[2] and Jayaram Sethuraman[3]

Florida State University

Hardy, Littlewood, and Pólya (1952) introduced the notion of one function being convex with respect to a second function and developed some inequalities concerning the means of the functions. We use this notion to establish a partial order called convex-ordering among functions. In particular, the distribution functions encountered in many parametric families in reliability theory are convex-ordered. We have formulated some inequalities which can be used for testing whether a sample comes from F or G, when F and G are within the same convex-ordered family. Performance characteristics of different coherent structures can also be compared with respect to this partial ordering. For example, we will show that the reliability of a $k+1$-*out-of-n* system is convex with respect to the reliability of a k-*out-of-n* system.

When F is convex with respect to G, the tail of the distribution F is heavier than that of G; therefore, our convex-ordering implies stochastic ordering. Convex-ordering is also related to total positivity and monotone likelihood ratio families. This provides us a tool to obtain some useful results in reliability and mathematical statistics.

1. Introduction. Notions of partial ordering among survival distributions have played a useful role in providing numerous inequalities in reliability. The notion of a random variable X with distribution F being stochastically larger than another random variable Y with distribution G is well known in the literature. Van

---

[1]Now at Digital Equipment Corporation.

[2]Research supported by the Air Force Office of Scientific Research, System Command, USAF, under Grant AFOSR 88-0040.

[3]Research supported by the U.S. Army Research Office under Grant DAAL 03-86-K-0094.

AMS 1980 subject classifications. 60E15, 60N05.

Key words and phrases. Convex-ordering, total positivity, partial ordering of distributions.

Zwet (1964) defined F to be *convex-ordered* with respect to G (written $F \overset{<}{Z} G$) if $G^{-1}F(x)$ is a convex function in x on the support of F. When this ordering occurs, one can show that Y is a convex function of X. Barlow and Proschan (1966) have obtained tolerance limits for distributions satisfying this ordering.

Lee (1981) defined and analyzed another notion of convex-ordering: F is *convex-hazard ordered* with respect to G, written $F \overset{<}{CH} G$ if $R_F R_G^{-1}$ is convex, where $\bar{F} = 1 - F$ and $R_F = -\log \bar{F}$ is the hazard function of F. Lee used this ordering to generalize certain inequalities and preservation theorems in reliability. We give still another notion of convex-ordering in Definition 2.1, which is different from those proposed by Van Zwet (1964) and Lee (1981).

2. Convex-Ordering Among Distributions. Throughout this paper, we define the inverse function h^{-1} of a nondecreasing function h by $h^{-1}(t) = \inf\{x : h(x) \geq t\}$. When h is nonincreasing, we define $h^{-1}(t) = \inf\{x : h(x) \geq t\}$. We use "$\ll$" to symbolize "absolutely continuous."

DEFINITION 2.1. Let G be any continuous distribution and F be absolutely continuous with respect to G. We say that F is *more convex than* G, written $F \overset{c}{>} G$, if $FG^{-1}(t)$ is a convex function in the interval (0,1).

Throughout this paper, we refer to the above as convex-ordering. Other notions of convex-ordering will be referred to with their authors' names like Van Zwet, etc. This definition of convex-ordering checks directly whether one distribution function can be expressed as a convex transformation of another distribution function, in contrast to that of Van Zwet which checks if the random variables can be so transformed. Thus, the distribution function x^3 is more convex than the distribution function x^2 on the interval (0,1). This concept coincides with that of Hardy, Littlewood, and Pólya (1952, p. 65). Although the above definition applies to the class of all monotonic functions, we shall generally restrict our attention to life distributions. (For an exception, see Theorem 2.8.)

The following lemma gives useful properties of FG^{-1}.

LEMMA 2.2. *Let G be any continuous distribution and F be absolutely continuous with respect to G. Then*

(i) $FF^{-1}(t) = t, 0 < t < 1$; $F^{-1}F(x) \leq x, x \geq 0$.

(ii) $FG^{-1}G(x) = F(x), x \geq 0$.

(iii) $(FG^{-1})^{-1} = GF^{-1}$.

(iv) FG^{-1} *is nondecreasing and continuous. If G is also absolutely continuous with respect to F, then*

(v) FG^{-1} *is strictly increasing.*

(vi) $F \overset{c}{>} G$ *implies that* GF^{-1} *is concave.*

PROOF.

(i) The facts that $F^{-1}F(x) \leq x$ and $FF^{-1}(t) \geq t$ are easily seen from the definition of F^{-1} and do not require the continuity of F. Suppose now that $FF^{-1}(t) > t$. Then by the continuity of F, there is an $x < F^{-1}(t)$ such that $F(x) > t$, which contradicts the definition of F^{-1}. Hence $FF^{-1}(t) = t$.

(ii) From (i), we have $FG^{-1}G(x) \leq F(x)$. Suppose $FG^{-1}G(x) < F(x)$, then $F\{(G^{-1}G(x), x)\} > 0$ which implies that $G(G^{-1}G(x)) < G(x)$, since $F << G$. This leads to a contradiction because $G(G^{-1}G(x)) = G(x)$.

(iii) $FG^{-1}(GF^{-1}(t)) = FF^{-1}(t) = t$ implies that $GF^{-1}(t) \geq (FG^{-1})^{-1}(t)$.

Conversely, if $FG^{-1}(x) \geq t$, then $G^{-1}(x) \geq F^{-1}FG^{-1}(x) \geq F^{-1}(t)$, which in turn implies that $x = GG^{-1}(x) \geq GF^{-1}(t)$. Thus $(FG^{-1})^{-1}(t) \geq GF^{-1}(t)$.

(iv) Let $t_n \to t$. Then $t_n = GG^{-1}(t_n) \to GG^{-1}(t) = t$. Since $F << G$, this implies that $FG^{-1}(t_n) \to FG^{-1}(t)$.

(v) For $t_1 < t_2$, $GG^{-1}(t_1) = t_1 < t_2 = GG^{-1}(t_2)$. Since $G << F$, this implies $FG^{-1}(t_1) < FG^{-1}(t_2)$.

(vi) Let $\phi = FG^{-1}$, then ϕ is convex and strictly increasing. We need to show that ϕ^{-1} is concave. Let $0 \leq \lambda \leq 1$, then for any x and Y, we have

$$\begin{aligned}\phi\phi^{-1}[\lambda x + (1-\lambda)y] &= \lambda x + (1-\lambda)y \\ &= \lambda\phi\phi^{-1}(x) + (1-\lambda)\phi\phi^{-1}(y) \\ &\geq \phi(\lambda\phi^{-1}(x) + (1-\lambda)\phi^{-1}(y)).\end{aligned}$$

Since ϕ is strictly increasing, this implies $\phi^{-1}(\lambda x + (1-\lambda)y) \geq \lambda\phi^{-1}(x) + (1-\lambda)\phi^{-1}(y)$. Therefore $\phi^{-1} = GF^{-1}$ is concave.

Convex-ordering represents a partial ordering in the class of continuous distributions as indicated below:

(a) Reflexivity: $F \overset{c}{>} F$. (Since F is continuous, $FF^{-1}(t) = t$ and this is a convex function.)

(b) Transitivity: $F \overset{c}{>} G$ and $G \overset{c}{>} H$ imply that $F \overset{c}{>} H$. (Since $FH^{-1} = FG^{-1}(GH^{-1})$ and FG^{-1}, GH^{-1} are convex nondecreasing functions on $(0,1)$, it follows that FH^{-1} is convex.)

(c) Antisymmetry: $F \overset{c}{>} G$ and $G \overset{c}{>} F$ imply that $F = G$. (Since $FG^{-1} = (GF^{-1})^{-1}$, FG^{-1} and GF^{-1} are both convex and concave. Thus $FG^{-1}(t) = t = GF^{-1}(t)$, and $F(x) = FG^{-1}[G(x)] = G(x)$.)

A very useful way to characterize convex-ordering is by using the Radon-Nikodym derivative $\frac{dF}{dG}$ and is given in the following theorem.

THEOREM 2.3. *$F \overset{c}{>} G$ if and only if $f = \frac{dF}{dG}$ is nondecreasing almost everywhere with respect to G.*

PROOF. Let λ denote the Lebesgue measure on (0,1). Since G is continuous, $GG^{-1}(t) = t$. For each measureable set E, we have $\lambda(E) = G(G^{-1}(E))$.

This shows that the condition f is nondecreasing almost everywhere with respect to G is equivalent to fG^{-1} being nondecreasing almost everywhere with respect to λ. Now

$$FG^{-1}(t) = \int_{-\infty}^{G^{-1}(t)} f(x)dG(x) = \int_0^t fG^{-1}(y)dy$$

since G is continuous.
Consequently, FG^{-1} is convex if and only if f is nondecreasing almost everywhere with respect to G. ||

REMARK. If G is absolutely continuous with respect to λ, then $f = \frac{dF}{dG} = \frac{dF}{d\lambda} / \frac{dG}{d\lambda}$ a.e. G. Thus $F \overset{c}{>} G$ is equivalent to the property of monotone increasing likelihood ratio of F with respect to G; i.e., $\frac{dF}{d\lambda} / \frac{dG}{d\lambda}$ is nondecreasing on the support of G. This is the likelihood ratio ordering of Ross (1983).

REMARK. If $f = \frac{dF}{dG}$ is continuous and G is a continuous distribution, then $F \overset{c}{>} G$ if and only if f is monotone nondecreasing.

We now define the notion of a convex-ordered family.

DEFINITION 2.4. A family of distributions $\{F_\alpha\}$ is said to be a *convex-ordered family*, or simply a *convex family* if $\alpha_2 > \alpha_1$ implies that $F_{\alpha_2} \overset{c}{>} F_{\alpha_1}$.

The following families of distributions are convex-ordered with respect to α for $\alpha > 0$.

EXAMPLES.

(1) Exponential: $F_\alpha(t) = 1 - e^{-t/\alpha}$, $t > 0$.

(2) Gamma: $F_\alpha(t) = \frac{1}{\Gamma(\alpha)} \int_0^t x^{\alpha-1} e^{-x} dx$, $t > 0$.

(3) Truncated Normal: $F_\alpha(t) = \frac{1}{a_\alpha \sigma\sqrt{2\pi}} \int_0^t e^{-(x-\alpha)^2/2\sigma^2} dx$ for $t > 0$, where $\sigma > 0$ is fixed and $a_\alpha = \int_0^\infty \frac{1}{\sigma\sqrt{2\pi}} e^{-(x-\alpha)^2/2\sigma^2} dx$.

(4) Weibull: $F_\alpha(t) = 1 - e^{-(t/\alpha)^\lambda}$ for $t > 0$, where $\lambda > 0$ is fixed.

(5) Proportional hazards: $F_\alpha(t) = 1 - e^{-\frac{1}{\alpha}R(t)}$, $t > 0$, where $R(t) = -\log \bar{F}(t)$ is the hazard function of some life distribution F.

DEFINITION 2.5. A nonnegative function $f_\alpha(x)$ on $R \times R$ is *totally positive of order 2* (TP_2) in (α, x) if

$$\begin{vmatrix} f_{\alpha_1}(x_1) & f_{\alpha_1}(x_2) \\ f_{\alpha_2}(x_1) & f_{\alpha_2}(x_2) \end{vmatrix} \geq 0$$

for all $\alpha_1 < \alpha_2$ and $x_1 < x_2$ (also called the monotone likelihood ratio property.)

The following theorems relate total positivity to convex-ordering. The theory of total positivity has been fruitful in obtaining many new results in reliability and life testing. We can also make use of this powerful tool in studying convex-ordered families. Karlin (1968) is an excellent source for results on total positivity.

THEOREM 2.6. *F_α is a convex family if and only if the corresponding density $f_\alpha(t)$ with respect to some dominating measure λ is TP_2 in (α, t).*

PROOF. By Theorem 2.3, we have that $F_{\alpha_2} \overset{c}{>} F_{\alpha_1}$ for $\alpha_2 > \alpha_1$ if and only if $\frac{f_{\alpha_2}}{f_{\alpha_1}}$ is increasing. Thus for $\alpha_2 > \alpha_1$, $t_2 > t_1$,

$$\begin{vmatrix} f_{\alpha_1}(t_1) & f_{\alpha_1}(t_2) \\ f_{\alpha_2}(t_1) & f_{\alpha_2}(t_2) \end{vmatrix} \geq 0,$$

which is the defining condition that $f_\alpha(t)$ is TP_2 in (α, t). $\|$

THEOREM 2.7. *If $\{F_\alpha\}$ is a convex family, then $F_\alpha(t)$ is TP_2 in (α, t).*

PROOF. For $\alpha_2 > \alpha_1$, $F_{\alpha_2}(F_{\alpha_1}^{-1})$ is convex on (0,1). This implies that $F_{\alpha_2}(F_{\alpha_1}^{-1}(t))/t$ is nondecreasing in t. Since f_{α_1} is continuous, $F_{\alpha_1}F_{\alpha_1}^{-1}(t) = t$. We have $F_{\alpha_2}(F_{\alpha_1}^{-1}(t))/F_{\alpha_1}\ (F_{\alpha_1}^{-1}(t))$ is nondecreasing in t. By noting that $F_{\alpha_1}^{-1}$ is increasing, we conclude that $F_\alpha(t)$ is TP_2 in (α, t). $\|$

Another characterization of convex-ordering is given by:

THEOREM 2.8. *$F \overset{c}{>} G$ if and only if $\bar{G} \overset{c}{>} \bar{F}$.*

PROOF. Since $F^{-1}(1-t) = \bar{F}^{-1}(t)$ on (0,1), we have $\bar{G}F^{-1}(t) = 1 - GF^{-1}(1-t)$. Thus FG^{-1} is convex if and only if $\bar{G}\bar{F}^{-1}$ is convex. $\|$

REMARK. An immediate consequence of Theorems 2.7 and 2.8 is that $F \overset{c}{>} G$ implies both $\frac{F}{G}$ and $\frac{\bar{F}}{\bar{G}}$ are nondecreasing.

We end this section with comparisons of our convex-ordering and other partial orderings. We note that the Weibull family is not convex-ordered with respect to the shape parameter λ, but it is convex-ordered in the sense of Van Zwet. For an example showing that convex-ordering of distribution does not imply convex-ordering in the sense of Van Zwet, consider $F_1(t) = t^2$ and $F_2(t) = 1 - \sqrt{1-t^2}$

for t in $[0,1]$. Then F_2 is more convex than F_1, i.e., $F_2 \overset{c}{>} F_1$. Since $F_1^{-1}F_2(t) = (1-\sqrt{1-t^2})^{\frac{1}{2}}$ is not convex, F_2 is not convex-ordered with respect to F_1 in Van Zwet's sense.

When F is absolutely continuous with respect to the Lebesgue measure λ, the failure rate function of F is defined to be $\tau_F = \frac{f}{\bar{F}}$, where $f = \frac{dF}{d\lambda}$. F has a larger failure rate function than G if $\tau_F(t) \geq \tau_G(t)$ for all $t \geq 0$. The next theorem compares convex-ordering with failure rate ordering.

THEOREM 2.9. *If F and G are absolutely continuous distributions with respective densities f and g, then $F \overset{c}{>} G$ implies $\tau_F(t) \leq \tau_G(t)$ for all t.*

PROOF. By Theorem 2.3, $f(t_1)g(t) \leq f(t)g(t_1)$ for all $t_1 \leq t$. Integrating this over $[t_1, \infty)$, we have $f(t_1)\bar{G}(t_1) \leq \bar{F}(t_1)g(t_1)$ for all t_1. ||

Comparing this result with the convex-hazard order of Lee (1981), which requires that $\frac{\tau_F}{\tau_G}$ be a nondecreasing function of t, we see that convex-ordering neither implies nor is implied by the convex hazard function ordering of Lee.

3. Preservation of Convex-Ordering Under Operations. In this section, we show that our notion of convex-ordering is preserved under various standard statistical operations.

First, we show that convex-ordering is preserved under mixture of distributions.

THEOREM 3.1. *If $F_\alpha \overset{c}{>} G_\beta$ for each pair (α, β), then $\int F_\alpha d\mu(\alpha) \overset{c}{>} \int G_\beta d\nu(\beta)$ for any mixing distribution μ and ν.*

PROOF. The proof can be split into two parts. Suppose that $F_\alpha \overset{c}{>} G$ for each α. We will show that $\int F_\alpha d\mu(\alpha) \overset{c}{>} G$ for any mixing distribution μ. Let $f_\alpha = \frac{dF_\alpha}{dG}$, then by Theorem 2.3, f_α is nondecreasing for each α. Thus $\int f_\alpha d\mu(\alpha)$ is nondecreasing, and this implies that $\int F_\alpha d\mu(\alpha) \overset{c}{>} G$.

A similar proof shows that if $F \overset{c}{>} G_\beta$ for each β, then $F \overset{c}{>} \int G_\beta d\nu(\beta)$ for any mixing distribution ν.

These two results establish Theorem 3.1. ||

It should be noted that the condition in Theorem 3.1 cannot be weakened to $F_\alpha \overset{c}{>} G_\alpha$ for each α, as shown in the following example.

EXAMPLE 3.2. Let

$$\bar{F}_1(t) = e^{-t/1.1},\ \bar{F}_2(t) = e^{-t/5.1},\ \bar{G}_1(t) = e^{-t},\ \bar{G}_2(t) = e^{-t/5}.$$

Then $F_1 \overset{c}{>} G_1$ and $F_2 \overset{c}{>} G_2$. However, $G_2 \overset{c}{>} F_1$. Let $F = \frac{1}{2}(F_1 + F_2)$, $G = \frac{1}{2}(G_1 + G_2)$. To check whether $F \overset{c}{>} G$, consider the ratio of the derivatives of F and G,

$$h(t) = \frac{dF}{dG}(t) = \frac{\frac{1}{1.1}e^{-t/1.1} + \frac{1}{5.1}e^{-t/5.1}}{e^{-t} + \frac{1}{5}e^{-t/5}}$$

We note that h is continuous, but h is not increasing because $h(4.6) = 1.04 > 1.026 = h(6)$. Thus the ordering $F \overset{c}{>} G$ is shown to be false in view of Theorem 2.3.

Next, we show that convex-ordering is preserved under formation of certain coherent structures. We begin with some basic definitions and notations from reliability.

Consider n independent components, each of which is either functioning or not. We use the binary variable x_i to indicate the state of the i-th component:

$$x_i = \begin{cases} 1 & \text{if component } i \text{ is functioning} \\ 0 & \text{otherwise.} \end{cases}$$

The state of a system composed of these components is determined by the states of the components. The function $\phi(x_1, \ldots, x_n)$ is called the *structure function* of the system and is defined by

$$\phi(x_1, \ldots, x_n) = \begin{cases} 1 & \text{if system is functioning} \\ 0 & \text{otherwise.} \end{cases}$$

EXAMPLE. A k-out-of-n system functions if and only if at least k out of the n components function. The structure function is given by

$$\phi(x_1, \ldots, x_n) = \begin{cases} 1 & \text{if } \sum_{i=1}^{n} x_i \geq k \\ 0 & \text{otherwise.} \end{cases}$$

The i-th component is *irrelevant* to the structure ϕ if ϕ is constant in x_i. We consider *monotone* systems, that is, systems for which $\phi(x_1, \ldots, x_n) \geq \phi(y_1, \ldots, y_n)$ whenever $x_i \geq y_i$ for all $i = 1, \ldots, n$. If a monotone system has no irrelevant components, it is said to be a *coherent* system.

Let $P(X_i = 1) = p_i$ denote the reliability of the i-th component; the system reliability is given by

$$h_\phi(p_1, \ldots, p_n) = P(\phi(x_1, \ldots, x_n) = 1).$$

Denote the life distribution of the i-th component by F_i; then the life distribution F_ϕ of the system is given by

$$F_\phi(t) = 1 - h_\phi(\bar{F}_1(t), \ldots, \bar{F}_n(t)).$$

As a special case, we will consider a parallel structure of n components, i.e., a 1-*out-of-n* system. The life distribution of the system is given by the product $\Pi_{i=1}^n F_i(t)$. The following theorem shows that convex-ordering is preserved under formation of parallel systems.

THEOREM 3.3. *Suppose $F_i \overset{c}{>} G_j$ for each pair (i, j). Then $\Pi_{i=1}^n F_i \overset{c}{>} \Pi_{i=1}^n G_i$.*

PROOF. It suffices to prove the theorem for $n = 2$. We first establish that $F_1 \cdot F_2$ is absolutely continuous ($<<$) with respect to $G_1 \cdot G_2$. From our assumption it follows that $F_i << G_j$, $i = 1, 2$, $j = 1, 2$. Note that $(F_1 \cdot F_2)(x) = F_1(x)F_2(x)$ is a distribution function. The distribution function $G_1 \cdot G_2$ is similarly defined. It is easy to establish that

$$G_1(E)G_2(E) \leq (G_1 \cdot G_2)(E) \leq G_1(E) + G_2(E)$$

and

$$F_1(E)F_2(E) \leq (F_1 \cdot F_2)(E) \leq F_1(E) + F_2(E),$$

for intervals E and unions of intervals E and extending it for Borel sets E. Now, if $(G_1 \cdot G_2)(E) = 0$, then either $G_1(E)$ or $G_2(E) = 0$. Either case implies that both $F_1(E) = 0$ and $F_2(E) = 0$ and thus $(F_1 \cdot F_2)(E) = 0$. Thus $F_1 \cdot F_2 << G_1 \cdot G_2$. Let λ be a measure dominating all the G_i's. Let $f_i = \frac{dF_i}{d\lambda}$ and $g_i = \frac{dG_i}{d\lambda}$. Then

$$\frac{d(F_1 \cdot F_2)}{d(G_1 \cdot G_2)} = \frac{f_1 \cdot F_2 + f_2 \cdot F_1}{g_1 \cdot G_2 + g_2 \cdot G_1} = \left[\frac{g_1 \cdot G_2}{f_1 \cdot F_2} + \frac{g_2 \cdot G_1}{f_1 \cdot F_2}\right]^{-1} + \left[\frac{g_1 \cdot G_2}{f_2 \cdot F_1} + \frac{g_2 \cdot G_1}{f_2 \cdot F_1}\right]^{-1},$$

is a nondecreasing function, since $\frac{f_i}{g_i}, \frac{F_i}{G_j}$ are nondecreasing. Hence $F_1 \cdot F_2 \overset{c}{>} G_1 \cdot G_2$ by Theorem 2.3. ||

It is natural to ask whether we can relax the assumption of Theorem 3.3 to $F_i \overset{c}{>} G_i$, all i? A counterexample paralleling Example 3.2 is given below.

EXAMPLE 3.4. Let F_1, F_2, G_1, G_2 be defined as in Example 3.2. Then

$$k(t) = \frac{d(F_1 \cdot F_2)}{d(G_1 \cdot G_2)}(t) = \frac{f_1(t)F_2(t) + f_2(t) \cdot F_1(t)}{g_1(t)G_2(t) + g_2(t) \cdot G_1(t)}$$

is continuous and $k(4.5) = 1.01944 > 1.01935 = k(4.7)$. Theorem 2.3 shows that the ordering $F_1 \cdot F_2 \overset{c}{>} G_1 \cdot G_2$ does not hold.

Another important coherent structure is the k-out-of-n system. We now show that convex-ordering is preserved under formation of such systems with independent components.

THEOREM 3.5. *Let $F_{n,k}$ be the life distribution of a k-out-of-n system with independent components with life distributions $F_1, \ldots, F_n$. Similarly let $G_{n,k}$ be the life distribution of a k-out-of-n system with independent components with life distributions $G_1, \ldots, G_n$. Suppose that $F_i \overset{c}{>} G_j$ for each pair (i, j). Then $F_{n,k} \overset{c}{>} G_{n,k}$.*

PROOF. Suppose $G_{n,k}(E) = 0$. Then $G_j(E) = 0$ for some j. Hence $F_i(E) = 0$ for all i. Thus $F_{n,k}(E) = 0$, so that $F_{n,k} << G_{n,k}$.

Let λ be a measure which dominates all the G_i's. Let $f_i = \frac{dF_i}{d\lambda}$ and $g_i = \frac{dG_i}{d\lambda}$, $i = 1, \ldots, n$. Then for a k-out-of-n system, the densities of $F_{n,k}$ and $G_{n,k}$ with respect to λ are

$$\begin{aligned} f_{n,k} &= \frac{n!}{(k-1)!(n-k)!}\Sigma f_{\alpha_1}\bar{F}_{\alpha_2}\ldots\bar{F}_{\alpha_k}F_{\alpha_{k+1}}\ldots F_{\alpha_n}, \\ g_{n,k} &= \frac{n!}{(k-1)!(n-k)!}\Sigma g_{\alpha_1}\bar{G}_{\alpha_2}\ldots\bar{G}_{\alpha_k}G_{\alpha_{k+1}}\ldots G_{\alpha_n}, \end{aligned}$$

where the summation is taken over all permutations $(\alpha_1, \ldots, \alpha_n)$ of the integers $1, 2, \ldots n$. The ratio of derivatives of $F_{n,k}$ with respect to $G_{n,k}$ is

$$\frac{dF_{n,k}}{dG_{n,k}} = \frac{\Sigma f_{\alpha_1}\cdot\bar{F}_{\alpha_2}\cdot\ldots\cdot\bar{F}_{\alpha_k}\cdot F_{\alpha_{k+1}}\cdot\ldots\cdot F_{\alpha_n}}{\Sigma g_{\alpha_1}\cdot\bar{G}_{\alpha_2}\cdot\ldots\cdot\bar{G}_{\alpha_k}\cdot G_{\alpha_{k+1}}\cdot\ldots\cdot G_{\alpha_n}}.$$

To complete the proof that this is a nondecreasing function, we show that the ratio of every term in the numerator to any term in the denominator is non-decreasing. A typical term is of the form

$$\frac{f_1\cdot\bar{F}_2\cdot\ldots\cdot\bar{F}_k\cdot F_{k+1}\cdot\ldots\cdot F_n}{g_{\alpha_1}\cdot\bar{G}_{\alpha_2}\cdot\ldots\cdot\bar{G}_{\alpha_k}\cdot G_{\alpha_{k+1}}\cdot\ldots\cdot G_{\alpha_n}},$$

which is nondecreasing since $\frac{f_i}{g_j}, \frac{\bar{F}_i}{\bar{G}_j}, \frac{F_i}{G_j}$ are nondecreasing. By Theorem 2.3, we conclude that $F_{n,k} \overset{c}{>} G_{n,k}$. ||

When all components are identical, the following theorem shows among other results, that the reliability of a k-out-of-n system is more convex than the reliability of a $(k-1)$-out-of-n system.

THEOREM 3.6. *Let $h_{n,k}(p)$ be the reliability function of a k-out-of-n system with identical components. Then $h_{n,k+1} \overset{c}{>} h_{n+1,k+1} \overset{c}{>} h_{n,k} \overset{c}{>} h_{n+1,k}$.*

PROOF.

$$\begin{aligned} h_{n,k}(p) &= \Sigma_{i=1}^{n}\binom{n}{i}p^i(1-p)^{n-i} \\ &= \frac{\Gamma(n+1)}{\Gamma(k)\Gamma(n-k+1)}\int_0^p t^{k-1}(1-t)^{n-k}dt, \end{aligned}$$

the incomplete Beta function. Taking the derivative of $h_{n,k}$, we have

$$h'_{n,k}(p) = \frac{n!}{(n-k)(k-1)!}p^{k-1}(1-p)^{n-k}.$$

Thus,

$$\frac{h'_{n,k}(p)}{h'_{n+1,k}(p)} = \frac{n-k+1}{n+1} \cdot \frac{1}{1-p}$$

is increasing in p, establishing that $h_{n,k} \overset{c}{>} h_{n+1,k}$. The remaining inequalities can be proved similarly. ||

Since the distribution of the $(n-k+1)$-th order statistic corresponds to the life distribution of a k-out-of-n system of identical components, the following corollary is essentially a restatement of Theorem 3.6.

COROLLARY 3.7. *Let $F_{n,k}$ be the distribution of the $(n-k+1)$-th order statistic in a sample of size n from F. Then $F_{n,k+1} \overset{c}{>} F_{n+1,k+1} \overset{c}{>} F_{n,k} \overset{c}{>} F_{n+1,k}$.*

In view of Theorem 3.5, one might ask whether convex-ordering is preserved under formation of coherent systems. The example below shows that this is not true in general.

EXAMPLE. Consider the coherent structure of identical components presented in the following diagram.

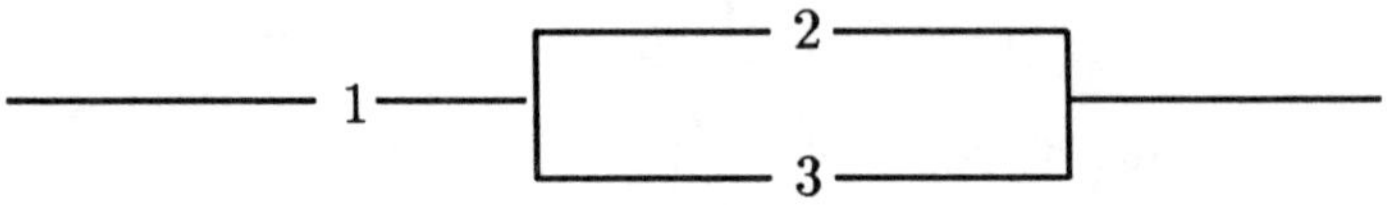

The system reliability is $h_\phi(p) = p^2(2-p)$. For $\bar{F}(t) = e^{-t}$ and $\bar{G}(t) = e^{-2t}$, we have $F \overset{c}{>} G$; but

$$\frac{f_\phi(t)}{g_\phi(t)} = \frac{4-3e^{-t}}{2e^{-2t}(4-3e^{-2t})}$$

which is continuous and not monotone nondecreasing in t.

In order to show that convex-ordering is preserved under convolution, we need to consider the class of Pólya frequence densities of order 2 (PF_2).

DEFINITION 3.8. f is a *Pólya frequency function of order 2* (PF_2) if for all $\triangle > 0$, $f(x+\triangle)/f(x)$ is decreasing in x, $-\alpha < x < \alpha$.

An equivalent definition is that $\log f(x)$ is concave. Note that each PF_2 function $f(x)$ defines a TP_2 function, $h(x,y) = f(x-y)$.

The following theorem, due to Ghurye and Wallace (1959), gives a sufficient condition on convex families for preservation of convexity under convolution.

THEOREM 3.9. *Let $\{F_\alpha\}$ and $\{G_\alpha\}$ be convex families with PF_2 densities. Then $(F_\alpha * G_\alpha)$ is a convex family.*

Denote the n-fold convolution of F by $F^{(n)}$. Then for life distribution F with log concave density, $F^{(n+1)}$ is more convex than $F^{(n)}$. This is a special case of Theorem 1 in Karlin and Proschan (1960).

THEOREM 3.10. *Let F be a life distribution with PF_2 density, then $\{F^{(n)}\}$ is a convex family with respect to n.*

4. Application of Convex-Ordering. Very often in life testing we do not know the exact form of the distribution, but based on physical evidence, we know something about the properties of the distribution. For example, in situations where a normal distribution is assumed, we might suspect that the tail of the underlying distribution is, in fact, heavier than that of the normal distribution. Therefore, we want to test the normal assumption against convex-ordered alternatives. In this section, we will present an inequality for convex families and apply this inequality to develop tests of such a hypothesis.

THEOREM 4.1. *(Hardy, Littlewood, and Pólya (1952), p. 75.) $F \overset{c}{>} G$ if and only if*

$$F^{-1}\left(\Sigma_{i=1}^{n}\lambda_i F(x_i)\right) \geq G^{-1}\left(\Sigma_{i=1}^{n}\lambda_i G(x_i)\right)$$

for all x_i and $\lambda_i \geq 0$, $i = 1, \ldots, n$, such that $\Sigma_{i=1}^{n}\lambda_i = 1$.

Note that $\Sigma_{i=1}^{n}\lambda_i F(x_i)$ is a weighted average. We now apply this result to hypothesis testing.

APPLICATION 4.2. Let $X_1, \ldots, X_n$ be a random sample. Suppose we wish to test:

$$H_0 : X_1, \ldots, X_n \sim G \ \text{(known)}$$

against the alternative

$$H_1 : X_1, \ldots, X_n \sim F, \quad F \overset{c}{>} G \ \text{ but otherwise unknown.}$$

Notice that when H_1 holds, $G^{-1}\left(\frac{1}{n}\Sigma_{i=1}^{n}G(X_i)\right) \leq F^{-1}\left(\frac{1}{n}\Sigma_{i=1}^{n}F(X_i)\right)$.

The right hand side can be estimated by $F_n^{-1}\left(\frac{1}{n}\Sigma_{i=1}^{n}F_n^{-1}(X_i)\right) = F_n^{-1}((n+1)/2n) \approx$ median of the sample where F_n is the empirical distribution function. Therefore, our test procedure is to reject H_0 if $G^{-1}\left(\frac{1}{n}\Sigma_{i=1}^{n}G(X_i)\right)$ is sufficiently smaller than $F_n^{-1}(\frac{1}{2})$. Recall that G is known.

5. Convex-Ordering for Symmetric Distribution Functions. In this section, we consider convex orderings between continuous symmetric distribution functions: $\bar{F}(x) = F(-x)$ for all x.

DEFINITION 5.1. $F \overset{sc}{>} G$ if F and G are continuous symmetric distribution functions and $F \overset{c}{>} G$ on $[0, \infty)$; i.e., $F << G$ and FG^{-1} is concave on $(-\infty, 0]$ and convex on $[0, \infty)$.

Examples of such ordered distributions are:

1) Normal.
Let F_α be the distribution function on $N(0, \alpha^2)$, $\alpha > 0$. Then $\alpha_2 > \alpha_1 \Rightarrow F_{\alpha_2} \overset{sc}{>} F_{\alpha_1}$

2) Double Exponential.
Let F_α be the density function given by

$$f_\alpha(x) = \frac{1}{2\alpha} e^{-|x|/\alpha}, \ \alpha > 0, \ -\infty < x < \infty.$$

Then $\alpha_2 > \alpha_1 \Rightarrow F_{\alpha_2} \overset{sc}{>} F_{\alpha_1}$.

A characterization of this ordering is given in the next theorem.

THEOREM 5.2. *$F \overset{sc}{>} G$ if and only if $f(t) = \frac{dF}{dG}(t)$ is nondecreasing in $|t|$ for almost every t.*

PROOF. By Theorem 2.3, f is increasing on $[0, \infty)$ and decreasing on $(-\infty, 0]$. ||

Thus, if we wish to show $F \overset{sc}{>} G$, we need only to consider f on the positive axis. As an immediate consequence of this and Theorem 3.1, we have:

THEOREM 5.3. *Let $F_\alpha \overset{sc}{>} G_\beta$ for each pair of (α, β). Then $\int F_\alpha d\mu(\alpha) \overset{sc}{>} \int G_\beta d\nu(\beta)$ for any mixing distributions μ and ν.*

When F is more convex than G, then G is more peaked about the origin than F, as shown in Theorem 5.5 below. We now compare this notion of relative peakedness to the following definition given by Birnbaum (1948).

DEFINITION 5.4. Y is more peaked than X if

$$P(|Y| \geq t) \leq P(|X| \geq t) \ \text{ for all } t \geq 0.$$

If X and Y have symmetric distribution functions F and G respectively, then this is equivalent to $G(t) \geq F(t)$ for all $t \geq 0$.

THEOREM 5.5. *If $F \overset{sc}{>} G$, then G is more peaked than F.*

PROOF. Since $F \overset{c}{\geq} G$ on $[0, \infty)$, it follows that $FG^{-1}(u) - \frac{1}{2}$ is convex on $[0, \frac{1}{2})$, and hence $(FG^{-1}(u) - \frac{1}{2})/(u - \frac{1}{2})$ is increasing on (0,1). This implies that $(F(t) - \frac{1}{2})/(G(t) - \frac{1}{2})$ is increasing on $(0, \infty)$ and is less than or equal to 1. This shows that $F(t) \leq G(t)$ for $t > 0$. Since F and G are symmetric, it follows that G is more peaked than F. ||

Since the product of symmetric distribution functions need not be symmetric, we do not have a result analogous to Theorem 3.3. It can also be shown that this ordering is not necessarily preserved under convolution. If $F \overset{sc}{>} G$, then we can show that the even central moments of F are greater than those of G. To prove this we need the following result.

LEMMA 5.6. *(Barlow and Proschan, 1975, p. 120.)*
Let $W(x)$ be a Lebesgue-Stieltjes measure, not necessarily positive for which $\int_t^\infty dW(x) \geq 0$ for all t, and let $h \geq 0$ be increasing. Then $\int_{-\infty}^{\infty} h(x)dW(x) \geq 0$.

THEOREM 5.7. $F \overset{sc}{>} G \Rightarrow \mu_{2n}(F) \geq \mu_{2n}(G)$ *for all* n.

PROOF. $F \overset{sc}{>} G \Rightarrow F(t) \leq G(t)\ \forall\, t \geq 0$. Let

$$W(x) = \begin{cases} F(x) - G(x) & \text{if } x \geq 0 \\ 0 & \text{otherwise,} \end{cases}$$

and

$$h(x) = \begin{cases} x^{2n} & \text{if } x \geq 0 \\ 0 & \text{otherwise.} \end{cases}$$

Then, by Lemma 5.6, $\mu_{2n}(F) \geq \mu_{2n}(G)$. ||

COROLLARY 5.8. *If $F \overset{sc}{>} G$, then Var $F \geq$ Var G.*

We conclude this section with the following theorem.

THEOREM 5.9. *Let the distribution functions F and G be absolutely continuous with densities f and g. If $F \overset{sc}{>} G$ and F is unimodal, then G is unimodal.*

PROOF. Both g/f and f are nonnegative and decreasing on $[0, \infty)$; thus g is decreasing on $[0, \infty)$. ||

REFERENCES

BARLOW, R.E. and PROSCHAN, F. (1981). *Statistical Theory of Reliability and Life Testing.* To Begin With, Silver Spring, MD.

BARLOW, R.E. and PROSCHAN, F. (1966). Tolerance and confidence limits for classes of distributions based on failure rate. *Ann. Math. Statist.* **37** 1593–1601.

BIRNBAUM, Z.W. (1948). On random variables with comparable peakedness. *Ann. Math. Stat.* **19** 76–81.

GHURYE, S.G. and WALLACE, D.L. (1959). A convolution class of monotone likelihood ratio families. *Ann. Math. Stat.* **30** 1158–1164.

HARDY, G.H., LITTLEWOOD, J.E., and PÓLYA, G. (1952). *Inequalities.* Cambridge University Press.

KARLIN, S. (1968). *Total Positivity, Vol. 1.* Stanford University Press.

KARLIN, S. and PROSCHAN, F. (1960). Pólya type distributions of convolutions. *Ann. Math. Stat.* **31** 721–736.

LEE, L. (1981). Distributions ordered by their hazard functions. VPI Technical Report.

ROSS, S.M. (1980). *Stochastic Processes.* Wiley, New York.

VAN ZWET, W.R. (1964). *Convex Transformations of Random Variables.* Mathematical Centrum.

DEPARTMENT OF STATISTICS
FLORIDA STATE UNIVERSITY
TALLAHASSEE, FL 32306–3033

UNBIASEDNESS OF TESTS OF HOMOGENEITY WHEN ALTERNATIVES ARE ORDERED

BY ARTHUR COHEN,[1] MICHAEL D. PERLMAN,[2] AND HAROLD B. SACKROWITZ[1]

Rutgers University, University of Washington, and Rutgers University

Let $X_1, X_2, \ldots, X_k$ be independent random variables whose densities are from an exponential family with parameters $\theta_1, \theta_2, \ldots, \theta_k$, respectively. That is, the densities are $f(x_i \mid \theta_i) = c(\theta_i)e^{x_i\theta_i}g(x_i)$. Assume that g is a Polya frequency function of order two (PF_2). Consider testing the null hypothesis $H : \theta_1 = \theta_2 = \ldots = \theta_k$ vs. the alternative $K : \theta_1 \geq \theta_2 \geq \ldots \geq \theta_k$. Write $\mathbf{x} = (x_1, x_2, \ldots, x_k)$ and define a partial ordering $>>^*$ on $\Re^k$ by $\mathbf{x} >>^* \mathbf{y}$ if and only if $\Sigma_{i=1}^j x_i \geq \Sigma_{i=1}^j y_i$ for $j = 1, 2, \ldots, k-1$ and equality for $j = k$. A function $\varphi(\mathbf{x})$ is said to be ISO* if $\mathbf{x} >>^* \mathbf{y}$ implies $\varphi(\mathbf{x}) \geq \varphi(\mathbf{y})$. We prove that if $\varphi(\mathbf{x})$ is a similar test which is ISO* then φ is unbiased. In fact if $\varphi(\mathbf{x})$ is ISO* the power function of the test is conditionally monotone nondecreasing along rays orthogonal to the equiangular line. For cases where the distribution satisfies the semi-group property the power function is unconditionally monotone along these rays. Furthermore a way to generate unbiased tests with monotone power is given.

The result contrasts with and complements the result of Robertson and Wright (1982). They prove that when the density has the semi-group property (normal and Poisson, for example) the tests which are ISO* have ISO* power functions. Such a finding is different from ours. The class of distributions for which our result holds is larger than the class in Robertson and Wright.

Applications for particular distributions and particular tests are given. Also some admissibility results are given for particular distributions. For example, it is proven that Bartholomew's test is admissible for the normal case.

[1]Research supported by NSF Grant DMS–84–18416.

[2]Research supported by NSF Grant DMS–86–03489.

AMS 1980 subject classification. Primary 62F03.

Key words and phrases. Homogeneity, unbiasedness, similar test, Neyman structure, Polya frequency function of order two (PF_2), log concave, multivariate total positivity of order two (MTP_2), FKG inequality, admissibility, Bartholomew's test.

The authors would like to thank Professor Ester Samuel-Cahn for helpful remarks that led to Corollary 2.4.

1. Introduction and Summary. Let $X_1, X_2, \ldots, X_k$ be independent continuous or integer-valued random variables distributed according to a one-parameter exponential family with parameters θ_i, $i = 1, 2, \ldots, k$. That is, the joint density of the X_i is

$$f(\mathbf{x}, \theta) = \left(\Pi_{i=1}^k \beta(\theta_i)\right) e^{\Sigma_{i=1}^k x_i \theta_i} \left(\Pi_{i=1}^k g(x_i)\right), \tag{1}$$

where $\mathbf{x} = (x_1, x_2, \ldots, x_k)$, $\theta = (\theta_1, \theta_2, \ldots, \theta_k)$. The dominating measure for each X_i is Lebesgue measure on $(-\infty, \infty)$ for the continuous case and counting measure on $\{0, \pm 1, \pm 2, \ldots\}$ for the case where the X_i are integer-valued. Assume that g is a Polya frequency function of order two (PF_2); that is, $g(\geq 0)$ is log concave on $(-\infty, \infty)$ or $\{0, \pm 1, \ldots\}$, respectively. The problem is to test

$$H : \theta_1 = \theta_2 = \ldots = \theta_k \text{ vs. } K : \theta_1 \geq \theta_2 \geq \ldots \geq \theta_k,$$

with at least one strict inequality under K. We study the unbiasedness and admissibility of tests.

Robertson and Wright (1982) study the problem of testing H vs. K when either (i) the distributions of the X_i come from a translation family, (ii) the distributions of the X_i satisfy the semi-group property (normal and Poisson, for example), or the distribution of $\mathbf{X} = (X_1, \ldots, X_k)$ is multinomial. They prove a result which yields a monotonicity property for the power functions of certain tests. This monotonicity property, called ISO*, implies unbiasedness of similar tests. More precisely, let $t_j(\mathbf{x}) = \Sigma_{i=1}^j x_i$, $j = 1, 2, \ldots, k$. We define the partial ordering $\mathbf{x} >>^* \mathbf{y}$ as follows: The vector $\mathbf{x} >>^* \mathbf{y}$ if and only if $t_j(\mathbf{x}) \geq t_j(\mathbf{y})$ for $j = 1, 2, \ldots, k-1$ and $t_k(\mathbf{x}) = t_k(\mathbf{y})$. A function h is said to be ISO* if $\mathbf{x} >>^* \mathbf{y}$ implies $h(\mathbf{x}) \geq h(\mathbf{y})$. For the distributions they study, Robertson and Wright (1982) prove that if $\varphi(\mathbf{x})$ is a test function which is ISO* then the power function of that test is ISO*.

One of the main results of this paper (Theorem 2.3) is that for the distributions in (1) with g PF_2, if $\varphi(\mathbf{x})$ is a test function which is similar and ISO* then φ is unbiased. Furthermore φ is such that its conditional power function (conditioned on $T_k = \Sigma_{i=1}^k X_i$) is monotone nondecreasing along rays orthogonal to the equiangular line. Still further if the distribution in (1) has the semi-group property then φ has a power function that is unconditionally monotone along rays orthogonal to the equiangular line.

In Theorem 2.3 we actually prove that the conditional power function of a similar ISO* test function satisfies a monotonicity property with respect to a certain partial ordering. The partial ordering is different from the one considered by Robertson and Wright (1982). Neither partial ordering implies the other. Points on rays orthogonal to the equiangular line will be ordered by both partial orderings mentioned above. Thus a power function which is conditionally monotone with respect to either partial ordering will be conditionally monotone along rays orthogonal to the equiangular line. What is important however is that the class of distributions for which Theorem 2.3 holds is larger than the class studied by

Robertson and Wright. Our class of distributions (1) includes the normal family with common known variance and means θ_i, the Poisson family with means related to θ_i, the binomial family with common sample size n and probabilities related to θ_i, the gamma family with common shape parameter (≥ 1) and scale parameters related to θ_i (which includes the chi-square distribution with two or more degrees of freedom), and many others. The binomial family is *not* covered by the results of Robertson and Wright (1982). In fact, at the end of Section 3 we present a counterexample to show that in the binomial case an ISO* test function need *not* have an ISO* power function, at least conditionally.

Hereafter, when there is no confusion it is convenient to let monotone power function mean a power function which is conditionally monotone along rays orthogonal to the equiangular line.

In Cohen and Sackrowitz (1987b), the class of distributions in (1) were considered for the problem of testing H vs. K': not H. In that study, test functions which were similar and Schur convex were shown to be unbiased. The method of proof used here will be different (and simpler) than the method used there.

In addition to providing a sufficient condition for a test to be unbiased and have a monotone power function for H vs. K, we give a method of generating such tests in Theorem 2.6.

Whereas the model of the paper is stated in terms of a single observation for each population it will be seen that all results remain true when we have n observations from each population, provided that each X_i is replaced by the sample mean $\bar{X}_i$ from the ith population. This follows as a special case of Theorem 2.7, a result on unbiasedness of tests for the important case of unequal numbers of observations from each population.

Applications to particular tests and to particular distributions will be made in Section 3. For gamma distributions with common shape parameter (≥ 1), unbiasedness is established for a certain natural class of tests for equality of the scale parameters.

For the statistical model of this paper one can easily derive an essentially complete class of test procedures from the result of Eaton (1970). For the normal case we prove that the likelihood ratio test derived by Bartholomew (1959) is admissible. We indicate other admissible tests for the normal case and indicate admissible tests for the binomial and Poisson cases as well.

Unbiasedness and monotone power results are given in Section 2. Section 3 contains applications to specific distributions and tests, while admissibility of tests is discussed in Section 4.

2. Unbiasedness of Tests. For the statistical model described near (1) we note first that any unbiased test for H vs. K must be similar and therefore must have Neyman structure with respect to $T \equiv \Sigma_{i=1}^{k} X_i$ (see Lehmann (1986), Theorem 2, p. 144). Hence any unbiased test of size α must have conditional size α (given $T = t$). It is clear that tests which are conditionally unbiased of size α (for every t) are unbiased of size α. Our plan will therefore be to show that the similar tests under study will be conditionally unbiased.

We now discuss the notion of multivariate totally positive distribution as done in Karlin and Rinott (1980). This notion is closely related to the FKG inequality. Let $f(x)$ be a nonnegative function defined on $\mathcal{X}^{(k)} \equiv \mathcal{X}_1 \times \mathcal{X}_2 \times \ldots \times \mathcal{X}_k$, where each $\mathcal{X}_i$ is a totally ordered subset of $\Re^1$, satisfying

$$f(\mathbf{x} \vee \mathbf{y}) f(\mathbf{x} \wedge \mathbf{y}) \geq f(\mathbf{x}) f(\mathbf{y}), \tag{2}$$

where $\vee$ and $\wedge$ are the corresponding lattice operations on $\mathcal{X}^{(k)}$, i.e.,

$$\begin{aligned} \mathbf{x} \vee \mathbf{y} &= (\max(x_1, y_1), \max(x_2, y_2), \ldots, \max(x_k, y_k)), \\ \mathbf{x} \wedge \mathbf{y} &= (\min(x_1, y_1), \min(x_2, y_2), \ldots, \min(x_k, y_k)). \end{aligned}$$

A function f with the property (2) is called multivariate totally positive of order 2 (MTP_2) on $\mathcal{X}^{(k)}$. In this paper, either $\mathcal{X}_i = (-\infty, \infty)$ for $i = 1, \ldots, k$ or $\mathcal{X}_i = \{0, \pm 1, \ldots\}$ for $i = 1, \ldots, k$.

From Karlin and Rinott (1980) we note that if $f(\mathbf{x})$ and $g(\mathbf{x})$ are MTP_2 on $\mathcal{X}^{(k)}$ then $f(\mathbf{x})g(\mathbf{x})$ is MTP_2 on $\mathcal{X}^{(k)}$. Also, if $f(\mathbf{x}) = g(x_i, x_j)$ where g is TP_2 on $\mathcal{X}_i \times \mathcal{X}_j$, then f is MTP_2 on $\mathcal{X}^{(k)}$, hence products of such functions are MTP_2 on $\mathcal{X}^{(k)}$.

Now define $T_j = \Sigma_{i=1}^{j} X_i$, $j = 1, 2, \ldots, k$, $\mathbf{T} = (T_1, \ldots, T_k)$, and $\mathbf{T}^{(k-1)} = (T_1, \ldots, T_{k-1})$. The range of $\mathbf{T}$ is again $\mathcal{X}^{(k)}$ while the range of $\mathbf{T}^{(k-1)}$ is $\mathcal{X}^{(k-1)}$. Let $f_\theta(\mathbf{t}^{(k-1)} \mid t_k)$ denote the conditional density of $\mathbf{T}^{(k-1)}$ given $T_k = t_k$.

LEMMA 2.1. *Assume (1) with g PF_2. Then for any θ, $f_\theta(\mathbf{t}^{(k-1)} \mid t_k)$ is MTP_2 on $\mathcal{X}^{(k-1)}$.*

PROOF. The density of $f_\theta(\mathbf{t}^{(k-1)} \mid t_k)$ satisfies

$$\begin{aligned} f_\theta(\mathbf{t}^{(k-1)} \mid t_k) &= C_{t_k}(\theta) e^{(\theta_1 - \theta_2)t_1 + \cdots + (\theta_{k-1} - \theta_k)t_{k-1}} \cdot \\ &\qquad g(t_1) g(t_2 - t_1) \cdots g(t_k - t_{k-1}), \end{aligned} \tag{3}$$

for $\mathbf{t}^{(k-1)} \in \mathcal{X}^{(k-1)}$. Since g is log concave on $\mathcal{X}_i \Leftrightarrow g(x_{i+1} - x_i)$ is TP_2 on $\mathcal{X}_i \times \mathcal{X}_{i+1}$, it follows from (3) that for fixed t_k, $f_\theta(\mathbf{t}^{(k-1)} \mid t_k)$ is MTP_2 on $\mathcal{X}^{(k-1)}$. ||

Now let $r_{\theta^{(2)}, \theta^{(1)}}(\mathbf{t}^{(k-1)} \mid t_k) = f_{\theta^{(2)}}(\mathbf{t}^{(k-1)} \mid t_k) / f_{\theta^{(1)}}(\mathbf{t}^{(k-1)} \mid t_k)$ where $\theta^{(2)}, \theta^{(1)}$ lie in the parameter space and let H_k denote the family of componentwise nondecreasing functions on $\mathcal{X}^{(k)}$ i.e. if $h(\mathbf{x}) \varepsilon H_k$, then $h(x_1, x_2, \cdots x_i, \cdots x_k)$ is nondecreasing in x_i while $x_1, x_2, \cdots, x_{i-1}, x_{i+1}, \cdots, x_k$ are fixed, for $i = 1, 2, \ldots k$. Also for $\theta^{(2)}$ lying in the alternative space and $\theta^{(1)}$ lying in the null space or the alternative space let $\xi = \theta^{(2)} - \theta^{(1)}$.

LEMMA 2.2. *Suppose $\xi' = (\xi_1, \xi_2, \cdots \xi_k)$ is such that $\xi_1 \geq \xi_2 \geq \cdots \geq \xi_k$. Then for fixed t_k the ratio $r_{\theta^{(2)}, \theta^{(1)}}(\mathbf{t}^{(k-1)} \mid t_k)$ lies in H_{k-1}.*

PROOF. From (3) we find

$$r_{\theta^{(2)},\theta^{(1)}}(\mathbf{t}^{(k-1)} \mid t_k) = C_{t_k}(\theta^{(2)},\theta^{(1)})e^{(\xi_1-\xi_2)t_1+\ldots+(\xi_{k-1}-\xi_k)t_{k-1}},$$

where $C_{t_k}(\theta^{(2)},\theta^{(1)}) > 0$ whenever the marginal density of T_K at t_k is positive. The result is immediate. ||

The well-known FKG inequality for $w_1, w_2 \in H_k$ and an MTP$_2$ density function f on $\mathcal{X}^{(k)}$ states that

$$Ew_1(\mathbf{T})w_2(\mathbf{T}) \geq Ew_1(\mathbf{T})Ew_2(\mathbf{T}). \tag{4}$$

See, for example, Karlin and Rinott (1980).

THEOREM 2.3. *Let $\theta^{(2)}$ be a parameter point in the alternative space and let $\theta^{(1)}$ be a parameter point in the null space or alternative space. Let $\xi = \theta^{(2)} - \theta^{(1)}$. Let $\varphi(\mathbf{x})$ be a similar size α test which is ISO* in $\mathbf{x}$. If $\xi_1 \geq \xi_2 \geq \cdots \geq \xi_k$, then $E_{\theta^{(2)}}(\varphi(\mathbf{x}) \mid T) \geq E_{\theta^{(1)}}(\varphi(\mathbf{x}) \mid T)$. Also for any $\theta^{(1)}$ lying in the null space $E_{\theta^{(2)}}\varphi(\mathbf{x}) \geq \alpha = E_{\theta^{(1)}}\varphi(\mathbf{x})$, which implies $\varphi(\mathbf{x})$ is unbiased.*

PROOF. It suffices to show that φ satisfies the above inequalities conditionally given $T_k = t_k$. Define $\psi(\mathbf{t}) = \varphi(t_1, t_2-t_1, \cdots, t_k-t_{k-1})$. Since φ is ISO*, $\varphi \in H_{k-1}$ as a function of $\mathbf{t}^{(k-1)}$ for fixed t_k. Then by Lemma 2.2 and (4),

$$\begin{aligned}
E_{\theta^{(2)}}(\varphi(\mathbf{x}) \mid T_k = t_k) &= E_{\theta^{(2)}}(\psi(\mathbf{t}) \mid T_k = t_k) \\
&= \int \psi(\mathbf{t})r_{\theta^{(2)},\theta^{(1)}}(\mathbf{t}^{(k-1)} \mid t_k)f_{\theta^{(1)}}(\mathbf{t}^{(k-1)} \mid t_k)d\mu(\mathbf{t}^{(k-1)}) \\
&\geq \int \psi(\mathbf{t})f_{\theta^{(1)}}(\mathbf{t}^{(k-1)} \mid t_k)d\mu(\mathbf{t}^{(k-1)}) \\
&\times \int f_{\theta^{(1)}}(\mathbf{t}^{(k-1)} \mid t_k)d\mu(\mathbf{t}^{(k-1)}) \\
&= E_{\theta^{(1)}}(\psi(\mathbf{t}) \mid T_k = t_k) \\
&= E_{\theta^{(1)}}(\varphi(\mathbf{x}) \mid T_k = t_k). \quad ||
\end{aligned}$$

Now let $\theta^{(2)}$ be any point in the alternative space and let $\bar{\theta}^{(2)} = (\bar{\theta}^{(2)}, \bar{\theta}^{(2)}, \ldots, \bar{\theta}^{(2)})$ be the projection of $\theta^{(2)}$ onto the equiangular line. Let $\theta^{(1)} = \lambda\theta^{(2)} + (1-\lambda)\bar{\theta}^{(2)}$, for $0 < \lambda < 1$; i.e., $\theta^{(1)}$ lies in the alternative space on the ray orthogonal to the equiangular line connecting $\theta^{(2)}$ and $\bar{\theta}^{(2)}$.

COROLLARY 2.4. *Let $\varphi(\mathbf{x})$ be a similar size α test which is ISO* in $\mathbf{x}$. Then the power function of $\varphi(\mathbf{x})$ is conditionally monotone nondecreasing along rays orthogonal to the equiangular line. If the density in (1) satisfies the semi-group property, then the power function is unconditionally monotone.*

PROOF. Let $\theta^{(2)}$ be any point in the alternative space and let $\theta^{(1)}$ be defined as above, i.e. $\theta^{(1)} = \lambda\theta^{(2)} + (1-\lambda)\bar{\theta}^{(2)}$. Then $\theta^{(2)} - \theta^{(1)} = (1-\lambda)(\theta^{(2)} - \bar{\theta}^{(2)}) = \xi$.

Since $\xi_1 \geq \xi_2 \geq \cdots \geq \xi_k$, the conditional monotonicity property follows from Theorem 2.3. If the density in (1) satisfies the semigroup property then the marginal distribution of T_k under $\theta^{(2)}$ is the same as the marginal distribution under $\theta^{(1)} = \lambda\theta^{(2)} + (1-\lambda)\bar{\theta}^{(2)}$. Thus Theorem 2.3 implies that unconditionally the power function is monotone.

REMARK 2.5. Theorem 2.3 can be proved by the arguments used in Cohen and Sackrowitz (1987b). The argument there is more difficult because the FKG inequality approach fails in that problem.

The next theorem identifies a general class of unbiased tests for H vs. K. Let $D = \{\mathbf{x} \mid x_1 \geq x_2 \geq \ldots \geq x_k\}$ and let $P(\mathbf{x} \mid D)$ be the unique point $\mathbf{z} \equiv (x_1, \ldots, z_k)$ in D which minimizes $\Sigma_1^k (x_i - z_i)^2$, i.e., $P(\mathbf{x} \mid D) \equiv P(\mathbf{x})$ is the projection of $\mathbf{x}$ onto D. See Brunk (1965).

THEOREM 2.6. *Let $\varphi(\mathbf{x})$ be a test function such that φ is Schur convex in $\mathbf{x}$. Let $\varphi^*(\mathbf{x}) = \varphi(P(\mathbf{x}))$. Then if $\varphi^*(\mathbf{x})$ is similar, $\varphi^*(\mathbf{x})$ is unbiased and has monotone power for H vs. K.*

PROOF. Suppose that $\mathbf{x} >>^* \mathbf{y}$. Then by Corollary 2.3 of Robertson and Wright (1982), $P(\mathbf{x}) >>^* P(\mathbf{y})$. Since $P(\mathbf{x})$ and $P(\mathbf{y})$ both lie in D it follows that $P(\mathbf{x})$ majorizes $P(\mathbf{y})$. Also since φ is Schur convex we have that

$$\varphi^*(\mathbf{x}) = \varphi(P(\mathbf{x})) \geq \varphi(P(\mathbf{y})) = \varphi^*(\mathbf{y}).$$

Thus φ^* is ISO* and the result follows from Theorem 2.3. ||

Theorem 2.6 provides a method of constructing unbiased tests for H vs. K. Suppose that $h(\mathbf{x})$ is a Schur convex test statistic used to test H vs. K': not H. Let φ be the size α test function based on h and suppose that φ is similar for H, hence has Neymann structure. Therefore

$$\begin{aligned} \varphi(\mathbf{x}) &= 1 \quad \text{if } h(\mathbf{x}) > C_\alpha(t_k) \\ &= \gamma(\mathbf{x}) \quad \text{if } h(\mathbf{x}) = C_\alpha(t_k) \\ &= 0 \quad \text{if } h(\mathbf{x}) < C_\alpha(t_k), \end{aligned}$$

where $C_\alpha(t_k)$ and $\gamma(\mathbf{x})$ are such that φ has conditional size α given $T_k = t_k$. Next define $h^*(\mathbf{x}) = h(P(\mathbf{x}))$ and now let

$$\begin{aligned} \varphi^*(\mathbf{x}) &= 1 \quad \text{if } h^*(\mathbf{x}) > C_\alpha^*(t_k) \\ &= \gamma^*(\mathbf{x}) \quad \text{if } h^*(\mathbf{x}) = C_\alpha^*(t_k) \\ &= 0 \quad \text{if } h^*(\mathbf{x}) < C_\alpha^*(t_k), \end{aligned}$$

where $C_\alpha^*(t_k)$ and $\gamma^*(\mathbf{x})$ are such that φ^* has conditional size α. Then φ^* is ISO* and by Theorem 2.6, φ^* is unbiased and has monotone power for H vs. K. We note that the critical values change from the φ test to the φ^* test.

The last result in this section concerns the case where n_i observations are taken from population i. That is, let r, $n_1, \ldots, n_r$ be positive integers such that $n_1 + n_2 + \ldots + n_r = k$ and suppose in (1) that

$$\theta_{n_1+\ldots+n_{i-1}+1} = \ldots = \theta_{n_1+\ldots+n_i} \equiv \theta_{(i)}, \quad 1 \le i \le r,$$

where $n_0 = 0$. Consider the problem of testing

$$H : \theta_{(1)} = \ldots = \theta_{(r)} \text{ vs. } \tilde{K} : \theta_{(1)} \ge \ldots \ge \theta_{(r)}$$

with at least one strict inequality under $\tilde{K}$. (Note that H is the same null hypothesis as before.) Clearly $\bar{\mathbf{X}} \equiv (\bar{X}_1, \ldots, \bar{X}_r)$ is a sufficient statistic for $(\theta_{(1)}, \ldots, \theta_{(r)})$, where

$$\bar{X}_i = \frac{1}{n_i} \Sigma_{q=1}^{n_i} X_{n_1+\ldots+n_{i-1}+q}, \quad 1 \le i \le r.$$

In Theorem 2.7 we present a condition under which a similar test $\varphi(\bar{\mathbf{x}})$ is unbiased with monotone power for H vs. $\tilde{K}$.

As in Robertson and Wright (1982), Section 5, define the weighted ordering $>>_W^*$ on $\Re^r$ as follows: $\mathbf{y} >>_W^* \mathbf{z}$ if and only if $\Sigma_{i=1}^j n_i y_i \ge \Sigma_{i=1}^j n_i z_i$, $j = 1, 2, \ldots, r-1$, with equality for $j = r$. A function f on $\Re^r$ is said to be ISO$_W^*$ if $\mathbf{y} >>_W^* \mathbf{z}$ implies $f(\mathbf{y}) \ge f(\mathbf{z})$. Clearly, if $n_1 = \ldots = n_r$, then $>>_W^*$ and ISO$_W^*$ reduce to the unweighted versions $>>^*$ and ISO*, respectively.

THEOREM 2.7. *Let $\varphi(\bar{\mathbf{x}})$ be a similar size α test function for H vs. $\tilde{K}$ which is ISO$_W^*$ in $\bar{\mathbf{x}}$. Then φ is unbiased and has monotone power.*

PROOF. If $\varphi(\bar{\mathbf{x}})$ is ISO$_W^*$ in $\bar{\mathbf{x}} \in \Re^r$, then it is easily seen that as a function of $\mathbf{x} \equiv (x_1, \ldots, x_k)$, φ is ISO* on $\Re^k$. The result then follows immediately from Theorem 2.3. ||

3. Applications. Robertson and Wright (1982), Section 4, describe a class of contrast statistics which are ISO*. Furthermore, they discuss a class of statistics based on an ℓ_2 distance between estimates satisfying the inequalities defining the alternative hypothesis K (or $\tilde{K}$). For example, in the case $n_1 = \ldots = n_r$ of equal sample sizes for each population (cf. the end of Section 2), let

$$S = \Sigma_{i=1}^r (\bar{\theta}_{(i)} - m(\bar{\mathbf{x}}))^2,$$

where $\bar{\theta} \equiv (\bar{\theta}_{(1)}, \ldots, \bar{\theta}_{(r)}) = P(\bar{\mathbf{x}} \mid D)$, and $m(\bar{\mathbf{x}}) \equiv \Sigma n_i \bar{x}_i / \Sigma n_i$ is the overall mean. Then S is ISO* by Theorem 2.6. In fact, as remarked after Theorem 2.6, we can generate an ISO* test function for H vs. K (or $\tilde{K}$) from any Schur convex test function for H vs. K'. In Cohen and Sackrowitz (1987b), test functions which were Schur convex were studied. These same test functions are ISO* when $\mathbf{x}$ is replaced by $P(\mathbf{x} \mid D)$.

To achieve unbiasedness and monotone power we recall that the critical values of the test statistics must be chosen conditionally for each value of $T_k = \Sigma_{i=1}^k X_i$. This is so since we prove unbiasedness by proving conditional unbiasedness and requiring that all conditional sizes are α. One exception is the normal case where the statistic may be chosen to be independent of T_k under H. Such is the case for Bartholomew's test ($\equiv$ the likelihood ratio test) which is based on the statistic S. It is known that the distribution of S under H is chi-bar squared and is independent of T_k. (A simple proof using Basu's theorem (see Lehmann, (1983), p. 46) establishes the independence.) Other statistics in the normal case could be chosen which are also ISO* and independent of T_k under H.

The binomial case is an important one because it arises frequently in practice. Ordered alternatives appear to be natural in experiments involving increased doses of a drug or increased learning of a subject. The current most popular approach is to use either the likelihood ratio test statistic for H vs. K or a χ^2-type statistic for H vs. K, comparing it with a critical value determined by the asymptotic distribution of the statistic under H. See Barlow, Bartholomew, Bremner, and Brunk (1972), pp. 192–193 for a discussion of these two tests. Use of the asymptotic distribution requires that the sample size in each population tends to ∞. Such a test not only requires large sample sizes but it is not unbiased. It is not unbiased because one critical value is used for all T_k and thus the test will not have Neyman structure. To find an unbiased test one can instead proceed as follows: Consider the conditional distribution of $\mathbf{X}$ (or $\bar{\mathbf{X}}$) given T_k. Test H vs. K' by a likelihood ratio test or a chi-square test. (The chi-square test is the one used for a $2 \times k$ contingency table.) Finally replace $\mathbf{x}$ (or $\bar{\mathbf{x}}$) by $P(\mathbf{x} \mid D)$ (or $P(\bar{\mathbf{x}} \mid D)$) to test H vs. K. It would be necessary to consider the conditional distribution of the statistic under H to determine critical values for each value of T_k.

Most remarks above concerning the binomial case also pertain to the Poisson case. Conditionally, in this latter case we deal with testing whether multinomial probabilities are equal to $(1/k, \ldots, 1/k)$ (or to $(1/r, \ldots, 1/r)$).

For the case of gamma random variables with a common shape parameter and unknown scale parameters $\theta_1, \ldots, \theta_k$, consider the following two-parameter family of tests for H vs. K' studied in Cohen and Strawderman (1971) and in Marshall and Olkin (1979), p. 387:

$$\varphi(\mathbf{x}) = \begin{cases} 1 & \text{if } R(\lambda, \eta) \equiv \left(\Sigma_{i=1}^k x_i^\lambda\right)^{1/\lambda} / \left(\Sigma_{i=1}^k x_i^\eta\right)^{1/\eta} > C_\alpha, \\ 0 & \text{otherwise,} \end{cases} \tag{5}$$

where C_α is a constant determined by the size α. (This test is similar because it is scale invariant.) In (5), replace $\mathbf{x}$ by $P(\mathbf{x} \mid D) \equiv P(\mathbf{x})$ to obtain the statistic $R^*(\lambda, \eta)$ and the test function $\varphi^*(\mathbf{x}) \equiv \varphi(P(\mathbf{x}))$ for H vs. K. $\|$

THEOREM 3.1. *For $\lambda \geq 0 \geq \eta$, the test φ^* based on $R^*(\lambda, \eta)$ is unbiased and has monotone power for H vs. K.*

PROOF. Let $Y_i = \log X_i$, so that the distribution of $\mathbf{Y} \equiv (Y_1, \ldots, Y_k)$ form a translation-parameter family with parameter $(\log\theta_1, \log\theta_2, \ldots, \log\theta_k)$. Theorem 3.1 of Robertson and Wright (1982) implies that if $R^*(\lambda, \eta)$ is ISO* in $\mathbf{y}$ then the test based on it is unbiased. But $R(\lambda, \eta)$ is Schur convex in $\mathbf{y}$ (cf. Marshall and Olkin (1979), p. 387) so the argument used in the proof of our Theorem 2.6 implies that $\varphi(P(\mathbf{x}))$ is ISO* in $\mathbf{y}$ and hence unbiased with monotone power. (Robertson and Wright's theorem establishes the property of an ISO* power function.) ||

Note that when $\lambda = 1$ and $\eta = 0$, the test φ in (5) is equivalent to the likelihood ratio test for H vs. K'. When $\mathbf{x}$ is replaced by $P(\mathbf{x} \mid D)$, the resulting test φ^* is the likelihood ratio test for H vs. K. When $\lambda = \infty$ and $\eta = -\infty$, φ is Hartley's test for H vs. K' and φ^* is its analog for the ordered alternative K. When $\lambda = 2$ and $\eta = 1$, the test φ is the locally most powerful unbiased test for H vs. K' (see Cohen and Sackrowitz (1987b)) while when $\lambda = \infty$ and $\eta = 1$, φ corresponds to Cochran's test. Although Theorem 3.1 does not apply to these latter two cases, Theorems 2.3 and 2.6 imply that they are in fact unbiased as long as the gamma density is PF_2, i.e., whenever the shape parameter is at least one.

We conclude this section by mentioning two counterexamples. The first already appears in Cohen, Sackrowitz, and Strawderman (1985), Example 5.4. It is an example where a test function φ for H vs. K is similar and ISO* and where the underlying density is of exponential type, but the test is <u>not</u> conditionally unbiased. Theorem 2.3 does not apply here because the function g in (1) is *not* PF_2 in this example.

The second counterexample presents a test function for H vs. K which is ISO* but which does not have (conditionally) an ISO* power function when the underlying distribution is binomial. In fact, take $k = 3$ and let $X_i \sim B(2, p_i)$, $i = 1, 2, 3$. Conditional on $T_3 \equiv X_1 + X_2 + X_3 = 2$, the six possible sample points are (2,0,0), (1,1,0), (0,2,0), (1,0,1), (0,1,1) and (0,0,2). The test function which rejects H for (2,0,0), (1,1,0) and (0,2,0) is conditionally ISO* with conditional size 0.4. The conditional power at $\mathbf{p}^{(1)} = (.88, .01, .01)$ is .8841, while the conditional power at $\mathbf{p}^{(2)} = (.45, .44, .01)$ is .9708, although $\mathbf{p}^{(1)} >>^* \mathbf{p}^{(2)}$. This example shows that the method of Robertson and Wright (1982) cannot work conditionally for the binomial case, while our method does establish unbiasedness. An open question is whether the power function is *unconditionally* ISO* for the binomial case when the test function is ISO*.

4. Admissibility of Tests. It follows from a theorem of Eaton (1970) that tests which conditionally are ISO* with convex acceptance sections for fixed T_k form an essentially complete class of tests. In the binomial and Poisson cases one can use the technique in Matthes and Truax (1967), Section 4(b) to establish that these tests are actually admissible. (The Poisson case requires only a bit more argument but an example of such an argument appears in Cohen and Sackrowitz (1987a).)

For the continuous cases, however, conditional admissibility does not imply unconditional admissibility except for the case $k = 2$. Nevertheless, we can prove the following theorem for the normal case.

THEOREM 4.1. *Let $X_1, \ldots, X_k$ be independent with $X_i \sim N(\theta_i, 1)$, $i = 1, \ldots, k$. The likelihood ratio test for H vs. K (Bartholomew's test) is admissible.*

PROOF. As discussed in Section 3 the likelihood ratio statistic S is independent of $T_k = \Sigma_{i=1}^k X_i$. Furthermore each acceptance section is convex by the argument in Birnbaum (1955), Section 9. Hence the overall acceptance region is convex. Since the test is ISO* as well as convex this permits application of the theorem of Stein (1956) to establish admissibility.

REMARK 4.2. It is clear that for the model (1) there are many ISO* tests for H vs. K that are unbiased but are inadmissible. Any similar test function $\varphi(\mathbf{x})$ which is ISO* but yet has non convex acceptance sections for fixed T_k is unbiased (by Theorem 2.3) and inadmissible (by the complete class theorem of Mattes and Truax (1967) or of Eaton (1970)). One simple example is to take $k = 3$, $T_3 = t_3$ and the acceptance region to be the union of $\{X_1 < C_1\}$ and $\{X_3 > C_3\}$.

REFERENCES

BARLOW, R.E., BARTHOLOMEW, D.J., BREMNER, J.M. and BRUNK, H.D. (1972). *Statistical Inference Under Order Restrictions.* Wiley, New York.

BARTHOLOMEW, D.J. (1959). A test for ordered alternatives. *Biometrika* **45** 36–48.

BIRNBAUM, A. (1955). Characterizations of complete classes of tests of some multiparametric hypotheses, with applications to likelihood ratio tests. *Ann. Math. Statist.* **26** 21–36.

BRUNK, H.D. (1965). Conditional expectation given a σ-lattice and applications. *Ann. Math. Statist.* **36** 1339–1350.

COHEN, A. and SACKROWITZ, H.B. (1987a). Admissibility of goodness of fit tests for discrete exponential families. *Statistics and Probability Letters* **5** 1–3.

COHEN, A. and SACKROWITZ, H.B. (1987b). Unbiasedness of tests for homogeneity. *Ann. Statist.* **15** 805–816.

COHEN, A., SACKROWITZ, H.B. and STRAWDERMAN, W.E. (1985). Multivariate locally most powerful unbiased tests. *Multivariate Analysis* VI, (P.R. Krishnaiah, ed.), Elsevier, 121–144.

COHEN, A. and STRAWDERMAN, W.E. (1970). Unbiasedness of tests for homogeneity of variances. *Ann. Math. Statist.* **42** 355–360.

EATON, M.L. (1970). A complete class theorem for multi-dimensional one sided alternatives. *Ann. Math. Statist.* **41** 1884–1888.

EFRON, B. (1965). Increasing properties of Polya frequency functions. *Ann. Math. Statist.* **36** 272–279.

KARLIN, S. and RINOTT, Y. (1980). Classes of orderings of measures and related correlation inequalities. I. Multivariate totally positive distributions. *Journal of Multivariate Analysis* **10** 457–498.

LEHMANN, E.L. (1983). *Theory of Point Estimation.* Wiley, New York.

LEHMANN, E.L. (1986). *Testing Statistical Hypotheses,* Second edition. Wiley, New York.

MARSHALL, A.W. and OLKIN, I. (1979). *Inequalities: Theory of Majorization and Its Applications.* Academic Press, New York.

MATTHES, T.K. and TRUAX, D.R. (1967). Tests of composite hypotheses for the multivariate exponential family. *Ann. Math. Statist.* **38** 681–697.

ROBERTSON, T. and WRIGHT, F.T. (1982). On measuring the conformity of a parameter set to a trend with applications. *Ann. Statist.* **10** 1234–1245.

STEIN, C. (1956). The admissibility of Hotelling's T^2–test. *Ann. Math. Statist.* **27** 616–623.

BOUNDS FOR TAIL PROBABILITIES OF WEIGHTED SUMS OF INDEPENDENT GAMMA RANDOM VARIABLES

BY PERSI DIACONIS[1] AND MICHAEL D. PERLMAN[2]

Harvard University and University of Washington

The tail probabilities of two weighted sums of independent gamma random variables are compared when the first vector of weights majorizes the second vector of weights. The conjecture that the two cumulative distribution functions cross exactly once is established in four special cases by means of the variation-diminishing property of totally positive kernels. Bounds are obtained for the location of the unique crossing point and its asymptotic behavior is determined.

1. Introduction. In this paper we continue the study of tail probabilities of weighted sums of independent, identically distributed (i.i.d.) gamma random variables begun by Diaconis (1976) and extended by Bock, Diaconis, Huffer, and Perlman (1987) [hereafter abbreviated as BDHP (1987)].

Let $Y_1, \ldots, Y_n$ be i.i.d. gamma random variables with common probability density function (pdf)

$$g_{\alpha,\beta}(y) = [\beta^\alpha \Gamma(\alpha)]^{-1} y^{\alpha-1} e^{-y/\beta}, \quad 0 < y < \infty, \tag{1}$$

where $\alpha > 0$ and $\beta > 0$ denote the shape and scale parameters, respectively. We denote this gamma distribution by $G(\alpha, \beta)$. For nonnegative weights $\theta_1, \ldots \theta_n$, the tail probabilities of the weighted sum $\sum \theta_i Y_i$ are denoted as follows:

$$\begin{aligned} F_{\boldsymbol{\theta}}(t) &= P\left[\sum \theta_i Y_i \le t\right] \\ \bar{F}_{\boldsymbol{\theta}}(t) &= P\left[\sum \theta_i Y_i \ge t\right] = 1 - F_{\boldsymbol{\theta}}(t), \end{aligned} \tag{2}$$

where $\boldsymbol{\theta} = (\theta_1, \ldots \theta_n)$ and $0 \le t < \infty$.

[1]Research supported in part by NSF Grant No. DMS 86-00235.
[2]Research supported in part by NSF Grant No. DMS 86-03489.

AMS 1980 *subject classifications.* Primary 60E15; secondary 62E15.

Key words and phrases. Tail probabilities, weighted sums, independent gamma random variables, Schur convexity, more dispersed, crossing point, total positivity.

We wish to thank Russell Millar, Ingram Olkin, and David Ragozin for many helpful ideas and discussions, and especially the referee who read the manuscript with great care.

Such weighted sums arise in many contexts in statistics and probability, for example as the distribution of quadratic forms $X'AX$ where X is an n-dimensional normal random vector and A is an arbitrary $n \times n$ positive semidefinite matrix. Such a quadratic form occurs, for example, as the limiting distribution of the chi-squared goodness-of-fit statistic when parameters are estimated on the basis of the ungrouped, rather than grouped, data—cf. Chernoff and Lehmann (1954). Weighted sums of exponential random variables also occur in the form $-\log(\Pi P_i^{\theta_i})$, a weighted version of the Fisher statistic for combining independent p-values $P_1, \ldots P_n$, where each P_i is uniformly distributed on (0,1) under the combined null hypothesis—cf. Good (1955) and Zelen and Joel (1959). See BDHP (1987) for additional examples.

Because the distribution of $\sum \theta_i Y_i$ cannot be expressed in a simple form, it is important to determine approximations or bounds for its tail probabilities. Much work concerning such approximations exists in the literature—cf. Johnson and Kotz (1970), Chapter 29—but little is known about bounds. One obvious question is the comparison of the tail probabilities of $\sum \theta_i Y_i$ and $\bar{\theta} \sum Y_i$, where $\bar{\theta} = n^{-1} \sum \theta_i$. This comparison is both natural (since $E(\sum \theta_i Y_i) = E(\bar{\theta} \sum Y_i))$ and potentially useful, since the tail probabilities of $\bar{\theta} \sum Y_i \sim G(n\alpha, \bar{\theta}\beta)$ are easily determined. For the reason mentioned in the next paragraph, it is appropriate to conjecture that the tail probabilities of $\bar{\theta} \sum Y_i$ provide lower bounds for those of $\sum \theta_i Y_i$.

Since $\boldsymbol{\theta} \equiv (\theta_1, \ldots, \theta_n)$ majorizes $\bar{\boldsymbol{\theta}} \equiv (\bar{\theta}, \ldots, \bar{\theta})$ [denoted by $\boldsymbol{\theta} \succ \bar{\boldsymbol{\theta}}$—cf. Marshall and Olkin (1979)], the above suggests a stronger conjecture, namely, that the tail probabilities of $\sum \theta_i Y_i$ exceed those of $\sum \eta_i Y_i$ whenever $\boldsymbol{\theta} \succ \boldsymbol{\eta} \equiv (\eta_1, \ldots, \eta_n)$. [Recall that $\boldsymbol{\theta} \succ \boldsymbol{\eta}$ requires that $\sum \theta_i = \sum \eta_i$, so that again $E(\sum \theta_i Y_i) = E(\sum \eta_i Y_i)$. Also, we shall adopt the convention that $\boldsymbol{\theta} \succ \boldsymbol{\eta}$ requires that $(\eta_1, \ldots, \eta_n)$ *not* be a permutation of $(\theta_1, \ldots, \theta_n)$.] Support for this conjecture is immediate: since

$$\text{(3)} \qquad \operatorname{Var}\left(\sum \theta_i Y_i\right) = \left(\sum \theta_i^2\right) \operatorname{Var} Y_1$$

and $\sum \theta_i^2$ is a strictly Schur-convex function of $(\theta_1, \ldots, \theta_n)$,

$$\text{(4)} \qquad \boldsymbol{\theta} \succ \boldsymbol{\eta} \Longrightarrow \operatorname{Var}\left(\sum \theta_i Y_i\right) > \operatorname{Var}\left(\sum \eta_i Y_i\right).$$

This states that if the weights $\theta_1, \ldots, \theta_n$ are more dispersed (in the sense of majorization) than $\eta_1, \ldots, \eta_n$ about their common average, then the random variable $\sum \theta_i Y_i$ is more dispersed than $\sum \eta_i Y_i$ about their common expected value, as measured by their variances. Our basic question is whether $\sum \theta_i Y_i$ is more dispersed than $\sum \eta_i Y_i$ as measured by the stronger criterion of their tail probabilities.

In this paper we investigate two aspects of this question, requiring two different techniques. First, in Section 2 we investigate the conjecture that if $\boldsymbol{\theta} \succ \boldsymbol{\eta}$, then $F_{\boldsymbol{\theta}}(\cdot)$ and $F_{\boldsymbol{\eta}}(\cdot)$ cross exactly once on $(0, \infty)$ at a unique point t^*. If true, this conjecture (called the *Unique Crossing Conjecture*, or UCC), implies that the probability distribution of $\sum \eta_i Y_i$ is more concentrated about t^* than that of $\sum \theta_i Y_i$. Although we believe that the UCC is true in general, we are able to verify it only in the following special cases:

(a) $n = 2$ (Proposition 2.1);

(b) $n = 3$, $\alpha = 1$ (Proposition 2.5);

(c) $n \geq 3$, $\alpha \geq 1$, $\boldsymbol{\theta}$ and $\boldsymbol{\eta}$ differ in only two components (Proposition 2.3);

(d) $n \geq 3$, $\boldsymbol{\eta} = \bar{\boldsymbol{\theta}}$ (Proposition 2.7).

Second, in Section 3 we investigate a conjecture regarding the *location* of the unique crossing point t^* when $\boldsymbol{\eta} = \bar{\boldsymbol{\theta}}$. It has been established by Diaconis (1976) and BDHP (1987) that when $\boldsymbol{\theta} \succ \boldsymbol{\eta}$ and $n = 2$, the unique crossing point of $F_{\boldsymbol{\theta}}$ and $F_{\boldsymbol{\eta}}$ lies in the interval

$$(2\alpha\bar{\theta}\beta,\ (2\alpha + 1)\bar{\theta}\beta) \tag{5}$$

(See also Proposition 2.2 in Section 2.) This implies that $F_{\boldsymbol{\theta}}(t)$ is Schur-convex in $\boldsymbol{\theta}$ when $t \leq 2\alpha\bar{\theta}\beta$ and that $\bar{F}_{\boldsymbol{\theta}}(t)$ is Schur-convex in θ when $t \geq (2\alpha + 1)\bar{\theta}\beta$. For $n \geq 3$, however, BDHP (1987) obtained only that $\bar{F}_{\boldsymbol{\theta}}(t)$ is Schur-convex in $\boldsymbol{\theta}$ when

$$t \geq n(n\alpha + 1)\bar{\theta}\beta \tag{6}$$

which, when $n = 2$, is a smaller interval than that implied by (5). Furthermore, they obtained no general result on the Schur-convexity of $F_{\boldsymbol{\theta}}(t)$ for $n \geq 3$; in fact, BDHP (1987, p. 394) presented a counterexample to show that no such result is possible.

In Section 3 of the present paper, we shall show that when $\boldsymbol{\eta} = \bar{\boldsymbol{\theta}}$, the unique crossing point t^* of $F_{\boldsymbol{\theta}}$ and $F_{\bar{\boldsymbol{\theta}}}$ in fact satisfies

$$t^*(\boldsymbol{\theta}, \bar{\boldsymbol{\theta}}) \sim n\alpha\bar{\theta}\beta \quad \text{as } n \to \infty \tag{7}$$

uniformly in $\boldsymbol{\theta}$ for fixed $\bar{\theta}$. In the course of this demonstration, we derive approximate bounds of the form

$$n\alpha\bar{\theta}\beta e^{-\frac{1}{2}} \dot{\leq} t^*(\boldsymbol{\theta}, \bar{\boldsymbol{\theta}}) \dot{\leq} (n\alpha + 1)\bar{\theta}\beta \log[n\alpha(n-1)] \tag{8}$$

valid for all $n \geq 2$ (cf. Propositions 3.1 and 3.2 and also (50) and (60)).

As in BDHP (1987), many of our methods also can be applied to obtain bounds for tail probabilities of weighted sums of independent Weibull random variables. Furthermore, it is likely that part of our results extend to the case where some of the weights θ_i may be negative, and also to the case where $Y_1, \ldots, Y_n$ are not i.i.d. but are exchangeable with pdf of the form

$$(\Pi y_i)^{\alpha-1} h\left(\sum y_i\right)$$

for suitable functions h.

2. The Unique Crossing Conjecture. By (4), $F_{\boldsymbol\theta}$ and $F_{\boldsymbol\eta}$ cannot be identical when $\boldsymbol\theta \succ \boldsymbol\eta$. Since

$$\boldsymbol\theta \succ \boldsymbol\eta \Longrightarrow E\left(\sum \theta_i Y_i\right) = E\left(\sum \eta_i Y_i\right), \tag{9}$$

$F_{\boldsymbol\theta} - F_{\boldsymbol\eta}$ must change sign *at least once* on $(0, \infty)$. In this section we investigate the

Unique Crossing Conjecture (UCC): If $\boldsymbol\theta \succ \boldsymbol\eta$, then $F_{\boldsymbol\theta} - F_{\boldsymbol\eta}$ changes sign *exactly once* on $(0, \infty)$. This crossing occurs at a unique point $t^* \equiv t^*(\boldsymbol\theta, \boldsymbol\eta)$, which is the only zero of $F_{\boldsymbol\theta} - F_{\boldsymbol\eta}$ on $(0, \infty)$.

If t^* exists, then necessarily

$$F_{\boldsymbol\theta}(t) - F_{\boldsymbol\eta}(t) \begin{cases} > 0 & \text{for } 0 < t < t^*, \\ < 0 & \text{for } t^* < t < \infty. \end{cases} \tag{10}$$

To see this, note that (4) and (9) imply that

$$\begin{aligned} E\left(\sum \theta_i Y_i - t^*\right)^2 &= \operatorname{Var}\left(\sum \theta_i Y_i\right) + \left[E\left(\sum \theta_i Y_i\right) - t^*\right]^2 \\ &> \operatorname{Var}\left(\sum \eta_i Y_i\right) + \left[E\left(\sum \eta_i Y_i\right) - t^*\right]^2 \\ &= E\left(\sum \eta_i Y_i - t^*\right)^2, \end{aligned} \tag{11}$$

hence

$$\begin{aligned} 0 &< E\left(\sum \theta_i Y_i - t^*\right)^2 - E\left(\sum \eta_i Y_i - t^*\right)^2 \\ &= \int_0^\infty \left\{P\left[\left(\sum \theta_i Y_i - t^*\right)^2 \geq u\right] - P\left[\left(\sum \eta_i Y_i - t^*\right)^2 \geq u\right]\right\} du \\ &= \int_0^\infty \{[F_{\boldsymbol\eta}(t^* + \sqrt{u}) - F_{\boldsymbol\theta}(t^* + \sqrt{u})] + [F_{\boldsymbol\theta}(t^* - \sqrt{u}) - F_{\boldsymbol\eta}(t^* - \sqrt{u})]\} du. \end{aligned}$$

If the inequalities in (10) were reversed, then (11) would be violated, so (10) must hold. (See also Remark 2.4 for the case $\alpha \geq 1$.) The result (10) shows that if the UCC is true, then the distribution of $\sum \theta_i Y_i$ is indeed more dispersed about t^* than that of $\sum \eta_i Y_i$ in the strong sense of tail probabilities.

Without loss of generality, we may set the scale parameter $\beta = 1$ for the remainder of this section. We shall establish the UCC in four special cases by means of the representation (13) below for $F_{\boldsymbol\theta}$. First, define

$$\begin{aligned} S_n &= \sum_{i=1}^n Y_i \\ W_i &= Y_i / S_n, \qquad i = 1, \ldots, n, \\ \mathbf{W} &= (W_1, \ldots, W_n). \end{aligned}$$

Then $\mathbf{W}$ and S_n are independent,

$$\begin{aligned} S_n &\sim G(n\alpha, 1), \\ \mathbf{W} &\sim \text{Dirichlet}(\alpha, \ldots, \alpha), \end{aligned}$$

i.e., $\mathbf{W}$ has the (exchangeable) Dirichlet distribution on the simplex

$$\Sigma^{(n)} \equiv \left\{ \mathbf{w} | w_1 \geq 0, \ldots, w_n \geq 0, \sum w_i = 1 \right\} \tag{12}$$

with pdf proportional to $(\Pi W_i)^{\alpha-1}$. Thus

$$\begin{aligned} F_{\boldsymbol{\theta}}(t) &= E\left\{ P\left[\sum \theta_i W_i \leq t S_n^{-1} | S_n \right] \right\} \\ &= \int_0^\infty H_{\boldsymbol{\theta}}(ts^{-1}) g_{n\alpha,1}(s) ds \\ &= t \int_0^\infty H_{\boldsymbol{\theta}}(u) g_{n\alpha,1}(tu^{-1}) u^{-2} du, \end{aligned} \tag{13}$$

where

$$H_{\boldsymbol{\theta}}(u) = P\left[\sum \theta_i W_i \leq u \right]. \tag{14}$$

Note that the support of $\sum \theta_i W_i$ is the interval $(\theta_{\min}, \theta_{\max}) \subseteq (0, n\bar{\theta})$.

From (13) it follows that

$$\begin{aligned} F_{\boldsymbol{\theta}}(t) - F_{\boldsymbol{\eta}}(t) &= t \int_0^\infty [H_{\boldsymbol{\theta}}(u) - H_{\boldsymbol{\eta}}(u)] g_{n\alpha,1}(tu^{-1}) u^{-2} du \\ &= \frac{t^{n\alpha}}{\Gamma(n\alpha)} \int_0^\infty [H_{\boldsymbol{\theta}}(u) - H_{\boldsymbol{\eta}}(u)] e^{-t/u} u^{-n\alpha-1} du. \end{aligned} \tag{15}$$

Because the kernel

$$K_1(t, u) \equiv e^{-t/u}$$

is *strictly* totally positive (STP) [cf. Karlin (1968), p. 15, eqn. (9)], it is strictly variation-diminishing, i.e., the number of sign changes of $F_{\boldsymbol{\theta}} - F_{\boldsymbol{\eta}}$ on $(0, \infty)$ cannot exceed the number of sign changes of $H_{\boldsymbol{\theta}} - H_{\boldsymbol{\eta}}$ provided this latter number is finite, the sign changes of $F_{\boldsymbol{\theta}} - F_{\boldsymbol{\eta}}$ must occur at isolated crossing points, and these crossing points are the only zeroes of $F_{\boldsymbol{\theta}} - F_{\boldsymbol{\eta}}$ on $(0, \infty)$ [apply Theorem 3.1(b) on p. 21 of Karlin (1968)]. We make use of these facts to establish the UCC in several special cases.

PROPOSITION 2.1. *If $n = 2$, the UCC is valid.*

PROOF. Without loss of generality, assume that $\theta_1 > \theta_2$. Since $(\theta_1, \theta_2) \succ (\eta_1, \eta_2)$, it follows that

$$\begin{aligned}(\theta_1, \theta_2) &= (\bar{\theta} + a, \bar{\theta} - a) \\ (\eta_1, \eta_2) &= (\bar{\theta} + b, \bar{\theta} - b),\end{aligned}$$

where $\bar{\theta} \geq a > |b| \geq 0$. Since $W_1 + W_2 = 1$, it follows that

$$\begin{aligned}H_{\boldsymbol{\theta}}(u) &= P[\bar{\theta} + a(W_1 - W_2) \leq u] \\ H_{\boldsymbol{\eta}}(u) &= P[\bar{\theta} + |b|(W_1 - W_2) \leq u],\end{aligned} \tag{16}$$

where we use the fact that $W_1 - W_2$ is symmetrically distributed about 0 on $(-1, 1)$ since $\mathbf{W}$ is exchangeable. Also, since W_1 has a Beta distribution, $W_1 - W_2 \equiv 2W_1 - 1$ assigns positive probability to every open subinterval of $(-1, 1)$. Therefore

$$H_{\boldsymbol{\theta}}(u) - H_{\boldsymbol{\eta}}(u) \begin{cases} > 0 & \text{if } \bar{\theta} - a < u < \bar{\theta} \\ < 0 & \text{if } \bar{\theta} < u < \bar{\theta} + a \\ = 0 & \text{if } u \leq \bar{\theta} - a,\ u = \bar{\theta}, \text{ or } u \geq \bar{\theta} + a, \end{cases} \tag{17}$$

hence has exactly one sign change (at $\bar{\theta}$) on $(0, \infty)$. Thus $F_{\boldsymbol{\theta}} - F_{\boldsymbol{\eta}}$ can have at most one sign change, hence by (9), must have exactly one sign change on $(0, \infty)$. Furthermore, this sign change must occur at a unique point t^* which must be the only zero of $F_{\boldsymbol{\theta}} - F_{\boldsymbol{\eta}}$ on $(0, \infty)$. ||

For the case $n = 2$, a closer examination of (15) in fact yields the upper bound in (5) for $t^* \equiv t^*(\boldsymbol{\theta}, \boldsymbol{\eta})$, the unique crossing point of $F_{\boldsymbol{\theta}} - F_{\boldsymbol{\eta}}$ on $(0, \infty)$.

PROPOSITION 2.2. *(Diaconis and Perlman (1976), BDHP (1987)). When* $n = 2$ *and* $\boldsymbol{\theta} \succ \boldsymbol{\eta}$, $t^*(\boldsymbol{\theta}, \boldsymbol{\eta}) < (2\alpha + 1)\bar{\theta}$.

PROOF. From (16) and the symmetry of $W_1 - W_2$, $\Lambda \equiv H_{\boldsymbol{\theta}} - H_{\boldsymbol{\eta}}$ is antisymmetric about $\bar{\theta}$, i.e.,

$$\Lambda(u) = -\Lambda(2\bar{\theta} - u). \tag{18}$$

Thus, if we define

$$\varphi_t(u) = e^{-t/u} u^{-2\alpha - 1}, \tag{19}$$

we obtain from (15) and (18) that

$$\begin{aligned}\frac{\Gamma(2\alpha)}{t^{2\alpha}}[F_{\boldsymbol{\theta}}(t) - F_{\boldsymbol{\eta}}(t)] &= \int_0^{\bar{\theta}} \Lambda(u)\varphi_t(u)du + \int_{\bar{\theta}}^{2\bar{\theta}} \Lambda(u)\varphi_t(u)du \\ &= \int_0^{\bar{\theta}} \Lambda(u)[\varphi_t(u) - \varphi_t(2\bar{\theta} - u)]du.\end{aligned} \tag{20}$$

When $t \geq (2\alpha+1)\bar{\theta}$ and $0 < u < \bar{\theta}$, it follows from Lemma 2.8 (at the end of this section) that $\varphi_t(u) - \varphi_t(2\bar{\theta} - u) < 0$, so by (17) we have that $F_{\boldsymbol{\theta}}(t) - F_{\boldsymbol{\eta}}(t) < 0$. By (10), this implies that $t^* < (2\alpha+1)\bar{\theta}$ as claimed. ||

Unfortunately, this method does not yield the lower bound for t^* in (5), since it is not true that $\varphi_t(u) - \varphi_t(2\bar{\theta} - u) > 0$ for every $u \in (0, \bar{\theta})$ when $t \leq 2\alpha\bar{\theta}$. Furthermore, when $n \geq 3$, $\Lambda(\cdot)$ need not be antisymmetric so the method does not immediately yield useful information about the location of t^*. An alternate approach is presented in Section 3 which does provide upper and lower bounds (though not sharp) for t^* when $\boldsymbol{\eta} = \bar{\boldsymbol{\theta}}$.

We now return to the UCC for the case $n \geq 3$. At present we cannot establish the UCC in general, so must content ourselves with four propositions (2.3, 2.5, 2.7, 2.7a) dealing with special cases of interest.

PROPOSITION 2.3. *Suppose that $n \geq 3$ and $\alpha \geq 1$. If $\boldsymbol{\theta}$ and $\boldsymbol{\eta}$ differ in exactly two components, then the UCC is valid.*

PROOF. When $n = 3$, we may assume without loss of generality that $\theta_3 = \eta_3 > 0$, so that $\boldsymbol{\theta} \succ \boldsymbol{\eta} \iff (\theta_1, \theta_2) \succ (\eta_1, \eta_2)$. Now

$$\begin{aligned} F_{\boldsymbol{\theta}}(t) &= E\{P[\theta_1 Y_1 + \theta_2 Y_2 \leq t - \theta_3 Y_3 | Y_3]\} \\ &= \int_0^\infty F_{(\theta_1,\theta_2)}(t - v) g(v) dv, \end{aligned}$$

where $g = g_{\alpha,\theta_3}$. Thus

$$F_{\boldsymbol{\theta}}(t) - F_{\boldsymbol{\eta}}(t) = \int_0^t \Delta(u) g(t-u) du, \tag{21}$$

where

$$\Delta(u) = F_{(\theta_1,\theta_2)}(u) - F_{(\eta_1,\eta_2)}(u).$$

We shall apply (21) to show that for $0 < t_1 < t_2 < \infty$,

$$F_{\boldsymbol{\theta}}(t_1) - F_{\boldsymbol{\eta}}(t_1) = 0 \implies F_{\boldsymbol{\theta}}(t_2) - F_{\boldsymbol{\eta}}(t_2) < 0, \tag{22}$$

$$F_{\boldsymbol{\theta}}(t_2) - F_{\boldsymbol{\eta}}(t_2) = 0 \implies F_{\boldsymbol{\theta}}(t_1) - F_{\boldsymbol{\eta}}(t_1) > 0. \tag{23}$$

These implications, together with the facts that $F_{\boldsymbol{\theta}} - F_{\boldsymbol{\eta}}$ is continuous and has at least one zero crossing on $(0, \infty)$, imply that $F_{\boldsymbol{\theta}}$ and $F_{\boldsymbol{\eta}}$ satisfy the UCC, hence establish the proposition when $n = 3$.

By Proposition 2.1 and (10), there exists $t_0 \in (0, \infty)$ such that

$$\Delta(u) \begin{cases} > 0 & \text{if } 0 < u < t_0 \\ = 0 & \text{if } u = t_0 \\ < 0 & \text{if } t_0 < u < \infty. \end{cases}$$

If $F_{\boldsymbol{\theta}}(t_1) - F_{\boldsymbol{\eta}}(t_1) = 0$, then since $g > 0$ on $(0, \infty)$, (21) implies that $t_1 > t_0$. Thus

$$\begin{aligned} F_{\boldsymbol{\theta}}(t_2) - F_{\boldsymbol{\eta}}(t_2) &< \int_0^{t_0} \Delta(u) g(t_2 - u) du + \int_{t_0}^{t_1} \Delta(u) g(t_2 - u) du \\ &\le \frac{g(t_2 - t_0)}{g(t_1 - t_0)} \left[\int_0^{t_0} \Delta(u) g(t_1 - u) du + \int_{t_0}^{t_1} \Delta(u) g(t_1 - u) du \right] \\ &= \frac{g(t_2 - t_0)}{g(t_1 - t_0)} [F_{\boldsymbol{\theta}}(t_1) - F_{\boldsymbol{\eta}}(t_1)] \\ &= 0. \end{aligned}$$

The second inequality follows because $\alpha \ge 1 \Longrightarrow g_{\alpha,\beta}(\cdot)$ is log concave $\Longrightarrow g(t-u)$ is totally positive of order two (TP_2) [cf. Karlin (1968), p. 32]. Thus (22) is valid, and (23) is established in similar fashion.

For $n \ge 4$, the proposition is established by a similar argument, using induction on n. ||

We remark that the kernel

$$K_2(t, u) \equiv I_{[0,t]}(u) g(t - u)$$

is TP_2 [cf. Karlin (1968), p. 16], hence Theorem 3.1(a) of Karlin (1968), p. 21, together with (21), shows that $F_{\boldsymbol{\theta}} - F_{\boldsymbol{\eta}}$ has at most one sign change on $(0, \infty)$. However, $K(t, u)$ is not *strictly* TP_2, so Karlin's Theorem 3.1(b) cannot be applied to conclude that the crossing point is *unique*, hence the need for a direct demonstration of this fact.

REMARK 2.4. If $\boldsymbol{\theta} \succ \boldsymbol{\eta}$ and $\alpha \ge 1$, we may apply Proposition 2.3 to deduce that $F_{\boldsymbol{\theta}}(t) - F_{\boldsymbol{\eta}}(t)$ is positive for sufficiently small $t > 0$ and negative for sufficiently large t. This follows from the fundamental majorization result that if $\boldsymbol{\theta} \succ \boldsymbol{\eta}$, then there exists a finite sequence

$$\boldsymbol{\theta} \equiv \boldsymbol{\psi}_0 \succ \boldsymbol{\psi}_1 \succ \ldots \succ \boldsymbol{\psi}_k \equiv \boldsymbol{\eta}$$

such that $\boldsymbol{\psi}_i$ and $\boldsymbol{\psi}_{i+1}$ differ in exactly two components [cf. Marshall and Olkin (1979), p. 21]. By Proposition 2.3 and (10), each difference $F_{\boldsymbol{\psi}_i}(t) - F_{\boldsymbol{\psi}_{i+1}}(t)$ must be positive for small $t > 0$ and negative for large t, hence the same must be true of the sum

$$\sum_{i=1}^{k-1} \left[F_{\boldsymbol{\psi}_i}(t) - F_{\boldsymbol{\psi}_{i+1}}(t) \right] \equiv F_{\boldsymbol{\theta}}(t) - F_{\boldsymbol{\eta}}(t).$$

Finally, we note that this implies that if the number of sign changes of $F_{\boldsymbol{\theta}} - F_{\boldsymbol{\eta}}$ on $(0, \infty)$ is finite, then this number must be *odd*. ||

REMARK 2.4A. The referee has pointed out that the argument for Proposition 2.3 also shows the following fact when $\alpha \ge 1$: if the UCC is valid for the vectors $\boldsymbol{\theta}$

and $\boldsymbol{\eta}$ in $\mathbb{R}^n$, then it remains valid for the vectors $(\boldsymbol{\theta}, \mathbf{c})$ and $(\boldsymbol{\eta}, \mathbf{c})$ in $\mathbb{R}^{n+k}$, where $\mathbf{c} = (c_1, \ldots, c_k)$ with $c_i > 0$. $\|$

PROPOSITION 2.5. *If $n = 3$ and $\alpha = 1$, the UCC is valid.*

Proof. Since $\alpha = 1$, $\mathbf{W} \equiv (W_1, W_2, W_3)$ is uniformly distributed on the 3-simplex $\Sigma^{(3)}$ (cf. (12)) with vertices (1,0,0), (0,1,0), (0,0,1). Let $h_{\boldsymbol{\theta}}$ denote the pdf of $\sum \theta_i W_i$, i.e.,

$$h_{\boldsymbol{\theta}}(u) = \frac{d}{du} H_{\boldsymbol{\theta}}(u)$$

(cf. (14)). Then $h_{\boldsymbol{\theta}}$ is a continuous triangular density function with support $(\theta_{\min}, \theta_{\max})$; $h_{\boldsymbol{\theta}}$ increases linearly on $(\theta_{\min}, \theta_{\mathrm{med}})$ and decreases linearly on $(\theta_{\mathrm{med}}, \theta_{\max})$, where $\theta_{\min}$, θ_{med}, and $\theta_{\max}$ denote the minimum, median, and maximum of $(\theta_1, \theta_2, \theta_3)$. Similarly, $h_{\boldsymbol{\eta}}$ is a triangular density function that increases linearly on $(\eta_{\min}, \eta_{\mathrm{med}})$ and decreases linearly on $(\eta_{\mathrm{med}}, \eta_{\max})$. Since

$$(24) \qquad \boldsymbol{\theta} \succ \boldsymbol{\eta} \Longrightarrow \theta_{\min} \leq \eta_{\min} \leq \eta_{\max} \leq \theta_{\max},$$

it follows that $h_{\boldsymbol{\theta}} - h_{\boldsymbol{\eta}}$ changes sign at most twice. But

$$(25) \qquad H_{\boldsymbol{\theta}}(u) - H_{\boldsymbol{\eta}}(u) = \int_0^\infty [h_{\boldsymbol{\theta}}(v) - h_{\boldsymbol{\eta}}(v)] I_{[0,u]}(v) dv$$

and the kernel

$$K_3(u, v) \equiv I_{[0,u]}(v)$$

is totally positive of every order [Karlin (1968), p. 16] hence is TP_3. Thus by (25) and Theorem 3.1(a) of Karlin (1968), p. 21, $H_{\boldsymbol{\theta}} - H_{\boldsymbol{\eta}}$ changes sign at most twice, so $F_{\boldsymbol{\theta}} - F_{\boldsymbol{\eta}}$ changes sign at most twice and has at most two zeroes on $(0, \infty)$, which must coincide with the crossing points (cf. the discussion following (15)). But the final sentence in Remark 2.4 implies that the number of sign changes of $F_{\boldsymbol{\theta}} - F_{\boldsymbol{\eta}}$ must be odd, hence cannot exceed one. Thus $F_{\boldsymbol{\theta}} - F_{\boldsymbol{\eta}}$ must have exactly one sign change and exactly one zero, i.e., the UCC is valid in this case. $\|$

It seems likely that Proposition 2.5 remains valid for $n \geq 4$. When $\alpha = 1$, both $h_{\boldsymbol{\theta}}$ and $h_{\boldsymbol{\eta}}$ are univariate B-splines (cf. Karlin, Micchelli, and Rinott (1986)) with knots $\theta_1, \ldots, \theta_n$ and $\eta_1, \ldots, \eta_n$, respectively. *We conjecture that the integrated B-splines $H_{\boldsymbol{\theta}}$ and $H_{\boldsymbol{\eta}}$ cross exactly once whenever $\boldsymbol{\theta} \succ \boldsymbol{\eta}$.* By the argument following (15), the validity of this conjecture would imply the validity of the UCC when $\alpha = 1$. We have been able to establish this conjecture when $\boldsymbol{\theta}$ and $\boldsymbol{\eta}$ differ in exactly two components. In fact, for *every* $0 < \alpha < \infty$, $H_{\boldsymbol{\theta}}$ and $H_{\boldsymbol{\eta}}$ cross exactly once when $n = 2$ (recall (17)), while when $\alpha \geq 1$ and $n \geq 3$ the following result obtains:

PROPOSITION 2.6. *(Diaconis and Perlman (1976)). Suppose that $n \geq 3$ and $\alpha \geq 1$. If $\boldsymbol{\theta} \succ \boldsymbol{\eta}$ and differ in exactly two components, then $H_{\boldsymbol{\theta}}$ and $H_{\boldsymbol{\eta}}$ cross exactly once.* $\|$

Our proof of this result is similar to that of Proposition 2.3 though somewhat longer, hence is omitted. Note that Proposition 2.3 follows from Proposition 2.6 by the argument following (15).

Now fix $\alpha = 1$ and consider the above conjecture regarding the single crossing of the integrated B-splines $H_{\boldsymbol{\theta}}$ and $H_{\boldsymbol{\eta}}$ when $\boldsymbol{\theta} \succ \boldsymbol{\eta}$ and differ in *more* than two components. *This conjecture is valid when* $n = 3$. To see this, one first shows that (4), (9), and (10) remain true with $Y_1, \ldots, Y_n$ replaced by $W_1, \ldots, W_n$ and F replaced by H. Then, by applying Proposition 2.6 rather than Proposition 2.3, it may be shown as in Remark 2.4 that the number of crossings of $H_{\boldsymbol{\theta}}$ and $H_{\boldsymbol{\eta}}$ must be *odd*. But it has been established in the proof of Proposition 2.5 that $H_{\boldsymbol{\theta}}$ and $H_{\boldsymbol{\eta}}$ can cross at most twice, hence they must cross exactly once as conjectured. (Note again that Proposition 2.5 follows from this result by the argument following (15).)

D.L. Ragozin has obtained some convincing numerical evidence that the above conjecture is valid when $n = 4$.

In the final proposition of this section, we return to the UCC for $F_{\boldsymbol{\theta}}$ and $F_{\boldsymbol{\eta}}$ when $\boldsymbol{\eta} = \bar{\boldsymbol{\theta}} \equiv (\bar{\theta}, \ldots \bar{\theta})$.

PROPOSITION 2.7. *(Diaconis and Perlman (1976), Shaked (1980)). If* $\boldsymbol{\eta} = \bar{\boldsymbol{\theta}}$, *the UCC is valid for every* $n \geq 2$.

PROOF. When $\boldsymbol{\eta} = \bar{\boldsymbol{\theta}}$, $\sum \eta_i W_i = \bar{\theta} \sum W_i = \bar{\theta}$, so

$$H_{\bar{\boldsymbol{\theta}}}(u) = \begin{cases} 0 & \text{if } u < \bar{\theta}, \\ 1 & \text{if } u > \bar{\theta}. \end{cases}$$

Thus $H_{\boldsymbol{\theta}} - H_{\bar{\boldsymbol{\theta}}}$ has exactly one sign change, so by (15), $F_{\boldsymbol{\theta}} - F_{\bar{\boldsymbol{\theta}}}$ has at most one sign change, hence exactly one sign change and exactly one zero on $(0, \infty)$. ||

We are grateful to the referee for pointing out the following extension of Proposition 2.7:

PROPOSITION 2.7a. *If* $\boldsymbol{\eta} = (1 - \lambda)\bar{\boldsymbol{\theta}} + \lambda \boldsymbol{\theta}$ *for* $0 \leq \lambda < 1$, *the UCC is valid for every* $n \geq 2$.

PROOF. Since $H_{\boldsymbol{\eta}}(u) = H_{\boldsymbol{\theta}}(\lambda^{-1}(u - (1 - \lambda)\bar{\theta}))$ when $\lambda > 0$, $H_{\boldsymbol{\theta}} - H_{\boldsymbol{\eta}}$ has exactly one sign change (at $u = \bar{\theta}$), so the argument for Proposition 2.7 remains applicable. ||

The following lemma was needed in the proof of Proposition 2.2.

LEMMA 2.8. *Define* $\varphi_t(u)$ *by (19). If* $t \geq (2\alpha + 1)\bar{\theta}$ *and* $0 < u < \bar{\theta}$, *then* $\varphi_t(u) < \varphi_t(2\bar{\theta} - u)$.

PROOF. As u increases from 0 to $\bar{\theta}$, $b \equiv \bar{\theta} u^{-1} - 1$ decreases from ∞ to 0. Since $t \geq (2\alpha + 1)\bar{\theta}$, the desired inequality will follow from the inequality

$$\frac{2\bar{\theta} - u}{u} < \exp\left\{\frac{2\bar{\theta}(\bar{\theta} - u)}{u(2\bar{\theta} - u)}\right\},$$

which is equivalent to

$$1 + 2b < \exp\left\{\frac{2b(1+b)}{1+2b}\right\}$$

and therefore to the inequality

$$f(b) \equiv \frac{2b(1+b)}{1+2b} - \log(1+2b) > 0 \tag{26}$$

for $b > 0$. But $f(0) = 0$ while

$$\frac{1}{4}(1+2b)^2 f'(b) = b^2 > 0,$$

hence (26) holds. ||

3. Location of the Unique Crossing Point When $\eta = \bar{\theta}$. It is a consequence of Proposition 2.7 that F_{θ} and $F_{\bar{\theta}}$ have a unique crossing point $t^* \equiv t^*(\theta, \bar{\theta}) \in (0, \infty)$ and that (10) holds when $\eta = \bar{\theta}$. In this section we present partial results regarding the location of t^*. Once again, without loss of generality we may assume that $\beta = 1$.

Our results are based on the following alternate representation of F_{θ} (compare to (13)):

$$\begin{aligned} F_{\theta}(t) &= E\left\{P[S_n \le t\left(\sum \theta_i W_i\right)^{-1} | \mathbf{W}]\right\} \\ &\equiv E\{G\left(t^{-1}\sum \theta_i W_i\right)\} \end{aligned} \tag{27}$$

where, for $0 < u < \infty$, $G \equiv G_{n\alpha}$ is defined by

$$G(u) = P[S_n \le u^{-1}] = \int_0^{u^{-1}} g_{n\alpha,1}(s)ds. \tag{28}$$

Since

$$G'(u) = -\frac{1}{\Gamma(n\alpha)} u^{-n\alpha-1} e^{-u^{-1}}, \tag{29}$$

$$G''(u) = \frac{1}{\Gamma(n\alpha)} u^{-n\alpha-2} e^{-u^{-1}}[(n\alpha+1) - u^{-1}], \tag{30}$$

it is immediate that G is strictly decreasing on $[0, \infty)$, strictly concave on $[0, (n\alpha + 1)^{-1}]$, and strictly convex on $[(n\alpha+1)^{-1}, \infty)$. Thus, for every $u \in [0, \infty)$ there exists a unique line L_u tangent to the graph of G at $(u, G(u))$, the equation of which is given by

$$L_u(v) = G(u) + G'(u)(v - u), \quad 0 \le v < \infty. \tag{31}$$

When $u < (n\alpha+1)^{-1}$, L_u is tangent to the graph of G from above, while when $u > (n\alpha+1)^{-1}$, L_u is tangent from below.

Because G is strictly concave on $[0,(n\alpha+1)^{-1}]$, clearly $L_u(0) > G(0) \equiv 1$ when $0 < u \le (n\alpha+1)^{-1}$, while

$$L_\infty(0) = G(\infty) - \lim_{u\to\infty}[uG'(u)] = 0.$$

Thus since

$$\frac{d}{du}[L_u(0)] = -uG''(u) < 0$$

for $u > (n\alpha+1)^{-1}$, there exists a unique positive number

$$\hat{u} \equiv \hat{u}_{n\alpha} > (n\alpha+1)^{-1} \tag{32}$$

such that $L_{\hat{u}}(0) = 1 = G(0)$. From (31) the point $\hat{u}$ is the unique positive solution to

$$G(\hat{u}) = 1 + \hat{u}G'(\hat{u}). \tag{33}$$

The line $L_{\hat{u}}$ is tangent to the graph of G at $(\hat{u}, G(\hat{u}))$ and elsewhere lies strictly below this graph, except at the point (0,1) where they coincide. In fact, for every point $u \ge \hat{u}$, L_u is tangent to the graph at $(u, G(u))$ and elsewhere lies strictly below the graph, i.e.,

$$u \ge \hat{u} \Longrightarrow L_u(v)\begin{cases} = G(v) & \text{if } v = u, \\ < G(v) & \text{if } v \ne 0,\ u. \end{cases} \tag{34}$$

If we set $t = \bar{\theta}u^{-1}$ with $u \ge \hat{u}$, it follows from (27), (34), and the linearity of $L_u(\cdot)$ that

$$\begin{aligned} F_{\boldsymbol{\theta}}(t) &= E\left\{G\left(u\bar{\theta}^{-1}\sum\theta_i W_i\right)\right\} \\ &> E\left\{L_u\left(u\bar{\theta}^{-1}\sum\theta_i W_i\right)\right\} \\ &= L_u\left(E(u\bar{\theta}^{-1}\sum\theta_i W_i)\right) \\ &= L_u\left(u\bar{\theta}^{-1}\left(\sum\theta_i\right)EW_1\right) \\ &= L_u(u) \\ &= G(u) \\ &= P[\bar{\theta}S_n \le t] \\ &= F_{\bar{\boldsymbol{\theta}}}(t). \end{aligned} \tag{35}$$

Here we have used the exchangeability of $(W_1,\ldots,W_n)$ and the fact that $\sum W_i = 1$. Since $u \ge \hat{u}$ iff $t \le \bar{\theta}\hat{u}^{-1}$, we have derived the following result:

PROPOSITION 3.1. *Let $\hat{u} \equiv \hat{u}_{n\alpha}$ denote the unique positive solution to (33), where $G \equiv G_{n\alpha}$ is given by (28). Then $F_{\boldsymbol{\theta}}(t) > F_{\bar{\boldsymbol{\theta}}}(t)$ whenever $t \leq \bar{\theta}\hat{u}_{n\alpha}^{-1}$, i.e.,*

(36) $$t^*(\boldsymbol{\theta}, \bar{\boldsymbol{\theta}}) > \bar{\theta}\hat{u}_{n\alpha}^{-1}$$ ||

Thus the *lower* tail probabilities of $\sum \theta_i Y_i$ are bounded below by those of $\bar{\theta} \sum Y_i$ for sufficiently small t. In order to show that the same is true of the *upper* tail probabilities when t is sufficiently large, we must modify (34) to show that $L_u(v)$ lies *above* the graph of G, at least for v in the support of $t^{-1} \sum \theta_i W_i$, for sufficiently small u.

For each $u \in (0, (n\alpha+1)^{-1})$ there exists a unique point $v(u) > (n\alpha+1)^{-1}$ such that $L_u(v(u)) = G(v(u))$, i.e., such that

(37) $$G(v(u)) = G(u) + (v(u) - u)G'(u).$$

For each u it is clear that

(38) $$L_u(v) \begin{cases} = G(v) & \text{if } v = u \text{ or } v = v(u) \\ > G(v) & \text{if } 0 < v < v(u) \text{ and } v \neq u. \end{cases}$$

For $u \in (0, (n\alpha + 1)^{-1})$ the function $v(u)$ is strictly decreasing and satisfies

$$v(0) = \infty, \quad v((n\alpha + 1)^{-1} = (n\alpha + 1)^{-1}.$$

Therefore, $v(u)/u$ strictly decreases from ∞ to 1, so there exists a unique point

(39) $$\tilde{u} \equiv \tilde{u}_{n\alpha,n} < (n\alpha + 1)^{-1}$$

such that

(40) $$v(\tilde{u}) = n\tilde{u}.$$

Hence

(41) $$u \leq \tilde{u} \Longrightarrow u\bar{\theta}^{-1} \sum \theta_i W_i \leq n\tilde{u} = v(\tilde{u}) \leq v(u),$$

so for $t = \bar{\theta}u^{-1}$ it follows from (38) and (41) that

$$\begin{aligned} F_{\boldsymbol{\theta}}(t) &= E\left\{G\left(u\bar{\theta}^{-1} \sum \theta_i W_i\right)\right\} \\ &< E\left\{L_u\left(u\bar{\theta}^{-1} \sum \theta_i W_i\right)\right\} \\ &= F_{\bar{\boldsymbol{\theta}}}(t) \end{aligned} \tag{42}$$

as in (35). Because $u \leq \tilde{u}$ iff $t \geq \bar{\theta}\tilde{u}^{-1}$, we thus have the following result:

PROPOSITION 3.2. *Let $\tilde{u} \equiv \tilde{u}_{n\alpha,n}$ denote the unique solution to (40) in the interval $(0,(n\alpha+1)^{-1})$, where $v(u)$ is defined by (37). Then $\bar{F}_{\boldsymbol{\theta}}(t) > \bar{F}_{\bar{\boldsymbol{\theta}}}(t)$ whenever $t \geq \bar{\theta}\tilde{u}_{n\alpha,n}^{-1}$, i.e.,*

$$t^*(\boldsymbol{\theta},\bar{\boldsymbol{\theta}}) < \bar{\theta}\tilde{u}_{n\alpha,n}^{-1}. \tag{43}$$ ||

To be useful, of course, Propositions 3.1 and 3.2 require estimates for $\hat{u}_{n\alpha}^{-1}$ and $\tilde{u}_{n\alpha,n}^{-1}$. To estimate the former, set $\nu = n\alpha$ and $x = \hat{u}_{\nu}^{-1}$, then use (29) to rewrite (33) as

$$\int_0^x s^{\nu-1}e^{-s}ds = \Gamma(\nu) - x^{\nu}e^{-x} \tag{44}$$

or, equivalently,

$$\int_x^{\infty} s^{\nu-1}e^{-s}ds = x^{\nu}e^{-x} \tag{45}$$

The substitution $w = s - x$ converts (45) to

$$\int_0^{\infty}\left(1+\frac{w}{x}\right)^{\nu} e^{-w}\frac{dw}{x+w} = 1, \tag{46}$$

then integration by parts yields

$$\int_0^{\infty}\left(1+\frac{w}{x}\right)^{\nu} e^{-w}dw = \nu + 1. \tag{47}$$

This integral strictly decreases from ∞ to 1 as x increases from 0 to ∞, hence x is the unique solution to (47).

A rough lower bound for $x \equiv \hat{u}_{\nu}^{-1}$ is obtained by expressing (47) as

$$\Gamma(\nu+1)E(W^{-1}+x^{-1})^{\nu} = \nu + 1 \tag{48}$$

where $W \sim G(\nu+1,1)$, then applying Jensen's inequality to obtain

$$\hat{u}_{\nu}^{-1} > \left[\left(\frac{\nu+1}{\Gamma(\nu+1)}\right)^{1/\nu} - \frac{1}{\nu+1}\right]^{-1} \sim (e-1)^{-1}\nu \doteq 0.58\nu \text{ as } \nu \to \infty. \tag{49}$$

A sharper bound may be obtained when $\nu > 1$ (in fact, equality holds when $\nu = 1$):

$$\hat{u}_{\nu}^{-1} > \left[\left(\frac{\nu+1}{\Gamma(\nu+1)}\right)^{1/\nu} - \frac{1}{\nu}\right]^{-1} \sim (e-1)^{-1}\nu \doteq 0.58\nu \text{ as } \nu \to \infty. \tag{49a}$$

A lengthier argument yields a better bound when $\nu > 1$:

$$\hat{u}_{\nu}^{-1} > e^{-(\nu-1)/2\nu}\nu^{\nu/(\nu+1)} \sim e^{-1/2}\nu \doteq 0.61\nu \text{ as } \nu \to \infty. \tag{50}$$

These bounds are crude, however, as seen from the following result and tabulation. (We warmly thank Russell Millar for suggesting this approach.)

PROPOSITION 3.3. *$\hat{u}_\nu^{-1} \sim \nu$ as $\nu \to \infty$.*

PROOF. By (45), $x \equiv \hat{u}_\nu^{-1}$ is the unique solution to the equation

$$f_\nu(x) \equiv \int_1^\infty y^{\nu-1} e^{-x(y-1)} dy - 1 = 0. \tag{51}$$

Since $f_\nu(\cdot)$ is strictly decreasing on $(0, \infty)$,

$$f_\nu(a) > 0 \Longrightarrow x > a. \tag{52}$$

Now define

$$\begin{aligned} f_\nu^*(x) &= x^\nu e^{-x} f_\nu(x)/\Gamma(\nu) \\ &= \frac{1}{\Gamma(\nu)} \int_x^\infty s^{\nu-1} e^{-s} ds - \frac{x^\nu e^{-x}}{\Gamma(\nu)} \\ &= P[G(\nu,1) \geq x] - \frac{x^\nu e^{-x}}{\Gamma(\nu)}. \end{aligned} \tag{53}$$

Then for any $0 < \delta < 1$, it follows from the Law of Large Numbers and Stirling's formula that

$$\begin{aligned} f_\nu^*((1-\delta)\nu) &= P[G(\nu,1)/\nu \geq 1-\delta] - \frac{((1-\delta)\nu)^\nu e^{-(1-\delta)\nu}}{\Gamma(\nu)} \\ &\sim 1 - \sqrt{\frac{\nu}{2\pi}}((1-\delta)e^\delta)^\nu \\ &\to 1 \end{aligned} \tag{54}$$

as $\nu \to \infty$, since $(1-\delta)e^\delta < 1$. Thus, by (52) and (54),

$$\liminf_{\nu \to \infty} (\nu \hat{u}_\nu)^{-1} \geq 1,$$

while $(\nu \hat{u}_\nu)^{-1} < (\nu+1)/\nu$ by (32), hence the asserted result follows. ||

The following tabulation obtained by Russell Millar indicates that the convergence of $(\nu \hat{u}_\nu)^{-1}$ to 1 is slow:

ν	$(\nu\hat{u}_\nu)^{-1}$
1	1.000
2	.809
5	.728
10	.730
20	.756
100	.840
1,000	.930
10,000	.973
100,000	.990

By Propositions 3.1 and 3.3, the crossing point $t^*(\boldsymbol{\theta},\bar{\boldsymbol{\theta}})$ satisfies

$$\liminf_{n\alpha\to\infty} \frac{t^*(\boldsymbol{\theta},\bar{\boldsymbol{\theta}})}{n\alpha\bar{\theta}} \geq 1, \quad \text{uniformly in } \boldsymbol{\theta}. \tag{55}$$

In order to estimate $\tilde{u}_{n\alpha,n}^{-1}$, set $\nu = n\alpha$ and $z = \tilde{u}_{\nu,n}^{-1}$, then combine (37) and (40) to obtain

$$\int_{z/n}^{z} s^{\nu-1}e^{-s}ds = (n-1)z^\nu e^{-z}. \tag{56}$$

The substitution $y = s/z$ converts (56) to

$$\int_{1/n}^{1} y^{\nu-1}e^{z(1-y)}dy = n-1. \tag{57}$$

Since the integral strictly increases from $\nu^{-1}(1-n^{-\nu})(< n-1)$ to ∞ as z increases from 0 to ∞, z is the unique solution to (57).

To apply Jensen's inequality, express (57) as

$$\frac{n^\nu - 1}{\nu n^\nu} E e^{z(1-Y)} = n-1, \tag{58}$$

where Y is a random variable with pdf $p(y)$ given by

$$p(y) = \begin{cases} \frac{\nu n^\nu}{n^\nu - 1} y^{\nu-1} & \text{if } \frac{1}{n} < y < 1 \\ 0 & \text{otherwise.} \end{cases} \tag{59}$$

Then

$$EY = \frac{\nu(n^{\nu+1}-1)}{(\nu+1)(n^{\nu+1}-n)},$$

hence

$$\begin{aligned} n-1 &> \frac{n^\nu - 1}{\nu n^\nu} \exp\left\{ z \left[1 - \frac{\nu(n^{\nu+1}-1)}{(\nu+1)(n^{\nu+1}-n)} \right] \right\} \\ &= \frac{n^\nu - 1}{\nu n^\nu} \exp\left\{ \frac{z}{\nu+1} \left[1 - \frac{\nu(n-1)}{(n^{\nu+1}-n)} \right] \right\}. \end{aligned}$$

It can be verified that the term in square brackets is positive, hence by taking logarithms we obtain the following upper bound for $z \equiv \tilde{u}_{\nu,n}^{-1}$:

$$\begin{aligned} \tilde{u}_{\nu,n}^{-1} &< (\nu+1)\left[1 - \frac{\nu(n-1)}{n^{\nu+1}-n}\right]^{-1} \log\left[\frac{\nu(n-1)}{1-n^{-\nu}}\right] \\ &\sim (\nu+1)\log[\nu(n-1)] \end{aligned} \tag{60}$$

as $n \to \infty$ or $\nu \to \infty$. But (60) is not sharp, as the next result indicates.

PROPOSITION 3.4. *$\tilde{u}_{\nu,n}^{-1} \sim \nu$ as $n \to \infty$ with α fixed.*

PROOF. By (57), $z \equiv \tilde{u}_{\nu,n}^{-1}$ is the unique positive solution to the equation

$$f_{\nu,n}(z) \equiv \int_{\frac{1}{n}}^{1} y^{\nu-1} e^{z(1-y)} dy - (n-1) = 0. \tag{61}$$

Since $f_{\nu,n}(\cdot)$ is strictly increasing,

$$f_{\nu,n}(a) > 0 \Longrightarrow z < a. \tag{62}$$

Now define

$$\begin{aligned} f_{\nu,n}^*(z) &= z^\nu e^{-z} f_{\nu,n}(z)/\Gamma(\nu) \\ &= P\left[\frac{z}{n} \le G(\nu,1) \le z\right] - \frac{(n-1)z^\nu e^{-z}}{\Gamma(\nu)} \end{aligned} \tag{63}$$

Then for any $0 < \delta < 1$ and $n \ge 2$,

$$\begin{aligned} f_{\nu,n}^*((1+\delta)\nu) &= P\left[\frac{1+\delta}{n} \le \frac{G(\nu,1)}{\nu} \le 1+\delta\right] - \frac{(n-1)((1+\delta)\nu)^\nu e^{-(1+\delta)\nu}}{\Gamma(\nu)} \\ &\sim 1-(n-1)\sqrt{\frac{\nu}{2\pi}}((1+\delta)e^{-\delta})^\nu \\ &\to 1 \end{aligned} \tag{64}$$

as $n \to \infty$, since $\nu = n\alpha$ and $(1+\delta)e^{-\delta} < 1$. Thus (62) and (64) together imply that

$$\limsup_{n\to\infty} (\nu \tilde{u}_{\nu,n})^{-1} \le 1,$$

while $(\nu\tilde{u}_{\nu,n})^{-1} > (\nu+1)/\nu$ by (39), so the result follows. ||

Once again, the convergence of $(\nu\tilde{u}_{\nu,n})^{-1}$ to 1 is slow. The following tabulations are given for the case $\alpha = 1$, so $n = \nu$:

ν	$(\nu\tilde{u}_{\nu,\nu})^{-1}$
2	1.88
5	1.92
10	1.87
20	1.72
100	1.39
1,000	1.14
10,000	1.05
100,000	1.02

Propositions 3.2 and 3.4 imply that for fixed α the crossing point $t^*(\boldsymbol{\theta}, \bar{\boldsymbol{\theta}})$ satisfies

(65) $$\limsup_{n\to\infty} \frac{t^*(\boldsymbol{\theta},\bar{\boldsymbol{\theta}})}{n\alpha\bar{\theta}} \le 1, \quad \text{uniformly in } \boldsymbol{\theta}.$$

Thus, by (55) and (65), we have derived the following approximation for the unique crossing point of $F_{\boldsymbol{\theta}}$ and $F_{\bar{\boldsymbol{\theta}}}$:

PROPOSITION 3.5. *$t^*(\boldsymbol{\theta}, \bar{\boldsymbol{\theta}}) \sim n\alpha\bar{\theta}$ as $n \to \infty$, uniformly in $\boldsymbol{\theta}$ for fixed $\bar{\theta}$ and α.* ||

The relation (7) in Section 1 follows immediately from this proposition.

The normal approximation to the distribution of $\sum Y_i$ suggests the conjecture that

(66) $$t^*(\boldsymbol{\theta}, \bar{\boldsymbol{\theta}}) = n\alpha\bar{\theta} + 0((n\alpha)^{1/2}) \quad \text{as } n\alpha \to \infty$$

uniformly in $\boldsymbol{\theta}$ for fixed $\bar{\theta}$. To support this conjecture, consider the behavior of $t^*(\boldsymbol{\theta}, \bar{\boldsymbol{\theta}})$ when

(67) $$\boldsymbol{\theta} = \boldsymbol{\theta}(n,k) \equiv (\underbrace{\frac{n}{k}, \ldots, \frac{n}{k}}_{k}, \underbrace{0, \ldots, 0}_{n-k}),$$

where $k \in \{1, \ldots, n-1\}$. For each real number c, define

(68) $$t_{n\alpha}(c) = n\alpha + c(n\alpha)^{1/2}.$$

Since $\bar{\theta} = 1$ for each k,

$$\begin{aligned} F_{\bar{\boldsymbol{\theta}}}(t_{n\alpha}(c)) &= P\left\{(n\alpha)^{-1/2}[G(n\alpha,1)-n\alpha] \le c\right\} \\ &\to \Phi(c) \end{aligned} \tag{69}$$

as $n\alpha \to \infty$, where Φ denotes the standard normal distribution function.

Now consider the following two special cases:

(i) $k\alpha \to k_0 < \infty$: here, as $n\alpha \to \infty$,

$$\begin{aligned} F_{\boldsymbol{\theta}}(t_{n\alpha}(c)) &= P[G(k\alpha,1) \le k\alpha + ck\alpha(n\alpha)^{-1/2}] \\ &\to P[G(k_0,1) \le k_0]. \end{aligned} \tag{70}$$

(ii) $k\alpha \to \infty$, $k/n \to \gamma \in [0,1)$: here,

$$\begin{aligned} F_{\boldsymbol{\theta}}(t_{n\alpha}(c)) &= P\left\{(k\alpha)^{-1/2}[G(k\alpha,1)-k\alpha] \le c(k/n)^{1/2}\right\} \\ &\to \Phi(c\gamma^{1/2}). \end{aligned} \tag{71}$$

Thus, if we define

$$c_0 = \begin{cases} \Phi^{-1}(P[G(k_0,1) \le k_0]) & \text{in case (i)}, \\ 0 & \text{in case (ii)}, \end{cases} \tag{72}$$

it follows from (69)–(71) that

$$\lim_{n\alpha\to\infty}[F_{\boldsymbol{\theta}}(t_{n\alpha}(c)) - F_{\bar{\boldsymbol{\theta}}}(t_{n\alpha}(c))] \begin{cases} > 0 & \text{if } c < c_0 \\ = 0 & \text{if } c = c_0 \\ < 0 & \text{if } c > c_0. \end{cases} \tag{73}$$

This implies that when (67) holds, then

$$t^*(\boldsymbol{\theta},\bar{\boldsymbol{\theta}}) = t_{n\alpha}(c_0) + o((n\alpha)^{1/2}) \quad \text{as } n\alpha \to \infty \tag{74}$$

in both special cases (i) and (ii), hence the conjecture (66) holds in these cases. In fact, in case (ii) it is true that

$$t^*(\boldsymbol{\theta},\bar{\boldsymbol{\theta}}) = n\alpha + o((n\alpha)^{1/2}) \quad \text{as } n\alpha \to \infty.$$

(This approach gives no information for the third special case where $\gamma = 1$, e.g., $k = n - 1$.)

The conjecture (66) likely follows from an appropriate uniform exponential bound for the tail probabilities of the weighted sums $\sum \theta_i Y_i$, but we do not pursue this here.

REFERENCES

BOCK, M.E., DIACONIS, P., HUFFER, F.W., and PERLMAN, M.D. (1987). Inequalities for linear combinations of gamma random variables. *Canadian J. Statist.* **15** 387–395.

DIACONIS, P. (1976). On general upper bounds for sums of the form $\sum \theta_i X_i$. Unpublished manuscript.

DIACONIS, P. and PERLMAN, M.D. (1976). Tail probabilities of sums of gamma random variables. Unpublished manuscript.

CHERNOFF, H. and LEHMANN, E.L. (1954). The use of maximum likelihood estimates in the χ^2 test for goodness of fit. *Ann. Math. Statist.* **25** 579–586.

GOOD, I.J. (1955). On the weighted combination of significance tests. *J. Roy. Statist. Soc. B* **17** 264–265.

JOHNSON, N.L. and KOTZ, S. (1970). *Continuous Univariate Distributions-2.* John Wiley and Sons, New York.

KARLIN, S. (1968). *Total Positivity.* Stanford University Press, Stanford, CA.

KARLIN, S., MICCHELLI, C.A., and RINOTT, Y. (1986). Multivariate splines: A probabilistic perspective. *J. Multiv. Anal.* **20** 69–90.

MARSHALL, A.W. and OLKIN, I. (1979). *Inequalities: Theory of Majorization and Its Applications.* Academic Press, New York.

SHAKED, M. (1980). On mixtures from exponential families. *J. Roy. Statist. Soc. B* **42** 192–198.

ZELEN, M. and JOEL, L.S. (1959). The weighted compounding of two independent significance tests. *Ann. Math. Statist.* **30** 885–895.

DEPARTMENT OF MATHEMATICS
HARVARD UNIVERSITY
CAMBRIDGE, MA 02138

DEPARTMENT OF STATISTICS GN–22
UNIVERSITY OF WASHINGTON
SEATTLE, WA 98195

AN OVERVIEW OF DEPENDENCY MODELS FOR CROSS-CLASSIFIED CATEGORICAL DATA INVOLVING ORDINAL VARIABLES

By Ruth Douglas[1] and Stephen E. Fienberg[1]

Carnegie Mellon University

In the late 1970's the popularity of loglinear and logistic model techniques for cross-classified categorical data led to a resurgence of interest in models and methods which directly incorporate information about the ordinal structure of the categories corresponding to the classification variables. In this paper we present an overview of some of the models for dependence that have been the focus of interest in this recent literature. In particular, we consider a class of association models extensively developed by Goodman and we examine order restrictions on parameters corresponding to the ordinal structure of the underlying variables. We attempt to summarize what is known about how these order restrictions for association and other models characterize monotonicity constraints on the underlying cross-classification probabilities or marginal totals. The principle context for our discussion is the dependency structure for two-dimensional ordinal contingency tables, but extensions to multi-dimensional tables that build on loglinear model ideas are relatively direct.

1. Introduction. The study of dependence among continuous random variables has a long history in statistics. The corresponding issue regarding dependency for categorical random variables also has a long history going back to the work of Yule (1900) and Pearson (1900); it has only been since the 1960's that a coherent and elaborate literature has developed. Much of the emphasis before this period was on the development of measures of association (e.g., see Goodman and

[1]Supported in part by NSF Grant SES–8701606 to Carnegie Mellon University.

AMS 1980 subject classifications. Primary 62H20; secondary 62H17.

Key words and phrases. Association models; goodness-of-fit test statistics; likelihood estimation; monotonicity constraints; odds ratios.

A version of this paper was presented at the *Symposium on Dependence in Statistics and Probability* August 1-5, 1987. Zvi Gilula, Michael Meyer, and Allan Sampson provided ideas and materials that helped to shape our perspective on this topic.

Kruskal, 1954, 1979; Kruskal, 1958). In the mid-1960's attention of researchers interested in categorical data shifted to the study of loglinear models and most of the resulting literature was focused on the dependency or interaction structures for nominal categorical variables, i.e., with unordered categories. That literature does not exclude ordinal variables, and as was noted in Fienberg (1982), the ordinal nature of some categorical variables is often crucial to the structural organization of categorical data subjected to loglinear analysis (e.g., triangular arrays, social mobility tables, and tables representing age-period-cohort structures). The key feature of these models involving ordinal structures is that they are not permutation invariant, i.e., the categories of the variables cannot be permuted in an arbitrary way without affecting the parameters describing the dependency structures.

The standard loglinear model approach to two-dimensional $I \times J$ contingency tables represents the probability, P_{ij}, of an observation falling into the ith row and jth column as

$$\log P_{ij} = u + u_{1(i)} + u_{2(j)} + u_{12(ij)}, \tag{1}$$

where

$$\sum_i u_{1(i)} = \sum_j u_{2(j)} = \sum_i u_{12(ij)} = \sum_j u_{12(ij)} = 0, \tag{2}$$

and

$$\sum_i \sum_j P_{ij} = 1. \tag{3}$$

The model may be rewritten in multiplicative form as

$$P_{ij} = \alpha_i \beta_j \exp\{u_{12(ij)}\}. \tag{4}$$

The traditional approach for two-dimensional tables has been to treat the interaction terms, $u_{12(ij)}$, as being unrestricted (the so-called *saturated model*) or to set them equal to zero, thereby assuming that row classification is *independent* of column classification. Neither approach reflects any ordinal structure that may be present in the row or column categories.

In the 1970's two separate approaches to the study of dependency structures for ordinal variables emerged. The first of these was linked to the correspondence analysis approach developed by Benzécri and his associates (e.g. see Benzécri, et al. 1973; Greenacre, 1984) and focused on correlational-like ideas. This work was later picked up by Goodman (1981, 1985) and Gilula (1982). The second approach, proposed independently in the 1960's by Rasch (see Christiansen, 1966) and by Fienberg (1968), was developed extensively by Haberman (1974a, 1974b) and Goodman (1979, 1981, 1984, 1985, 1986) and linked in a formal way to ordinal variables by Agresti (1984), Agresti and Chuang (1985), and Fienberg (1982).

The association-model approach focuses attention on the interaction parameters, $u_{12(ij)}$, in model (1) and models them in terms of a reduced number of parameters that may be chosen to reflect ordinal structure. In this review, we concentrate our attention on these models and describe aspects of them that explicitly incorporate montonicity constraints corresponding to ordinal structures.

In Section 2 we describe the class of association models developed by Goodman, and in Section 3 we briefly outline the related correlation models. Then in Section 4 we turn to order restrictions imposed upon the parameters of association-models from Section 2.

In the final section of this paper, we briefly consider extensions to multidimensional tables, some of which are relatively direct.

2. Association Models. As we mentioned in the introduction, the class of association models was proposed by Rasch (see Christiansen, 1966) and Fienberg (1968) as a categorical analogue to the Tukey's one-degree-of-freedom model for nonadditivity and its generalizations. The first careful development of these models and their formal linkage to ideas on cross-product or odds ratios were given by Goodman (1979). Later elaborations by Goodman (1981, 1985) led to the general model described below. We use Goodman's notation wherever possible. Equivalent models and special cases have been formulated by several other authors (see for example the stereotype model by Anderson (1984), and the general base-comparison logit model by Cox and Chuang (1984)).

2.1. Model Formulation. Goodman (1985) puts forth the following reparametrization of the standard loglinear model which he refers to as the *saturated RC association model*:

$$P_{ij} = \alpha_i \beta_j \exp\{\sum_{m=1}^{M} \phi_m \mu_{im} \nu_{jm}\}, \tag{5}$$

where

- $\{\mu_{im}\}$ and $\{\nu_{jm}\}$ are standardized row scores and standardized column scores for row category i and column category j, respectively (these are parameters to be estimated from the data);
- the $\{\phi_m\}$ are measures of the "intrinsic association" (if $\phi_m = 0$ for $m = 1, 2, \ldots, M$ then the table exhibits independence of rows and columns);

(6)•
$$M = \min(I-1, J-1).$$

Furthermore for m and $m^* = 1, 2, \ldots, M$, with $m \neq m^*$ we have the following identifying restrictions:

$$\sum_{i=1}^{I} \mu_{im} P_{i+} = \sum_{j=1}^{J} \nu_{jm} P_{+j} = 0, \tag{7}$$

$$\sum_{i=1}^{I} \mu_{im}^2 P_{i+} = \sum_{j=1}^{J} \nu_{jm}^2 P_{+j} = 1, \tag{8}$$

$$\sum_{i=1}^{I} \mu_{im}\mu_{im^*} P_{i+} = \sum_{j=1}^{J} \nu_{jm}\nu_{jm^*} P_{+j} = 0. \tag{9}$$

These are the marginal-weight versions of Becker and Clogg's (1989) generalized identification restrictions

$$\sum_{i=1}^{I} h_i \mu_{im} = \sum_{j=1}^{J} g_j \nu_{jm} = 0, \tag{10}$$

$$\sum_{i=1}^{I} h_i \mu_{im}^2 = \sum_{j=1}^{J} g_j \nu_{jm}^2 = 1, \tag{11}$$

where the $\{h_i\}$ and $\{g_j\}$ are row and column category weights, respectively, and some restrictions are applied to the cross dimension correlations

$$\rho_{m,n} = \sum_{i=1}^{I} h_i \mu_{im} \mu_{in} \tag{12}$$

and

$$\sigma_{m,n} = \sum_{j=1}^{J} g_j \nu_{jm} \nu_{jn}. \tag{13}$$

Becker and Clogg (1989) point out the importance of weighting systems both in measuring association and in comparing sets of contingency tables, and suggest some other possible choices for the weights.

This saturated RC model is essentially just an explicit rewriting of the original loglinear model of expressions (1) to (4) where the interaction terms, $\{u_{12(ij)}\}$ and the ANOVA-like constraints have been replaced by the sums

$$\sum_{m=1}^{M} \phi_m \mu_{im} \nu_{jm}$$

and somewhat different constraints.

If $\phi_m = 0$ for $m = M^* + 1, \ldots, M$, then we get the *unsaturated RC model* of order M^*. When $M^* = 1$, expression (5) reduces to

$$P_{ij} = \alpha_i \beta_j \exp\{\phi \mu_i \nu_j\}, \tag{14}$$

which is model II of Goodman (1979) expressed in terms of identifiable parameters. Expression (14) is what Goodman refers to as the general multiplicative row-and-column-effects (RC) model when the $\{\mu_i\}$ and $\{\nu_j\}$ are unspecified. Two special cases are (i) the row-effects (R) model when the $\{\mu_i\}$ are parameters and the $\{\nu_j\}$ are known, or they are ordered and the spacing between them is specified; (ii) the column-effects (C) model when the $\{\nu_j\}$ are parameters and the $\{\mu_i\}$ are known, or they are ordered and the spacing between them is specified. Gilula, Krieger, and Ritov (1988) provide an interpretation of ϕ as a measure of *stochastic order entropy.*

When the *orders* of the row and column categories are specified, and the rows and columns are appropriately ordered, the U association model may be considered. In the case in which the rows are equally spaced and the columns are equally spaced, we have (in Goodman's (1985) notation):

$$(15)\qquad \mu_i - \mu_{i+1} = \Delta', \qquad i = 1, 2, \ldots, I-1,$$

$$(16)\qquad \nu_j - \nu_{j+1} = \Delta'', \qquad j = 1, 2, \ldots, J-1.$$

The *local log odds ratios*, for the cells in the 2×2 subtables formed from the cells in adjacent rows and columns, are

$$\begin{aligned} \log \theta_{ij} &= \log \frac{P_{ij} P_{i+1,j+1}}{P_{i,j+1} P_{i+1,j}}, \\ (17) \qquad &= \phi(\mu_i - \mu_{i+1})(\nu_j - \nu_{j+1}), \ i = 1, 2, \ldots, I-1, \ j = 1, 2, \ldots, J-1. \end{aligned}$$

For the equal spacing model of (15) and (16), we can rewrite expression (17) as

$$\begin{aligned} \log \theta_{ij} &= \log \frac{P_{ij} P_{i+1,j+1}}{P_{i,j+1} P_{i+1,j}} \\ &= \phi \Delta' \Delta'' \\ (18) \qquad &= \text{constant}. \end{aligned}$$

The term *linear-by-linear association* is used to denote the generalized form of this model, for which

$$(19)\qquad \mu_i - \mu_{i+1} = \Delta_i' \ i = 1, 2, \ldots, I-1,$$

$$(20)\qquad \nu_j - \nu_{j+1} = \Delta_j'' \ j = 1, 2, \ldots, J-1,$$

but where the spacing values, $\{\Delta_i'\}$ and $\{\Delta_j''\}$ are known. For these association models there is an explicit (known) monotonic structure introduced by the fixed spacings.

The various models with their degrees of freedom are as follows:

Model	d.f.
Independence	$(I-1)(J-1)$
U	$(I-1)(J-1)-1$
R	$(I-1)(J-2)$
C	$(I-2)(J-1)$
RC	$(I-2)(J-2)$

2.2. Estimation of Parameters. Let the observed value in the ijth cell be x_{ij}. Then the maximum likelihood estimate $\hat{m}_{ij}$ of the expected frequency $m_{ij} = NP_{ij}$ for the RC association model will satisfy

$$\hat{m}_{i+} = x_{i+},\ i = 1, 2, \ldots, I, \tag{21}$$
$$\hat{m}_{+j} = x_{+j},\ j = 1, 2, \ldots, J, \tag{22}$$
$$\sum_j \hat{\nu}_j \hat{m}_{ij} = \sum_j \hat{\nu}_j x_{ij},\ i = 1, 2, \ldots, I, \tag{23}$$
$$\sum_i \hat{\mu}_i \hat{m}_{ij} = \sum_i \hat{\mu}_i x_{ij},\ j = 1, 2, \ldots, J, \tag{24}$$

when the x_{ij} follow any of the standard contingency table sampling models, i.e., Poisson, multinomial, product-multinomial (see, for example, Bishop, Fienberg, and Holland, 1975).

Goodman (1979) suggests the following iterative procedure for solving the likelihood equations. Let

$$\rho_i = \mu_i - \bar{\mu}, \tag{25}$$
$$\sigma_j = \nu_j - \bar{\nu}, \tag{26}$$
$$\bar{\mu} = \sum_i \mu_i / I, \tag{27}$$
$$\bar{\nu} = \sum_j \nu_j / J. \tag{28}$$

We denote the values of the estimates at any given stage in the iterative procedure by α_i^*, β_j^*, μ_i^*,and ν_j^*. Then we update the estimates one at a time, each update being followed by a recalculation of the values of m_{ij}. Denote the current expected frequencies by m_{ij}^*, then we replace $(m_{ij}^*, \alpha_i^*, \beta_j^*, \mu_i^*, \nu_j^*)$ by $(m_{ij}^{**}, \alpha_i^{**}, \beta_j^{**}, \mu_i^{**}, \nu_j^{**})$ where

$$\alpha_i^{**} = \alpha_i^* x_{i+} / m_{i+}^*, \tag{29}$$
$$\beta_j^{**} = \beta_j^* x_{+j} / m_{+j}^*, \tag{30}$$
$$\mu_i^{**} = \mu_i^* (\sum_j \sigma_j^* (x_{ij} - m_{ij}^*)) / (\sum_j \sigma_j^{*2} m_{ij}^*), \tag{31}$$

$$(32) \qquad \nu_j^{**} = \nu_j^*(\sum_i \rho_i^*(x_{ij} - m_{ij}^*))/(\sum_i \rho_i^{*2} m_{ij}^*).$$

Here ρ_i^*, σ_j^*, ρ_i^{**}, σ_j^{**}, have the obvious meanings. The algorithm appears to work in practice but to our knowledge there is no formal proof for its convergence. Becker and Clogg (1989) extend this algorithm to deal with the case in which K two way tables are considered and compared.

Alternative approaches to maximum likelihood estimation involve the use of general nonlinear maximization routines and other variants of standard Newton Raphson algorithms (e.g., see Chuang, 1980). See also Gilula (1982) who discusses the use of the singular value decomposition of the local odds ratio matrix to estimate the model parameters. It should, however, be noted that thus far in the literature, necessary and sufficient conditions for the existence of maximum likelihood estimates for the RC model have not been formulated.

2.3. Ordering Properties Implied by the Association Models. At this stage, we defer discussion of estimation under order restrictions. It is worth noting that tables fit by these models exhibit ordering properties which relate directly to traditional notions of dependence. We summarize some of these properties as they appear in the categorical data literature, and re-express them in the language of the dependence literature.

Let A and B denote the variables corresponding to rows and columns in the contingency table. Under the RC model, the association is isotropic, that is, rows and columns can be ordered in such a way that the *local odds ratios* have the property:

$$(33) \qquad \theta_{ij} = \frac{P_{ij}P_{i+1,j+1}}{P_{i+1,j}P_{i,j+1}} \geq 1 \quad i = 1, \ldots, I-1, \quad j = 1, \ldots, J-1.$$

Tables possessing the property (33) are *totally positive of order 2,* denoted TP_2 (see Schriever, 1986, or Gill and Schriever, 1987).

Agresti (1984) defines the local-global odds ratio

$$(34) \qquad \theta_{ij}^r = \frac{(\sum_{k\leq j} P_{ik})(\sum_{k>j} P_{i+1\,k})}{(\sum_{k>j} P_{ik})(\sum_{k\leq j} P_{i+1\,k})}, \quad j = 1, 2, \ldots, J-1.$$

These odds ratios are local in the row variable, but "global" in the column variable, and may be defined for any two rows, a and b, rather than adjacent rows i and $i+1$. Analogous global-local odds ratios, global in the row variable and local in the column variable may also be defined for columns j and $j+1$ i.e.,

$$(35) \qquad \theta_{ij}^c = \frac{(\sum_{k\leq i} P_{kj})(\sum_{k>i} P_{k\,j+1})}{(\sum_{k>i} P_{kj})(\sum_{k\leq i} P_{k\,j+1})}, \quad i = 1, 2, \ldots, I-1.$$

Then $\theta_{ij}^r > 1$ for each j implies that the conditional distribution in row $i+1$ is stochastically larger than the conditional distribution in row i, while $\theta_{ij}^c > 1$ for each i implies that the conditional distribution in row $j+1$ is stochastically larger than the conditional distribution in row j.

In fact, when the rows and columns are ordered appropriately such that (33) holds, we have the following stochastic ordering relationships (Goodman, 1981):

(i) The conditional distribution for the ith row of the table, P_{ij}/P_{i+} is stochastically smaller than the conditional distribution for the i'th row of the table if $i' > i$, i.e.,

$$1 \leq i < i' \leq I \Longrightarrow P(B \leq j|A = i) \geq P(B \leq j|A = i') \qquad j = 1, 2, \ldots, J. \tag{36}$$

Barlow and Proschan (1981) denote this property by *SI(B|A)*.

(ii) The conditional distribution for the jth column of the table, P_{ij}/P_{+j} is stochastically smaller than the conditional distribution for the j'th row of the table if $j' > j$, i.e.,

$$1 \leq j < j' \leq J \Longrightarrow P(A \leq i|B = j) \geq P(A \leq i|B = j') \qquad i = 1, 2, \ldots, I. \tag{37}$$

Barlow and Proschan denote this property by *SI(A|B)*.

Schriever (1983, 1986) refers to the variables A and B as being *double regression dependent of order 1 (DR_1)* where the order relationships (34) and (35) hold. Thus tables satisfying the RC association models are double regression dependent up to a permutation of rows and columns. This property Schriever refers to as *order dependence of order 1.* Schriever (1983, 1986) carries these arguments further noting that, for the RC association model with rows and columns reordered so that μ_i and ν_j are increasing in their indices, the table of probabilities is totally positive of order M where $M = min(I-1, J-1)$, i.e., the table is TP_M.

While the RC association model can always lead to a reordering of rows and columns such that TP_2 implies conditions (34) and (35), the reverse is not true. Thus there exist tables exhibiting order dependence for which the RC association model does not hold. This is a special case of results due to Schriever (1986) i.e.:

$$TP_k \Longrightarrow DR_{k-1}, \tag{38}$$

Thus for k=2, we get the implication that TP_2 implies DR_1 but the reverse is not necessarily true (except for $I \times 2$ and $J \times 2$ tables) since any rank 2 probability table can have its rows and columns permuted such that it is DR_1 (Schriever, 1986).

At this point, we note that there is clearly a hierarchical relationship between the dependence notions that have emerged thus far. In fact, we have identified the upper levels of Barlow and Proschan's (1975) tree of notions of bivariate dependence. These, and other dependence concepts forming nodes of the tree may be re-expressed in terms of conditions on the odds ratios of the contingency table or various subtables thereof. See Douglas et al. (1990) for a more detailed treatment of these linkages in tree-like form. It is relatively straightforward to elicit the implication structure once we view the dependence concepts in the contingency table framework.

Agresti, Chuang, and Kezouh (1987) show that for the row (columns) association model, since the estimated conditional distributions in the rows (columns) are stochastically ordered according to the values of the $\{\hat{\mu}_i\}$ and the $\{\hat{\nu}_j\}$, then the $\{\hat{\mu}_i\}$ and the $\{\hat{\nu}_j\}$ have the same ordering as the sample row and column means. This result follows from a rewriting of the likelihood equations from expressions (21), (22), (23), and (24).

3. Correlation Models. For completeness, we describe briefly the correlation models corresponding to the association models in Section 2 above. Goodman (1985, 1986) presents a much more careful treatment of these models and their relationships to the association models. He is quick to point out that although the association and correlation approaches are related not only may they yield different results, but also one approach may be preferable over the other in specific settings. The models are identical under independence of the row and column variables, and turn out to be reasonable approximations of each other when the intrinsic association (ϕ_m above) and correlation parameters (λ_m below) are close to zero in value.

The *saturated RC canonical correlation model* may be formulated as

$$P_{ij} = P_{i+}P_{+j}(1 + \sum_{m=1}^{M} \lambda_m x_{im} y_{jm}) \tag{39}$$

where

- x_{im} and y_{jm} are standardized row scores and standardized column scores respectively, to be estimated from the data.
- λ_m measures the correlation between x_{im} and y_{jm} and

$$\lambda_1 \geq \ldots \geq \lambda_M, \tag{40}$$

  and

$$M = \min(I-1, J-1). \tag{41}$$

- Furthermore for $m = 1, 2, \ldots, M$, we have the following identifying restrictions:

(42) $$\sum_{i=1}^{I} x_{im} P_{i+} = \sum_{j=1}^{J} y_{im} P_{+j} = 0,$$

(43) $$\sum_{i=1}^{I} x_{im}^2 P_{i+} = \sum_{j=1}^{J} y_{im}^2 P_{+j} = 1,$$

(44) $$\sum_{i=1}^{I} x_{im} x_{im^*} P_{i+} = \sum_{j=1}^{J} y_{jm} y_{jm^*} P_{+j} = 0.$$

Maximum likelihood estimation is direct for these saturated correlation models; not so for their unsaturated counterparts. Estimation for the unsaturated models is described in Goodman (1985).

The saturated RC canonical correlation model is simply a reparametrization of the *saturated RC correspondence analysis* model (see, e.g., Greenacre, 1984), i.e.

(45) $$P_{ij} = P_{i+} P_{+j} (1 + \sum_{m=1}^{M} x'_{im} y'_{jm} / \lambda_m),$$

where $x'_{im} = \lambda_m x_{im}$ and $y'_{jm} = \lambda_m y_{jm}$.

Unsaturated RC correlation models, with $\lambda_m = 0$ for $m = m^* + 1, \ldots, M$ and *U correlation models* may be formulated as in the association approach. Their order restriction properties have been studied in depth by Schriever (1983, 1984). For example, the RC correlation model with $m^* = 1$ also demonstrates order dependence of order 1 (i.e. expressions (34) and (35) hold after a suitable reordering of rows and columns). Moreover, Schriever (1983, 1986) proves the following result:

THEOREM. *Suppose that the row and column variables, A and B are double regression dependent. Then the first set of correspondence analysis row and column scores , that is , for m=1, can be chosen to satisfy*

(46) $$x_{11} \leq x_{21} \ldots \leq x_{I1}, \quad y_{11} \leq y_{21} \leq \cdots \leq y_{J1}.$$

Strict inequalities in (30) and (31) imply strict inequalities in (46).

Schriever also generalizes this theorem to higher order sets of scores (i.e. $m > 1$). Gilula, Krieger, and Ritov (1988) provide an interpretation of λ_1 for the RC model with $m^* = 1$, as a measure of *stochastic order extremity.* They also note the link to Kimeldorf and Sampson's (1978) coefficient of monotonic dependence.

Canonical correlation analysis has also been used to develop tests for independence of rows and columns. Haberman (1981) assigns scores ϕ_i and ψ_j to row category i and column category j respectively, and maximizes the correlation

$$R_1 = \sum_{i=1}^{I}\sum_{j=1}^{J} p_{ij}\phi_i\psi_j \tag{47}$$

where $p_{ij} = n_{ij}/N$, the observed relative frequency in cell j. NR_1^2 is shown to be approximately asymptotically distributed according to the distribution of the maximum eigenvalue of a central Wishart matrix with $J-1$ degrees of freedom. Approximate critical values for this distribution are available from existing tables.

In the case in which we consider the $I \times J$ table to represent observations on a discrete bivariate random vector (X, Y) taking values (i, j), $i = 1, \ldots, I$, $j = 1, \ldots, J$, Sethuraman (1977) has shown the above distribution to be the limiting distribution of nR_n^*. Here R_n^* is the sample Renyi maximum correlation

$$R_n^* = R(x, y) = \max_{f(x), g(y)} \rho(f(x), g(y)) \tag{48}$$

where

- the sample consists of n observations on the bivariate variable (X, Y),
- the maximum is over all functions f of X and g of Y such that $\mathrm{E}f^2$ and $\mathrm{E}g^2$ are finite,
- ρ denotes correlation, and
- the following conditions are satisfied:
    1. if $p_i = P(X = i)$, $I \leq J$, the $I \times I$ matrix with diagonal elements $p_i - p_i^2$, and off diagonal elements $-p_i p_j$ is of rank $(I-1)$, and
    2. $R(X, Y) = 0$, i.e., X and Y are independent.

4. Order-Restricted Association Models. In this section we consider association models of the form (5) in which one or both of the following constraints are assumed to hold:

$$\mu_1 \leq \mu_2 \leq \ldots \leq \mu_I, \tag{49}$$

$$\nu_1 \leq \nu_2 \leq \ldots \leq \nu_J. \tag{50}$$

Before detailing parameter estimation and goodness of fit procedures, we present three important illustrations of these models in order to better familiarize the reader with their structure.

4.1. The Stereotype Model. Anderson's (1984) stereotype model is a special case of the qualitative logistic regression model

$$\Pr(y = y_s \mid z) = \frac{\exp(\eta_{0s}^* + \eta_s^T z)}{\sum_{t=1}^{k} \exp(\eta_{0t}^* + \eta_t^T z)}, \quad s = 1, \ldots, k, \tag{51}$$

where

- y is an ordered categorical response with categories $y_1 \dots y_k$.
- $\eta_s^T = (\eta_{s1} \dots \eta_{sk})$ gives the regression coefficients for the odds of $y = y_s$ relative to $y = y_k$.

In the stereotype model, the $\{\eta_s\}$ are taken to be parallel, i.e.,

$$\eta_s = -\xi_s \eta, \quad s = 1, \dots, k, \tag{52}$$

and the $\{\xi_s\}$ are taken to be monotone decreasing, i.e.,

$$1 = \xi_1 > \xi_2 > \dots > \xi_k = 0. \tag{53}$$

The resulting model is expressible as

$$\Pr(y = y_s \mid z) = \frac{\exp(\eta_{0s}^* - \xi_s \eta^T z)}{\sum_{t=1}^k \exp(\eta_{0t}^* - \xi_t \eta^T z)} \quad s = 1, \dots, k. \tag{54}$$

This model is one dimensional in that only one linear function, viz. $\eta^T z$, is required to describe the relationship between y and z for all categories y_s. The importance of this model lies not only in the fact that it is ordered, but also that it can be extended to incorporate a multidimensional regression relationship. Anderson applies this model to the $2 \times k$ contingency table, showing that the stereotype model can be applied if and only if the cross product ratios are monotone increasing or decreasing.

We can apply the stereotype model to the $I \times J$ table if we consider the ordered categorical response y with categories $y_1, \dots, y_J$ and the single predictor, z, with values $z_1, \dots, z_I$. Denoting the cell probabilities by P_{ij}, we get

$$\frac{P_{ij}}{P_{iJ}} = \frac{\Pr(y = y_j \mid z = z_i)}{\Pr(y = y_J \mid z = z_i)} = \exp(\eta_{0j}^* - \xi_j \eta z_i). \tag{55}$$

Thus

$$\log(\frac{P_{ij}}{P_{iJ}}) = \eta_{0j}^* - \xi_j \eta z_i. \tag{56}$$

The RC association model of expression (5) gives

$$\log(\frac{P_{ij}}{P_{iJ}}) = \log(\eta_j - \eta_J) + \xi \mu_i (\nu_j - \nu_J). \tag{57}$$

Thus we have that Anderson's model is really just the RC association model with the monotonicity constraint of expression (53) and the following correspondence of component parameters:

Stereotype Model	RC model
β_{0j}^*	$\beta_j - \beta_J$
ϕ_j	$\nu_j - \nu_J$
z_i	μ_i
β	ϕ

The stochastic ordering properties of expressions (34) and (35) described in Section 2.3 above are noted by Anderson for his model.

4.2. The Proportional Odds Model. Another regression-like model for the analysis of ordinal data is proposed by McCullagh (1980), who distinguishes clearly between explanatory and response variables. The response has k ordered categories, with probabilities $\pi_1(\underline{x}), \pi_2(\underline{x}), \ldots \pi_k(\underline{x})$, where $\underline{x}$ is the vector of covariates. The proportional odds model is expressed as

$$\log \frac{\gamma_j(\underline{x})}{1-\gamma_j(\underline{x})} = \theta_j - \underline{\beta}^T \underline{x} \quad 1 \leq j < k \tag{58}$$

where

- $\gamma_j(\underline{x}) = \pi_1(\underline{x}) + \ldots \pi_j(\underline{x})$
- $\underline{\beta}$ is the vector of unknown parameters.

The model is thus a cumulative logit model and induces the stochastic ordering properties of expressions (34) and (35) as follows. The difference: $\text{logit}(\gamma_j(\underline{x}_1))-\text{logit}(\gamma_j(\underline{x}_2))$ is equal to $\underline{\beta}^T(\underline{x}_2 - \underline{x}_1)$, thus constant over j. McCullagh denotes this difference by Δ. Since the logit function is monotonic, the sign of Δ determines whether $\gamma_j(\underline{x}_1) > \gamma_j(\underline{x}_2)$ or $\gamma_j(\underline{x}_1) < \gamma_j(\underline{x}_2)$ for all j.

4.3. The Monotone Scores Association Model. Chuang and Agresti (1986) give a detailed treatment of a model in which the row variable is nominal, and the column variable is regarded as an ordinal response, with score parameters for the columns constrained to be monotone increasing. The model is

$$\log P_{ij} = \mu + a_i + b_j + \mu_i \nu_j, \tag{59}$$

where

$$\sum_{i=1}^{I} a_i = \sum_{j=1}^{J} b_j = \sum_{i=1}^{I} \mu_i = 0, \tag{60}$$

and

$$1 = \nu_1 \leq \ldots \leq \nu_J = J, \tag{61}$$

Model (58) is simply the RC association model of Section 2, with the added ordering constraint (61). This constraint produces a stochastic ordering among the response distributions of the rows. Suppose that rows correspond to two drugs a and b, with $\mu_a > \mu_b$. Then, because of the ordering of the $\{\nu_j\}$, the log odds ratios for adjacent responses,

$$\log \frac{P_{aj} P_{b,j+1}}{P_{a,j+1} P_{bj}} = (\mu_b - \mu_a)(\nu_{j+1} - \nu_j), \tag{62}$$

are nonnegative, and the response distribution for the bth drug, $\{P_{bj}/P_{b+}\}$ is stochastically greater than that for the ath drug, $\{P_{aj}/P_{a+}\}$. Chuang and Agresti (1986) use the model to analyze a data set in which the column variable is the response (on an ordinal scale) to treatment with different drugs (the row variables).

4.4. Parameter Estimation. In the monotone scores case we are essentially concerned with fitting the RC association model of expression (5) under a monotonicity constraint on the $\{\nu_j\}$. Here we consider estimation for the order restricted R, C, and RC models i.e. the R model under the restriction $\mu_1 \leq \ldots \leq \mu_I$, the C model under the restriction $\nu_1 \leq \ldots \leq \nu_J$, and the RC model under one or both of the preceding restrictions. Results will be stated for the most general case when available, i.e., in terms of the RC model; if not, then in terms of the R model (in which case it is implied that they hold analogously for the C model) or the C model (in which case it is implied that they hold analogously for the R model).

If the true parameters are strictly monotone, i.e. they do not lie on the boundary of the parameter space, then the estimates for the order-restricted RC model and the RC model are asymptotically equivalent.

For estimation under order restrictions, Goodman (1985) discusses an approximation yielding an ordered solution for the R or C models with $\phi > 0$ when the actual maximum likelihood estimates for the unrestricted model, are not in the correct order. He notes that for each violated restriction the likelihood is maximized on the boundary where these adjacent values are equal to each other. Then in his iterative procedure, he combines rows (or columns) corresponding to pairs of scores violating the order constraints, and refits the model. This, in effect, enforces an equality constraint on the score parameters for the rows (columns) concerned. Thus the order restricted solution is the same as the ordinary ML solution for the appropriately collapsed table.

For the R model, the following necessary and sufficient conditions that completely characterize this collapsing are given in Agresti, Chuang, and Kezouh (1987):

$$\hat{m}^*_{i+} = x_{i+} \quad i = 1, \ldots, I, \tag{63}$$

$$\hat{m}^*_{+j} = x_{+j} \quad j = 1, \ldots, J, \tag{64}$$

$$\sum_{i \leq b}(\sum_j \nu_j \hat{m}^*_{ij}) \leq \sum_{i \leq b}(\sum_j \nu_j x_{ij}) \quad b = 1, \ldots, I, \tag{65}$$

$$\sum_{i \leq r_k}(\sum_j \nu_j \hat{m}^*_{ij}) = \sum_{i \leq r_k}(\sum_j \nu_j x_{ij}) \quad k = 1, \ldots, a, \tag{66}$$

where $\{r_1, \ldots, r_a\}$ are such that

$$\hat{\mu}^*_1 = \ldots = \hat{\mu}^*_{r_1} < \hat{\mu}^*_{r_1+1} = \ldots = \hat{\mu}^*_{r_2} < \ldots < \hat{\mu}^*_{r_{a-1}+1} = \ldots = \hat{\mu}^*_{r_a}. \tag{67}$$

A heuristic argument for conditions (65) and (66) may be formulated as follows: In the unrestricted case the inequality (65) is an equality (23). When order constraints are applied, the restricted maximum is lower than the unrestricted maximum; this is manifested by inequalities in the likelihood equations with equalities only on the boundaries of the level sets. For the RC model, however, there are no sufficient conditions. This is due to the fact that, because the RC model is not loglinear, the log likelihood is not necessarily concave.

Agresti and Chuang (1985) and Agresti, Chuang, and Kezouh (1987) note that the partition for the order restricted ML solution $R_1, \ldots, R_a$ above, where $R_k = \{r_{k-1}+1, \ldots, r_k\}$, is identical to the partition of level sets obtained in using the *Pool Adjacent Violators* algorithm *(PAV)* to obtain the regression of the sample row means (weighted by the row totals) in the class of functions isotonic with respect to the simple order in the rows (see Barlow, et al., 1972).

Gilula (personal communication) has pointed out the kinds of problems that come with the *PAV* approach. He considers the following 3 by 3 table:

1	1	1
1	2	1
1	1	1

and notes what happens when we fit the row-effects model of expression (14) with the following fixed values of $\{\nu_j\}$ under constraint (61):

$$\nu_1 = 0, \quad \nu_2 = 1, \quad \nu_1 = 1.$$

Then the unrestricted MLE's of $\{\mu_i\}$ are:

$$\hat{\mu}_1 = -0.2582, \quad \hat{\mu}_2 = 0.3873, \quad \hat{\mu}_3 = -0.2582$$

and

$$\hat{\phi} = 0.6424.$$

Now, if we impose a monotonicity constraint on the $\{\mu_i\}$ and apply the *PAV* algorithm we get

$$\hat{\mu} = (-0.4833, 0.2065, 0.2065)$$

and

$$\hat{\phi} = 0.3193$$

But the same maximum value of the likelihood is achieved by

$$\hat{\mu} = (0.4833, -0.2065, -0.2065)$$

and

$$\hat{\phi} = -0.3193.$$

Ritov and Gilula (1987) point out that the problem with direct application of *PAV* to the score parameters lies in the fact that these parameters are mutually dependent. Thus reordering a pair of μ_i's may result in a violation of the order constraint on the ν_j's. A solution is to apply the *PAV* algorithm to specific functions of the parameters, viz.

$$\tilde{E}_i(\hat{\nu}) = \sum_{j=1}^{J} \tilde{P}_{ij}\hat{\nu}_j / \tilde{P}_{i.} \tag{68}$$

and

$$\tilde{F}_j(\hat{\mu}) = \sum_{i=1}^{I} \tilde{P}_{ij}\hat{\mu}_i / \tilde{P}_{.j} \tag{69}$$

where $\tilde{P}_{ij}$ represents the empirical distribution in the observed contingency table. The quantities $(\tilde{E}_i - \tilde{E}_{i-1})$ and $(\tilde{F}_j - \tilde{F}_{j-1})$ are asymptotically uncorrelated and amalgamation can be done separately (independently) for rows and columns. The unrestricted maximum likelihood estimates for the collapsed table, in which the $\tilde{E}_i$ and $\tilde{F}_j$ follow the desired order, are asymptotically equivalent to the order restricted maximum likelihood estimates.

Dykstra and Lemke (1988) discuss this restricted maximization problem, and its dual I-projection problem. They note the applicability of a general algorithm from Dykstra (1985) to solve the I-projection problem, and thus the maximum likelihood problem here.

4.5. Goodness-of-Fit Statistics. When the true parameters for the order restricted model are in fact monotone, then the Pearson chisquare and Likelihood Ratio chisquare are an asymptotic χ^2. When there are equalities between adjacent parameters, the goodness of fit statistics are distributed as mixtures of chisquares. For example, it follows from Agresti, Chuang, and Kezouh (1987) that in the RC model, under monotonicity constraints on the $\{\mu_i\}$, when the $\{\mu_i\}$ are strictly monotone, except for one identical adjacent pair, the likelihood ratio statistic has an asymptotic distribution that is an equal mixture of the one for the RC model and one with an additional degree of freedom. This is because when one pair is equal then

- with limiting probability 0.5, asymptotically, the ordinary MLE's will follow the order restriction, and G^2 under the order restriction will be the same as the unrestricted G^2 with $\chi^2_{(I-2)(J-2)}$ distribution.
- with limiting probability 0.5, the ordinary MLE's will have the estimates for the pair of parameters concerned out of order. The order restricted solution will be the same as the ordinary solution for the table with the two rows concerned, collapsed.

From Agresti, Chuang, and Kezouh (1987), Theorem 4, the likelihood ratio chi-square statistic in this case can be decomposed as:

$$G^2(RC^*) = G^2(RC') + G^2(I), \tag{70}$$

where

- $G^2(RC^*)$ denotes the fit of the order restricted model to the original table,
- $G^2(RC')$ denotes the fit of the ordinary RC model to the collapsed table, and
- $G^2(I)$ denotes the fit of the model of independence to the $2 \times J$ table formed from the pair of rows with parameters out of order.

The first of the quantities on the right hand side has an asymptotic $\chi^2_{(I-3)(J-2)}$ distribution, and the second a $\chi^2_{(J-1)}$ distribution, and they are asymptotically independent. Thus their sum is asymptotically distributed as $\chi^2_{(I-2)(J-2)+1}$.

In general, however, the foregoing result does not tell us the asymptotic distribution of G^2 when the order constraints are included in the model because we do not know the values of the true parameters and hence the equalities/inequalities that may hold among them.

Ritov and Gilula (1987) show that the chisquare statistic for testing H_0: Order restricted RC model against H_1: Unrestricted RC model has the following asymptotic distribution:

$$P(X^2 > c) = \sum_{k=1}^{I'+J'-2} \beta_k P(\chi^2_{(k)} > c) \qquad (71)$$

where I' and J' denote the maximum number of row and column parameters that can be equal and the mixture probabilities β_k depend on the marginals $P_{i.}$ and $P_{.j}$.

The general problem of inference under order restrictions has been treated in detail by Barlow et al. (1972), and, most recently by Robertson, Wright, and Dykstra (1988) which includes some of the material in Barlow et al. and also covers subsequent developments in the area of isotonic methods. Raubertas, Lee, and Nordheim (1986) consider hypothesis tests for linearly constrained normal means, and Robertson (1978) and Lee (1987) deal with tests for order restrictions on multinomial parameters.

Robertson considers the following three hypotheses:

- $H_0: \quad p = q,$ where $p = (p_1, \ldots, p_k)$ is a probability vector of unknown values and q is a known probability vector.
- $H_1: \quad p \neq q,$ but p satisfies some order restriction O on its components.
- $H_2: \quad p \epsilon R^k, \ \sum p_i = 1.$

The asymptotic distribution of the test statistic for testing H_0 versus H_1 is of the $\bar{\chi}^2$ form (see Barlow et al., 1972, ch. 3), i.e., a weighted sum of standard chi-squares, depending on the alternate hypothesis through the weights. The test of H_1 versus H_2 is not similar in that it depends on the particular p satisfying H_1; however, asymptotically, an upper bound on the significance level of the test of H_1 versus H_2 can be found. Raubertas, Lee, and Nordheim (1986) give similar results for the normal means case. In both cases, the maximum likelihood estimates that appear in the test statistics turn out to be projections onto polyhedral cones.

5. Parameter Constraints in Terms of Table Margins. Yet another way to approach information on ordering is to incorporate it into the model in the form of order restrictions on the marginal totals of the cross classification. This has been explored by Eddy, Fienberg, and Meyer (1982). The motivation behind this approach is similar to the motivation behind the structural zero method for modelling contingency tables in which some cell frequencies are known to be zero (see Bishop, Fienberg and Holland, 1975). There appears to be an interesting link between this approach and ideas associated with loglinear models.

6. Association Models for Multi-Dimensional Tables. Natural extensions of the association models from Section 2 to three and higher-dimensional contingency tables are reasonably direct and have been explored previously by Chuang (1980), Clogg (1982), Fienberg (1982), Agresti (1983), and Goodman (1986). Rather than attempting an exhaustive treatment of the topic we present some illustrative examples and features of these extensions, noting the links to the association models for $I \times J$ tables and the kinds of stochastic ordering features described in Section 2.

For the $I \times J \times K$ cross-classification involving variables A, B, and C, let P_{ijk} be the probability of an observation falling into the ith row, jth column and kth layer. The saturated loglinear model for this 3-dimensional table is usually written in the form:

$$\text{(72)}\quad \log P_{ijk} = u + u_{1(i)} + u_{2(j)} + u_{3(k)} + u_{12(ij)} + u_{13(ik)} + u_{23(jk)} + u_{123(ijk)}$$

with identifying constraints requiring the sum of any subscripted u-term over each subscript to be equal to zero. Extensions of the association models to this situation typically involve

1. setting one or more subscripted u-term equal to zero;

2. representing some of the remaining u-terms with 2 or more subscripts as products of parameters.

For both (1) and (2) the choices must be in accord with the generalized hierarchy principle (e.g., see Fienberg, 1980, p. 43 and p. 100) associated with ANOVA-like models wherein restriction terms must be compatible with restrictions or structures for the lower-order terms which are marginal to them. Thus representing u_{12} in multiplicative form implies related multiplication forms for u_{123}. In particular setting $u_{12} = 0$ implies that variables A and B not be linked for any multiplicative component used to model related higher-order terms, e.g. u_{123}.

Chuang (1980), for example, described several choices for association models based on (72) where the 3-factor term is represented as

(73) $$\text{(a) } u_{123(ijk)} = \lambda v_{1(i)}\, v_{2(j)}\, v_{3(k)},$$

(74) $$\text{(b) } u_{123(ijk)} = \lambda v_{1(i)}\, v_{23(jk)}.$$

In the multiplicative notation of Goodman (1985) these models are representable as

(75) $$\text{(a) } P_{ijk} = \alpha_{ij}^{AB} \alpha_{ik}^{AC} \alpha_{jk}^{BC} \exp\,(\phi \mu_i^A \mu_j^B \mu_k^C),$$

(76) $$\text{(b)} P_{ijk} = \lambda_{ij}^{AB} \lambda_{ik}^{AC} \lambda_{jk}^{BC} \exp\,(\phi \mu_i^A \mu_j^{BC}).$$

Some other examples of loglinear-association models are:

$$\begin{aligned} \text{(c) } u_{123(ijk)} &= 0, \\ u_{12(ij)} &= \lambda_1 v_{1(i)} v_{2(j)}, \\ u_{13(ik)} &= \lambda_2 v'_{1(i)} v'_{2(j)}, \\ u_{23(jk)} &= \lambda_3 v^*_{2(j)} v^*_{3(j)}, \end{aligned} \tag{77}$$

$$\begin{aligned} \text{(d) } u_{12(ij)} &= 0, \\ u_{123(ijk)} &= \lambda v_{1(i)} v_{23(jk)}, \end{aligned} \tag{78}$$

which are representable in Goodman's notation as

(79) $$\text{(c) } P_{ijk} = \alpha_i^A \alpha_j^B \alpha_k^C \exp(\phi_1 \mu_i^A \mu_j^B + \phi_2 \nu_i^A \nu_k^C + \phi_3 \alpha_j^B \alpha_k^C),$$

(80) $$\text{(d) } P_{ijk} = \alpha_{ik}^{AC} \alpha_{jk}^{BC} \exp(\phi \mu_i^A \mu_{jk}^{BC}),$$

(see also Clogg, 1982).

Maximum likelihood estimates for these loglinear-association models require some form of iterative procedure. Chuang (1980) proposes a variant on the Newton-Raphson method and direct generalizations of Goodman's iterative method for two-way tables are available.

The various mixed loglinear-association models described above can all be reformulated as models for the key parameters for three-way arrays, i.e. ratios of odds-ratios of the form

(81) $$\theta_{ijk} = \frac{P_{ijk} P_{i+1,jk}}{P_{i,j+1,k} P_{i+1,j+1,k}} \Big/ \frac{P_{ij,k+1}\; P_{i+1,j,k+1}}{P_{i,j+1,k}\; P_{i+1,j+1,k+1}}.$$

The (θ_{ijk}) do not readily lend themselves to any total postivity order restrictions and, to date, no-one has considered appropriate generalizations of the stochastic ordering properties discussed in Section 2.2.

No one in the categorical data analysis literature has yet presented methods for estimation of the higher dimensional loglinear-association models subject to order on the multiplicative parameters such as in expression (49) and (50). The estimation problems for the RC model alluded to in Section 4.4 come home in spades when we move to higher dimensions and more than two sets of order restructures. Again the log likelihood is not necessarily concave and approaches modelled after the Pooled Adjacent Violators algorithm will not necessarily work.

Schriever (1983, 1986) considers some multi-dimensional generalizations of the correlation models of Section 3, and related stochastic ordering properties. His description allows a two-dimensional matrix representation involving submatrices which are two-dimensional marginals. Thus his approach, while far more restrictive than the one suggested here for loglinear-association models, does allow for the direct examination of total-positivity and order-dependence properties.

REFERENCES

AGRESTI, A. (1984). *Analysis of Ordinal Categorical Data.* Wiley, New York.

AGRESTI, A. (1983). A survey of strategies for modeling cross-classifications having ordinal variables. *J. Amer. Statist. Assoc.* **78** 184–198.

AGRESTI, A. and CHUANG, C. (1985). Bayesian and maximum likelihood approaches to order-restricted inference for models for ordinal categorical data. *Proceedings of the Symposium on Order Restricted Statistical Inference* held in Iowa City, Iowa, Sept. 11–13, 1985, 6–27.

AGRESTI, A., CHUANG, C., and KEZOUH, A. (1987). Order restricted score parameters in association models for contingency tables. *J. Amer. Statist. Assoc.* **82** 619–623.

ANDERSON, J.A. (1984). Regression and ordered categorical variables. *J. Roy. Statist. Soc. B* **46** 1–30.

BARLOW, R.E., BARTHOLOMEW, D.J., BREMNER, J.M., and BRUNK, H.D. (1972). *Statistical Inference Under Order Restrictions.* John Wiley and Sons, New York.

BARLOW, R.E. and PROSCHAN, F. (1981). *Statistical Theory of Reliability and Life Testing: Probability Models.* To Begin With, Silver Spring, Maryland.

BECKER, M.P. and CLOGG, C.C. (1989). Analysis of sets of two-way contingency tables using association models. *J. Amer. Statist. Assoc.* **84** 142–151.

BENZÉCRI, J.P., et al. (1973). *L'Analyse des Données, Vol. 1: La Taxinomie, Vol. 2: L'Analyse des Correspondances.* Dunod, Paris (Second Edition 1976).

BISHOP, Y., FIENBERG, S.E., and HOLLAND, P.W. (1975). *Discrete Multivariate Analysis.* M.I.T. Press, Cambridge.

CHRISTIANSEN, U. (1966). *The Lectures by Professor G. Rasch over the Theory of Statistics* (in Danish). University Foundation for the Printing of Textbooks, Copenhagen.

CHUANG, C. and AGRESTI, A. (1986). A new model for an ordinal pain data set from a pharmaceutical study. *Statist. in Medicine* **5** 15–20.

CHUANG, C. (1980). Analysis of categorical data with ordered categories. Ph.D. Dissertation, School of Statistics, University of Minnesota.

CLOGG, C.C. (1982). Some models for the analysis of association in multiway cross-classifications having ordered categories. *J. Amer. Statist. Assoc.* **77** 803–815.

COX, C. and CHUANG, C. (1984). A comparison of chi-square partitioning and two logit analyses of ordinal pain data from a pharmaceutical study. *Statist. in Medicine* **3** 273–285.

DOUGLAS, R.B., FIENBERG, S.E., LEE, M.-L.T., SAMPSON, A.R., and WHITAKER, L. (1990). Positive dependence concepts and contingency table parametrizations. In *Topics in Statistical Dependence* (H.W. Block, A.R. Sampson, T.H. Savits, eds.), *IMS Lecture Notes/Monograph Series.* Hayward, CA.

DYKSTRA, R.L. (1985). An iterative procedure for obtaining I-projections onto the intersection of convex sets. *Ann. Prob.* **13**, 975–984.

DYKSTRA, R.L. and LEMKE, J.H. (1988). Duality of I-projections and maximum likelihood estimation for log-linear models under cone constraints. *J. Amer. Statist. Assoc.* **83** 546–554.

EDDY, WM. F., FIENBERG, S.E., and MEYER, M.M. (1982). Contingency table estimation with order restrictions on the margins. Unpublished manuscript.

FIENBERG, S.E. (1968). The estimation of cell probabilities in two-way contingency tables. Ph.D. Dissertation, Department of Statistics, Harvard University.

FIENBERG, S.E. (1980). *The Analysis of Cross-classified Categorical Data.* (2nd ed.) M.I.T. Press, Cambridge.

FIENBERG, S.E. (1982). Using information on ordering for loglinear model analysis of multidimensional contingency tables. In *Proceedings of XIIth International Biometrics Conference.* Tonteruse, France, 133–139.

GILL, R.D. and SCHRIEVER, B.F. (1986). Comment on "Some useful extensions of the usual correspondence analysis approach and the usual log-linear models approach in the analysis of contingency tables." *Int. Statist. Rev.* **54** 289–291.

GILULA, Z. (1982). A note on the analysis of association in cross-classifications having ordered categories. *Commun. Statist.-Theor. Meth.* **11**(11) 1233–1240.

GILULA, Z., KRIEGER, A.M., and RITOV, Y. (1988). Ordinal association in contingency tables: Some interpretive aspects. *J. Amer. Statist. Assoc.* **83** 540–545.

GOODMAN, L.A. (1979). Simple models for the analysis of association in cross-classifications having ordered categories. *J. Amer. Statist. Assoc.* **74** 537–552.

GOODMAN, L.A. (1981). Association models and canonical correlation in the analysis of cross-classifications having ordered categories. *J. Amer. Statist. Assoc.* **76** 320–334.

GOODMAN, L.A. (1984). *The Analysis of Cross-Classified Data Having Ordered Categories.* Harvard University Press, Cambridge.

GOODMAN, L.A. (1985). The analysis of cross-classified data having ordered and/or unordered categories: Association models, correlation models, and asymmetry models for contingency tables with or without missing entries. *Ann. Statist.* **13** 10–69.

GOODMAN, L.A. (1986). Some useful extensions of the usual correspondence analysis approach and the usual log-linear models approach in the analysis of contingency tables (with discussion). *Int. Statist. Rev.* **54** 243–309.

GOODMAN, L.A. and KRUSKAL, W.H. (1954). Measures of association for cross classification. *J. Amer. Statist. Assoc.* **49** 732–764.

GOODMAN, L.A. and KRUSKAL, W.H. (1979). *Measures of Association for Cross Classifications.* Springer-Verlag, New York.

GREENACRE, M.J. (1984). *Correspondence Analysis.* Academic Press, New York.

HABERMAN, S.J. (1974a). Log-linear models for frequency tables with ordered classifications. *Biometrics* **30** 589–600.

HABERMAN, S.J. (1974b). *The Analysis of Frequency Data.* University of Chicago Press, Chicago.

HABERMAN, S.J. (1981). Tests for independence in two-way contingency tables based on canonical correlation and on linear-by-linear interaction. *Ann. Statist.* **9** 1178–1186.

KIMELDORF, G. and SAMPSON, A. (1978). Monotone dependence. *Ann. Statist.* **6** 895–903.

KRUSKAL, W.H. (1958). Ordinal measures of association. *J. Amer. Statist. Assoc.* **53** 814–861.

LEE, C.C. (1987). Chi-squared tests for and against an order restriction in multinomial parameters. *J. Amer. Statist. Assoc.* **82** 611–618.

MCCULLAGH, P. (1980). Regression models for ordinal data. *J. Roy. Statist. Soc. B* 109–142 (with discussion).

PEARSON, K. (1900). On a criterion that a given system of deviations from the probable in the case of a correlated system of variables is such that it can be reasonably supposed to have arisen from random sampling. *Philos. Mag.* **50** No. 5, 157–175.

RAUBERTAS, R.F., LEE, C.C., and NORDHEIM, E.V. (1986). Hypothesis tests for normal means constrained by linear inequalities. *Commun. Statist.-Theor. Meth.* **15**(9) 2809–2833.

RITOV, Y. and GILULA, Z. (1987). The order restricted RC model for ordered contingency tables: Estimation and testing for fit. Manuscript.

ROBERTSON, T. (1978). Testing for and against an order restriction on multinomial parameters. *J. Amer. Statist. Assoc.* **73** 197–202.

ROBERTSON, T., WRIGHT, F.T., and DYKSTRA, R.L. (1988) *Order Restricted Statistical Inference.* Wiley.

SCHRIEVER, B.F. (1983). Scaling of order dependent categorical variables with correspondence analysis. *Int. Statist. Rev.* **51** 225–238.

SCHRIEVER, B.F. (1986). *Order Dependence.* CWI-tract 20. Centre for Mathematics and Computer Science, Amsterdam.

SETHURAMAN, J. (1977). The limit distribution of the Renyi maximum correlation, with applications to contingency tables and correspondence analysis. In *Proceedings of the 41st Session of the International Statistical Institute*, Book 4, 701–703.

YULE, G.U. (1900). On the association of attributes in statistics: with illustration from the material of the childhood society, c. *Philos. Trans. Roy. Soc. Ser. A* **194** 257–319.

DEPARTMENT OF STATISTICS
CARNEGIE MELLON UNIVERSITY
PITTSBURGH, PA 15213

POSITIVE DEPENDENCE CONCEPTS FOR ORDINAL CONTINGENCY TABLES

By Ruth Douglas, Stephen E. Fienberg, Mei-Ling T. Lee, Allan R. Sampson,[1] and Lyn R. Whitaker[2]

Carnegie Mellon University, Carnegie Mellon University, Boston University, University of Pittsburgh, Naval Postgraduate School

This paper provides a connection for cross-classified variables between ordinal contingency tables and positive dependence concepts. Various bivariate positive dependence properties are reinterpretted in terms of odd ratios for certain subtables of the two-way array.

1. Introduction. The primary purpose of this paper is to provide a bridge between two literatures: ordinal contingency tables and positive dependence concepts. Both focus on relationships among discrete random variables. Contingency table research has concentrated on statistical analysis of parametric models for cell frequencies or probabilities. Well known examples of such models are Bishop, Fienberg, and Holland's (1975) log-linear models, and Goodman's (1979, 1985) association models. In the modeling of ordinal contingency tables, the ordered structure of the levels of the cross-classifying variables is usually translated into a related ordering constraint on the corresponding model parameters, e.g., Douglas and Fienberg (1990).

The dependence literature, on the other hand, has focused on probabilistic characterizations, which lead to the study of properties associated with the cell probabilities, rather than with the parameters in models. For a discussion of positive dependence properties and their interrelationships, see Barlow and Proschan (1981).

In this paper, we study in depth the connection between ordinal contingency table parametrizations and positive dependence properties of cross-classified variables. The relationships we examine are established through generalized odds ratios in a contingency table. A few of these connections have been considered previously by Agresti (1984), Grove (1984), and Yanagimoto (1972). Throughout our presentation, we pursue the dual goals of promoting potentially new models for contingency table researchers and bringing ordinal statistical models and techniques to researchers in positive dependence.

---

[1]Research supported by the National Security Agency under Grant MDA904–90–H–4036.

[2]Research supported in part by the NPS Research Foundation.

AMS 1980 subject classification. Primary 62H17; secondary 62H05.

Key words and phrases. Positive dependence property, two-way contingency table, odds-ratio, one-parameter family of distributions.

To illustrate the connections between the two literatures, we begin by considering a two-way, $I \times J$ ordinal table, with cross-classifying variables X and Y. We suppose that X takes values $\{x_1 < x_2 < \cdots < x_I\}$, and Y takes values $\{y_1 < y_2 < \cdots < y_J\}$. We denote the joint probability $P(X = x_i, Y = y_j)$ by p_{ij}, and we assume that $p_{ij} > 0$, $i = 1, \ldots, I$, $j = 1, \ldots, J$. It is well known that the property requiring the positivity of all log (local) odds ratios:

$$\log \theta_{ij} \equiv \log \frac{p_{ij}p_{i+1,j+1}}{p_{ij+1}p_{i+1,j}} > 0, \tag{1}$$

$i = 1, \ldots, I-1$, $j = 1, \ldots, J-1$, is equivalent to the positive dependence concept that the distribution of (X, Y) is totally positive of order 2 (TP_2) (e.g., see Barlow and Proschan (1981)). Moreover, if the RC model of order 1 (e.g., see Douglas and Fienberg (1990)) holds in this table, i.e., if

$$p_{ij} = \alpha_i \beta_j \exp(\phi \mu_i \nu_j), \tag{2}$$

where μ_i and ν_j are, respectively, the row and column parameters, and the ordering of the row and column parameters is the same as that of the rows and columns, then log local odds ratios are all positive and, hence, the parametrized distribution of expression (2) must be TP_2.

In Section 2 of this paper we explore, in detail, connections between positive dependence concepts and odds ratios, and their suitable generalizations. There is no discussion of the links between parametric models such as in expression (2) and related concepts, but the connections should be kept in mind. In Sections 3 and 4, we present the basic results linking various positive dependence concepts in terms of inequality restrictions on generalized odds ratios.

2. Generalized Odds Ratios. While detailed discussions of the positive dependence properties can be found in many sources, e.g., Barlow and Proschan (1981), Grove (1984), Shaked (1977), or Yanagimoto (1972), virtually every author uses different nomenclature and notation. Here we develop a unified nomenclature and notation that is linked directly to odds ratios for certain subtables of the two-way array. In the next section we apply the nomenclature to establish a hierarchy among various notions of positive dependence.

Agresti (1984) uses the following nomenclature and notation for three basic classes of odds ratios in a two-way table:

1. *Local odds ratios* (see Bishop, Fienberg, and Holland (1975) or Fienberg (1980))

$$\theta_{ij} = \frac{p_{ij}p_{i+1,j+1}}{p_{i+1,j}p_{i,j+1}}, \quad i = 1, \ldots, I-1, \;\; j = 1, \ldots, J-1.$$

Figure 2.1

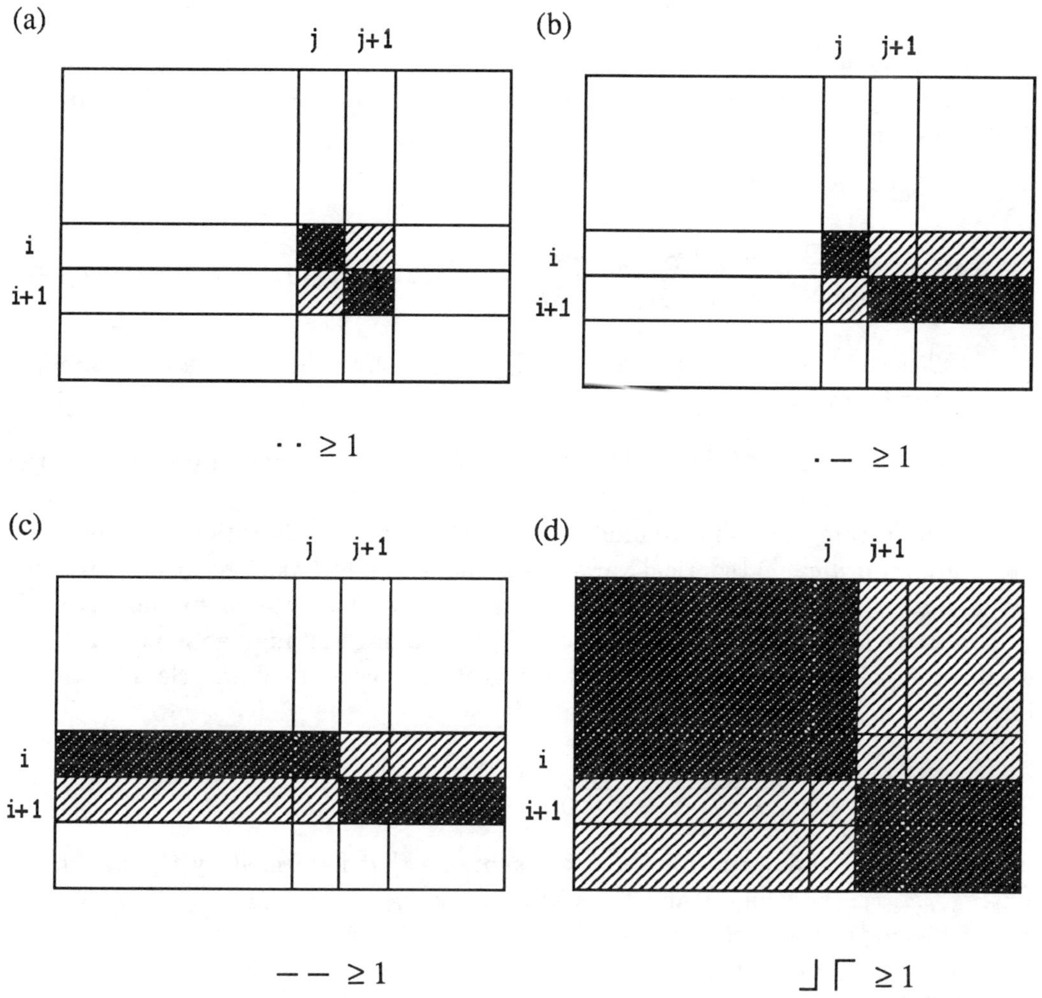

2. *Local-global odds ratios*

$$\theta^{c}_{ij} = \frac{(\Sigma_{k\le j} p_{ik})(\Sigma_{k>j} p_{i+1,k})}{(\Sigma_{k>j} p_{ik})(\Sigma_{k\le j} p_{i+1,k})}, \quad i = 1,2,\ldots,I-1, \;\; j = 1,2,\ldots,J-1.$$

(These odds ratios are local in the row variable, and global in the column variable as indicated by the superscript c. The notion of global-local odds ratios, θ^{r}_{ij} can be defined in a similar fashion.).

3. *Global odds ratios*

$$\theta^{rc}_{ij} = \frac{(\Sigma_{k\le i}\Sigma_{\ell\le j} p_{k\ell})(\Sigma_{k>i}\Sigma_{\ell>j} p_{k\ell})}{(\Sigma_{k\le i}\Sigma_{\ell>j} p_{k\ell})(\Sigma_{k>i}\Sigma_{\ell\le j} p_{k\ell})}, \quad i = 1,2,\ldots,I-1, \;\; j = 1,2,\ldots,J-1.$$

(These odds ratios are global in both the row and column variables, as indicated by the superscript rc.

We illustrate the sets of cells making up three types of odds ratios in Figure 2.1, parts (a), (c), and (d).

The literature of log-linear models is based on local odds ratios, whereas the literature on ordinal categorical variables is more closely linked to local-global or global odds ratios. All three types of quantities can be viewed as special cases of *generalized odds ratios* which are formed by partitioning rectangular subarrays into 4 sets of adjacent cells based on a collapsing of the row and column classifications into dichotomies. For example, the quantity

$$\frac{P\{X = x_i, Y \le y_j\} P\{X < x_i, Y > y_j\}}{P\{X < x_i, Y \le y_j\} P\{X = x_i, Y > y_j\}}$$

is a generalized odds ratio for the $i \times J$ subtable that is formed by the partition $\{x_1, x_2, \ldots, x_{i-1}\}$ and $\{x_i\}$ and the partition $\{y_1, y_2, \ldots, y_j\}$ and $\{y_{j+1}, y_{j+2}, \ldots, y_J\}$. This odds ratio is represented diagrammatically in Figure 2.2.

For our purposes here it suffices to restrict attention to those rectangular subarrays that are either 2×2 subtables or that include at least one extremal value of X or Y. The resulting class of odds ratios includes those based on the continuation ratios described by Fienberg (1980), i.e.,

$$\frac{p_{ij}(\Sigma_{k>j} p_{i+1,k})}{(\Sigma_{k>j} p_{ik}) p_{i+1,j}} \text{ for } i = 1,2,\ldots,I-1, \;\; j = 1,2,\ldots,J-1,$$

as depicted in Figure 2.1 (b), in addition to the basic odds ratios of Agresti (1984) described above.

Rather than use these verbal descriptions of different classes of odds ratios, in this paper we adopt a hieroglyphic-like symbolic notation which describes these sets of cells geometrically, and also symbolically gives relationships among concepts. This symbolic notation readily handles generalizations of these odds-ratio concepts.

Figure 2.2.

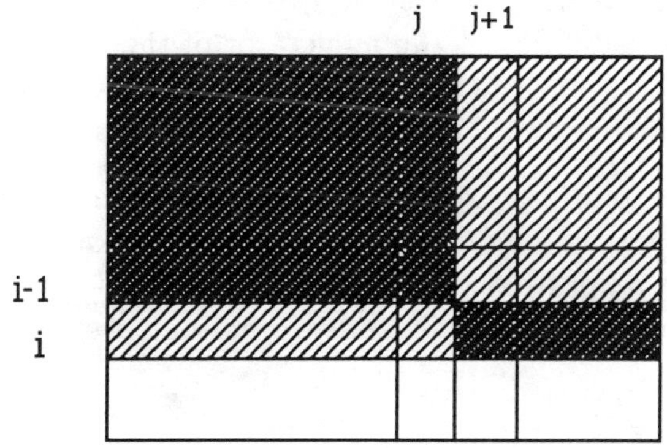

⌋ − ≥ 1

We will focus on four types of *sets* of cells:

1. individual cells, denoted by ·,
2. horizontal strips, denoted by −,
3. vertical strips, denoted by |,
4. rectangular sections, denoted by ⌋, ⌊, ⌉, and ⌈ for upper-left, upper-right, lower-left, and lower-right corners of the table, respectively.

Our goal is to use inequalities on the $(I-1)(J-1)$ odds ratios of a given type to describe different dependence properties. In general, requiring a collection of odds ratios to be ≥ 1 typically corresponds to some form of positive dependence. Using the hieroglyphic-like notation allows us to express such inequalities in a simple notational form that can be related to relationships among dependence properties. We illustrate by example.

We can describe the property "all local odds ratios are greater than or equal to 1," by the notation "·· ≥ 1". The first "·" refers to the fact that the "upper-left hand" set of the local odds ratio is a single cell, and the second "·" to the fact that the "lower-right hand" set is also a single cell. Thus this use of the set notation describes the sets of cells corresponding to the two probabilities in the numerator of the odds ratio and the symbols for the sets of cells corresponding to the two probabilities in the denominator are uniquely defined by implication. Similarly, the property that the "local-global odds ratios are ≥ 1" (i.e., odds ratios described by Figure 2.1 (c)) is described by "−− ≥ 1", and "⌋⌈≥ 1" means that "all global odds ratios are ≥ 1".

Two set symbols are said to be *comparable* if they uniquely describe the cells corresponding to the probabilities that are used to construct a generalized odds ratio for a rectangular subarray. Since the two symbols are used to describe the sets of cells for the probabilities in the numerator of the odds ratios, this means that they must uniquely define the complementary sets of cells for the probabilities in the denominator. Thus $\rfloor$ and $\lceil$ are comparable, but $\rfloor$ and $\lfloor$ are not.

3. Relationships Among Odds Ratio Inequalities. There are 16 possible odds ratios and corresponding dependence properties expressed by appropriate choices of comparable pairs from the four types of sets. To establish the hierarchies among these various positive dependence notions, we employ the following theorem:

THEOREM 3.1.

$$
\text{(a) } S\cdot \geq 1 \quad \begin{matrix} \Nearrow & S| \geq 1 & \Searrow \\ & & \\ \Searrow & S- \geq 1 & \Nearrow \end{matrix} \quad S\lceil \geq 1
$$

and

$$
\text{(b) } \cdot T \geq 1 \quad \begin{matrix} \Nearrow & -T \geq 1 & \Searrow \\ & & \\ \Searrow & |T \geq 1 & \Nearrow \end{matrix} \quad \rfloor T \geq 1
$$

where S and T are any sets which are appropriately comparable.

PROOF OF THEOREM 3.1. A complete proof requires examination of each inequality for each possible type of set S or T. We illustrate the essence of the complete proof by focusing on two inequalities for part (b).

Suppose that $T = \lceil$. The following diagram depicts the relevant structures of the inequalities $\cdot\lceil \geq 1$ for two neighboring points (cells) a and b and a residual horizontal strip c along with the remainder of the table, a lower rectangle d and two vertical strips, e and f:

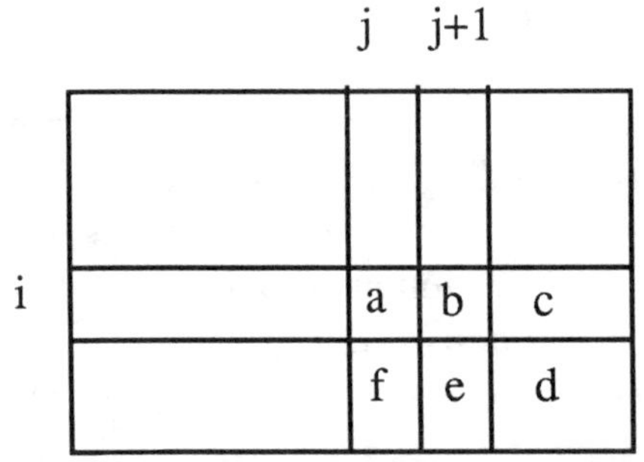

We shall refer to the probability associated with a set of cells using a capital letter corresponding to lower case letter used to denote the set, e.g., the probability corresponding to the set "a" is denoted by "A."

To prove that $\cdot\lceil \geq 1 \Rightarrow -\lceil \geq 1$, it suffices for us to prove that

$$BD \geq CE \tag{3}$$

and

$$A(E+D) \geq F(B+C) \tag{4}$$

imply

$$(A+B)D \geq (F+E)C, \tag{5}$$

where the capital letter represents the probability of the corresponding lower case letter's region. We consider two possible cases. Suppose that $FB \geq AE$. Then it follows from expression (4) that $AD \geq FC$. Combining this inequality with that of expression (3) yields expression (5). Alternatively, suppose that $FB < AE$. We can rewrite expression (3) as $(CE)^{-1} \geq (BD)^{-1}$. Then

$$AE(CE)^{-1} > FB(BD)^{-1}$$

or

$$AD > FC.$$

Combining this inequality with that of expression (3) yields expression (5) once again. Clearly, the proof of $\cdot\lceil \geq 1 \Rightarrow |\lceil \geq 1$ has exactly the same form.

Next suppose that $T = |$. The following diagram depicts the relevant tabular structure:

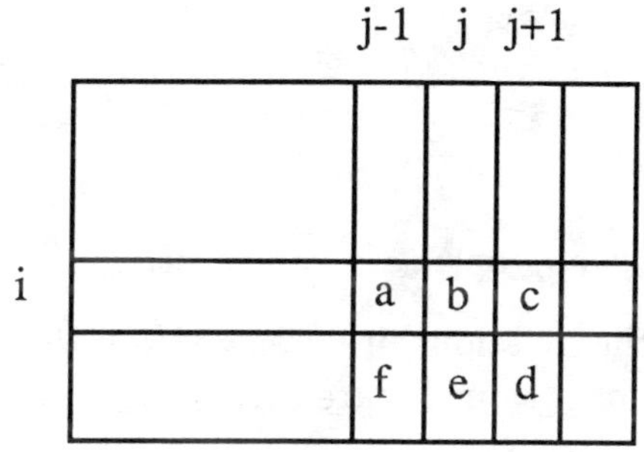

To prove that $\cdot| \geq 1 \Rightarrow -| \geq 1$, it suffices for us to prove that

$$BD \geq CE \tag{6}$$

and

$$AE \geq BF \tag{7}$$

imply

$$(A+B)D \geq (E+F)C. \tag{8}$$

We rewrite expression (6) as $(CE)^{-1} \geq (BD)^{-1}$ and combine this through multiplication with expression (7) to get $AD \geq CF$. Adding this inequality to that of expression (6) yields (8). The proof of the remaining components of part (b) of Theorem 3.1 take on one of the two preceding forms.

The proof of part (a) of Theorem 3.1 uses essentially the same form of proof applied to similarly constructed tabular arrays.

An interesting consequence of Theorem 3.1 is that the same relationships hold when the inequalities "≥ 1" are replaced by the equalities "$\equiv 1$." We write this result as:

COROLLARY 3.1.

$$\text{(a) } S\cdot \equiv 1 \quad \begin{matrix} \nearrow & S| \equiv 1 & \searrow \\ & & \\ \searrow & S- \equiv 1 & \nearrow \end{matrix} \quad S\lceil \equiv 1.$$

and

$$\text{(b) } \cdot T \equiv 1 \quad \begin{matrix} \nearrow & -T \equiv 1 & \searrow \\ & & \\ \searrow & |T \equiv 1 & \nearrow \end{matrix} \quad \rfloor T \equiv 1.$$

for any sets S and T that are appropriately comparable.

We can display all of the relationships among the 16 different dependence properties in a hierarchical form as in Figure 3.1.

4. Equivalent Dependence Concepts. Thirteen of the 16 concepts in Figure 3.1 have been multiply studied by various previous authors in their specific context. See Shaked (1977) for the definition of DTP; Yanagimoto (1972) for $\mathbf{P}(\cdot,\cdot)$, Barlow and Proschan (1981) for TP_2, SI, RCSI, LTD, RTI, and PQD; and Grove (1984) for his various equations.

Figure 3.1.

Hierarchy Among Bivariate Dependence Properties

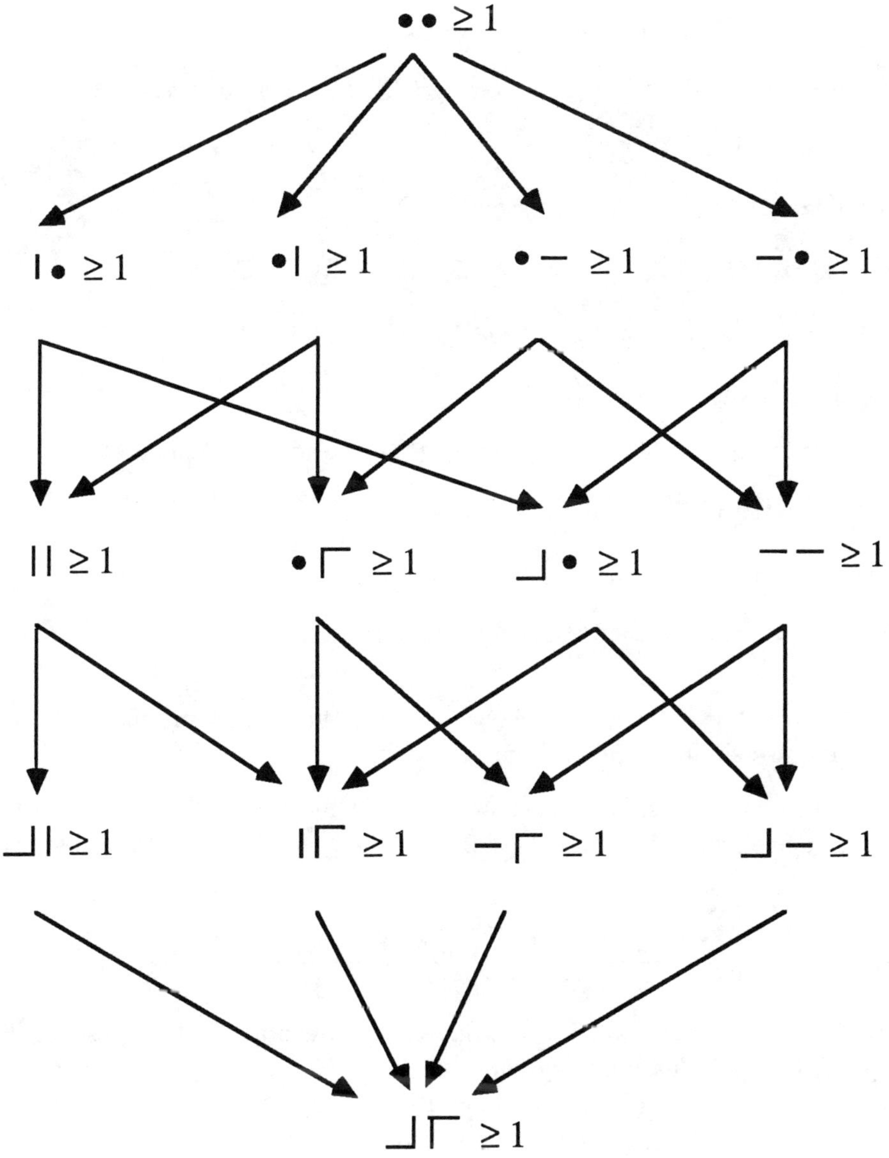

THEOREM 4.1. For each subpart below, all dependence concepts are equivalent:

(i) ·· ≥ 1, TP_2, DTP(0,0), **P**(3,3), Grove (1.1).

(ii) |· ≥ 1, **P**(3,2″).

(iii) ·| ≥ 1, DTP(0,1), **P**(3,2′), Grove (3.1).

(iv) ·− ≥ 1, DTP(1,0), **P**(2′,3), Grove (3.1, x and y interchanged).

(v) −· ≥ 1, **P**(2″,3).

(vi) || ≥ 1, **P**(3,1), SI($Y|X$), Grove (3.2).

(vii) ·⌈≥ 1, DTP(1,1), **P**(2′,2′), X and Y are RCSI, Grove (3.3).

(viii) −− ≥ 1, **P**(1,3), SI($X|Y$), Grove (3.2, x and y interchanged).

(ix) ⌋| ≥ 1, **P**(2″,1), LTD($Y|X$).

(x) |⌈≥ 1, **P**(2′,1), RTI($Y|X$), Grove (3.4, x and y interchanged).

(xi) −⌈≥ 1, **P**(1,2′), RTI($X|Y$), Grove (3.4).

(xii) ⌋− ≥ 1, **P**(1,2″), LTD($X|Y$).

(xiii) ⌋⌈≥ 1, **P**(1,1), X and Y are PQD, Grove (3.5).

PROOF. We only indicate how to prove (vii). The proofs of the remaining parts require similar type of manipulations.

Shaked (1977) essentially shows that RCSI <=> DTP(1,1) <=> **P**(2′,2′). Thus, we need only show that **P**(2′,2′) <=> ·⌈≥ 1. The condition **P**(2′,2′) requires that

$$\frac{P\{x_\beta < X, y_\beta < Y\}P\{x_\alpha < X \le x_\beta, y_\alpha < Y \le y_\beta\}}{P\{x_\beta < X, y_\alpha \le Y \le y_\beta\}P\{x_\alpha < X \le x_\beta, y_\beta < Y\}} \ge 1, \tag{9}$$

for all $x_\alpha < x_\beta$, $y_\alpha < y_\beta$. Fix a point x_i, y_j and note that expression (9) clearly implies ·⌈≥ 1 at this point by choosing $x_\beta = x_i$, $x_\alpha = x_{i-1}$, $y_\beta = y_i$, $y_\alpha = y_{i-1}$. Now view expression (9) as X, Y conditional on $X > x_\alpha$, $Y > y_\beta$ having property ⌋⌈≥ 1 at the point x_i, y_j. We observe that ·⌈≥ 1 holding for X, Y at x_i, y_j implies ·⌈≥ 1 holds for the preceding conditional random variables at this point. Next, we apply Theorem 3.1 (b) with T = ⌈. This shows that ·⌈≥ 1⇒⌋⌈≥ 1, and, hence, ·⌈≥ 1 ⇒ expression (9) holds at x_i, y_j for fixed x_α, y_β. Finally, vary x_α, y_β.

Comments

(i) X and Y being independent implies that ·· ≡ 1.

(ii) All implications in Figure 3.1 are strict, in that there exist counter-examples that show that the properties are not equivalent.

(iii) No other implications among these dependence notions than those in Figure 3.1 hold for all possible $I \times J$ tables of probabilities.

(iv) $\cdot\lceil \geq 1$ is equivalent to $P\{X = x_i, Y = y_j | X \geq x_i, Y \geq y_j\}$ being L-superadditive in i and j. (A real-valued function $h(i,j)$ is L-superadditive in i and j if for all $i < i'$, $j < j'$, $h(i,j) + h(i',j') - h(i,j') - h(i',j) \geq 0$.)

(v) $\cdot| \geq 1$ is equivalent to Prob $\{Y = y_j | X = x_i\}$/Prob $\{Y > y_j | X = x_i\}$ is nonincreasing in x_i for each fixed y_j. Shaked (1977) refers to this property as *conditional hazard rate decreasing* of Y given X (HRD($Y|X$)). The property $\cdot - \geq 1$ is equivalent to HRD($X|Y$).

(vi) $|\cdot \geq 1$ is equivalent to Prob $\{Y = y_j | X = x_i\}$/Prob $\{Y < y_j | X = x_i\}$ is nonincreasing in x_i for all y_j. An analogous equivalence exists for $-\cdot \geq 1$.

(vii) $\rfloor\cdot \geq 1$ is equivalent to Prob $\{(X,Y) \leq (x_i, y_j) | (X,Y) \leq (x_i', y_j')\} \geq$ Prob $\{(X,Y) \leq (x_i, y_j) | (X,Y) \leq (x_i^*, y_j^*)\}$ for $x_i' \leq x_i^*$, $y_j' \leq y_j^*$, and for any choice of (x_i, y_j). In the spirit of the terminology for Harris (1970), we refer to this property as *lower corner set decreasing* (LCSD (X,Y)).

(viii) The dependence properties $-| \geq 1$ and $|- \geq 1$ appear not to have been studied previously.

Comment (i) can be further amplified upon in light of Corollary 3.1.

THEOREM 4.2. *The random variables X and Y are independent if and only if $ST \equiv 1$ holds for any pair of comparable sets, S and T.*

PROOF. Independence is easily shown to be equivalent to $\cdot\cdot \equiv 1$, and also equivalent to $\rfloor\lceil \equiv 1$. The result then follows from Corollary 3.1.

One possible interpretation of Theorem 4.2 is that different odds ratios parametrizations provide a variety of positive dependence alternatives to independence by comparing the ratios to 1. This notion has been used in testing by Haberman (1974) who tests $H_0 : X, Y$ are independent (i.e., $\cdot\cdot \equiv 1$) against $H_A : X, Y$ are strictly TP_2 (i.e., X, Y are $\cdot\cdot \geq 1$, but not $\cdot\cdot \equiv 1$).

There are various other positive dependence properties leading to interesting ordinal contingency table parametrizations which are obtained by combining two or more dependence properties where there are inequality asymmetries. For instance, the situation where $|\cdot \geq 1$ and $\cdot| \geq 1$ both hold is equivalent to P(3,2) of Yanagimoto (1972). Another approach is to combine notions which treat X and Y asymmetrically. For example, Schriever (1983) described the situation where $|| \geq 1$ and $-- \geq 1$ hold together as double regression dependent of order 1.

Other positive dependence properties appear not able to be treated using the notational approach in this paper. For instance, X and Y being *associated* (Esary, Proschan and Walkup (1967)) requires that $P\{(X,Y) \in U|(X,Y) \in V\} \geq P\{(X,Y) \in U\}$ for all possible upper sets U, V (see also Sampson and Whitaker (1989)). We note that each of $\rfloor| \geq 1$, $|\lceil \geq 1$, $-\lceil \geq 1$, and $\rfloor- \geq 1$, implies association which, in turn, implies $\rfloor\lceil \geq 1$.

We underscore the fact that our approach for positive dependence properties permits an easy expression for each of the properties in terms of the parameters $\{p_{ij}\}$. For example, the concept $\cdot| \geq 1$, i.e., HRD($Y|X$), is equivalent to

$$p_{ij} \sum_{\ell=j+1}^{J} p_{i+1,\ell} \Big/ [p_{i+1,j} \sum_{\ell=j+1}^{J} p_{i\ell}] \geq 1,$$

for $i = 1, \ldots, I-1$ and $j = 1, \ldots, J-1$; and $\rfloor| \geq 1$ is equivalent to

$$(\sum_{k=1}^{i} \sum_{\ell=1}^{j} p_{k\ell})(\sum_{\ell=j+1}^{J} p_{i+1,\ell}) \Big/ [(\sum_{k=1}^{i} \sum_{\ell=j+1}^{J} p_{k\ell})(\sum_{\ell=l}^{j} p_{i+1,\ell}] \geq 1,$$

for $i = 1, \ldots, I-1$ and $j = 1, \ldots, J-1$. Similarly, our approach permits an easy representation for properties for resulting from intersections of other properties.

5. Discussion. Virtually all of the bivariate concepts that we consider, by reversing inequalities, can be made into negative dependence properties. For example, requiring the local odds ratios to be less than or equal to one, i.e., $\cdot\cdot \leq 1$, is equivalent to reverse regular rule of order 2 (RR$_2$), and $\rfloor\lceil \leq 1$, is equivalent to negative quadrant dependence. For a further discussion of negative dependence, see Block, Savits, and Shaked (1982).

For at least three of the concepts, requiring the odds ratios all to be equal to the same parameter θ produces an interesting one-parameter family of distributions. For example, $\cdot\cdot = \theta$ corresponds to the uniform association model (Goodman, (1979)), $\rfloor\lceil = \theta$ generates the Plackett family of distributions (Plackett (1965)), and $\cdot\lceil = \theta$ is considered in Clayton (1978). Which, if any of the other positive dependence properties produces interesting one-parameter families of distributions remains an open question.

To date, little attention has been given to the estimation of cell probabilities, p_{ij}, under the various odds ratios ≥ 1, except for models based on local odds ratios (e.g., see Douglas and Fienberg (1990)). Some limited testing results have been considered primarily for the $\cdot\cdot$ and $\rfloor\lceil$ concepts in Haberman (1974), Nguyen and Sampson (1987), and Krishnaiah, Rao, and Subramanyam (1987). Related results have been obtained by Grove (1980), Gilula (1986), and Gilula, Krieger, and Ritov (1988).

We note the interesting but little explored problem of comparing bivariate ordered contingency tables using positive dependence conceptualizations. Results concerning positive dependence orderings are recent and developed primarily for comparing two bivariate distributions. Three of the best known orderings are more positive quadrant dependent (Tchen (1980)), more associated (Schriever (1985)), and more totally positive of order 2 (Kimeldorf and Sampson (1987)).

REFERENCES

AGRESTI, A. (1984). *Analysis of Ordinal Categorical Data.* John Wiley, NY.

BARLOW, R. and PROSCHAN, F. (1981). *Statistical Theory of Life Testing: Probability Models.* To Begin With, Silver Spring, MD.

BISHOP, Y.M.M., FIENBERG, S.E., and HOLLAND, P.H. (1975). *Discrete Multivariate Analysis. Theory and Practice.* MIT Press, Cambridge, MA.

BLOCK, H.W., SAVITS, T.H., and SHAKED, M. (1982). Some concepts of negative dependence. *Ann. Prob.* **10** 765–772.

CLAYTON, D.G. (1978). A model for association in bivariate life tables and its application in epidemiological studies of familial tendency in chronic disease incidence. *Biometrika* **65** 141–151.

DOUGLAS, R. and FIENBERG, S.E. (1990). An overview of dependency models for cross-classified categorical data involving ordinal variables. In *Topics in Statistical Dependence* (H.W. Block, A.R. Sampson, T.H. Savits, eds.), *IMS Lecture Notes/Monograph Series.* Hayward, CA.

ESARY, J., PROSCHAN, F., and WALKUP, D. (1967). Association of random variables with applications. *Ann. Math. Statist.* **38** 1466–1474.

FIENBERG, S.E. (1980). *The Analysis of Cross-Classified Categorical Data, 2nd Ed..* MIT Press, Cambridge, MA.

GOODMAN, L.A. (1979). Simple models for the analysis of association in cross-classifications having ordered categories. *J. Amer. Statist. Assoc.* **74** 537–552.

GOODMAN, L.A. (1985). The analysis of cross-classified data having ordered and/or unordered categories: association models, correlation models and assymmetry models for contingency tables with or without missing entries. *Ann. Statist.* **13** 10–69.

GILULA, Z. (1986). Grouping and association in contingency tables: An exploratory canonical correlation approach. *J. Amer. Statist. Assoc.* **81** 773–779.

GILULA, Z., KRIEGER, A., and RITOV, Y. (1988). Ordinal association in contingency tables: Some interpretive aspects. *J. Amer. Statist. Assoc.* **83** 540–545.

GROVE, D.M. (1980). A test of independence against a class of ordered alternatives in a $2 \times C$ contingency table. *J. Amer. Statist. Assoc.* **75** 454–459.

GROVE, D.M. (1984). Positive association in two-way contingency tables: Likelihood ratio tests. *Comm. Stat. Th. Meth.* **13** 931–945.

HABERMAN, S.J. (1974). Log-linear models for frequency tables with ordered classification. *Biometrics* **35** 589–600.

HARRIS, R. (1970). A multivariate definition for increasing hazard rate distribution functions. *Ann. Math. Statist.* **41** 713–717.

KIMELDORF, G. and SAMPSON, A.R. (1987). Positive dependence orderings. *Ann. Inst. Statist. Math. A* **39** 113–128.

KIMELDORF, G. and SAMPSON, A.R. (1989). A framework for positive dependence. *Ann. Inst. Statist. Math. A* **41** 31–45.

KRISHNAIAH, P., RAO, M.B., and SUBRAMANYAM, K. (1987). A structure theorem on bivariate positive quadrant dependent distributions and tests for independence in two-way contingency tables. *J. Multi. Anal.* **23** 93–118.

NGUYEN, T. and SAMPSON, A.R. (1987). Testing for positive quadrant dependence in ordinal contingency tables. *Naval Res. Logist. Quar.* **34** 859–877.

PLACKETT, R.L. (1965). A class of bivariate distributions. *J. Amer. Statist. Assoc.* **60** 516–522.

SAMPSON, A.R. and WHITAKER, L. (1988). Positive dependence, upper sets, and multidimensional partitions. *Math. Oper. Res.* **3** 254–264.

SAMPSON, A.R. and WHITAKER, L. (1989). Computational aspects of association for bivariate discrete distributions. In *Contributions to Probability and Statistics. Essays in Honor of Ingram Olkin* (L.J. Gleser, M.D. Perlman, S.J. Press, A.R. Sampson, eds.), Springer-Verlag, NY, 288–299.

SCHRIEVER, B.F. (1985). Order Dependence. Ph.D. Dissertation, Free University of Amsterdam.

SCHRIEVER, B.F. (1983). Scaling of order dependent categorical variables with correspondence analysis. *Int. Statist. Rev.* **51** 225–238.

SHAKED, M. (1977). A family of concepts of dependence for bivariate distributions. *J. Amer. Statist. Assoc.* **72** 642–650.

TCHEN, A. (1980). Inequalities for distributions with given marginals. *Ann. Prob.* **8** 814–827.

YANAGIMOTO, T. (1972). Families of positive dependent random variables. *Ann. Inst. Statist. Math. A* **24** 559–573.

DEPARTMENT OF STATISTICS
CARNEGIE MELLON UNIVERSITY
PITTSBURGH, PA 15213

DEPARTMENT OF STATISTICS
CARNEGIE MELLON UNIVERSITY
PITTSBURGH, PA 15213

DEPARTMENT OF MATHEMATICS
BOSTON UNIVERSITY
BOSTON, MA

DEPARTMENT OF MATHEMATICS AND STATISTICS
UNIVERSITY OF PITTSBURGH
PITTSBURGH, PA 15260

DEPARTMENT OF OPERATIONS RESEARCH
NAVAL POSTGRADUATE SCHOOL
MONTEREY, CA 93940

DEPENDENCE RELATIONSHIPS FOR RAYS IN A CONVEX BODY IN $\Re^n$

By E.G. Enns[1]

University of Calgary

Line segments within a convex body, one of whose endpoints lies on the surface of the body are called rays. A forward and backward ray are complementary extensions of one another to form a secant of the body. This article illustrates that when these rays are generated by ν-randomness they display a negative orthant dependence when the convex body is a sphere in $\Re^n$. However, the forward and backward rays in any convex body in $\Re^n$ do not always display this dependence.

1. Introduction. In the field of geometrical probability, one of the topics which is extensively discussed is that of the distributions of the lengths of random rays and secants within and through convex bodies. See, for example, Coleman (1969, 1973), Enns and Ehlers (1978, 1980, 1981), Kellerer (1971, 1984), Kingman (1969), and Santalo (1976, 1986).

In most cases one is concerned with the distribution of a single quantity. This article will illustrate that for a sphere in $\Re^n$, the bivariate distribution of two related rays displays a negative orthant dependence (see Block, Savits, and Shaked (1982) and Joag-Dev and Proschan (1983)). It will also be illustrated that this dependence relationship does not hold for these rays in all convex bodies in $\Re^n$.

2. Definitions and Notation. In the notation of Enns and Ehlers (1978), the normalized average overlap volume of a convex body K with its translate in $\Re^n$ is defined as:

$$\Omega(\ell) = \mathcal{E}_\theta[V(K \cap K(\ell, \theta))]/V(K) \tag{1}$$

where $K(\ell, \theta)$ is the translate of a convex body K a distance ℓ in direction θ. Also $V(\cdot)$ and $\mathcal{E}_\theta(\cdot)$ are respectively the volume and expected value of $(\cdot)$ when uniformly averaged over all θ.

Line segments may be entirely in the interior of K or they may have one endpoint (rays) or both endpoints (secants) in the surface of K. These rays and

[1]The author acknowledges the support of the National Science and Engineering Council of Canada.

Key words and phrases. Convex body, random rays, negative orthant dependence.

secants may be generated by a variety of measures or types of randomness, for example, Coleman (1969) or Ehlers and Enns (1981).

The measure considered here is that of ν-randomness, namely a point P is chosen at random in K, according to the uniform distribution (Lebesgue measure) on K. A direction θ is then selected independently of P from a uniform distribution over all possible directions in $\Re^n$. The length of the forward ray L_1 is the distance from P to the surface of K in the direction θ. Similarly, the backward ray extends from P to the surface of K in the opposite direction to θ and has length L_2. Obviously L_1 and L_2 have the same marginal distributions and from Enns and Ehlers (1980) these are:

$$P(L_1 > \ell) = P(L_2 > \ell) = \Omega(\ell). \tag{2}$$

The secant traverses the whole body K and has length $L = L_1 + L_2$. It has been shown in Kingman (1969) and Enns and Ehlers (1978) that the probability density function of L is:

$$f(\ell) = \ell \frac{d^2\Omega(\ell)}{d\ell^2}.$$

The bivariate distribution of L_1 and L_2 derived in Enns and Ehlers (1980) is:

$$P(L_1 > \ell_1, L_2 > \ell_2) = \Omega(\ell_1 + \ell_2). \tag{3}$$

3. The Negative Dependence Condition. The bivariate distribution of the lengths of the forward and backward rays (3) displays a negative orthant dependence (NOD) if (2) has an increasing hazard rate. This may be shown by considering the form of the bivariate distribution (3). For NOD one requires:

$$P(L_1 > \ell_1, L_2 > \ell_2) \leq P(L_1 > \ell_1)P(L_2 > \ell_2), \tag{4}$$

see, for example, Block, Savits, and Shaked (1982) or Joag-Dev and Proschan (1983). Incorporating (2) and (3) into (4) yields our condition for NOD, namely:

$$\Omega(\ell_1 + \ell_2) \leq \Omega(\ell_1) \cdot \Omega(\ell_2). \tag{5}$$

If $r(\ell)$ is the hazard rate of the distribution of L_1 or L_2, then:

$$\Omega(\ell) = \left[\exp - \int_0^\ell r(x)dx\right]. \tag{6}$$

Condition (5) then implies that

$$\int_0^{\ell_1+\ell_2} r(x)dx \geq \int_0^{\ell_1} r(x)dx + \int_0^{\ell_2} r(x)dx$$

or equivalently:

(7) $$\int_{\ell_1}^{\ell_1+\ell_2} r(x)dx \geq \int_0^{\ell_2} r(x)dx, \text{ for all positive } \ell_1 \text{ and } \ell_2.$$

Now (7) is the condition for an increasing hazard rate average (IHRA). Hence if L_1 and L_2 have an increasing hazard rate (IHR) then they are (IHRA) and hence the pair (L_1, L_2) is NOD.

In the following section it will be shown that if L_1 and L_2 are ν-random rays in a sphere in $\Re^n$ (n-sphere), then (L_1, L_2) are NOD. It is intuitive that if a forward ray in a sphere is large, then it is likely that the backward ray is small. However (L_1, L_2) are not NOD for all convex bodies K. An example where NOD breaks down is a sufficiently elongated rectangle. Here it is intuitive that a long forward ray will have to be in the direction of the elongation and will hence most likely have a correspondingly long backward ray.

4. Negative Orthant Dependence of the Forward and Backward Rays in an n-Sphere. From Enns and Ehlers (1978), it has been shown that for a unit n-sphere

(8) $$P(L_1 > \ell) = \Omega(\ell) = \frac{2C_{n-1}}{C_n} \int_{\ell/2}^{1} (1-x^2)^{\frac{n-1}{2}} dx, \quad 0 \leq \ell \leq 2,$$

where $C_n = \pi^{n/2}/\left(\frac{n}{2}\right)!$ = the volume of a unit n-sphere.

From Section 3 it is now sufficient to show that the hazard rate of $\Omega(\ell)$, namely $r(\ell) = -\frac{d}{d\ell}\ell n\Omega(\ell)$ is increasing in ℓ. This then implies (L_1, L_2) are NOD.

Let

(9) $$\xi(\ell) = 2\Omega^2(\ell)\,[C_n/C_{n-1}]^2\,\frac{dr(\ell)}{d\ell}.$$

If $\xi(\ell) > 0$, then $r'(\ell) > 0$ and the result is shown.

From (8) one obtains

$$\xi(\ell) = 2\left[1 - \frac{\ell^2}{4}\right]^{n-1} - (n-1)\ell\left[1 - \frac{\ell^2}{4}\right]^{\frac{n-3}{2}} \int_{\ell/2}^{1} (1-x^2)^{\frac{n-1}{2}} dx$$

or equivalently:

(10) $$\xi(\ell) = \left[1 - \frac{\ell^2}{4}\right]^{\frac{n-3}{2}} \int_{\ell/2}^{1} (1-x^2)^{\frac{n-1}{2}} \{2(n+1)x - (n-1)\ell\} dx.$$

Let the partial integrand $\alpha(\ell, x) = 2(n+1)x - (n-1)\ell$.

For x in the range of the integral one obtains:

$$\alpha(\ell, x) \geq 2\ell \geq 0$$

and it follows that $\xi(\ell) > 0$ and the result is proved.

A recent article by Enns and Ehlers (1988) shows that the dimensional moment of the extended ray length does not depend on the generating body K. The extended ray is the ray projected to the boundary of the convex body G where the convex body $K \subset G$. Redefine L_1 and L_2 as the extended forward and backward rays when K is any convex body and G is an n-sphere. I conjecture that (L_1, L_2) will be NOD. Again intuitively it does not matter where the generating point is chosen within G. If the forward ray is long, then the backward ray will tend to be short. Hence the generating point may be chosen in any interior region K.

Negative orthant dependence between the two rays does not always hold for some convex bodies. It can be shown for a sufficiently elongated rectangle, that NOD is violated. This is a messy but straightforward algebraic exercise and will not be reproduced here.

REFERENCES

BLOCK, H.W., SAVITS, T.H., and SHAKED, M. (1982). Some concepts of negative dependence. *Ann. Prob.* **10** 765–772.

COLEMAN, R. (1969). Random paths through convex bodies. *J. Appl. Prob.* **6** 430–441.

COLEMAN, R. (1973). Random paths through rectangles and cubes. *Metallography* **6** 103–114.

ENNS, E.G. and EHLERS, P.F. (1978). Random paths through a convex region. *J. Appl. Prob.* **15** 144–152.

ENNS, E.G. and EHLERS, P.F. (1980). Random paths originating within a convex region and terminating on its surface. *Austral. J. Statist.* **22**(1) 60–68.

ENNS, E.G. and EHLERS, P.F. (1988). Chords through a convex body generated from within an embedded body. *J. Appl. Prob.* **25** 700–707.

EHLERS, P.F. and ENNS, E.G. (1981). Random secants of a convex body generated by surface randomness. *J. Appl. Prob.* **18** 157–166.

JOAG-DEV, K. and PROSCHAN, F. (1983). Negative association of random variables with applications. *Ann. Statist.* **11** 286–295.

KELLERER, A.M. (1971). Considerations on the random transversal of convex bodies and solutions for general cylinders. *Radiation Research* **47** 359–376.

KELLERER, A.M. (1984). Chord-length distributions and related quantities for spheroids. *Radiation Research* **98** 425–437.

KINGMAN, J.F.C. (1969). Random secants of a convex body. *J. Appl. Prob.* **6** 660–672.

SANTALO, L.A. (1976). *Integral Geometry and Geometric Probability.* Addison-Wesley, Reading, MA.

SANTALO, L.A. (1986). On the measure of line segments entirely contained in a convex body. *Aspects of Mathematics and Its Applications*, North-Holland.

DEPARTMENT OF MATHEMATICS AND STATISTICS
UNIVERSITY OF CALGARY
CALGARY, AB
CANADA T2N 1N4

BIVARIATE MARKOV CHAINS CONTAINING A FAILURE PROCESS

By Nancy Flournoy

The American University

This paper is directed toward the challenge to model dependencies among discrete-state processes. In an earlier motivating application, we used proportional hazards regression models with time-dependent covariates to examine the relationship between relapse following treatment for leukemia, internal biological processes fighting the leukemia, and interventions intended to fix defects in these processes or to stimulate them to behave more aggressively. Complexities in this application led to the introduction of new dependency measures derived from extending Kolmogorov's differential equations. In this paper these dependency measures are interpreted for a bivariate Markov chain where one process is a failure process and another evolves concurrently. Likelihood construction using these dependency measures and others currently used in the context of a failure process are discussed.

1. Introduction. Our interest in measures of association that evolve in time has its genesis in an application involving multiple parallel, non-independent, discrete-state random processes evolving in time. A motivating application involving certain random events and the relationship between these events is described briefly in Section 2 to provide a context for the theoretical development. The principal issue is how to define and model the dependency of a failure process on discrete-state continuous-time random covariates. In Flournoy (1990) we develop measures of dependency among multivariate discrete state stochastic processes by extending the concept of intensity functions using multivariate extensions of Kolmogorov's differential equations. Now these dependency measures are made explicit and interpreted for the situation where one process is a failure process and another (possibly vector-valued) process evolves concurrently. Also we describe

AMS subject classifications. 60J27; 15A51.

Key words and phrases. Intensity functions, Kolmogorov's differential equations, point processes, proportional hazards regression, time-dependent covariates.

The author wishes to acknowledge the comments and suggestions of the reviewer that lead to an improved manuscript.

the incorporation of these dependency measures into the likelihood function for purposes of estimation.

It is important to note that there may not be a unique natural multivariate and conditional extension of Markovian processes or of univariate intensity functions. Indeed a variety of extensions can be created depending on different needs and different criteria. There are two approaches to incorporating the Markov property of lack of memory. One approach assumes that each process is marginally memoryless and characterizes multivariate processes with this property. However, the resulting multivariate processes are not memoryless. This approach was used, for example, by Yadin and Syski (1979) to explore the randomization of intensities in a Markov chain, and by Cogburn (1980) in describing random environments. In contrast to this approach, we examine measures of dependency based on the assumption that two processes are jointly memoryless, yet inhomogeneous.

The rationale for a model that is jointly memoryless, yet inhomogeneous, arises from medical applications such as the one described in the next section. Section 3 provides the necessary notation and a discussion of likelihood construction for one general class of continuous time Markov chains, and for failure processes in particular. In Section 4 a bivariate derivation of intensity functions is obtained that is analogous to Kolmogorov's (1931) univariate derivation. The likelihood is given for the situation in which one process is a failure process. In Section 5 dependencies within the bivariate chain are described and incorporated into the joint likelihood. These dependencies are motivated by one proposed by Cox (1972).

2. A Motivating Scientific Problem. A description of the original motivating application should serve to provide a context to our development. The comparison of several proportional hazards regression models analyzing this application is reported by Weiden, Flournoy, Thomas, Fefer, and Storb (1981) with conclusions reported earlier by Weiden, Flournoy, Thomas, Prentice, Fefer, Buckner, and Storb (1979). To be brief and retain focus, the present description is restricted to the situation surrounding two primary events of interest without regard to competing causes of failure or other covariates.

A bone marrow transplant is preceded by a dose of radiation that destroys the bone marrow that produces leukemic cells. Originally, the effect of this treatment was hypothesized to be dual: the first hypothesis has been demonstrated effectively, namely, that extremely high doses of radiation can be given, increasing the destruction of leukemic cells with the transplant rescuing the patient; the second hypothesis is the subject of inquiry, that if the body's failure to mount an effective fight against the leukemic cells results from a defect in its ability to identify the leukemic cells as foreign objects, and if a healthy immune system is transplanted, one that matches the patient's own immune system so closely that it will accept the patient as itself, the transplanted immune system will recognize that the leukemic cells are strangers and wage an aggressive battle against any that survived the radiation treatment.

Assume that two random processes initiate realizations on the day a patient receives a bone marrow transplant to treat leukemia. One process governs the recurrence of the leukemia, the relapse rate. It is this process that we seek to control and model. A second process governs the emergence of side-effects caused by the bone marrow graft that resemble lupus, an auto-immune disease. If the second hypothesis holds, then an appearance of this side-effect, called graft-versus-host disease, indicates that the transplanted immune system recognizes the foreign cells, and is engaging in a battle that reduces relapse probabilities. However, these battles are frequently fatal. If the hypothesis is true, the battle between the transplanted immune system and its new host, the patient, should be encouraged rather than stifled. But accepting this second hypothesis implies risking the lives of patients believed to be at high risk of later relapse in order to learn how to control the transplant's attack on residual leukemia cells without killing the patient.

The second hypothesis is that the onset of graft-versus-host disease reduces the probability of relapse. Yet if graft-versus-host disease does affect relapse, it is hard to imagine that its effect is constant or even proportional. The need to model inhomogeneous effects led us to consider the Chapman-Kolmogorov equations as a tool for developing multivariate measures of dependency.

3. Notation and the Likelihood. Now we present the necessary notation in the context of a single discrete state process. Then we specialize to a two-state failure process such as is commonly called a death process (see, e.g., Karlin and Taylor (1981)), and establish a correspondence with the notation commonly used in survival analysis. The likelihood is given for the situation in which independent censoring mechanisms terminate the observation of sample paths. A method now used to incorporate dependencies into failure models is described and used to motivate alternative approaches. One such approach in which dependencies are defined by straightforward extensions of classical univariate Markov chain theory is described in subsequent sections. Connections to the theory of counting processes are not pursued at this time.

3.1. The Likelihood for a Univariate Markov Chain. To introduce notation and elementary definitions, consider a continuous time Markov chain $U(t)$, or simply U, with a finite discrete-state space $\theta = 1, 2, \ldots, \Omega_u$ and continuous time parameter $t \in T = [0, \infty)$. Throughout, lower case Roman letters are used to denote times and Greek letters are used to denote states. When considering functions of a process U and the context is clear, we omit the 'U'. Let $H^{t|s} \equiv H^{t|s}(U)$ be a $\Omega_u \times \Omega_u$ matrix of *transition probabilities* with elements

$$h_{\theta|\sigma}^{t|s} \equiv h_{\theta|\sigma}^{t|s}(U) \equiv P\{U(t) = \theta | U(s) = \sigma\}, \quad \sigma, \theta = 1, 2, \ldots, \Omega_u, \tag{1}$$

each of which is a probability that the process is in state θ at time t given it was in state σ at time s. Note that in this notation the current time t and state θ are given before the vertical line, and the prior time s and state σ are given after the vertical line. Of course, some transitions may have probability zero.

We adopt the assumption that the process U cannot return to a state once it has been there and left, which implies that $H^{t|s}$ is an upper triangular matrix for all $0 < s \leq t$. Also note that, in general, $H^{r|s}$ and $H^{t|r}$ do not commute. These features are significant for solving matrix versions of Kolmogorov's differential equations. Furthermore, these features hold for the motivating application where graft-versus-host disease (the event), relapse of leukemia (the failure), and death from causes other than leukemia together with termination of observation (the censorings) are not reversible.

Assuming the limits exist, let Q^t be a $\Omega_u \times \Omega_u$ dimensional matrix with elements the univariate *intensity functions* $q^t_{\theta|\sigma}(U)$ defined as by Kolmogorov (1931) to be the derivatives of inhomogeneous transition probabilities with time s evaluated at the later time t. The derivative of $H^{t|s}$ taken elementwise yields the upper triangular matrix

$$Q^t \equiv \frac{\partial}{\partial t} H^{t|s}(U)|_{s=t}. \tag{2}$$

Since the diagonal elements of $H^{t|s}$ are positive except in degenerate cases, the inverse $(H^{t|s})^{-1}$ generally exists and so will solutions to Kolmogorov's differential equations

$$\frac{\partial}{\partial t} H^{t|s} = H^{r|s} \frac{\partial}{\partial t} H^{t|r}|_{r=t} \equiv H^{t|s} Q^t \tag{3}$$

for $s \leq r \leq t$.

As pointed out by Chang and Yang (1990), standard reference books on stochastic processes (e.g. Chang (1980)) discuss solutions only of homogeneous Markov transition probabilities, and solutions for inhomogeneous Markov transition probabilities (e.g. for (3)) remain very much application dependent. This remains true although Feller (1940) proved existence and uniqueness theorems for the solutions of inhomogeneous Markov chains; and although more recently the solutions of inhomogeneous Markov chains have been further characterized by Getz (1976) and Hartfiel (1985). Therefore, it is of practical significance that maximum likelihood estimates can be obtained without requiring a general solution to (3). This is the case in the present model.

In fact, solutions are required only for the diagonal elements of $H^{t|s}$ as we now explain. Since $H^{t|s}$ is an upper triangular matrix, both the derivative of $H^{t|s}$ and its inverse are upper triangular matrices, and consequently, the product $(H^{t|s})^{-1}\partial H^{t|s}/\partial t$ is upper triangular. In particular, each diagonal element $q^t_{\theta|\theta}$, $\theta = 1, 2, \ldots, \Omega_u$ in the solution of (3) equals the product of the respective diagonal elements in the matrices $(H^{t|s})^{-1}$ and $\partial H^{t|s}/\partial t$ yielding a system of Ω_u differential equations for each s and t with solutions

$$h^{t|s}_{\theta|\theta} = \exp\{\int_s^t q^r_{\theta|\theta} \mathrm{dr}\}, \quad \theta = 1, 2, \ldots, \Omega_u. \tag{4}$$

Given $n = 1, \ldots, N$ independent realizations of $U(t)$, a likelihood can be constructed using probability laws for sample paths. For the nth realization, let τ_{nm} be the time of the mth transition for $m = 1, \ldots, M_n$ and let $\{\theta_{n0}, \theta_{n1}, \ldots, \theta_{nM_n}\}$ denote the distinct states visited. For notational convenience, we assume that the realizations are of finite duration so that the M_nth transition is to an absorbing state. Karlin and Taylor (1981, pp. 145–149) describe the process of determining probability laws for sample paths from homogeneous Markov chains and point out that the process is analogous for sample paths from general continuous time Markov chains. Following their argument, let k be fixed and $K > 0$ be an arbitrary positive integer. Then for the nth realization $[U_n(t),\ 0 \leq t \leq \tau_{M_n}]$,

$$h_{\theta_{nm}|\theta_{nm}}^{\tau_{n(m+1-(k/K))}|\tau_{nm}} = P\left\{U_n(\tau) = \theta_{nm},\ \tau_{nm} \leq \tau \leq \tau_{n(m+1-(k/K))}\right\}$$

and (assuming $U(t)$ is separable),

$$\lim_{K \to \infty} h_{\theta_{nm}|\theta_{nm}}^{\tau_{n(m+1-(k/K))}|\tau_{nm}}$$

may be considered as just

$$h_{\theta_{nm}|\theta_{nm}}^{\tau_{n(m+1)}|\tau_{nm}} = P\left\{U_n(\tau) = \theta_{nm},\ \tau_{nm} < \tau < \tau_{n(m+1)}\right\} = \exp\left\{\int_{\tau_{nm}}^{\tau_{n(m+1)}} q_{\theta|\theta}^{\tau} \mathrm{d}\tau\right\},$$

where the last equality follows from (4). Also the probability that $U(t)$ remains at θ_{nm} from τ_{nm} to $\tau_{n(m+1)}$ and then jumps to $\theta_{n(m+1)} \neq \theta_{nm}$ in time $\mathrm{d}\tau$ is

$$h_{\theta_{nm}|\theta_{nm}}^{\tau_{n(m+1)}|\tau_{nm}} q_{\theta_{n(m+1)}|\theta_{nm}}^{\tau_{n(m+1)}} \mathrm{d}\tau.$$

Therefore, assuming the probability of more than one jump in the time span from $\tau_{n(m+1)}$ to $\tau_{n(m+1)} + \mathrm{d}\tau$ goes to zero as $\mathrm{d}\tau \to 0^+$ and given the initial states $\{\theta_{n0},\ n = 1, \ldots, N\}$, the likelihood of $[U_n(t),\ 0 \leq t \leq \tau_{M_n}]$ is

$$\begin{aligned}(5) \qquad \ell_n &= \left[\prod_{m=0}^{M_n-1} h_{\theta_{nm}|\theta_{nm}}^{\tau_{n(m+1)}|\tau_{nm}} q_{\theta_{n(m+1)}|\theta_{nm}}^{\tau_{n(m+1)}}\right] \\ &= \prod_{m=0}^{M_n-1} \exp\left\{\int_{\tau_{nm}}^{\tau_{n(m+1)}} q_{\theta_{nm}|\theta_{nm}}^{\tau} \mathrm{d}\tau\right\} q_{\theta_{n(m+1)}|\theta_{nm}}^{\tau_{n(m+1)}},\end{aligned}$$

where the last equality follows from (4). Note in the first expression in (5) that the likelihood depends on diagonal elements from $H^{t|s}$ that correspond to the time spent in those states that were actually visited and off-diagonal elements from Q^t that correspond to the actual state changes at observed transition times. The second expression in (5) shows that, by solving for only the diagonal elements of $H^{t|s}$, the likelihood can be written strictly in terms of elements of Q^t. This fact facilitates the modeling dependencies on failure processes.

3.2. The Likelihood of a Failure Process. A simple *failure process*, namely, a two state process with one absorbing state is now described, and its likelihood is derived assuming that independent censoring mechanisms terminate observation of the sample paths. Let V be a failure process with state space $\Omega_v = \{0, 1\}$, where state 1 denotes failure. Then the matrix $H^{t|s}$ of transition probabilities simplifies because the transition matrix depends on only one element, i.e.,

$$H^{t|s}(V) = \begin{bmatrix} h_{0|0}^{t|s} & h_{1|0}^{t|s} \\ h_{0|1}^{t|s} & h_{1|1}^{t|s} \end{bmatrix} = \begin{bmatrix} 1 - h_{1|0}^{t|s} & h_{1|0}^{t|s} \\ 0 & 1 \end{bmatrix}, \tag{6}$$

with the consequence that Q^t also simplifies:

$$Q = \begin{bmatrix} -q_{1|0}^t & q_{1|0}^t \\ 0 & 0 \end{bmatrix}. \tag{7}$$

Clearly, the failure process is completely specified by $q_{1|0}^t$ and the system of differential equations (3) reduces to the single one-to-one correspondence familiar in survival analysis with the solution:

$$S(t) = \exp\left\{-\int_0^t q_{1|0}^t \mathrm{dr}\right\}, \tag{8}$$

where $S(t) \equiv h_{0|0}^{t|s}|_{s=0}$, the probability that no state changes occur up to time t, is called the *survival function* and the *hazard function*, commonly denoted by $\lambda(t)$, is the intensity of failure $q_{1|0}^t$.

The failure process we consider is extended to a three state process to accommodate censoring; the initial state θ_{n0} is survival, and transitions are possible to the absorbing states, θ_{n1} and θ_{n2}, of failure and censoring, respectively. Since only the first transition is observable, realizations are described by the observations $\{(\tau_n, \theta_n),\ n = 1, \ldots, N\}$, where $\tau_n \equiv \min\{\tau_{n1}, \tau_{n2}\}$ is the minimum of the failure and censoring times for the nth observation, and θ_n equals one if the failure state θ_{n1} is observed and zero if the censoring state θ_{n2} is observed. Using (7) and assuming an independent censoring mechanism, Kalbfleisch and Prentice (1980) show that the likelihood (5) for the nth observation is

$$\ell_n = \left(q_{1|0}^{\tau_n}\right)^{\theta_n} \exp\left\{-\int_0^{\tau_n} q_{1|0}^{r} \mathrm{dr}\right\}. \tag{9}$$

One approach for incorporating dependencies into a Poisson survival model is becoming popular in applications. If $\mathbf{U}_n(t)$ is a row vector of covariate processes for the nth observation and $\boldsymbol{\beta}$ is a row vector of unknown parameters, then a model

$$q_{[1|0]_n}^t = h(\mathbf{U}_n(t), \boldsymbol{\beta}) \tag{10}$$

relating the nth observed hazard function $q_{[1|0]_n}^t$ to covariates $\mathbf{U}_n(t)$ is selected. Then dependencies are evaluated by replacing $q_{[1|0]_n}^t$ with $h(\mathbf{U}_n(t), \boldsymbol{\beta})$ in (9) or in Cox's (1972) partial likelihood and evaluating $\boldsymbol{\beta}$ by maximum likelihood methods. The most famous example of this approach to evaluating the dependency of a covariate process $U(t)$ on a failure process $V(t)$ is the use of Cox's (1972) proportional hazards regression model

$$q^t_{[1|0]_n} = h(t) \exp\left\{\mathbf{U}_n(t)\boldsymbol{\beta}^T\right\},$$

where $h(t) \geq 0$ is an arbitrary function of time and each covariate acts proportionally on the failure intensity $q^t_{1|0}$, in Cox's (1975) partial likelihood. The partial likelihood ℓ_{pn} for the nth observation is

$$\ell_{pn} = \left[\left(q^{\tau_n}_{[1|0]_n}\right) \Big/ \sum_{j=1}^{R_n} q^{\tau_n}_{[1|0]_j}\right]^{\theta_n}, \tag{11}$$

where R_n is the set $\{j : \tau_j \geq \tau_n\}$ of realizations that have not failed or been censored at the time τ_n that the nth observation fails. Heuristically, $\theta_n = 1$ for each failure, and the corresponding ratio in the partial likelihood is the probability that, of all the realizations at risk at τ_n, it was the nth realization that failed.

When $U(t)$ is deterministic, the insertion of (10) into (9) or (11) can be justified using conditionality arguments. However, when $U(t)$ is a random function of time, insertion of (10) into (9) or into (11) does not take the Jacobian of the transformation into account and clearly their probabilistic content is altered (see Flournoy (1980) and Yashin and Arjas (1988)). The increasing number of applications that use (10) in (9) or (11) when covariates $U(t)$ are random is one motivation for exploring alternative approaches.

Another motivation is to establish a framework which extends the types of dependencies that can be modeled beyond those that can be expressed in the form of (10). Note that in (10) as written, the covariate process(es) maps into the failure intensity at parallel times t. Meaningful dependencies between $U(t)$ and $V(t)$ may involve lag or lead times, or it may be other functions such as the rate at which states change in $U(t)$ that most directly affect $V(t)$. An obvious extension of (10) is $q^t_{[1|0]_n} = h(\mathbf{U}_n(s), \boldsymbol{\beta},\ 0 \leq s \leq t)$. But this extension is so general that forms of $h(\cdot)$ that are meaningful in applications must be determined for the extension to be useful.

4. A Failure Process in a Bivariate Markov Chain. A common practice in modeling a Markov chain with a multivariate state space is to reparameterize to a univariate Markov chain by expressing the state space as a list of all possible points in the multidimensional space. Such an approach yields a parameter for every possible transition which makes the model intractable for most realistic applications and almost certainly intractable if the processes are taken to be inhomogeneous. One way to reduce the complexity of inhomogeneous Markov chains is to assume that functional dependencies exist between the transitions. This is the role of the proportional hazards assumption in Cox's (1972) regression model for failure data. We characterize a multivariate representation of the Markov chain in order to facilitate the conceptualization of other types of functional dependencies within a more general probability framework.

Let $[U(t),\ V(t)]$ be a two-dimensional row vector of discrete-state stochastic processes, where $V(t)$ is a univariate failure process and $U(t)$, without loss of generality, might be vector-valued. Note that the elements of $[U(t),\ V(t)]$ are jointly indexed by a single continuous time parameter $t \in T = [0, \infty)$. When 't' is not critical to the context, we write $[U, V] \equiv [U(t),\ V(t)]$.

Denote the bivariate state space of $[U(t),\ V(t)]$ by $\Omega = \Omega_u \times \Omega_v$. For some t, the probability of being in certain states may be zero. For example, when a failure in $V(t)$ is fatal to $U(t)$, the combined sample space might degenerate to a single point representing the absorbing state of the entire system. Let $[\theta^t_{ui},\ \theta^t_{vj}]$ denote an element in the bivariate state space Ω at time t. Rather than maintain the superscript 't' in the state space notation, we indicate the states at different times by another letter σ. That is, $[\theta_{ui},\ \theta_{vj}]$ and $[\sigma_{ui},\ \sigma_{vj}]$ denote the ith state of process U and the jth state of process V at two different times.

The state space Ω_u of U is determined by the application, whereas there are only two states for the failure process V. Let θ_{v1} or '$V = 1$' denote the state of failure while θ_{v0} or '$V = 0$' denotes no failure. Thus in the bivariate state space, $[\theta_{ui},\ \theta_{v0}] \equiv [\theta_{ui},\ 0]$ and $[\theta_{ui},\ \theta_{v1}] \equiv [\theta_{ui},\ 1]$ denote $U(t) = \theta_i$ jointly with survival or failure, respectively. When it is not necessary to reference a specific state, we let $(\theta_u,\ \sigma_u)$ and $(\theta_v,\ \sigma_v)$ denote arbitrary states of U in Ω_u and of V in Ω_v at times t, and s, respectively. The nth realization of U and V is denoted by U_n and V_n, respectively, and the mth state visited by the nth realization is denoted by $[\theta_u,\ \theta_v]_{nm}$.

DEFINITION 1. Now let $H^{t|s} \equiv H^{t|s}(U, V)$ denote a $2\Omega_u \times 2\Omega_u$ dimensional matrix of *joint transition probabilities*, where $H^{t|s}$ can be partitioned into four $\Omega_u \times \Omega_u$ upper triangular submatrices,

$$H^{t|s} = \begin{bmatrix} H^{t|s}_{0|0} & H^{t|s}_{1|0} \\ 0 & H^{t|s}_{1|1} \end{bmatrix}, \tag{12}$$

where the 0 submatrix reflects the assumption that the state of failure is absorbing. The matrix $H^{t|s}_{1|0}$ is upper triangular with elements defined by the probability of transition in U from state i at time s to state j at time t $(i, j \in \Omega_u)$ jointly with the transition of V from survival (0) into failure (1):

$$\begin{aligned} h_{\theta_u 1|\sigma_u 0} &\equiv h^{t|s}_{\theta_u 1|\sigma_u 0}(U, V) \\ &\equiv P\{U(t) = \theta_{uj},\ V(t) = 1 | U(s) = \sigma_{ui},\ V(s) = 0\}; \end{aligned} \tag{13}$$

$H^{t|s}_{0|0}$ and $H^{t|s}_{1|1}$ are also upper triangular matrices with elements defined to be probability transitions function of U joint with continued survival $(V = 0)$ from s to

t and with continued failure ($V = 1$), respectively. In referencing the transition probabilities (13), we omit the notation '(U, V)' for simplicity when it does not play a role. The superscript '$t|s$' is also omitted for notational simplicity whenever it is not essential to the context.

The nature of a failure's impact on U is expressed through $H_{1|1}$. For example, $H_{1|1}$ may be the identity matrix I_{Ω_u} implying $U(t)$ is frozen in its current state at the time of failure, or $U(t)$ may enter a single absorbing state regardless of its current state at the time of failure. Often when we focus on how failures depend on a coprocess, it is natural to treat $H_{1|1}$ as ancillary and restrict consideration of joint transition probabilities to the upper submatrices in (12), namely $H_{1|0}$, in which failure occurs, and $H_{0|0}$, in which failure does not occur. $H_{1|1}$ is ancillary when it does not contain parameters in common with $H_{0|0}$ or $H_{1|0}$ which are conditioned on prior survival.

DEFINITION 2. Two discrete state continuous time random processes U and V form a *bivariate Markov chain* if $[U, V]$ have joint loss of memory, that is, if for any s, r, and t such that $0 \le s \le r \le t < \infty$,

$$\begin{aligned} P\{[U(t), V(t)] = [\theta_u, \theta_v] | [U(s), V(s)], \forall s : s \in [0, r]\} \\ = P\{[U(t), V(t)] = [\theta_u, \theta_v] | [U(r), V(r)]\}. \end{aligned} \tag{14}$$

A bivariate extension of the Chapman-Kolmogorov equations follows directly:

$$H^{t|s}(U, V) = H^{r|s}(U, V) H^{t|r}(U, V). \tag{15}$$

Note that the bivariate Chapman-Kolmogorov equation (15) implies that $[U, V]$ is a bivariate Markov chain and does not imply that either marginal process U or V is a Markov chain, that is, (15) does not imply that the univariate Chapman-Kolmogorov equations hold for U or V. Indeed, the bivariate loss of memory property (15) will hold also for marginal processes only under very restrictive conditions (see Yadin and Syski (1979)).

DEFINITION 3. Let two random processes U and V form a bivariate Markov chain and assume that the derivatives of (13) exist with respect to t, then their $\Omega \times \Omega$-dimensional *joint intensity matrix* is defined analogous to (2):

$$Q^t \equiv Q^t(U, V) \equiv \frac{\partial}{\partial t} H^{t|s}(U, V)|_{s=t} \equiv \frac{\partial}{\partial t} H^{t|s}|_{s=t}. \tag{16}$$

The elements of $Q^t(U, V)$ are denoted by $q^t_{\theta_u \theta_v | \sigma_u \sigma_v}$, and are called *joint intensity functions.*

Extending the forward differential equations by taking the derivative of both sides of (15) elementwise with respect to t, and evaluating the result at $r = t$, gives bivariate forward equations for all s and t analogous to (3). When $[U, V]$ is a bivariate Markov chain and V is a failure process, the forward equations are

(17) $$\frac{\partial}{\partial t}H^{t|s} \equiv \begin{bmatrix} \frac{\partial}{\partial t}H^{t|s}_{0|0} & \frac{\partial}{\partial t}H^{t|s}_{1|0} \\ 0 & \frac{\partial}{\partial t}H^{t|s}_{1|1} \end{bmatrix} = \begin{bmatrix} H^{t|s}_{0|0} & H^{t|s}_{1|0} \\ 0 & H^{t|s}_{1|1} \end{bmatrix} \begin{bmatrix} Q^{t}_{0|0} & Q^{t}_{1|0} \\ 0 & Q^{t}_{1|1} \end{bmatrix}$$
$$= \begin{bmatrix} H^{t|s}_{0|0}Q^{t}_{0|0} & H^{t|s}_{0|0}Q^{t}_{1|0} + H^{t|s}_{1|0}Q^{t}_{1|1} \\ 0 & H^{t|s}_{1|1}Q^{t}_{1|1} \end{bmatrix},$$

and since $H^{t|s}$ is upper triangular with positive diagonal elements (except in degenerate cases), the solution of (17) for the diagonal elements in $H^{t|s}$ is analogous to (4):

(18) $$h^{t|s}_{\theta_j|\theta_j} = \exp\left\{\int_s^t q^{r}_{\theta_j|\theta_j}\,\mathrm{dr}\right\}; \quad j = 0,1,;\ \theta = 1,2,\ldots,\Omega_u.$$

Recall that in the case where a failure in V freezes the coprocess U, $H_{1|1}$ is the identity matrix and hence (17) simplifies considerably because $Q^{t}_{1|1} \equiv 0$.

Assuming for notational convenience that the M_nth transition is to an absorbing state, the likelihood for the nth realization is of the same form as (5) except that the realized states $[\theta_u, \theta_v]_{nm}$ are now binary with $\theta_v = 0,1$; $m = 1,\ldots,M_n$:

(19) $$\ell_n = \prod_{m=0}^{M_n-1} h^{\tau_{n(m+1)}|\tau_{nm}}_{[\theta_u\theta_v]_{nm}|[\theta_u\theta_v]_{nm}} q^{\tau_{n(m+1)}}_{[\theta_u\theta_v]_{n(m+1)}|[\theta_u\theta_v]_{nm}}$$
$$= \prod_{m=0}^{M_n-1} \exp\left\{\int_{\tau_{nm}}^{\tau_{n(m+1)}} q^{r}_{[\theta_u\theta_v]_{nm}|[\theta_u\theta_v]_{nm}}\,\mathrm{dr}\right\} q^{\tau_{n(m+1)}}_{[\theta_u\theta_v]_{n(m+1)}|[\theta_u\theta_v]_{nm}},$$

where the last equality follows from (18). Hence the likelihood can be constructed using the diagonal elements of $H^{t|s}$, modeled in terms of the diagonal elements of Q^t using (18), together with the off-diagonal elements of Q^t as was the case for the univariate model. If a failure in $V(s)$ freezes $U(t)$ for $t \geq \inf\{s : V(s) = 1\}$ so that $Q^{t}_{1|1} = 0$, (19) becomes an extension of (9):

(20) $$\ell_n = \left(\prod_{m=0}^{M_n-1} \exp\left\{\int_{\tau_{nm}}^{\tau_{n(m+1)}} q^{r}_{[\theta_u 0]_m|[\theta_u 0]_m}\,\mathrm{dr}\right\}\right)\left(\prod_{m=0}^{M_n-2} q^{\tau_{n(m+1)}}_{[\theta_u 0]_{(m+1)}|[\theta_u 0]_m}\right)$$
$$\times \left(q^{\tau_{n(m+1)}}_{[\theta_u 0]M_n|[\theta_u 0]_{(M_n-1)}}\right)^{1-\delta_{nM_n}} \left(q^{\tau_{n(m+1)}}_{[\theta_u 1]M_n|[\theta_u 0]_{(M_n-1)}}\right)^{\delta_{nM_n}},$$

where $\delta_{nM_n} = 1$ if the M_nth transition is to failure and 0 otherwise.

5. Dependencies Within the Bivariate Chain. We have established the analogy between likelihood construction for univariate and bivariate discrete state

Markov chains and shown that their respective likelihoods can be modeled strictly in terms of the intensity matrix Q^t. However, the joint intensities in Q^t constitute a limited class of dependencies. We now describe two other classes of dependencies defined in Flournoy (1990) in the context of a failure process; other dependencies in time and space could also be considered.

First we consider the process $U(t)$ as it depends on prior realizations of $U(s)$ and $V(s)$ jointly. There are a variety of applications in which the dependency of $U(t)$ on $U(s)$, given $V(s) = 0$ for $0 \leq s \leq t$, is of interest. For example, consider a gross simplification of an educational application developed by Debanne, Rowland, Eielefeld, and Maw (1989) in which U has states that are school grades (i.e. freshman, sophomore, junior, senior), and V is equal to 0 if a student remains in high school and is equal to 1 if the student drops out. One dependency of interest involves the transition through grades of school among students who remain in school.

The second class of dependency we consider is one in which the failure process $V(t)$ at time t depends on itself $V(s)$ at a prior time s together with the coprocess at both times, namely $U(s)$ and $U(t)$. One might wish to model this type of dependency, for example, to study the way in which failure ($V(t) = 1$) depends on transitions in the coprocess ($U(s)$ to $U(t)$) given prior survival ($V(s) = 0$). In the next section, we show how these two classes of dependencies relate to each other and to the joint process and thereby their role in the likelihood.

5.1. Conditional Dependencies. Let $H^{t|s}(U\|[U,V]) \equiv H(U(t)\|[U(s),V(s)])$ be a $\Omega_u \times 2\Omega_u$ matrix of *conditional transition probabilities* with U at time t conditioned on itself together with V at time s, that is, with elements

$$h^{t|s}_{\theta_u|\sigma_u\sigma_v} \equiv P\left\{U(t) = \theta_u \| [U(s), V(s)] = [\sigma_u, \sigma_v]\right\}, \qquad \theta_u,\ \sigma_u = 1, \ldots, \Omega_u;\ \sigma_v = 0, 1. \tag{21}$$

The matrix of conditional transition probabilities is partitioned with respect to $V(s)$ into two upper triangular submatrices in order to focus on the conditional transition probabilities at time t given $U(s)$ joint with survival ($V(s) = 0$) separately from conditional transition probabilities given $U(s)$ joint with failure ($V(s) = 1$):

$$\begin{aligned} H^{t|s}(U\|[U,V]) &\equiv [H(U\|[U,0]), H(U\|[U,1])] \\ &\equiv \left[H^{t|s}(U\|[U,0]), H^{t|s}(U\|[U,1])\right]. \end{aligned} \tag{22}$$

Similar to the simplification that is possible in the joint transition matrix, $H^{t|s}(U\|[U, 1])$ may equal I_{Ω_u} or $U(t)$ may enter a single absorbing state when a failure (i.e. $V(t) = 1$) is fatal to the entire system.

The *conditional intensity matrix* is defined, analogous to (2), through the derivatives of (21) to be

$$(23) \qquad Q^t(U\|[U,V]) \equiv \left[\frac{\partial}{\partial t} H^{t|s}(U\|[U,V])\right]_{s=t},$$

with elements $q^t_{\theta_u|\sigma_u\sigma_v}$. Partition $Q^t(U\|[U,V])$ in (23) compatibly with $H^{t|s}(U\|[U, V])$ in (22) to obtain:

$$(24) \qquad \begin{aligned} Q^t(U\|[U,V]) &\equiv \left[\frac{\partial}{\partial t} H^{t|s}(U\|[U,0]),\ \frac{\partial}{\partial t} H^{t|s}(U\|[U,1])\right]_{s=t} \\ &\equiv \left[Q^t(U\|[U,0]),\ Q^t(U\|[U,1])\right]. \end{aligned}$$

The conditional dependencies are aggregations of the joint transition probabilities with

$$(25) \qquad \begin{aligned} H(U\|[U,0]) &= H_{0|0} + H_{1|0}, \qquad & H(U\|[U,1]) &= H_{1|1}, \\ Q(U\|[U,0]) &= Q_{0|0} + Q_{1|0}, \qquad & Q(U\|[U,1]) &= Q_{1|1}. \end{aligned}$$

Using (18) in (25), the diagonal elements of $H(U\|[U,0])$ are

$$h^{t|s}_{\theta_u|\theta_u\sigma_v} = \exp\left\{\int_s^t q^r_{\theta_u 0|\theta_u 0} \mathrm{dr}\right\} + \exp\left\{\int_s^t q^r_{\theta_u 1|\theta_u 0} \mathrm{dr}\right\}.$$

In the situation where $Q^t_{1|0} = -Q^t_{0|0}$, the number of parameters to be modeled in the joint likelihood (20) is reduced and the analogy between (20) and the univariate likelihood (9) for a failure process is strengthened. This simplification occurs when $Q^t(U\|[U,0]) = 0$, that is, when

$$(26) \qquad \frac{\partial}{\partial t} P\{U(t) = j|U(s) = i,\ V(s) = 0\}\,|_{s=t} = 0, \qquad i,j = 1,\ldots,\Omega_u.$$

Equation (26) is attained when the conditional transition probabilities are constant functions of t for each i. An important special case in which (26) is attained is the case in which $H^{t|s}(U\|[U,0]) = I_{\Omega_u}$, that is, the case in which $U(r)$ is constant for $s \le r \le t$ almost everywhere. In many applications, there is no scientific basis for the assumption that $Q_{0|0} = -Q_{1|0}$. Consequently, the reduction of parameters in the likelihood (9) that resulted from such an equality (see 7) for the univariate failure process does not apply generally to the likelihood (20) for a bivariate chain that contains a failure process.

We now introduce a second class of dependencies before discussing the relationship of both classes to the bivariate likelihood function.

5.2. Cross-conditional Dependencies. Let $H(V|U,[U,V]) \equiv H^{t|t,s}(V|U,[U,V])$ be a $2\Omega_u \times 2\Omega_u$ dimensional matrix with elements

(27) $$h^{t|t,s}_{\theta_v|\theta_u\sigma_u\sigma_v} \equiv P\{V(t)=\theta_v|U(t)=\theta_u,\ [U(s),V(s)]=[\sigma_u,\sigma_v]\},$$
$$\theta_u,\sigma_u=1,\ldots,\Omega_u;\ \sigma_u,\sigma_v=0,1,$$

that are called *cross-transition probabilities.* Cross-transition probabilities are dependencies between the failure process V at time t conditioned on the coprocess at the same time t as well as on the joint process $[U,V]$ at a prior time s. Thus, for example, the elements of $H^{t|t,s}(1|U,[U,0])$ are the probabilities of having failed by time t given survival at time s and given the state of process U at s and at t with states $[\sigma_u,0]$ and $[\theta_u,1]$:

(28) $$h^{t|t,s}_{\theta_v|\theta_u\sigma_u\sigma_v}\Big|_{\substack{V(s)=0\\V(t)=1}} = h^{t|t,s}_{1|\theta_u\sigma_u0} = 1-h^{t|t,s}_{0|\theta_u\sigma_u0},$$
$$\theta_u,\sigma_u=1,\ldots,\Omega_u.$$

Through the derivatives of (27), define a *cross-intensity matrix* with V at time t conditioned on U at time t and itself together with U at time s to be

(29) $$Q^t(V|U,[U,V]) \equiv \left[\frac{\partial}{\partial t}H^{t|t,s}(V|U,[U,V])\right]_{s=t},$$

with elements $q^t_{\theta_v|\theta_u\sigma_u\sigma_v}$ that are called *cross-intensity functions.*

Now $H(V|U,[U,V])$ and $Q(V|U,[U,V])$ each can be partitioned into four $\Omega_u\times\Omega_u$ dimensional submatrices. But these matrices can be defined in terms of one submatrix as is seen from an expanded representation of $H(V|U,[U,V])$:

(30) $$H(V|U,[U,V]) = \begin{bmatrix} H(0|U,[U,0]) & J-H(0|U,[U,0]) \\ 0 & J \end{bmatrix},$$

where J is a $\Omega_u\times\Omega_u$ dimensional matrix with each element identically one reflecting the absorbing state of the failure process. Note that, for convenience, the configuration of the cross-transition functions in (30) deviates from the conventional univariate representation of transition probabilities in which the row probabilities sum to one. From (30) we have, by definition (29), that

(31) $$Q(0|U,[U,0]) = -Q(1|U,[U,0])$$

analogous to (7).

5.3. Dependency Measures and the Likelihood. Theorem 1 states that the joint likelihood can be written in terms of conditional and cross-conditional intensity functions. Let $\delta_{M_n}=1$ if the M_nth transition of the nth realization is to failure and 0 otherwise.

Theorem 1. If $Q^t_{1|1}=0$, then the likelihood (20) for the nth realization of a bivariate Markov chain $[U,V]$ in which V is a failure process can be rewritten as

$$(32)\quad \ell_n = \left[\prod_{m=0}^{M_n-1} \exp\left\{\int_{\tau_{nm}}^{\tau_{n(m+1)}} \left(q^{r}_{[\theta_u]_{n(m+1)}|[\sigma_u 0]_{nm}} - \delta^{r}_{\sigma_u\theta_u} q^{r}_{[1]_{n(m+1)}|[\theta_u\sigma_u 0]_{nm}}\right) \mathrm{dr}\right\}\right]$$

$$\prod_{m=0}^{M_n-2} \left(q^{\tau_{n(m+1)}}_{[\theta_u]_{n(m+1)}|[\sigma_u 0]_{nm}} - \delta^{\tau_{n(m+1)}}_{\sigma_u\theta_u} q^{\tau_{n(m+1)}}_{[1]_{n(m+1)}|[\theta_u\sigma_u 0]_{nm}}\right)$$

$$\times \left(q^{\tau_{nM_n}}_{[\theta_u]_{nM_n}|[\sigma_u 0]_{n(M_n-1)}} - \delta^{\tau_{n(m+1)}}_{\sigma_u\theta_u} q^{\tau_{nM_n}}_{[1]_{nM_n}|[\theta_u\sigma_u 0]_{n(M_n-1)}}\right)^{1-\delta_{nM_n}}$$

$$\times \left(q^{\tau_{nM_n}}_{[1]_{nM_n}|[\theta_u\sigma_u 0]_{n(M_n-1)}}\right)^{\delta_{M_n}},$$

where $\delta^{t}_{\sigma_u\theta_u}$ is the Kronecker delta function.

Proof. The joint transition probabilities factor into two components of corresponding elements from $H(U|[U,V])$ and $H(V|U,[U,V])$:

$$(33)\quad \begin{aligned} h^{t|s}_{\theta_u\theta_v|\sigma_u\sigma_v} &= P(V(t)=\theta_v|U(t)=\theta_u,\ V(s)=\sigma_v,\ U(s)=\sigma_u) \\ &\quad \times \frac{P(U(t)=\theta_u, V(s)=\sigma_v, U(s)=\sigma_u)}{P(V(s)=\sigma_v, U(s)=\sigma_u)}, \\ &= h^{t|t,s}_{\theta_v|\theta_u\sigma_u\sigma_v} h^{t|s}_{\theta_u|\sigma_u\sigma_v}. \end{aligned}$$

Apply the chain rule to (33) in taking the derivative with respect to t to obtain

$$(34)\quad \frac{\partial}{\partial t} h^{t|s}_{\theta_u\theta_v|\sigma_u\sigma_v} = \left[h^{t|t,s}_{\theta_v|\theta_u\sigma_u\sigma_v}\right]\left[\frac{\partial}{\partial t} h^{t|s}_{\theta_u|\sigma_u\sigma_v}\right] + \left[h^{t|s}_{\theta_u|\sigma_u\sigma_v}\right]\left[\frac{\partial}{\partial t} h^{t|t,s}_{\theta_v|\theta_u\sigma_u\sigma_v}\right].$$

Note that

$$(35)\quad \begin{aligned} h^{t|t,s}_{\theta_v|\theta_u\sigma_u\sigma_v} &= P(V(t)=\theta_v|U(t)=\theta_u,\ U(s)=\sigma_u,\ V(s)=\sigma_u) \\ &= \frac{P(V(t)=\theta_v, V(s)=\sigma_v|U(t)=\theta_u, U(s)=\sigma_u)}{P(V(s)=\sigma_v|U(t)=\theta_u, U(s)=\sigma_u)} \end{aligned}$$

and (35) evaluated at $s=t$ is equal to 0 if $\sigma_v \neq \theta_v$ and is equal to 1 if $\sigma_v = \theta_v$. Similarly,

$$(36)\quad \begin{aligned} h^{t|s}_{\theta_u|\sigma_u\sigma_v}\big|_{s=t} &= P(U(t)=\theta_u|U(s)=\sigma_u,\ V(s)=\sigma_v)\big|_{s=t} \\ &= \frac{P(U(t)=\theta_u, U(s)=\sigma_u|V(s)=\sigma_v)}{P(U(s)=\theta_u|V(s)=\sigma_v)}\Big|_{s=t} = \delta^{t}_{\sigma_u\theta_u}. \end{aligned}$$

Evaluating (34) at $s=t$, using (35) and (36), yields a joint intensity function that is a weighted sum of a conditional and a cross-intensity function:

$$(37) \qquad q^t_{\theta_u\theta_v|\sigma_u\sigma_v} = \delta^t_{\theta_v\sigma_v} q^t_{\theta_u|\sigma_u\sigma_v} + \delta^t_{\sigma_u\theta_u} q^t_{\theta_v|\theta_u\sigma_u\sigma_v}.$$

Evaluate the right hand side of (37) for $q^t_{\theta_u 0|\sigma_u 0}$ and $q^t_{\theta_u 1|\sigma_u 0}$. Insert the results into (20), and by (31), replace $q^t_{[0]nm|[\theta_u\sigma_u 0]_{n(m-1)}}$ with $-q^t_{[1]nm|[\theta_u\sigma_u 0]_{n(m-1)}}$ for each t and m to yield the theorem. ||

Cox defined the hazard function depending on a covariate value at t to be

$$(38) \qquad \lambda(t|U(t)) = \lim_{\Delta t\to 0} \frac{1}{\Delta t} P\left\{V(t+\Delta t) - V(t) = 1 | V(t) = 0,\ U(t)\right\},$$

which is similar to, but not the same as, the corresponding limit of the cross-transition probability (27) from survival to failure:

$$(39) \qquad \lim_{\Delta t\to 0} \frac{1}{\Delta t} h^{t+\Delta t|t+\Delta t, t}_{1|\theta_u\sigma_u 0}.$$

Both (38) and (39) are conditioned on survival to time t, but they differ in that $\lambda(t|U(t))$ is conditioned on only one value of U at time t, whereas (39) is conditioned on U at s and t. However, neither (38) nor (39) is equal to the conditional intensity function $q^t_{1|\theta_u\sigma_u 0}$. Note that the derivative of $H^{t|t,s}$ in (29) is taken with respect to 't' which appears in the conditioning event as well as the conditioned event. The number of parameters in the likelihood (32) can be reduced in two ways. First, models such as described in (10) can be used to provide structure to the cross-conditional intensities:

$$(40) \qquad \begin{aligned} q^t_{[1]nm|[\theta_u\sigma_u 0]_{n(m-1)}} &= h(U_n(t), \boldsymbol{\beta}), \\ m &= 0, 1, \ldots, M_n,\ \theta_u, \sigma_u = 1, \ldots, \Omega_u. \end{aligned}$$

Also the underlying science of an application or prior data on the covariate process U (which frequently exists conditional on survival) may suggest structural models for

$$(41) \qquad \begin{aligned} &q^{\tau_{n(m+1)}}_{[\theta_u]_{n(m+1)}|[\sigma_u 0]_{nm}}, \\ &m = 0, 1, \ldots, M_n,\ \theta_u, \sigma_u = 1, \ldots, \Omega_u. \end{aligned}$$

We expect that when the covariate process U is nonstationary, modeling the conditional (40) and cross conditional intensity functions (41) and using the joint likelihood (32) can lead to better estimates than can be obtained by using only the model (10) in the partial likelihood (11). Structural models to reduce the parameters in (40) and (41) will be proposed, and their performance analyzed elsewhere.

REFERENCES

CHANG, C.L. (1980). *An Introduction to Stochastic Processes in Biostatistics.* R.E. Krieger Publishing Company, New York.

CHANG, M.N. and YANG, G.L. (1990). A stochastic model for prevalence of hepatitis A antibody. *Math. Bio.* **98** 157–169.

COGBURN, R. (1980). Markov chains in random environments: The case of Markovian environments. *Ann. Prob.* **8** 908–916.

COX, D.R. (1972). Regression models and life-tables. *J. Roy. Statist. Soc.* Ser. B **34** 187–202.

COX, D.R. (1975). Partial likelihood. *Biometrika* **64** 269–276.

DEBANNE, S.M., ROWLAND, D.Y., EIELEFELD, R.A., and MAW, C.E. (1989). Adaption of the National Center for Education Statistics survey data for Markovian predicted models. P.215–219 in: *The 1989 Proceedings of the Social Statistics Section.* American Statistical Association, Alexandria, VA.

FELLER, W. (1940). On the integrodifferential equations of purely discontinuous Markov processes. *Trans. of the Amer. Math. Soc.* **48** 488–515.

FLOURNOY, N. (1980). On the survival and intensity functions. P.75 in: *1980 Abstracts. Summaries of papers presented at the joint statistical meeting of the American Statistical Association and the Biometric Society.* American Statistical Association: Alexandria, VA.

FLOURNOY, N. (1990). Dependency in multivariate Markov chains. *Linear Algebra and Its Applications* **127** 85–106.

GETZ, W.M. (1976). Stochastic equivalents of the linear and Lotka-Volterra system of equations—a general birth–and–death process formulation. *Math. Bio.* **29** 235–257.

HARTFIEL, D.J. (1985). On the solutions to $x' = A(t)x(t)$ over all $A(t)$, where $P \leq A(t) \leq Q$. *J. Math. Anal. and Appl.* **108** 230–240.

KALBFLEISCH, J.D. and PRENTICE, R.L. (1980). *The Statistical Analysis of Failure Time Data.* John Wiley and Sons, Inc., New York.

KARLIN, S. and TAYLOR, H.M. (1981). *A Second Course in Stochastic Processes.* Academic Press, Inc., New York.

KOLMOGOROV, A.N. (1931). Über die analytischen Methaden in der Wahrscheinlichkeitsrechnung. *Mathematische Annalen* **104** 415–458.

WEIDEN, P.L., FLOURNOY, N., THOMAS, E.D., FEFER, A., and STORB, R. (1981). Antitumor effect of marrow transplantation in human recipients of syngeneic or allogeneic grafts. Pages 11–23 in: *Graft-Versus-Leukemia in Man and Animal Models,* (ed. by J. Okunewick and R. Meredith), CRC Press, Inc.: Boca Raton, Florida.

WEIDEN, P.L., FLOURNOY, N., THOMAS, E.D., PRENTICE, R., FEFER, A., BUCKNER, C.D., and STORB, R. (1979). Antileukemic effect of graft-versus-host disease in human recipients of allogeneic grafts. *New Eng. J. Med.* **300** 1068–1073.

YADIN, M. and SYSKI, R. (1979). Randomization of intensities in a Markov chain. *Adv. in Appl. Prob.* **11** 397–421.

YASHIN, A. and ARJAS, E. (1988). A note on random intensities and conditional survival functions. *J. Appl. Prob.* **25** 630–635.

DEPARTMENT OF MATHEMATICS AND STATISTICS
AMERICAN UNIVERSITY
WASHINGTON, DC 20016-8050

A COMPARISON OF BONFERRONI-TYPE AND PRODUCT-TYPE INEQUALITIES IN PRESENCE OF DEPENDENCE

BY JOSEPH GLAZ[1]

University of Connecticut

Let $X_1, \ldots, X_n$ be a sequence of dependent random variables and for $j = 1, \ldots, n$, $A_j = (X_j \varepsilon I_j)$ where I_j's are infinite intervals of the same type, $I_j = (-\infty, a_j)$ or $I_j = (b_j, \infty)$. In this article we compare the performance of the Bonferroni-type and product-type inequalities in approximating the probabilities $P\{\cup_{i=1}^n A_i\}$ or $P\{\cap_{i=1}^n B_i\}$ where B_i is the complementary event of A_i.

The following results are proved. If $X_1, \ldots, X_n$ possess a positive dependence structure (MTP_2 or sub-Markov with respect to a sequence of infinite intervals of the same type) the product-type inequalities dominate the Bonferroni-type inequalities. If, on the other hand, the sequence of random variables is negatively dependent (S-MRR_2 or super-Markov with respect to a sequence of infinite intervals of the same type) the product-type inequalities complement the Bonferroni-type inequalities in approximating the probabilities mentioned above. Three examples are presented to illustrate the results obtained in this paper.

1. Introduction. Let $X_1, \ldots, X_n$ be a sequence of dependent random variables and for $j = 1, 2, \ldots, n$

$$A_j = (X_j \varepsilon I_j), \tag{1}$$

where I_j are infinite intervals of the same type; $I_j = (-\infty, a_j)$ or $I_j = (b_j, \infty)$. We are interested in studying the approximations for

[1]Supported in part by the Research Foundation of the University of Connecticut.

AMS 1980 subject classifications. Primary 60E15; secondary 62E10.

Key words and phrases. Probability inequalities, positive dependence, negative dependence, multivariate distributions.

I would like to thank the editors and the referee for their helpful comments on an earlier version of this paper.

$$(2) \qquad P_1 = P\{\cup_{i=1}^n A_i\},$$

or equivalently,

$$(3) \qquad 1 - P_1 = P\{\cap_{i=1}^n B_i\},$$

where $B_i = A_i^c$ is the complementary event of A. These approximations play an important role in many areas of statistics; to list just a few: multiple comparison analysis (Fuchs and Sampson, 1987; Games, 1977; Kenyon, 1986a; Sidak, 1971; and Tong, 1970), simultaneous prediction (Chew, 1968), location and scale shift detection (Bauer and Hackl, 1978, 1980, and 1985; Glaz, 1983; Glaz and Johnson, 1987; and Worsley, 1979), scan statistics (Berman and Eagleson, 1985; Gates and Westcott, 1984; Glaz, 1989; Glaz and Naus, 1983; Naus, 1982; and Samuel-Cahn, 1983), sequential testing (Bauer and Hackl, 1985; Glaz and Johnson, 1986; and Kenyon, 1986b), and outlier detection (Ellenberg, 1976; Galpin and Hawkins, 1981; and Joshi, 1972).

In Section 2 of this article, we briefly outline the up-to-date development in the area of Bonferroni-type inequalities. In Section 3 the product-type inequalities will be introduced along with the necessary dependence concepts. We then compare the Bonferroni-type and product-type inequalities for certain dependence structures for $X_1, \ldots, X_n$. In Section 4 three examples will be presented for the evaluation of Bonferroni-type and product-type inequalities. A brief discussion comparing these two classes of inequalities and evaluating the numerical results from Section 4 will be given in Section 5.

2. Bonferroni-Type Inequalities. The Bonferroni-type inequalities have been used by many authors to obtain bounds for P_1 given in equation (2):

$$(4) \qquad S_{1,n} - S_{2,n} \leq P_1 \leq S_{1,n},$$

where

$$(5) \qquad S_{1,n} = \Sigma_{i=1}^n p_i, \quad S_{2,n} = \Sigma_{i=1}^{n-1} \Sigma_{j=i+1}^n p_{i,j}$$

and

$$(6) \qquad p_i = P(A_i), \quad p_{i,j} = P(A_i \cap A_j).$$

As these bounds can be quite inaccurate, attempts have been made to improve their performance. Kwerel (1975) has shown that

$$(7) \qquad P_1 \geq aS_{1,n} + bS_{2,n},$$

where $a = 2/k$, $b = -2/k(k-1)$ and $k-2$ is the integer part of $2S_{2,n}/S_{1,n}$. The inequality of (7) is the tightest, given the probabilities (6). The computation of this lower bound, for large n, can be quite tedious and the performance unsatisfactory (Glaz, 1989).

The study of upper bounds for P_1 have received more attention, the reason being that it provides a conservative test or a confidence coefficient in a multiple comparison procedure (see references mentioned in the Introduction). Let $v_1, \ldots, v_n$ be the vertices of the graph G, representing the events $A_1, \ldots, A_n$, respectively. The vertices v_i and v_j are joined by an edge e_{ij} if and only if $A_i \cap A_j \neq \phi$. Hunter (1976) and Worsley (1982) proved that for a subgraph T of G

$$P_1 \leq S_{1,n} - \Sigma_{\{(i,j);e_{ij}\varepsilon T\}}\, p_{i,j}, \tag{8}$$

if and only if T is a tree. An important member of this class of upper bounds is

$$P_1 \leq S_{1,n} - \Sigma_{i=1}^{n}\, p_{i,i+1}, \tag{9}$$

which under certain conditions is the least upper bound in that class. The above statement is valid if the events $A_1, \ldots, A_n$ are exchangeable or are ordered in such a way that for $1 \leq i_1 < i_2 \leq n$, $P(A_{i_1} \cap A_{i_2})$ is maximized for $i_j - i_{j-1} = 1$ (see Worsley, 1982, Examples 3.1 and 3.2).

DEFINITION 2.1. An inequality for P_1 or $1 - P_1$ is of *order* k if it is given in terms of $P\{\cap_{j=1}^{m} A_{i_j}\}$ for $1 \leq m \leq k < n$, and contains the term $P\{\cap_{j=1}^{k} A_{i_j}\}$ for some $1 \leq i_1 < i_2 < \ldots < i_k \leq n$.

Recently, Hoover (1989) has derived a sequence of Bonferroni-type upper bounds of order k, $1 \leq k \leq n-1$:

$$P_1 \leq P\{\cup_{i=1}^{k} A_i\} + \Sigma_{j=k+1}^{n} P\{A_j \cap [\cap_{1\leq i_1<i_2<\ldots<i_k<n} (\cup_{j=1}^{k} A_{i_j})^c]\}, \quad i_1, i_2, \ldots, i_k \varepsilon S_j \tag{10}$$

where S_j is a subset of $\{1, 2, \ldots, j-1\}$ of size $k-1$ and $j \geq k+1$. For $k = 1$ and $k = 2$ the upper bounds in (10) reduce to the Bonferroni upper bound in (4) and the Hunter-Worsley upper bound in (8), respectively. In the case that $A_1, \ldots, A_n$ are naturally ordered in such a way that $P(\cap_{j=1}^{m} A_{i_j})$ is maximized for $i_j - i_{j-1} = 1$, $2 \leq j \leq m$ and $2 \leq m \leq n-1$, the natural ordering with $S_j = \{j-1, j-2, \ldots, j-k\}$ is recommended for the upper bound of order k. In this case (10) reduces to:

$$P_1 \leq S_{1,n} - \Sigma_{i=1}^{n-1}\, p_{i,i+1} - \Sigma_{j=2}^{k-l}\, \Sigma_{i=1}^{n-j}\, p^*_{i,i+1,\ldots,i+j}, \tag{11}$$

where

$$\Sigma_{j=2}^{1}\, d_j \equiv 0 \text{ and for } j \geq 2$$

$$p^*_{i,i+1,\ldots,i+j} = P(A_i \cap A^c_{i+1} \cap \ldots \cap A^c_{i+j-1} \cap A_{i+j}). \tag{12}$$

For $k = 2$ equation (11) reduces to equation (9). If the events $A_1, A_2, \ldots, A_n$ are exchangeable, a further simplification of (10) is obtained:

(13) $$P_1 \leq np_1 - (n-1)p_{1,2} - \Sigma_{j=2}^{k-1}(n-j)p^*_{1,2,\ldots,j+1},$$

where $p^*_{1,2,\ldots,j+1}$ is given by equation (12). In Section 4 three examples will be presented to evaluate these Bonferroni-type inequalities.

3. Product-Type Inequalities. Let $X_1,\ldots,X_n$ be a sequence of dependent random variables and let A_i be the events defined in equation (1). The so-called *product upper bound* for P_1 is given by:

(14) $$P_1 \leq 1 - \Pi_{i=1}^{n}(1-p_i),$$

where p_i is defined in equation (6). This inequality along with the conditions for its validity has been studied by Dunn (1958), Esary, Proschan and Walkup (1967), Jogdeo (1977), Khatri (1967), Sidak (1967, 1968, 1971, and 1973), and Scott (1967).

The following concept of positive dependence introduced by Esary, Proschan, and Walkup (1967) is useful in establishing the inequality (14). $X_1,\ldots,X_n$ are said to be *associated* if for every pair of coordinatewise increasing real valued functions f and g,

$$\mathrm{Cov}[f(\mathbf{X}), g(\mathbf{X})] \geq 0,$$

where $\mathbf{X} = (X_1,\ldots,X_n)$. Esary, Proschan, and Walkup (1967) proved that $X_1,\ldots,X_n$ being associated is a sufficient condition for the validity of (14). It is well-known that the product bound for P_1 is tighter than the Bonferroni upper bound in (4). On the other hand, the upper bound (9) outperforms the product bound (14) (Worsley, 1982, Example 3.1). According to the definition (2.1), the product bound (14) is a first order inequality. The rest of this section is devoted to presenting the product-type inequalities of order k and comparing them with the Bonferroni-type inequalities of corresponding order. In what follows we will assume that the events $A_1,\ldots,A_n$ are naturally ordered in such a way that $P\{\cap_{j=1}^{m} A_{i_j}\}$ is maximized for $i_j - i_{j-1} = 1$, $2 \leq j \leq m$ and $2 \leq m \leq n-1$.

To study the higher order product-type inequalities for P_1, the following concepts of dependence play an important role.

DEFINITION 3.1. (Karlin, 1968). A nonnegative real-valued function of two variables, $f(x,y)$, is *totally positive of order two*, TP_2 (*reverse rule of order two*, RR_2), if

$$f(x_1,y_1)f(x_2,y_2) - f(x_1,y_2)f(x_2,y_1) \geq (\leq) 0$$

for all $x_1 < x_2$ and $y_1 < y_2$.

DEFINITION 3.2. (Karlin and Rinott, 1980a, 1980b). A nonnegative real-valued function of n variables, $f(x_1,\ldots,x_n)$ is *multivariate totally positive of order two*, MTP_2 (*multivariate reverse rule of order two*, MRR_2), if for any pair of arguments x_i and x_j the function f, viewed as a function of x_i and x_j while the rest of the arguments are kept fixed, is TP_2 (RR_2). $f(x_1, \ldots, x_n)$ is said to be strongly MRR_2, S-MRR_2, if for any set of PF_2 functions $\{\phi_j\}$ (a function ϕ is PF_2 if and only if $\phi(x - y)$ is TP_2 in the variables $-\infty < x, y < \infty$), the marginals

$$g(x_{i_1}, \ldots, x_{i_k}) = \int \ldots \int f(x_1, \ldots, x_n) \Pi_{m=1}^{n-k} \phi(x_{j_m}) dx_{j_1} \ldots dx_{j_{n-k}}$$

are MRR_2 in the variables $(x_{i_1}, \ldots, x_{i_k})$, where the set $\{1, \ldots, n\} = \{i_1, \ldots, i_k\} \cup \{j_1, \ldots, j_{n-k}\}$. A sequence of random variables, $X_1, \ldots, X_n$, is said to be MTP_2 (S-MRR_2) if its joint density is MTP_2 (S-MRR_2).

The class of random variables with MTP_2 or S-MRR_2 densities is quite rich. For a listing of these densities, see Karlin and Rinott (1980a, 1980b). Barlow and Proschan (1975) defined the TP_2 in pairs property for $(X_1, \ldots, X_n)$. If the support of its distribution function is a product space, then TP_2 in pairs is equivalent to MTP_2.

We introduce the following concept of dependence that is closely related to the higher order product-type bounds.

DEFINITION 3.3. A sequence of random variables $X_1, \ldots, X_n$ is said to be *sub-Markov* (*super-Markov*) with respect to a sequence of intervals $I_1, \ldots, I_n$ if for any $1 \leq i < k \leq n$

$$P\{X_k \varepsilon I_k \mid \cap_{j=1}^{k-1}(X_j \varepsilon I_j)\} \geq (\leq) P\{X_k \varepsilon I_k \mid \cap_{j=i}^{k-1}(X_j \varepsilon I_j)\}.$$

In Glaz and Johnson (1984, Theorems 2.3 and 2.8) it is proved that if the joint density of $X_1, \ldots, X_n$ is MTP_2 (S-MRR_2), then $X_1, \ldots, X_n$ is sub-Markov (super-Markov) with respect to the intervals $I_j = (-\infty, a_j)$ or $I_j = (b_j, \infty)$, $j = 1, \ldots, n$. Moreover, if $X_1, \ldots, X_n$ are MTP_2 (S-MRR_2), we construct a decreasing (increasing) sequence of upper (lower) bounds for P_1:

$$\gamma_{k,n} = 1 - P\{\cap_{j=1}^{k} A_j^c\} \Pi_{m=k+1}^{n} P(A_m^c \mid \cap_{j=m-k+1}^{m-1} A_j^c), \tag{15}$$

where $1 \leq k \leq n-1$ and A_j^c is the complementary event of $A_j = (X_j \varepsilon I_j)$, $j = 1, \ldots, n$. Note that if $k = 1$, then $\gamma_{1,n}$ is the product bound given by the inequality (14) (in the positive dependence case). For $k \geq 1$, $\gamma_{k,n}$ is the kth order product-type bound for P_1. We now proceed to compare the product-type bounds with the Bonferroni-type bounds. For $k \geq 2$ let

$$\delta_{k,n} = S_{1,n} - \Sigma_{i=1}^{n-1} p_{i,i+1} - \Sigma_{j=2}^{k-1} \Sigma_{i=1}^{n-j} p^*_{i,i+1,\ldots,i+j} \tag{16}$$

denote the kth order Bonferroni-type bound where $S_{1,n}$, $p_{i,i+1}$, and $p^*_{i,i+1,\ldots,i+j}$ are given by equations (5), (6), and (12), respectively. The following result is true:

THEOREM 3.1. *Let $X_1, \ldots, X_n$ be a sequence of dependent random variables and $A_1, \ldots, A_n$ be the events defined in equation (1). Assume that* $0 < P_1 < 1$. *Then for* $k \geq 2$

$$\gamma_{k,n} \leq \delta_{k,n}, \tag{17}$$

where $\gamma_{k,n}$ and $\delta_{k,n}$ are given by equations (15) and (16), respectively. Moreover, if $n > k$ and for some $1 \leq m \leq k$ and $1 \leq j \leq n-k$

$$P\{A_n \cap (\cap_{i=1}^{k-1} A_{n-i}^c)\} > 0, \quad P\{A_j \mid \cap_{m=j+1}^{j+k} A_i^c)\} > 0 \tag{18}$$

then the inequality in (17) is sharp.

PROOF. We prove this result by induction on n, the number of the events A_j. For $n = k$,

$$\gamma_{k,k} = 1 - P\{\cap_{j=1}^{k} A_j^c\} = P\{\cup_{j=1}^{k} A_j\} = \delta_{k,k}.$$

Assume the conclusion of the theorem is true for $n-1$ events with a weak inequality in (17), and show that it holds for n events. Write for $k \geq 3$

$$\gamma_{k,n} = \gamma_{k,n-1} + (1 - \gamma_{k,n-1}) P(A_n \mid \cap_{j=n-k+1}^{n-1} A_j^c).$$

Then by the induction hypothesis, it follows that

$$\begin{aligned}
\gamma_{k,n} &\leq \delta_{k,n-1} + (1-\gamma_{k,n-1}) P(A_n \mid \cap_{j=n-k+1}^{n-1} A_j^c) \\
&= \delta_{k,n} - (\delta_{k,n} - \delta_{k,n-1}) + (1-\gamma_{k,n-1}) P(A_n \mid \cap_{j=n-k+1}^{n-1} A_j^c) \\
&= \delta_{k,n} - \{P(A_n) - P(A_{n-1} \cap A_n) \\
&\quad - \Sigma_{j=2}^{k-1} P[A_{n-j} \cap (\cap_{i=n-j+1}^{n-1} A_i^c) \cap A_n]\} \\
&\quad + (1-\gamma_{k,n-1}) P\{A_n \cap [\cap_{j=1}^{k-1} A_{n-j}^c]\} / P[\cap_{j=1}^{k-1} A_{n-j}^c] \\
&= \delta_{k,n} - P\{A_n \cap [\cap_{j=1}^{k-1} A_{n-j}^c]\} \\
&\quad + (1-\gamma_{k,n-1}) P\{A_n \cap [\cap_{j=1}^{k-1} A_{n-j}^c]\} / P[\cap_{j=1}^{k-1} A_{n-j}^c].
\end{aligned}$$

Since

$$(1-\gamma_{k,n-1}) / P[\cap_{j=1}^{k-1} A_{n-j}^c] = \Pi_{j=1}^{n-k} P\{A_j^c \mid \cap_{m=j+1}^{j+k-1} A_m^c\},$$

we get that

$$\gamma_{k,n} \leq \delta_{k,n} - P\{A_n \cap [\cap_{j=1}^{k-1} A_{n-j}^c]\}\{1 - \Pi_{j=1}^{n-k} P(A_j^c \mid \cap_{m=j+1}^{j+k-1} A_m^c)\}. \tag{19}$$

As the second term on the right-hand side of the inequality (19) is nonnegative, we obtain the inequality (17). It follows from the inequality (19) that if the conditions (18) hold, then the inequality in (17) is sharp. This concludes the proof of Theorem 3.1 for $k \geq 3$. For $k = 2$ the proof is similar, with equation (9) being used instead of (11). ||

The following two results are a direct consequence of Theorem 3.1 and Glaz and Johnson (1984, Theorem 2.3 and Theorem 2.8, respectively).

COROLLARY 3.2. *If $X_1, \ldots, X_n$ are MTP_2, then for $k \geq 1$*

$$P_1 \leq \gamma_k \leq \delta_k.$$

Moreover, γ_k and δ_k are nonincreasing sequences of k.

COROLLARY 3.3. *If $X_1, \ldots, X_n$ are S-MRR_2, then for $k \geq 1$*

$$\gamma_k \leq P_1 \leq \delta_k.$$

Moreover, the sequences γ_k and δ_k are nondecreasing and nonincreasing, respectively, in k.

REMARK. The condition of $X_1, \ldots, X_n$ being MTP_2 (S-MRR_2) in Corollary 3.2 (Corollary 3.3) can be relaxed to $X_1, \ldots, X_n$ being sub-Markov (super-Markov) with respect to the intervals $I_1^c, \ldots, I_n^c$.

In Section 4 we present three examples to evaluate the performance of the product-type inequalities.

4. Examples. To illustrate the inequalities discussed in Sections 2 and 3 and to compare their performance, we present three examples. A brief discussion will follow in Section 5.

4.1. Boundary Crossing Probabilities. Let $Z_1, \ldots, Z_n, \ldots$ be independent random variables from a normal distribution with mean 0 and variance 1 and $S_j = \Sigma_{i=1}^{j} Z_i$. Denote by

$$\tau = \inf\{j \geq 1; \mid S_j \mid > c_j\},$$

the first time that the sequence of partial sums cross a symmetric boundary given by the constants c_j. We are interested in approximations for

$$P(\tau > n) = P\{\cap_{j=1}^{n}(\mid S_j \mid \leq c_j)\},$$

$n = 1, 2, \ldots$. Based on these approximations, one can evaluate approximations for $E(\tau)$ and $\text{Var}(\tau)$, the expected time and the variance of the time for the first crossing of the boundary, respectively. It follows from Glaz and Johnson (1986, Theorem 2.1) and Karlin and Rinott (1982) that $\mid S_1 \mid, \ldots, \mid S_n \mid$ is MTP_2. Hence, Corollary 3.2 implies that for $k \geq 1$

$$P(\tau > n) \geq \gamma_{k,n}^* \geq \delta_{k,n}^*,$$

where

$$\gamma_{k,n}^* = 1 - \gamma_{k,n} \text{ and } \delta_{k,n}^* = 1 - \delta_{k,n}$$

are given in equations (15) and (16), respectively. Here, we will evalute the bounds in the case of a triangular boundary

$$c_j = a - bj,\ a > 0,\ b > 0,$$

that has been introduced by Anderson (1960) in the context of sequential tests of hypotheses. For a more elaborate discussion of this subject, the reader is referred to Glaz and Johnson (1986).

In Table 4.1 we compare the Bonferroni-type and product-type bounds of order $k \leq 3$ with the simulated values for $P(\tau > n)$. The triangular boundary in this example is given by $c_j = 7.5 - .2j$. The simulated values for $P(\tau > n)$ are denoted by $\hat{P}(\tau > n)$ and have been estimated from a simulation with 10,000 trials using IMSL (1975).

Table 4.1

Approximations for $P(\tau > n)$, $c_j = 7.5 - .2j$

n	$\delta_{1,n}^*$	$\gamma_{1,n}^*$	$\delta_{2,n}^*$	$\gamma_{2,n}^*$	$\delta_{3,n}^*$	$\gamma_{3,n}^*$	$\hat{P}(\tau > n)$
5	.9955	.9955	.9961	.9961	.9961	.9961	.9963
10	.7878	.8038	.8945	.8953	.8996	.8998	.9025
15	-	.3029	.6384	.6555	.6717	.6789	.6939
20	-	.0325	.2949	.3824	.3811	.4229	.4534
25	-	.0006	-	.1657	.0755	.2030	.2333
30	-	.0000	-	.0413	-	.0573	.0683
35	-	.0000	-	.0015	-	.0022	.0028

NOTE: The - in the table corresponds to values less than 0.

4.2. Moving Window Detection Probabilities. Let $Z_1, \ldots, Z_n, \ldots$ be independent observations from a normal distribution with mean 0 and variance one unit. For fixed $m \geq 2$, define

$$S_{j,m} = \Sigma_{i=j}^{j+m-1} Z_i,\ j \geq 1,$$

and

$$\tau_m = \inf\{j \geq 1;\ S_{j,m} > a\} + m - 1.$$

Then τ_m is the first time that the process of moving sums of length m crosses the straight line boundary specified by the constant a, $a > 0$. Applications to quality control are discussed in Bauer and Hackl (1980) and Lai (1974), who employ the first-order product bound $\gamma_{1,n}^*$ to approximate $P(\tau_m > n)$.

Note that in this example the sequence of moving sums, $\{S_{m,j}\}_{j=m}^n$, is associated but not MTP_2. Hence we cannot argue that $\gamma_{k,n}^*$ is a lower bound for $P(\tau_m > n)$. One can show (Glaz and Johnson, 1988) that

$$\lim_{n\to\infty} P(\tau_m > n \mid \tau_m > n-1) = \alpha,$$

where $0 < \alpha < 1$, and use this asymptotic stationarity property of $P(\tau_m > n \mid \tau_m > n-1)$ to justify the use of $\gamma_{k,n}^*$ as an approximation for $P(\tau_m > n)$. The quantity $\delta_{k,n}^*$ is still a lower bound for $P(\tau_m > n)$ and from Theorem 3.1 we have that $\gamma_{k,n}^* \geq \delta_{k,n}^*$.

In Table 4.2, for specified values of m, a, and n, we present the kth order Bonferroni-type bounds and product-type approximations, $k \leq 3$, and compare them with the simulated values $\hat{P}(\tau_m > n)$. $P(\tau_m > n)$ have been estimated from a simulation with 10,000 trials using IMSL (1975).

Table 4.2

Approximations for $P(\tau_{10} > n)$, $a = 2.0$

n	$\delta_{1,n}^*$	$\gamma_{1,n}^*$	$\delta_{2,n}^*$	$\gamma_{2,n}^*$	$\delta_{3,n}^*$	$\gamma_{3,n}^*$	$\hat{P}(\tau_{10} > n)$
15	-	.1595	.4436	.4866	.4976	.5148	.5278
20	-	.0346	.1508	.3216	.2723	.3650	.3866
25	-	.0075	-	.2125	.0470	.2588	.2785
30	-	.0016	-	.1404	-	.1835	.2002
35	-	.0004	-	.0928	-	.1300	.1443
40	-	.0000	-	.0613	-	.0922	.1039
45	-	.0000	-	.0405	-	.0654	.0762
50	-	.0000	-	.0268	-	.0463	.0572
60	-	.0000	-	.0112	-	.0233	.0236
70	-	.0000	-	.0051	-	.0117	.0159
80	-	.0000	-	.0022	-	.0059	.0072
90	-	.0000	-	.0010	-	.0029	.0036
100	-	.0000	-	.0004	-	.0015	.0019

NOTE: The - corresponds to values less than 0.

4.3. Multinomial Distribution. Let $\mathbf{X} = (X_1, \ldots, X_m)$ be a multinomial random variable with parameters $\mathbf{p} = (p_1, \ldots, p_m)$ and $\mathbf{n} = (n_1, \ldots, n_m)$, where $\Sigma_{i=1}^m p_i = 1$ and $\Sigma_{i=1}^m n_i = N$. It follows from Karlin and Rinott (1980b) that $\mathbf{X}$ is S-MRR_2. We are interested in approximations for $P(X_i \leq a_i;\ i = 1, \ldots, m)$ or

$P(X_i \geq b_i;\ i = 1, \ldots, m)$. We will assume that $p_1 = p_2 = \ldots = p_m = p$, in which case $X_1, \ldots, X_m$ are exchangeable. It follows from Corollary 3.3 that

$$\delta^*_{k,m} \leq P(X_i \leq a_i;\ i = 1, \ldots, m) \leq \gamma^*_{k,m} \tag{20}$$

and

$$\delta^*_{k,m} \leq P(X_i \geq b_i;\ i = 1, \ldots, m) \leq \gamma^*_{k,m}, \tag{21}$$

where $\gamma^*_{k,m} = 1 - \gamma_{k,m}$ and $\delta^*_{k,m} = 1 - \gamma_{k,m}$. Mallows (1968) has proved that $\gamma^*_{1,m}$ is an upper bound for the above probabilities.

An important special case of the approximations given in (20) and (21) is when $a_1 = a_2 = \ldots = a_m = a$ and $b_1 = b_2 = \ldots = b_m = b$. In these cases we obtain approximations for the distribution of

$$X_{(m)} = \max(X_1, \ldots, X_m) \text{ and } X_{(1)} = \min(X_1, \ldots, X_m),$$

respectively. We illustrate the performance of these approximations in the following example. Consider a roulette with $m = 38$ numbers. We would like to test the null hypothesis that $p_1 = p_2 = \ldots = p_{38} = 1/38$. Consider the test that rejects the null hypothesis for large values of $X_{(m)}$.

In Table 4.3 we present bounds for the P-values of this test when $N = 100$. The P-values are given by $P(X_{(38)} \geq n)$, where n is the largest observed cell count.

Table 4.3

Bounds for the P-Values for the Test of Equal Cell Probabilities

n	5	6	7	8	9	10	11
$\gamma_{3,38}$	.9944	.8562	.4758	.1744	.0496	.0121	.0026
$\delta_{3,38}$	> 1	> 1	.6200	.1894	.0507	.0121	.0026

5. Discussion. The Bonferroni-type and product-type inequalities, presented in Sections 2 and 3, have the same degree of complexity. In fact, one can show that both types of the kth order inequalities for $P\{\cup_{j=1}^n (X_j \varepsilon I_j)\}$ can be expressed in terms of $P\{\cap_{j=1}^i (X_j \varepsilon I_j)\}$, for $1 \leq i \leq k$.

If $X_1, \ldots, X_n$ possesses a positive dependence structure (MTP_2 or sub-Markov with respect to I_j's), the product-type inequalities dominate the Bonferroni-type inequalities (Corollary 3.2). In this case, Table 4.1 of Example 4.1 illustrates the amount of improvement achieved by the kth order product-type inequality over the kth order Bonferroni-type inequality for $k = 1, 2, 3$. The order of the inequality plays an important role in improving the approximations. Example 4.2 supports the use of product-type inequalities as approximations in cases when

$X_1, \ldots, X_n$ are positively dependent but does not necessarily satisfy the conditions of Corollary 2.2. In this situation, the Bonferroni-type inequalities along with the simulations provide a tool for evaluating the accuracy of the product-type approximations. Numerical results in Table 4.2 indicate that $\gamma_{3,n}$ can serve as a respectable approximation for the tail probabilities $P(\tau_m > n)$.

If $X_1, \ldots, X_n$ have a negative dependence structure (S-MRR$_2$ or super-Markov with respect to I_j's), the product-type inequalities complement the Bonferroni-type inequalities in approximating P_1 and $1 - P_1$, given by equation (2) and (3), respectively. This result is quite useful, as there are no tight lower (upper) bounds available for P_1 $(1 - P_1)$. In Example 4.3 both types of inequalities are utilized to approximate the P-value of the test for equal cell probabilities in a multinomial experiment. The numerical results in Table 4.3 indicate that these inequalities can provide us with quite accuracte approximations.

In conclusion, we would like to point out that the product-type bounds have the advantage of always having a value in the interval [0,1], while the Bonferroni-type bounds could have values outside the unit interval (see Tables 4.1–4.3).

REMARKS. Recently, Block, Costigan, and Sampson (1988) developed an optimized version of the second-order product-type inequality under conditions of positive dependence. As part of their work, they show that the second-order product-type inequality developed in Glaz and Johnson (1984) is superior to the corresponding second-order Bonferroni-type inequality, and both are based on the same spanning tree. Their proof of the result is analytical in nature. Hoover (1988) independently used a similar approach to the one used in this paper to derive the proof of Theorem 3.1.

REFERENCES

ANDERSON, T.W. (1960). A modification of the sequential probability ratio test to reduce sample size. *Ann. Math. Statist.* **31** 167–197.

BARLOW, R.E. and PROSCHAN, F. (1975). *Statistical Theory of Reliability and Life Testing Probability Models.* Holt, Rinehart and Winston, New York.

BAUER, P. and HACKL, P. (1978). The use of mosum's for quality control. *Technometrics* **20** 431–436.

BAUER, P. and HACKL, P. (1980). An extension of the mosum technique for quality control. *Technometrics* **22** 1–7.

BAUER, P. and HACKL, P. (1985). The application of Hunter's inequality in simultaneous testing. *Biomet. J.* **27** 25–38.

BERMAN, M. and EAGLESON, G.K. (1985). A useful upper bound for the tail probabilities of the scan statistic when the sample size is large. *J. Amer. Statist. Assoc.* **80** 886–889.

BLOCK, H.W., COSTIGAN, T. and SAMPSON, A.R. (1988). Optimal product-type probability bounds. Technical Report No. 88–07, Series in Reliability and Statistics, Department of Mathematics and Statistics, University of Pittsburgh.

CHEW, V. (1968). Simultaneous preduction intervals. *Technometrics* **10** 323–331.

DUNN, O.J. (1959). Confidence intervals for the means of dependent normally distributed variables. *J. Amer. Statist. Assoc.* **54** 613–621.

ELLENBERG, J.H. (1976). Testing for a single outlier from a general linear regression. *Biometrics* **21** 645–650.

ESARY, J.D., PROSCHAN, F. and WALKUP, D. (1967). Association of random variables with applications. *Ann. Math. Statist.* **38** 1466–1474.

FUCHS, C. and SAMPSON, A.R. (1987). Simultaneous confidence intervals for the general linear model. *Biometrics* **43** 457–469.

GALPIN, J.S. and HAWKINS, D.M. (1981). Rejection of a single outlier in two- or three-way layouts. *Technometrics* **23** 65–70.

GAMES, P. (1977). An improved t-table for simultaneous control on g contrasts. *J. Amer. Statist. Assoc.* **72** 531–534.

GATES, D.J. and WESTCOTT, M. (1984). On the distribution of scan statistics. *J. Amer. Statist. Assoc.* **79** 423–429.

GLAZ, J. (1983). Moving window detection for discrete data. *IEEE Trans. Inform. Theory* **IT-29** 457–462.

GLAZ, J. (1989) Approximations and bounds for the distribution of the scan statistics. *J. Amer. Statist. Assoc.* (in press).

GLAZ, J. and JOHNSON, B. MCK. (1984). Probability inequalities for multivariate distibutions with dependence structures. *J. Amer. Statist. Assoc.* **79** 436–441.

GLAZ, J. and JOHNSON, B. MCK. (1986). Approximating boundary crossing probabilities with application to sequential tests. *Seq. Anal.* **5** No. 1, 37–72.

GLAZ, J. and JOHNSON, B. MCK. (1988). Boundary crossing for moving sums. *J. Appl. Prob.* **25** 81–88.

GLAZ, J. and NAUS, J. (1983). Multiple clusters on the line. *Comm. in Stat.-Theor. Meth.* **12** 1961–1986.

HOOVER, D.R. (1988). Comparison of improved Bonferroni and Sidak/Slepain bounds with applications to normal Markov processes. Technical Report, Department of Statistics, University of South Carolina.

HOOVER, D.R. (1989). Component complement addition upper bounds—an improved inclusion/exclusion method. *J. Stat. Plan. Infer.* (in press).

HUNTER, D. (1976). An upper bound for the probability of a union. *J. Appl. Prob.* **13** 597–603.

International Mathematical and Statistical Library (1975). *IMSL Library 1.* Reference Manual (2 volumes). IMSL Corporation, Houston, Texas.

JOGDEO, K. (1977). Association and probability inequalities. *Ann. Stat.* **5** 495–504.

JOSHI, P.C. (1972). Some slippage tests of mean for a single outlier in linear regression. *Biometrika* **59** 109–120.

KARLIN, S. (1968). *Total Positivity.* Stanford University Press, Stanford, California.

KARLIN, S. and RINOTT, Y. (1980a). Classes of orderings of measures and related correlation inequalities–I: Multivariate totally positive distributions. *J. Mult. Anal.* **10** 467–498.

KARLIN, S. and RINOTT, Y. (1980b). Classes of orderings of measures and related correlation inequalities–II: Multivariate reverse rule distributions. *J. Mult. Anal.* **10** 499–516.

KARLIN, S. and RINOTT, Y. (1982). Total positivity properties of absolute value multinormal variables with applications to confidence interval estimates and related probabilistic inequalities. *Ann. Statist.* **9** 1035–1049.

KENYON, J.R. (1986a). Calculating improved bounds and approximations for multiple comparisons. *Comp. Sci. and Stat.: Proc. 18th Symp. Inter.* 367–371.

KENYON, J.R. (1986b). Calculating improved bounds and approximations for boundary crossing probabilities. *ASA 1986 Pro. Stat. Comp. Sec.* 202–205.

KHATRI, C.G. (1967). On certain inequalities for normal distributions and their applications to simultaneous confidence bounds. *Ann. Math. Statist.* **38** 1853–1867.

KWEREL, S.M. (1975). Most stringent bounds on aggregated probabilities of partially specified dependent probability systems. *J. Amer. Stat. Assoc.* **70** 472–479.

LAI, T.L. (1974). Control charts based on weighted sums. *Ann. Statist.* **2** 134–147.

MALLOWS, C.L. (1968). An inequality involving multinomial probabilities. *Biometrika* **55** 422–424.

NAUS, J.I. (1982). Approximations for distributions of scan statistics. *J. Amer. Stat. Assoc.* **77** 177–183.

SAMUEL-CAHN, E. (1983). Simple approximations to the expected waiting time for a cluster of any given size for point processes. *Adv. App. Prob.* **15** 21–38.

SCOTT, A. (1967). A note on conservative confidence regions for the mean of a multivariate normal. *Ann. Math. Statist.* **38** 278–280. Correction: *Ann. Math. Statist.* **39** 2161.

SIDAK, Z. (1967). Rectangular confidence regions for means of multivariate normal distributions. *J. Amer. Statist. Assoc.* **62** 626–633.

SIDAK, Z. (1968). On multivariate normal probabilities of rectangles. *Ann. Math. Statist.* **39** 1425–1434.

SIDAK, Z. (1971). On probabilities of rectangles in multivariate student distributions: Their dependence on correlations. *Ann. Math. Statist.* **42** 169–175.

SIDÁK, Z. (1973). A chain of inequalities for some types of multivariate distributions with nine special cases. *Aplikace Matematiky* **18** 110–118.

TONG, L.T. (1970). Some probability inequalities of multivariate normal and multivariate t. *J. Amer. Statist. Assoc.* **65** 1243–1247.

WORSLEY, K.J. (1979). On the likelihood ratio test for a shift in location of normal populations. *J. Amer. Statist. Assoc.* **74** 365–367.

WORSLEY, K.J. (1982). An improved Bonferroni inequality and applications. *Biometrika* **69** 297–302.

DEPARTMENT OF STATISTICS
UNIVERSITY OF CONNECTICUT
U–120, MSB 428
196 AUDITORIUM ROAD
STORRS, CT 06268

ROBUSTNESS OF MANN-WHITNEY-WILCOXON TEST FOR SCALE TO DEPENDENCE IN THE VARIABLES

By Z. Govindarajulu

University of Kentucky

The asymptotic efficiency of the Mann-Whitney-Wilcoxon (MWW) test for scale relative to the likelihood ratio test for equality of exponential scale parameters is evaluated. This efficiency is studied when the underlying variables have a bivariate exponential distribution of the form due to Morgenstern (1956), Gumbel (1960), Marshall and Olkin (1967), Downton (1970), Cowan (1987), and Sarkar (1987).

1. Introduction. Serfling (1968) studied the use of the Wilcoxon test statistic when there is some dependence among the X's and among the Y's. Hollander, Pledger, and Lin (1974) showed that the two-sample Wilcoxon test is asymptotically conservative when the X's and Y's having a bivariate distribution which is positively quadrant dependent. Govindarajulu (1975) studied the sensitivity of the Mann-Whitney-Wilcoxon (MWW) test for location alternatives when X and Y are dependent having an unknown bivariate distribution with continuous marginals. In the present paper we study the sensitivity of MWW test for scale alternatives when X and Y are dependent. In particular, we evaluate the Pitman efficiency of the MWW test relative to the likelihood ratio test for scale alternatives when (X, Y) has a bivariate exponential distribution. Several bivariate exponential distributions are available in the literature. See, for instance, Basu (1986) and Sarkar (1987) for a survey of these forms. Here we select a few of the bivariate exponential forms and evaluate the Pitman efficiencies of the MWW test.

2. An Asymptotically Distribution-free Test. Let $X[Y]$ be distributed as $F[G]$ where F and G are continuous. We wish to test the null hypothesis

$$H_0 : F(x) = G(x) \text{ for all } x$$

against the alternative

$$H_1 : F(x) \geq G(x) \text{ with strict inequality for some } x.$$

AMS Classification. 62G35.

Key words and phrases. Rank test for scale, dependent variables, Pitman efficiency.

The author thanks the referees for a careful reading of the manuscript.

Let (X_i, Y_i), $i = 1, 2, \ldots, n$ denote a random sample of size n drawn from $H(x, y)$. Let $H_n(x, y)$, $F_n(x)$ and $G_n(y)$ respectively denote the empirical distribution functions (e.d.f.'s) based on the samples $(X_i, Y_i) (i = 1, \ldots, n)$, $(X_1, \ldots, X_n)$ and $(Y_1, \ldots, Y_n)$. Let $Z_{ij} = 1$ or 0 according as $X_i \le Y_j$ or $X_i > Y_j$ respectively for $1 \le i, j \le n$.

Define

$$U = n^{-2} \sum_{i=1}^{n} \sum_{j=1}^{n} Z_{ij} = \int_{-\infty}^{\infty} F_n(x) dG_n(x).$$

Then we have the following result of Govindarajulu (1975), which was independently obtained by Hollander, Pledger, and Lin (1974).

RESULT 1. With the above notation, for all continuous F and G we have

$$\lim_{n \to \infty} P\{n^{\frac{1}{2}}(U - p)/\sigma \le z\} = \Phi(z), \quad \text{for all } z,$$

where

$$p = \int F dG. \tag{1}$$

$$\begin{aligned} \sigma^2 = 2 \iint_{x<y} F(x)[1 &- F(y)] dG(x) dG(y) \\ &+ 2 \iint_{x<y} G(x)[1 - G(y)] dF(x) dF(y) \\ &- 2 \iint_{-\infty}^{\infty} [H(x, y) - F(x)G(y)] dG(x) dF(y), \end{aligned} \tag{2}$$

and Φ denotes the standard normal distribution function.

PROOF. See Theorem 2.1 of Govindarajulu (1975).

One can rewrite σ^2 as

$$\sigma^2 = \int F^2 dG + \int G^2 dF - 2 \int \int H(x, y) dF(y) dG(x) - (1 - 2p)^2. \tag{3}$$

In order to test H_0 against H_1, we reject H_0 when U exceeds some $k_\alpha (\frac{1}{2} < k_\alpha < 1)$ where k_α is determined by α. Now since $F \ge G, E[F(Y) - G(Y)] = 0$ if and only if $F(Y) = G(Y)$ with probability one. Thus the test is consistent against H_1. To see this clearly, for large n we have $k_\alpha = (\frac{1}{2}) + \sigma(H_0) z_\alpha n^{-\frac{1}{2}}$ and the power of the test is $\Phi\left[\left\{(p - \frac{1}{2}) n^{\frac{1}{2}} - \sigma(H_0) z_\alpha\right\} / \sigma(H_1)\right]$ which tends to one as $n \to \infty$ since $p > \frac{1}{2}$ where $\sigma(H_0)$ and $\sigma(H_1)$ respectively denote the values of σ under H_0 and H_1. Also note that $\sigma^2(H_1) > \sigma^2(H_0) - (1 - 2p)^2$. Since σ^2 under H_0 is not free of $H(x, y)$, the test is not distribution-free. A consistent estimator of σ^2 under H_0 is given by

$$\hat{\sigma}^2 = \frac{2}{3} - 2n^{-2} \sum_{i=1}^{n} \sum_{j=1}^{n} H_n(Y_i, X_j). \tag{4}$$

Thus, an asymptotically distribution-free test of H_0 against H_1 is obtained by using $\hat{\sigma}^2$ in the place of σ^2. Thus

$$k_\alpha = \frac{1}{2} - \hat{\sigma} n^{-\frac{1}{2}} \Phi^{-1}(\alpha) \text{ for large } n.$$

Certain Remarks.

(i) Let $U^* = [n(n-1)]^{-1} \sum\sum_{i \neq j} Z_{ij}$. Consider

$$U - U^* = \{-1/n^2(n-1)\} \sum\sum_{i \neq j} Z_{ij} + n^{-2} \sum_{i=1}^{n} Z_{ii}.$$

Thus

$$n^{\frac{1}{2}} |U - U^*| \leq 2/n^{\frac{1}{2}} \to 0 \text{ as } n \to \infty.$$

Hence, U^* is asymptotically equivalent to U.

(ii) Suppose that n_N is random which is independent of (X_i, Y_i), $i = 1, 2, \ldots$, and there exists a positive integer N such that n_N/N converges to λ $(0 < \lambda < \infty)$ in probability. Then

$$P[n_N^{\frac{1}{2}} (U - p)/\sigma \leq z] \to \Phi(z) \text{ as } N \to \infty.$$

(This result is useful for handling the censored samples case.)

(iii) If one wishes to test H_0 against $H_2 : F \leq G$, one should interchange the roles of X and Y in the test procedure for H_0 against H_1.

3. Parametric Competitor. In this section we assume independent exponential marginals for the distributions of X and Y and derive the likelihood ratio test procedure for testing the null hypothesis of equality of the scale parameters. Since the scale parameters of the exponential marginals are the means of the distributions, it is not inappropriate to use the MWW test for testing the equality of the scale parameters.

Let X have the distribution $F(x) = 1 - \exp(-\lambda_1 x)$, $x > 0$ and Y have the distribution function $G(x) = 1 - \exp(-\lambda_2 x)$, $x > 0$. We wish to test $H_0 : \lambda_1 = \lambda_2$ against the alternative $H_1 : \lambda_2 \neq \lambda_1$. If $(X_1, \ldots, X_n)$ and $(Y_1, \ldots, Y_n)$ denote random samples from F and G respectively, one can easily show that the likelihood ratio criterion is given by Λ (since $\hat{\lambda}_1 = 1/\bar{X}, \hat{\lambda}_2 = 1/\bar{Y}$ and $\hat{\lambda} = 2/(\bar{X}+\bar{Y})$ when $\lambda_1 = \lambda_2 = \lambda$) where

$$\frac{1}{4}\Lambda^{\frac{1}{n}} = \bar{X}\bar{Y}/(\bar{X}+\bar{Y})^2,\ \bar{X} = n^{-1}\sum_1^n X_i,\ \bar{Y} = n^{-1}\sum_1^n Y_i. \tag{5}$$

Suppose we reject H_0 when

$$T = n\bar{X}\bar{Y}/(\bar{X}+\bar{Y})^2 < k_\alpha \tag{6}$$

where k_α is determined by the level of significance α. Notice that $\lambda_1\bar{X}$ and $\lambda_2\bar{Y}$ tend to unity in probability as n becomes large. If Z and W denote independent standard exponential random variables and $\bar{Z}$ and $\bar{W}$ denote sample means based on random samples of size n each, then by Slutsky's theorem, T has the same asymptotic distribution as T' where

$$T' = \lambda_1\lambda_2(\lambda_1+\lambda_2)^{-2} n\bar{Z}\bar{W}. \tag{7}$$

Next we compute the Pitman efficacy of T' (and hence that of T) assuming that Z and W have a bivariate exponential distribution with standard exponential marginals (that is, having scale parameters equal to unity) and correlation coefficient ρ^*. Note that for large n, $(\sqrt{n}\bar{Z}, \sqrt{n}\bar{W})$ is bivariate normal with means $(\sqrt{n}, \sqrt{n})$, unit variances and covariance $= E(ZW) - 1 = \text{corr}(Z, W) = \rho^*$, since

$$\text{Cov}(\sqrt{n}\bar{Z}, \sqrt{n}\bar{W}) = E(n\bar{Z}\bar{W}) - n = E(ZW) + n - 1 - n.$$

We need the following lemma.

LEMMA 3.1. *If (V_1, V_2) is bivariate normal with mean (0,0), unit variances and correlation ρ^*, then*

$$\begin{aligned} &(i)\quad E(V_1^2V_2) = E(V_1V_2^2) = 0, \text{ and} \\ &(ii)\quad E(V_1^2V_2^2) = 1 + 2\rho^{*2}. \end{aligned} \tag{8}$$

PROOF. (i) follows from the fact that all moments of odd orders are equal to zero and

$$\begin{aligned} E(V_1^2V_2^2) = E\{V_1^2E(V_2^2|V_1)\} &= E[V_1^2\{(1-\rho^{*2}) + \rho^{*2}V_1^2\}] \\ &= (1-\rho^{*2}) + 3\rho^{*2} = 1 + 2\rho^{*2}. \end{aligned}$$

For computing the Pitman efficacy of the test procedure based on T, we need to evaluate

$$E(n\bar{Z}\bar{W}|H_0) = n\ \mathrm{Cov}(\bar{Z},\bar{W}|H_0) + n = \rho^* + n, \tag{9}$$

$$E(n\bar{Z}\bar{W}|H_1) = n\ \mathrm{Cov}(\bar{Z},\bar{W}|H_1) + \frac{n}{\lambda_2} = (\rho^* + n)/\lambda_2, \quad \text{when } \lambda_1 = 1. \tag{10}$$

Also

$$\begin{aligned} E(n^2\bar{Z}^2\bar{W}^2|H_0) &= E\{n^2(\bar{Z}-1)^2(\bar{W}-1)^2\} + 4n^2E(\bar{Z}-1)(\bar{W}-1) + \\ &\qquad 2n^2E(\bar{Z}-1)^2 + n^2 \\ &= (1+2\rho^{*^2}) + 4n\rho^* + 2n + n^2. \end{aligned} \tag{11}$$

Hence

$$\begin{aligned} \mathrm{Var}(n\bar{Z}\bar{W}|H_0) &= 1 + 2\rho^{*^2} + 4n\rho^* + 2n + n^2 - (p^* + n)^2 = \\ &\qquad 1 + \rho^{*^2} + 2n\rho^* + 2n. \end{aligned} \tag{12}$$

Now letting $\theta = \lambda_2/\lambda_1$, with $\lambda_1 = 1$ and assuming that $\theta = 1 \pm \xi/n^{\frac{1}{2}}$, we obtain the Pitman efficacy of T to be

$$\begin{aligned} e(T) \stackrel{\text{def}}{=} \lim_{n\to\infty} \quad & \frac{[E(T'|H_1) - E(T'|H_0)]^2}{\xi^2\ \mathrm{var}(T'|H_0)} \\ &= \lim_{n\to\infty} \frac{E^2(n\bar{Z}\bar{W})(\frac{1-\lambda_2}{\lambda_2})^2}{\xi^2 \mathrm{var}(n\bar{Z}\bar{W})} \\ &= \lim_{n\to\infty} \frac{(\rho^*+n)^2(\xi^2)}{n\xi^2(1\pm\xi/n^{\frac{1}{2}})^2(1+\rho^{*^2}+2\rho^*n+2n)} \\ &= \frac{1}{2(1+\rho^*)}. \end{aligned} \tag{13}$$

4. Asymptotic Efficiency with Respect to Scale Alternatives. Govindarajulu (1975) studied the asymptotic efficiency of U relative to Student's t-test against location alternatives. Here, we will evaluate the asymptotic efficiency of U relative to the T-test against scale alternatives, especially when (X,Y) has a bivariate exponential distribution of the form due to Morgenstern (1956), Gumbel (1960), Marshall and Olkin (1967), Downton (1970), Cowan (1987), and Sarker (1987). Let us assume that $G(x) = F(\theta x)$. Then we can rewrite the null and alternative hypotheses as

$$H_0 : \theta = 1 \ \text{ versus } \ H_1 : \theta < 1.$$

Furthermore, the efficacy of the U-test is

$$e(U) = (\int yf^2(y)dy)^2/A \tag{14}$$

where

$$\begin{aligned} A &= (2/3) - 2I(\rho) \\ I(\rho) &= \int_0^1 \int_0^1 H(F^{-1}(u), F^{-1}(v))du\,dv. \end{aligned} \tag{15}$$

Hence, the asymptotic efficiency of U-test relative to T-test is given by

$$e(U,T) = 2(1+\rho^*)(\int yf^2(y)dy)^2/A. \tag{16}$$

Next we will evaluate (16) when $H(x,y)$ is a bivariate exponential distribution having standard exponential distribution for the marginals. Then, computations yield

$$\int_0^\infty yf^2(y)dy = 1/4. \tag{17}$$

Then

$$e(U,F) = (1+\rho^*)/8A. \tag{18}$$

The Bivariate Exponential Distribution of Morgenstern (1956).

The joint density (in the standard form) is given by

$$h(x,y) = e^{-x-y}[1+\rho(2e^{-x}-1)(2e^{-y}-1)], \quad x,y>0, \ -1<\rho<1.$$

Hence

$$H(x,y) = (1-e^{-x})(1-e^{-y})(1+\rho e^{-x-y})$$

and the correlation between X and $Y = \rho^* = \rho/4$. Hence

$$H(F^{-1}(u), F^{-1}(v)) = uv\{1+\rho(1-u)(1-v)\}.$$

Thus

$$\begin{aligned} I(\rho) &= \int_0^1 \int_0^1 H(F^{-1}(u), F^{-1}(v))du\,dv \\ &= (1/4) + (\rho/36); \\ A(\rho) &= (2/3) - 2I(\rho) = (3-\rho)/18. \end{aligned}$$

So

$$e(U,t) = 9(4+\rho)/16(3-\rho). \tag{19}$$

Table 4.1. Values of $e(U,T)$

ρ	-1	-0.5	0	0.5	1
$e(U,T)$	0.42	0.56	0.75	1.01	1.41

Bivariate Exponential Distribution of Gumbel (1960).
The joint density (in standard form) is

$$h(x,y) = e^{-x-y-\rho xy}\{(1+\rho x)(1+\rho y) - \rho\}, \quad x,y > 0,\ \rho > 0.$$

Then

$$H(x,y) = 1 - e^{-x} - e^{-y} + \exp(-x-y-\rho xy), \quad x,y > 0,\ 0 < \rho < 1.$$

Correlation between X and $Y = \rho^* = \frac{e^{1/\rho}}{\rho}E_1(\rho^{-1}) - 1$, where $E_1(x) = \int_x^\infty (e^{-u}/u)du$ stands for the exponential integral. Hence

$$\begin{aligned} I(\rho) &= \int_0^1\int_0^1 [1-(1-u)-(1-v)+(1-u)(1-v)e^{-\rho\ln(1-y)\ln(1-v)}]du\, dv \\ &= \int_0^1\int_0^1 [1-s-t+ste^{-\rho\ln s\ln t}]ds\, dt \\ &= \int_0^1\int_0^1 ste^{-\rho\ln s\ln t}ds\, dt. \end{aligned}$$

Let

$$a(t) = \int_0^1 se^{-\rho\ln s\ln t}ds = \int_0^1 s^{(1-\rho\ln t)}ds = (2-\rho\ln t)^{-1}.$$

Hence

$$\begin{aligned} I(\rho) &= \int_0^1 \{t/(2-\rho\ln t)\}dt \\ &= e^{4/\rho}\int_2^\infty \frac{e^{-2v/\rho}}{\rho v}dv \\ &= \frac{e^{4/\rho}}{\rho}\int_{4/\rho}^\infty \frac{e^{-w}}{w}dw = \frac{e^{4/\rho}}{\rho}E_1(4/\rho). \end{aligned} \tag{20}$$

So,

$$e(U,T) = (1+\rho^*)/4[(4/3) - (4/\rho)e^{4/\rho}E_1(4/\rho)]. \tag{21}$$

Computations yield the following table.

Table 4.2. Values of $e(U,T)$

ρ	0	0.1	0.2	0.25	0.3	0.4	0.5	0.75	0.8	0.9
$1+\rho^*$	1	0.92	0.85	0.82	0.80	0.876	0.72	0.65	0.64	0.62
$e(U,T)$	0.75	0.64	0.56	0.53	0.50	0.45	0.42	0.34	0.33	0.31

The Bivariate Exponential Distribution of Marshall and Olkin (1967). In standard form, we have

$$P(X > x, Y > y) = \exp[-x - y - \rho \max(x,y)], \quad \rho, x, y > 0$$

where

$$EX = EY = (1+\rho)^{-1}$$

and correlation between X and $Y = \rho^* = \rho/(2+\rho)$. Thus

$$\begin{aligned} H(x,y) &= P(X \le x) + P(Y \le y) - 1 + P(X > x, Y > y) \\ &= 1 - e^{-x(1+\rho)} - e^{-y(1+\rho)} + e^{-x-y-\rho\max(x,y)}. \end{aligned} \tag{22}$$

Hence

$$\begin{aligned} I(\rho) &= \int_0^1 \int_0^1 [u + v - 1 + \exp\{\frac{\ln(1-u) + \ln(1-v)}{(1+\rho)} - \frac{\rho}{(1+\rho)} \\ &\quad \times \max(-\ln(1-u), -\ln(1-v))\}] du\, dv \\ &= 2 \iint_{0<v<u<1} (1-u)(1-v)^{(1+\rho)^{-1}} dv\, du = (1+\rho)/(4+3\rho). \end{aligned}$$

So,

$$A = (2/3) - 2I(\rho) = 2/3(4+3\rho). \tag{23}$$

Consequently,

$$e(U,T) = 3(1+\rho)(4+3\rho)/8(2+\rho). \tag{24}$$

Table 4.3. Values of $e(U,T)$

ρ	0	0.5	1.0	1.5	2.0	2.5
$e(U,T)$	0.75	1.24	1.75	2.28	2.81	3.35

Bivariate Exponential Distribution of Downton (1970). The joint density (in standard form) is given by

$$h(x,y) = (1-\rho)^{-1} \exp\left\{-\frac{(x+y)}{1-\rho}\right\} I_0\left(\frac{2\sqrt{\rho x y}}{1-\rho}\right), \quad x, y, \rho > 0,$$

where I_0 denotes the modified Bessel function of order zero. Thus

$$H(x,y) = (1-\rho)^{-1} \int_0^x \int_0^y e^{-(u+v)/(1-\rho)} I_0\left(\frac{2\sqrt{\rho u v}}{1-\rho}\right) du\, dv.$$

Using the expansion $I_0(z) = \sum_{k=0}^{\infty} (\frac{1}{4}z^2)^k/(k!)^2$ we have

$$H(x,y) = \sum_{k=0}^{\infty} (1-\rho)\rho^k \left(\int_0^{x/(1-\rho)} \frac{u^k e^{-u}}{k!} du\right)\left(\int_0^{y/(1-\rho)} \frac{v^k e^{-v}}{k!} dv\right).$$

Hence

$$I(\rho) = \int_0^{\infty}\int_0^{\infty} H(x,y) e^{-(x+y)} dx\, dy \;=\; \sum_{k=0}^{\infty} (1-\rho)\rho^k \times \left[\int_{x=0}^{\infty} e^{-x}\left(\int_0^{x/(1-\rho)} \frac{u^k e^{-u}}{k!} du\right) dx\right]^2.$$

Consider

$$\begin{aligned} a(\rho) &= \int_0^{\infty} e^{-x}\left(\int_0^{x/(1-\rho)} \frac{u^k e^{-u}}{k!} du\right) \\ &= (1-\rho)\int_0^{\infty} e^{-(1-\rho)s}\left(\int_0^{s} \frac{u^k e^{-u}}{k!} du\right) ds \\ &= (1-\rho)\int_0^{\infty} \frac{u^k e^{-u}}{k!}\left(\int_u^{\infty} e^{-(1-\rho)s} ds\right) du \\ &= \int_0^{\infty} e^{-(2-\rho)u}\frac{u^k}{k!} du = (2-\rho)^{-(k+1)}. \end{aligned}$$

Substituting this in the expression for $I(\rho)$ we obtain

$$I(\rho) = \sum_{k=0}^{\infty} (1-\rho)\rho^k (2-\rho)^{-2(k+1)} = (1-\rho)/\{(2-\rho)^2 - \rho\}.$$

Consequently,

$$A(\rho) = \frac{2}{3} - 2I(\rho) = \frac{2(1-\rho)^2}{3\{(2-\rho)^2-\rho\}}.$$

Straightforward computations yield

$$\int_0^\infty \int_0^\infty xyh(x,y)dx\,dy = (1-\rho)^3 \sum_{k=0}^{\infty} (k+1)^2 \rho^k.$$

Next, writing $(k+1)^2 = (k+1)(k+2) - (k+1)$ and summing the right hand side series we obtain

$$E(ZW) = 1 + \rho.$$

Hence $\rho^* = \rho$ and consequently

$$\begin{aligned} e(U,T) = (1+\rho)/8A &= 3(1+\rho)\{(2-\rho)^2 - \rho\}/16(1-\rho)^2 = \\ &\quad 3(1+\rho)(4-\rho)/16(1-\rho). \end{aligned} \tag{25}$$

Table 4.4. Giving the Values of $e(U,T)$

ρ	0	0.25	0.5	0.75	0.9	1
$e(U,T)$	0.75	1.17	1.97	4.27	11.04	∞

The Bivariate Exponential Distribution of Cowan (1987).

Since

$$P[X > x, Y > y] = P[X \le x, Y \le y] - P[X \le x] - P[Y \le y] + 1,$$

one can write the general bivariate distribution function as

$$\begin{aligned} H(x,y) &= 1 - e^{-\lambda_1 x} - e^{-\lambda_2 y} + \\ &\quad \exp\left[-\frac{1}{2}\{\lambda_1 x + \lambda_2 y + (\lambda_1^2 x^2 + \lambda_2^2 y^2 - 2\lambda_1\lambda_2 xy\cos a)^{\frac{1}{2}}\}\right], \\ &\quad x, y > 0,\ 0 < a < \pi. \end{aligned}$$

Using the standard bivariate distribution

$$\begin{aligned} I(a) &= \int_0^\infty \int_0^\infty H(x,y)dF(x)dF(y) \\ &= \int_0^\infty \int_0^\infty [1 - e^{-x} - e^{-y} + e^{-\frac{1}{2}\{(x+y)+(x^2+y^2-2xy\cos a)^{\frac{1}{2}}\}}]e^{-x-y}dx\,dy \\ &= \int_0^\infty \int_0^\infty e^{-\frac{3}{2}(x+y)-\frac{1}{2}(x^2+y^2-2xy\cos a)^{\frac{1}{2}}}\,dx\,dy \\ &= \int_0^{\pi/2} \frac{4d\theta}{9(\cos\theta+\sin\theta)^2+1-\sin 2\theta\cos a+6(\cos\theta+\sin\theta)(1-\sin 2\theta\cos a)^{\frac{1}{2}}} \end{aligned}$$

after using a polar coordinate transformation. Hence

$$I(a) = 4\int_0^{\pi/2} \frac{d\theta}{10 + (9 - \cos a)\sin 2\theta + 6(\cos\theta + \sin\theta)(1 - \sin 2\theta \cos a)^{\frac{1}{2}}}$$

$$I(a) = 2\int_0^{\pi/2} \frac{d\theta}{5 + (5 - \eta)\sin 2\theta + 3(\cos\theta + \sin\theta)(1 + \sin 2\theta - 2\eta \sin 2\theta)^{\frac{1}{2}}}.$$

where $\eta = (1 + \cos a)/2$. One can compute (starting with the double integral)

$$I(\pi) = 1/4, \text{ and } I(0) = 1/3.$$

$$\begin{aligned} A &= \frac{2}{3} - 2I(a) \\ 2A &= 4(\frac{1}{3} - I(a)) \\ e(U,T) &= (1 + \rho^*)/8A. \end{aligned}$$

Also, Cowan (1987) gives

$$\begin{aligned} \text{Corr}(X,Y) &= 1 \quad \text{if } a = 0 \\ &= -1 + \frac{4}{1+\cos a}\left[1 - \frac{1-\cos a}{1+\cos a}\log(\frac{2}{1-\cos a})\right], \quad \text{for } 0 < a < \pi, \\ &= 0 \quad \text{if } a = \pi. \end{aligned}$$

Computations yield Table 4.5 giving values of $I(a)$, the correlation between X and Y and $e(U,T)$.

Table 4.5. Values of $I(a)$, the Correlations between X and Y and $e(U,T)$

a	0	30°	60°	90°	120°	105°	180°
Corr(X,Y)	1	.728	.434	.227	.096	.023	0
$I(a)$	.333	.318	.295	.276	.262	.253	.250
$e(U,T)$	∞	6.88	2.36	1.33	0.95	0.79	0.75

Bivariate Exponential Distribution of Sarkar (1987).

Sarkar (1987) obtains an absolutely continuous bivariate exponential distribution given by (for $\lambda_1, \lambda_2 > 0$ and $\lambda_{12} \geq 0$)

$$\begin{aligned} P[X \geq x, Y \geq y] &= \exp\{-(\lambda_2+\lambda_{12})y\}\{1-[B(\lambda_1 y)]^{-\gamma}[B(\lambda_1 x)]^{1+\gamma}\} \text{ if } 0 < x \leq y \\ &= \exp\{-(\lambda_1+\lambda_{12})x\}\{1-[B(\lambda_2 x)]^{-\gamma}[B(\lambda_2 y)]^{1+\gamma}\} \text{ if } x \geq y > 0, \end{aligned}$$

where $\gamma = \lambda_{12}/(\lambda_1+\lambda_2), B(z) = 1-\exp(-z)$ for $z > 0$. Notice that X and Y are independent if $\lambda_{12} = 0$. Considering the standard form, that is, when $\lambda_1 = \lambda_2 = 1$ and $\lambda_{12} = \rho$ we have $E(Z) = E(\lambda_1 X) = (1+\rho)^{-1}$, $EW = E(\lambda_2 Y) = (1+\rho)^{-1}$, Var Z = Var $W = (1+\rho)^{-2}$, and correlation between Z and W is

$$\text{Corr}(Z,W) = \rho^* = \frac{\rho}{2+\rho}\left[1+2\sum_{j=1}^{\infty}\frac{j!}{2+\rho+2j}\prod_{k=1}^{j}(2+\rho+k)^{-1}\right].$$

Since

$$P[Z\le z, W\le w] = 1-P(Z\ge z)-P(W\ge w)+P(Z\ge z, W\ge w).$$

We obtain

$$\begin{aligned}
I(\rho) &= \int_0^1\int_0^1 H(F^{-1}(u),F^{-1}(v))du\,dv \\
&= 1-\int_0^1(1-u)du-\int_0^1(1-v)dv \\
&\quad+\iint_{u\le v}(1-v)^{(1+\rho)}\{1-v^{-\rho/2}u^{1+(\rho/2)}\}du\,dv \\
&\quad+\iint_{u>v}(1-u)^{(1+\rho)}\{1-u^{-\rho/2}v^{1+(\rho/2)}\}du\,dv \\
&= 2\iint_{0<u\le v<1}(1-v)^{(1+\rho)}\{1-v^{-\rho/2}u^{1+(\rho/2)}\}du\,dv \\
&= 2\left[\int_0^1(1-v)^{(1+\rho)}v\,dv-(2+\rho/2)^{-1}\int_0^1 v^2(1-v)^{1+(\rho/2)}dv\right] \\
&= 2\left[\frac{1}{(3+\rho)(2+\rho)}-(2+\rho/2)^{-1}\frac{2}{(4+\rho/2)(3+\rho/2)(2+\rho/2)}\right] \\
&= 2\left[\frac{1}{(3+\rho)(2+\rho)}-\frac{32}{(8+\rho)(6+\rho)(4+\rho)^2}\right].
\end{aligned}$$

Hence

$$A(\rho) = 2\left[\frac{1}{3}-\frac{2}{(3+\rho)(2+\rho)}+\frac{64}{(8+\rho)(6+\rho)(4+\rho)^2}\right].$$

Then one can easily evaluate

$$e(U,T) = (1+\rho^*)/8A.$$

Table 4.6. Values of ρ^*, $A(\rho)$ and $e(U,T)$ for some selected values of ρ

ρ	0	0.5	1.0	1.5	2.0	2.5	3.0
$A(\rho)$	0.167	0.324	0.414	0.472	0.511	0.539	0.560
ρ^*	0	0.274	0.418	0.507	0.555	0.617	0.653
$e(U,T)$	0.75	0.49	0.43	0.40	0.38	0.38	0.37

Concluding Remarks. Mann-Whitney-Wilcoxon test is more robust to positive dependence in the X, Y variables while testing for scale alternative with certain bivariate exponential distributions for (X, Y). This is true for the bivariate exponential forms due to Morgenstern (1956), Gumbel (1960), Downton (1970), Cowan (1987), and Marshall and Olkin (1967). However, surprisingly for the bivariate form due to Sarkar (1987), the MWW test is sensitive to positive dependence between the variables X and Y.

REFERENCES

ABRAMOWICZ, M. and STEGUN, I.A. (1964). *Handbook of Mathematical Functions.* National Bureau of Standards, Applied Math. Series #55, 227–252.

BASU, A.P. (1988). Multivariate exponential distributions and their applications in reliability. *Handbook of Statistics* **7** (P.R. Krishnaiah and C.R. Rao, eds.), North Holland, 467–477.

COWAN, R. (1987). A bivariate exponential distribution arising in random geometry. *Ann. Inst. Statist. Math.* **39**, Part A, 103–111.

DOWNTON, F. (1970). Bivariate exponential distributions in reliability theory. *J. Roy. Statist. Soc. B* **32** 408–417.

GOVINDARAJULU, Z. (1975). Robustness of Mann-Whitney-Wilcoxon test to dependence in the variables. *Studia Scientiarum Mathematicarum Hungarica* **10** 39–45.

GOVINDARAJULU, Z., LECAM, L., and RAGHAVACHARI, M. (1967). Generalizations of theorems of Chernoff-Savage on the asymptotic normality of test statistics. *Proc. Fifth Berkeley Symp. Math. Statist. and Prob.* University of California Press, 609–638.

GUMBEL, E.J. (1960). Bivariate exponential distribution. *J. Amer. Statist. Assoc.* **55** 698–707.

HAWKES, A.G. (1972). A bivariate exponential distribution with applications to reliability. *J. Roy. Statist. Soc.* Ser. B **34** 129–131.

HOLLANDER, M., PLEDGER, G., and LIN, P.E. (1974). Robustness of the Wilcoxon test to a certain dependency between samples. *Ann. Statist.* **2** 177–181.

JOHNSON, N.L. and KOTZ, S. (1972). *Distributions in Statistics: Continuous Multivariate Distributions.* John Wiley, New York. Chapter 41, Section 3.

MARSHALL, A.W. and OLKIN, I. (1967). A generalized bivariate exponential distribution. *J. Appl. Prob.* **4** 291–302.

MORGENSTERN, D. (1956). Einfache beispiele zweidimensionalar verteilungen. *Mitteilungsblatt für mathematische Statistik* **8** 234–235.

NAGAO, M. and KADOYA, M. (1971). Two-variate exponential distribution and its numerical table for engineering application. *Bulletin of the Disaster Prevention Research Institute* **20**, No. 3, 183–215.

PAULSON, A.S. (1973). A characterization of the exponential distribution and a bivariate exponential distribution. *Sankhyā Ser. A.* **35** 69–78.

RAFERTY, A.C. (1984). A continuous multivariate exponential distribution. *Comm. Statist.*, Part A—*Theory and Methods* **13** 947–965.

SARKAR, S.K. (1987). A continuous bivariate exponential distribution. *J. Amer. Statist. Assoc.* **82** 667–675.

SERFLING, R.J. (1968). The Wilcoxon two-sample statistic on strongly mixing processes. *Ann. Math. Statist.* **39** 1202–1209.

DEPARTMENT OF STATISTICS
UNIVERSITY OF KENTUCKY
LEXINGTON, KY 40506-0027

RELATIVE ERRORS IN RELIABILITY MEASURES

BY PUSHPA L. GUPTA AND R.D. GUPTA[1]

University of Maine and University of New Brunswick

A common assumption, in reliability and lifetesting situations when the components are installed in series system, is that they are independent and are exponentially distributed. In this paper we study the relative error in reliability measures such as the reliability function, the failure rate and the mean residual life under the erroneous assumption of independence when in fact lifetimes follow a bivariate exponential model. The behavior of these errors is discussed to examine their structure as a function of time. Some of the existing results in the literature follow as special cases.

1. Introduction. Klein and Moeschberger (1986, 1987) have studied the relative error (defined in Section 3) in system reliability and system mean life when the components follow the bivariate exponential distributions of Marshall and Olkin (1967), Freund (1961), Gumbel (1960), Downton (1970), and Oakes (1982). Moeschberger and Klein (1984) have studied the relative error in the Gumbel II bivariate exponential model.

In this paper we consider a series system whose components follow bivariate exponential models. The joint distribution of the component lives may not be uniquely determined from the observable data on (T, I), where $T = \min(X_1, X_2)$ and $I = I_{\{X_1 < X_2\}}$, a problem of nonidentifiability as described by Tsiatis (1975) and others. If the data on T shows that T has an exponential distribution, then the component lives can be assumed to follow any one of the models described by Marshall and Olkin (1967), Freund (1961), and Block and Basu (1974). If the data shows simultaneous failure of both the components, the shock model developed by Marshall and Olkin will be more appropriate. If T is not exponential but the marginals are exponentials with the same parameters as for the independent case, then one may assume either one of Gumbel I or Gumbel II (1960). If in fact, the joint lifetimes follow any one of the five models mentioned above, the assumption of independence will lead to inappropriate conclusions.

[1]Partially supported by NSERC Research Grant A–4850.

AMS 1980 subject classification. Primary 62N, 62P.

Key words and phrases. Failure rate, reliability function, mean residual life, bivariate exponential distributions.

Section 2 describes the various bivariate models and Section 3 contains the definitions of the three reliability measures and the corresponding relative errors. The relative errors in the three measures for various models are given in Section 4. It also contains the analyses of these errors. The relative error in the system mean life studied by Klein and Moeschberger follows as a special case of the mean residual life at the origin.

2. The Models. Since the data on T is available, one can test to see whether the distribution of T is exponential. If it is found that the distribution of T is an exponential, then the joint distribution of X_1 and X_2 may be one of the following:

Independent:

$$\bar{F}_1(x_1, x_2) = e^{-\lambda_1 x_1 - \lambda_2 x_2}, \quad \lambda_i > 0, \quad x_i > 0, \quad i = 1, 2.$$

Marshall and Olkin model (1967):

$$\bar{F}_2(x_1, x_2) = e^{-\lambda_1 x_1 - \lambda_2 x_2 - \lambda_{12} \max(x_1, x_2)}, \quad \lambda_i > 0, \quad \lambda_{12} > 0, x_i > 0, \quad i = 1, 2.$$

Freund (1961):

$$\bar{F}_3(x_1, x_2) = \begin{cases} (\lambda_1/(\lambda_1 + \lambda_2 - \theta_2))\, e^{-(\lambda_1+\lambda_2-\theta_2)x_1 - \theta_2 x_2} \\ \quad + \; ((\lambda_2 - \theta_2)/(\lambda_1 + \lambda_2 - \theta_2))\, e^{-(\lambda_1+\lambda_2)x_2}, \; x_1 \le x_2, \\ (\lambda_2/(\lambda_1 + \lambda_2 - \theta_1))\, e^{-(\lambda_1+\lambda_2-\theta_1)x_2 - \theta_1 x_1} \\ \quad + \; ((\lambda_1 - \theta_1)/(\lambda_1 + \lambda_2 - \theta_1))\, e^{-(\lambda_1+\lambda_2)x_1}, \; x_1 > x_2, \\ \theta_i > 0, \; \lambda_i > 0, \; x_i > 0, \; i = 1, 2. \end{cases}$$

Block and Basu (1974):

$$\begin{aligned} \bar{F}_4(x_1, x_2) &= ((\lambda_1 + \lambda_2 + \lambda_{12})/(\lambda_1 + \lambda_2))\, e^{-\lambda_1 x_1 - \lambda_2 x_2 - \lambda_{12} \max(x_1, x_2)} \\ &- (\lambda_{12}/(\lambda_1 + \lambda_2))\, e^{-(\lambda_1 + \lambda_2 + \lambda_{12}) \max(x_1, x_2)} \\ &\quad \lambda_i > 0, \; \lambda_{12} \ge 0, \; x_i > 0, \; i = 1, 2. \end{aligned}$$

An error may occur by assuming the independent model when in fact the joint distribution of X_1 and X_2 is described by one of the other three models. In case, however, due to various difficulties and resources, the data on T is not available at the designing stage of a system, it is not unreasonable to assume that the marginal distributions of X_1 and X_2 are exponential with parameters λ_1 and λ_2, respectively. The following models satisfy this condition:

Independent:

$$\bar{F}_1(x_1, x_2) = e^{-\lambda_1 x_1 - \lambda_2 x_2}, \quad \lambda_i > 0, \quad x_i > 0, \quad i = 1, 2.$$

Gumbel I (1960):

$$\bar{F}_5(x_1,x_2) = e^{-\lambda_1 x_1-\lambda_2 x_2-\lambda_{12}x_1x_2}, \; \lambda_i > 0, \; \lambda_{12} \geq 0, \; x_i > 0, \; i = 1,2.$$

Gumbel II (1960):

$$\bar{F}_6(x_1,x_2) = \left[1+\alpha(1-e^{-\lambda_1 x_1})(1-e^{-\lambda_2 x_2})\right] e^{-\lambda_1 x_1-\lambda_2 x_2} \quad \lambda_i > 0, \; x_i > 0, \; |\alpha| < 1, \; i = 1,2.$$

Once again an error may occur due to the erroneous assumption of independence.

3. Definitions. Suppose T is a non-negative random variable denoting the life of a component having distribution function $F(t)$ and probability density function (pdf) $f(t)$. Then the survival function $\bar{F}(t)$, the mean residual life function (MRLF) $r(t)$ and the failure rate $\lambda(t)$ of T are defined as follows:

DEFINITION 3.1. The mean residual life function $r(t)$ of T is defined by

$$r(t) = E(T-t|T>t) = \int_t^\infty \bar{F}(x)dx/\bar{F}(t) = \left[\mu - \int_0^t \bar{F}(x)dx\right]/\bar{F}(t).$$

DEFINITION 3.2. The failure rate $\lambda(t)$ is defined by

$$\lambda(t) = f(t)/\bar{F}(t) = [1+r'(t)]/r(t)$$

DEFINITION 3.3. The survival function $\bar{F}(t)$ is given by

$$\bar{F}(t) = P(T>t) = (r(0)/r(t))\exp\left[-\int_0^t dx/r(x)\right].$$

For the series system with two components $T = \min(X_1, X_2)$, it is clear from the above definitions that $\bar{F}(t)$, $\lambda(t)$ and $r(t)$ are equivalent in the sense that given any one of them the other two can be determined. We also define the relative error, in various reliability measures, incurred by erroneously assuming the independent bivariate exponential model when in fact the models are dependent. The relative errors in reliability measures are defined as follows:

Relative error in reliability (survival) function is

$$\left(\bar{F}_D(t) - \bar{F}_I(t)\right)/\bar{F}_I(t).$$

Relative error in failure rate is

$$\left(\lambda_D(t) - \lambda_I(t)\right)/\lambda_I(t),$$

Table I

Reliability Measures for the Two Component Series System

Model	Reliability Function	Failure Rate	Mean-Residual Life
Independent	$\exp(-(\lambda_1+\lambda_2)t)$	$\lambda_1+\lambda_2$	$1/(\lambda_1+\lambda_2)$
Marshall & Olkin	$\exp(-\lambda t)$	λ	$1/\lambda$
Freund	$\exp(-(\lambda_1+\lambda_2)t)$	$\lambda_1+\lambda_2$	$1/(\lambda_1+\lambda_2)$
Block & Basu	$\exp(-\lambda t)$	λ	$1/\lambda$
Gumbel I	$\exp(-(\lambda_1+\lambda_2+\lambda_{12}t)t)$	$\lambda_1+\lambda_2+$ $2\lambda_{12}t$	$\sqrt{\pi/\lambda_{12}}(1-\phi(\sqrt{2t'}))$ $/\exp(-t')$
Gumbel II	$\exp(-(\lambda_1+\lambda_2)t)h(t)$	$\lambda_1+\lambda_2-$ $[h'(t)/h(t)]$	$[(1+\alpha)/(\lambda_1+\lambda_2)-$ $\alpha g(t)]/h(t)$

where

$$\begin{aligned}
\lambda &= \lambda_1+\lambda_2+\lambda_{12}, t' = \lambda_{12}(t+[(\lambda_1+\lambda_2)/2\lambda_{12}])^2, \phi(t) = \frac{1}{\sqrt{2\pi}}\int_{-\infty}^{t} e^{-x^2/2}dx \\
h(t) &= 1+\alpha(1-\exp(-\lambda_1 t))(1-\exp(-\lambda_2 t)) \\
g(t) &= \exp(-\lambda_1 t)/(2\lambda_1+\lambda_2)+\exp(-\lambda_2 t)/(\lambda_1+2\lambda_2)-\exp(-(\lambda_1+\lambda_2)t)/ \\
&\quad 2(\lambda_1+\lambda_2)
\end{aligned}$$

and the relative error in the mean-residual life is

$$(r_D(t)-r_I(t))/r_I(t),$$

where D stands for dependent and I for independent model.

Even though the three reliability measures described above are equivalent, the three relative errors do not exhibit such a property.

4. Relative Error in Reliability Measures. Table I lists the reliability measures for the two component series system. The relative errors in reliability measures under the assumption of independence are given in Table II.

Analysis of the errors.

(1) The relative error in all the three reliability measures is the same for Marshall and Olkin and Block and Basu models and there is no error for Freund model. For Marshall and Olkin, the relative error in reliability is negative and decreases from 0 to -1 as a function of t (λ_{12} fixed) or λ_{12} (t fixed). Thus in this case the independence assumption leads to an overassessment of reliability. For this model,

Table II

Relative Error in Reliability Measures Under the Assumption of Independence

True Model	Reliability Function	Failure Rate	Mean-Residual Life
Independent	0	0	0
Marshall & Olkin	$\exp(-\lambda_{12}t)-1$	$\lambda_{12}/(\lambda_1+\lambda_2)$	$-\lambda_{12}/\lambda$
Freund	0	0	0
Block & Basu	$\exp(-\lambda_{12}t)-1$	$\lambda_{12}/(\lambda_1+\lambda_2)$	$-\lambda_{12}/\lambda$
Gumbel I	$\exp(-\lambda_{12}t^2)-1$	$2\lambda_{12}t/(\lambda_1+\lambda_2)$	$[\sqrt{\pi/\lambda_{12}}(1-\phi(\sqrt{2t'}))$ $(\lambda_1+\lambda_2)/\exp(-t')]-1$
Gumbel II	$h(t)-1$	$-h'(t)/(\lambda_1+\lambda_2)$ $h(t)$	$([1+\alpha-\alpha(\lambda_1+\lambda_2)$ $g(t)]/h(t))-1$

where λ, t', $h(t)$, $\phi(t)$ and $g(t)$ are as above.

unlike the relative error in reliability, the other two relative errors are independent of t and are functions of all the three parameters λ_1, λ_2, and λ_{12}. Thus the wrong assumption leads to an underassessment in the case of failure rate and an overassessment in the case of mean residual life.

(2) For Gumbel I, the relative error in reliability is negative and decreases from 0 to -1 as a function of λ_{12} (t fixed) or $t(\lambda_{12}$ fixed), resulting in an overassessment under the wrong assumption of independence. In the case of Gumbel II, the relative error has the same sign as that of α and increases (decreases) from 0 to α if α is positive (negative).

(3) The relative error in failure rate for Gumbel I is positive and is a linear function of t (λ_{12} fixed). In this case the wrong assumption will lead to underassessment.

(4) For the case $\lambda_1 = \lambda_2$, the relative error in failure rate for Gumbel II is positive (negative) for negative (positive) values of α. If α is negative (positive) it increases (decreases) from 0 to $(1-\sqrt{1+\alpha})/2$. Thus the absolute maximum error in this case is $|1-\sqrt{\alpha+1}|/2$. The critical point is $t = -(1/\lambda_1)\ln[\sqrt{\alpha+1}(\sqrt{\alpha+1}-1)/\alpha]$.

(5) For the case $\lambda_1 = \lambda_2$, the relative error in the mean residual life for Gumbel II is positive (negative) for positive (negative) values of α. If α is positive (negative) it increases (decreases) from 0 to $(3\sqrt{\alpha+1}-\sqrt{\alpha+9})^2/12[\sqrt{(\alpha+9)(\alpha+1)}-(\alpha+3)]$. Thus the absolute maximum error is $(3\sqrt{\alpha+1}-\sqrt{\alpha+9})^2/12[\sqrt{(\alpha+9)(\alpha+1)}-(\alpha+3)]$.

(6) The relative error in the mean residual life for Gumbel I can be written as

$(\lambda_1+\lambda_2)$ (1/(failure rate of normal random variable with mean $= -(\lambda_1+\lambda_2)/2\lambda_{12}$ and variance $= 1/2\lambda_{12}$)) $- 1$. Since the failure rate of a normal random variable is increasing, the relative error under discussion is decreasing. It is always negative and decreases from $(\sqrt{\pi/\lambda_{12}})(1-\phi((\lambda_1+\lambda_2)/\sqrt{2\lambda_{12}}))(\lambda+\lambda_2)/e^{-(\lambda_1+\lambda_2)^2/4\lambda_{12}})-1$ to -1. Thus in this case the independence assumption leads to overassessment of the mean residual life.

REFERENCES

BLOCK, H.W. and BASU, A.P. (1974). A continuous bivariate exponential extension. *J. Amer. Statist. Assoc.* **69** (348) 1031–1037.

DOWNTON, F. (1970). Bivariate exponential distributions in reliability theory. *J. Roy. Statist. Soc.* **3(32)** 408–417.

FREUND, J.E. (1961). A bivariate extension of the exponential distribution. *J. Amer. Statist. Assoc.* **56** 971–977.

GUMBEL, E.J. (1960). Bivariate exponential distributions. *J. Amer. Statist. Assoc.* **55** 698–707.

KLEIN, J.P. and MOESCHBERGER, M.L. (1986). The independence assumption for a series or parallel system when component lifetimes are exponential. *IEEE Transactions on Reliability* R–**35**(3) 330–334.

KLEIN, J.P. and MOESCHBERGER, M.L. (1987). Independent or dependent competing risks: Does it make a difference? *Commun. Statist. Simul.* **16**(2) 507–533.

MARSHALL, A.W. and OLKIN, I. (1967). A multivariate exponential distribution. *J. Amer. Statist. Assoc.* **62** 30–40.

MOESCHBERGER, M.L. and KLEIN, J.P. (1984). Consequences of departures from independence in exponential series systems. *Technometrics* **26**(3) 277–284.

OAKES, D. (1982). A model for association in bivariate survival data. *J. Roy. Statist. Soc.* B **44** 414–422.

TSIATIS, A. (1975). A nonidentifiability aspect of the problem of competing risks. *Proc. Nat. Acad. Sci.* **72** 20–22.

DEPARTMENT OF MATHEMATICS
UNIVERSITY OF MAINE
ORONO, ME 04469

DIVISION OF MATHEMATICS,
ENGINEERING, COMPUTER SCIENCE
UNIVERSITY OF NEW BRUNSWICK
SAINT JOHN, N.B., CANADA E2L4L5

INFORMATION, CENSORING, AND DEPENDENCE

By Myles Hollander[1], Frank Proschan[1], and James Sconing[1]

Florida State University, Florida State University, and University of Iowa

Hollander, Proschan, and Sconing (1987) used the theory of majorization to develop and study various information measures in the randomly right-censored model where the basic observation is $\underline{Z} = (Z, \delta)$ where $Z = \min(X, Y)$, X is the survival time, Y is the censoring time, Y is assumed to be independent of X, and $\delta = 1$ if $X \leq Y$, $= 0$ otherwise. Here we use coefficients of divergence to derive measures of how dissimilar the joint distribution of $(X, \underline{Z})$ is from the product of its marginals. These measures contain some of the HPS information measures as special cases. We also introduce various concepts of bivariate dependence to measure the degree to which Y inhibits the ability to see X.

1. Introduction and Summary. Consider the randomly censored model where X is the survival time, Y is the censoring time, and where Y is assumed to be independent of X. We observe (Z, δ) where $Z = \min(X, Y)$, $\delta = I(X \leq Y)$, where $I(A)$ denotes the indicator of the event A. Hollander, Proschan, and Sconing (1987) [hereafter referred to as HPS (1987)] used the theory of majorization to develop and study various measures for this model.

One of the measures developed by HPS (1987) for the case where X and Y are discrete is a generalization of Shannon's (1948) information in the uncensored case.

Definition 1.1. For the censored model where X and Y have discrete distributions $p_i = \Pr(X = i), q_i = \Pr(Y = i)$, the information in the experiment (X, Y) is defined to be

$$(1) \qquad H(X,Y) = H(\underline{p}, \underline{q}) = -\sum_i q_i \left[\sum_{j \leq i} p_j \log p_j + \bar{P}_{i+1} \log \bar{P}_{i+1} \right]$$

[1]Research sponsored by the Air Force Office of Scientific Research, AFSC, USAF, under Grant AFOSR 88–0040. The U.S. Government is authorized to reproduce and distribute reprints for governmental purposes notwithstanding any copyright notation thereon.

AMS 1980 subject classifications. Primary 62N05; secondary 62B10.

Key words and phrases. Bivariate dependence, coefficients of divergence, association.

We are grateful to Professor Ian McKeague for helpful conversations about information. We also thank the referee and editors for helpful comments.

where $\underline{p} = (p_1, p_2, \ldots)$, $\underline{q} = (q_1, q_2, \ldots)$, and $\bar{P}_i = \sum_{j \geq i} p_j$. (The choice of the base of the logarithm is unimportant and henceforth will be defined as the base of the natural logarithm.)

When there is no censoring, (1) reduces to Shannon's (1948) measure

$$(2) \qquad H(X) = H(\underline{p}) = -\sum_i p_i \log p_i.$$

HPS (1987) showed that (1) is equivalent to Shannon's mutual information $H(X) - H(X|Z, \delta)$. Other properties concerning $H(X, Y)$ established by HPS (1987) include:

I: $H(X) \geq H(X, Y)$.

II: If $Y_1 \overset{st}{\leq} Y_2$, then $H(X, Y_1) \leq H(X, Y_2)$

III: If there exists a k such that for all $i > k$, $p_i > p_{i+1}$ and $\bar{P}_k < e^{-1}$, then $H(X, i+1) - H(X, i)$ is nonincreasing in $i, i > k$. Here $H(X, i)$ is an abbreviation for $H(X, Y)$ where $Y = i$ with probability one.

Property II essentially says that information increases as censoring decreases. However, there will be limits to such an increase and Barlow and Hsiung (1983) state "it would be interesting to see when this (information) gain is marginally decreasing." Property III gives a condition for that effect. HPS (1987) used majorization to prove I and II. Goel (1986) uses (2) and Blackwell's (1951) theory of comparison of experiments to prove I and II.

When X, Y are absolutely continuous with densities p, q, respectively, the analog of (1) is

$$(3) \quad H(X, Y) = H(p, q) = -\int_0^\infty q(y) \left[\int_0^y p(x) \log p(x) dx + \bar{P}(y) \log \bar{P}(y) \right] dy$$

where P is the distribution function of X and $\bar{P} = 1 - P$. This measure of information in the continuous case was introduced and considered in Sconing (1985) and Hollander, Proschan, and Sconing (1985). They noted that unlike (1), (3) is not scale-invariant. Baxter (1989) established analogues of properties I, II, III in the absolutely continuous case using measure $H(X, Y)$ defined by (3). Baxter does not view the lack of scale-invariance to be a serious limitation for the use of (3).

The original motivation for (1) and (3) was intuitive. Suppose in the discrete case the censoring variable assumes the value i. Then the information obtained is the full information $-p_j \log p_j$, if a death occurs prior to the censoring time. Otherwise we receive partial information, $-\bar{P}_{i+1} \log \bar{P}_{i+1}$. (If a death and a censorship occur at the same time we say that a death is observed.) The definition of (expected) information follows by averaging with respect to the censoring variable.

In Section 2 we derive the f-divergence [see (5)] of the Radon-Nikodym derivative of the joint distribution of X and $\underline{Z}$ with respect to the product of marginals.

This measures how dissimilar the joint distribution is from the product of the marginals. The measure derived is seen to contain some of the HPS (1987) information measures as special cases. As the censoring variable increases stochastically (the limiting case of $Y = \infty$ with probability one can be thought of as no censoring), Z and X become more similar and thus the divergence should decrease. Conditions for this to occur are given in Theorem 2.1.

In Section 3 we introduce various notions of bivariate dependence to measure the degree to which the censoring variable Y inhibits the ability to see the survival variable X. Let $Y_1 \overset{st}{\leq} Y_2$ and let $Z_i = \min(X, Y_i), i = 1, 2$. We compare the dependence between X and Z_1 to the dependence between X and Z_2. The notions introduced are "more positive quadrant dependent," "more associated," "more left-tail decreasing," "more right-tail increasing," and "more stochastically increasing." It is then shown that (X, Z_2) is more positive quadrant dependent than (X, Z_1). With the exception of "more associated," similar results are obtained for the other notions of dependence.

2. Coefficients of Divergence. When $X < Y$ we have $Z = X$. Since the variables X, Z are often equal, in some sense their underlying probabilistic structures should be similar. From Kullback (1959), coefficients which increase as two distributions become less similar and are called coefficients of divergence.

We define our information measure in the continuous case to be

$$I_g(p_X p_{\underline{Z}}, p_{X\times\underline{Z}}) = \int_0^\infty q(z)\left[\int_0^z p(x)g\{p(x)\}dx + \bar{P}(z)g\{\bar{P}(z)\}\right]dz. \tag{4}$$

This measure is equivalent to a measure of information in the discrete case developed in HPS (1987) and (with $g(x) = -\log x$) advocated in the absolutely continuous case by Baxter (1989).

Note that the information is defined as a relationship between $p_X p_{\underline{Z}}$, the product of the marginal distributions of X and $\underline{Z}$, and $p_{X\times\underline{Z}}$ the joint distribution. Our coefficent I_g is actually a measure of the distance of the joint distribution from the case where X and $\underline{Z}$ are independent. That (4) is actually a coefficient of divergence follows from the results of Csiszàr.

Csiszàr (1963, 1966) generalized the Kullback-Leibler information number in the following fashion. Let $f(x)$ be a convex function on R^+ satisfying $f(0) = \lim_{x\to 0} f(x), 0\cdot f(0/0) = 0, 0\cdot f(a/0) = \lim_{x\to\infty} af(x)/x, a > 0$. Let u_1 and u_2 be two probability distributions on some measurable space $(\mathcal{X}, \mathcal{A})$. Let λ be a measure on $(\mathcal{X}, \mathcal{A})$ such that u_i is absolutely continuous with respect to $\lambda, i = 1, 2$. Let p_i be the Radon-Nikodym derivative of u_i with respect to λ. Define

$$I_f(u_1, u_2) = \int p_1(x) f\left[\frac{p_2(x)}{p_1(x)}\right]\lambda(dx). \tag{5}$$

$I_f(u_1, u_2)$ is the $\underline{f\text{-divergence}}$ of u_1 and u_2.

From a completely different point of view, Ali and Silvey (1965a, 1965b, 1966) and independently Ziv and Zakai (1973) obtain an expression similar to (5). Both pairs of authors consider coefficients which quantify the distance between two probability measures. Their coefficient of divergence is defined as

$$d_f(P_1, P_2) = \int_{\phi<\infty} f(\phi)dP_1 + P_2(N) \lim_{\phi\to\infty} f(\phi)/\phi \tag{6}$$

where $f(x)$ is a convex function, $\phi = dP_2/dP_1$, and N is a P_1-null set where P_2 has positive measure. The only difference between (5) and (6) is the dominating measure λ. The two measures will be identical if P_1 and P_2 are mutually absolutely continuous. Note that the measures (5) and (6) are not symmetric in p_1 and p_2. However if $g(x) = xf(1/x)$ then $I_f(p_1, p_2) = I_g(p_2, p_1)$. Further g is convex if and only if f is convex. Define a new function $f^*(x) = f(x) + g(x)$; then the measure $I_{f^*}(p_1, p_2)$ will be symmetric.

Now we can derive the coefficient of (4) using the divergence measures in (5) or (6). Consider X and $\underline{Z}$ as the two variables of interest. We derive the f-divergence of the Radon-Nikodym derivative of the joint distribution of X and $\underline{Z}$ with respect to the product of their marginals. Note that the joint density of X and $\underline{Z}$ puts positive probability on the line where $X = Z$, the 45° line passing through the origin. This line has zero two-dimensional Lebesgue measure. Thus p_1 and p_2 defined as the joint distribution of X and $\underline{Z}$ and the product of the marginals are not mutually absolutely continuous. Hence the measures in (5) and (6) are no longer equivalent. Equation (6) is now useful only if $\lim_{x\to\infty} f(x)/x$ is finite. Equation (5) requires a measure $\lambda(x)$ which dominates both the joint density of X and $\underline{Z}$ and the product of the marginals. Let $\lambda(x)$ be the sum of two-dimensional Lebesgue measure and a measure u, which is Lebesgue measure on the 45° line $\{(x, y) : x = y, x > 0, y > 0\}$. For the joint probability measure of $(X, \underline{Z})$, we write $\Pr\{X = x, \underline{Z} = (z, 0)\} = p(x)q(z)$, for $x > z$, 0 otherwise, and $\Pr\{X = x, \underline{Z} = (z, 1)\} = p(x)\bar{Q}(x)$, for $x = z$, 0 otherwise. Then (5) becomes

$$\begin{aligned} I_f(p_X p_{\underline{Z}}, p_{X\times\underline{Z}}) &= \int_0^\infty p(x)p(x)\bar{Q}(x) f\left\{\frac{p(x)\bar{Q}(x)}{p(x)p(x)\bar{Q}(x)}\right\} dx + \\ &\quad \int_x \int_{z<x} p(x)q(z)\bar{P}(z) f\left\{\frac{p(x)q(z)}{p(x)q(z)\bar{P}(z)}\right\} dz dx, \end{aligned}$$

which reduces to,

$$\begin{aligned} I_f(p_X p_{\underline{Z}}, p_{X\times\underline{Z}}) &= \int_0^\infty p(x)p(x)\bar{Q}(x) f\{1/p(x)\} dx \\ &\quad + \int_0^\infty q(x)\bar{P}(x)\bar{P}(x) f\{1/\bar{P}(x)\} dx. \end{aligned} \tag{7}$$

Take $g(x) = xf(1/x)$; then (7) becomes

$$(8) \quad I_g(p_X p_{\underline{Z}}, p_{X\times\underline{Z}}) = \int_0^\infty p(x)\bar{Q}(x)g\{p(x)\}dx + \int_0^\infty q(x)\bar{P}(x)g\{\bar{P}(x)\}dx.$$

This can be rewritten to give the coefficient in (4). One would expect that (under reasonable conditions) I_g would decrease as censoring increases stochastically. Such a decrease is equivalent to the term $\psi(z) = \int_0^z p(x)g(p(x))dx + \bar{P}(z)g\bar{P}(z))$ being increasing. Assume g is differentiable; then

$$\psi'(z) = p(z)g\{p(z)\} - p(z)\bar{P}(z)g'\{\bar{P}(z)\} - p(z)g\{\bar{P}(z)\},$$

which is positive if and only if for every z

$$(9) \qquad g\{p(z)\} \geq \bar{P}(z)g'\{\bar{P}(z)\} + g\{\bar{P}(z)\}.$$

Unfortunately inequality (9) is not always satisfied. For example, take $g(x) = -\log x$ and $\bar{P}(x) = \exp\{-\lambda x\}$; then the direction of the inequality depends on λ. However some conditions can be found for $g(x)$ and $p(x)$ so that (9) is satisfied. Two such conditions are:

C1: g decreasing on [0,1] and $p(z)\{\bar{P}(z)\}^{-1} \leq 2$
C2: g increasing on [0,1] and $p(z)\{\bar{P}(z)\}^{-1} \geq 2$

The conditions C1 and C2 are introduced to keep the failure rate $p(z)\{\bar{P}(z)\}^{-1}$ from varying too much.

THEOREM 2.1. *If either C1 and C2 hold and $g'(x)$ is continuous on $[0, \infty]$, then $I_g(p_X p_{\underline{Z}}, p_{X\times\underline{Z}})$ is decreasing as censoring increases stochastically.*

PROOF. It is enough to show (9). Expand $g(p(z))$ in a Taylor series about $\bar{P}(z)$. Then

$$\begin{aligned} g\{p(z)\} &\geq g\{\bar{P}(z)\} + g'\{\bar{P}(z)\}\{p(z) - \bar{P}(z)\} \\ &\geq g\{\bar{P}(z)\} + \bar{P}(z)g'\{\bar{P}(z)\} + g'\{\bar{P}(z)\}(\{p(z) - 2\bar{P}(z)\} \\ &\geq g\{\bar{P}(z)\} + \bar{P}(z)g'\{\bar{P}(z)\}, \end{aligned}$$

if $g'\{\bar{P}(z)\}\{p(z) - 2\bar{P}(z)\} \geq 0$, which holds if C1 or C2 hold. ||

In terms of the original function $f(x)$, $g(x)$ decreasing is equivalent to $f(x)/x$ increasing, $1 \leq x < \infty$. Most of the functions $f(x)$ which are commonly used in f-divergence satisfy the necessary condition.

EXAMPLE 2.2.

1) $f(x) = x\log x$	$g(x) = -\log x$	Kullback-Leibler Information number
2) $f(x) = (1/2)(x^{1/2} - 1)^2$	$g(x) = (1/2)(x^{1/2} - 1)^2$	Hellinger metric
3) $f(x) = (1/2)\|x - 1\|$	$g(x) = (1/2)\|x - 1\|$	city-block distance
4) $f(x) = (x - 1)^2$	$g(x) = (x - 1)^2/x$	χ^2-distance

It is easy to verify that in the above four cases, $g(x)$ is decreasing. Note that the third function does not satisfy the conditions of Theorem 2.1. However the ordering still holds under slightly more restrictive conditions.

THEOREM 2.3. *If g is decreasing on (0,1) and $p(z)(\bar{P}(z))^{-1} \le 1$, then $I_g(p_X p_{\underline{Z}}, p_{X\times\underline{Z}})$ is decreasing as censoring increases stochastically.*

PROOF. If g is decreasing and $p(z)/\bar{P}(z) \le 1, g\{p(z)\} \ge g\{\bar{P}(z)\}$. Equation (9) follows since $g'(x) \le 0$ on (0,1). ||

These last two theorems use the divergence measure as defined in (5). As was stated previously (6) is not satisfactory unless $\lim_{x\to\infty} f(x)/x < \infty$. Of the four functions cited in Example 2.2 only the second and third functions fit this criterion. In particular the third function, $f(x) = (1/2)|x-1|$ is the one originally proposed by Ali and Silvey (1965a) for measuring dispersion between the joint distribution of two variables and the product of their marginals. In the censored model, the set N corresponds to the set where $X = Z$, or equivalently, where $X \le Y$. Then (6) becomes

$$(10) \quad d_f(p_X p_{\underline{Z}}, p_{X\times\underline{Z}}) = \int_0^\infty q(x)\bar{P}^2(x)f\{1/\bar{P}(x)\}dx + c\int_0^\infty p(x)\bar{Q}(x)dx$$

where $c = \lim_{x\to\infty} f(x)/x$.

THEOREM 2.4. *If f is such that $\lim_{x\to\infty} f(x)/x = c < \infty$ and $f(x)/x$ is increasing for $1 < x < \infty$, then $d_f(p_X p_{\underline{Z}}, p_{X\times\underline{Z}})$ increases as censoring decreases stochastically.*

PROOF. Consider (10) as an expected loss over the variable Z with loss $\bar{P}(x)f\{1/\bar{P}(x)\}$ when $Z = x$ and $Y < X$, and loss c when $X \le Y$. So the loss function can be written as $\bar{P}(x)f\{1/\bar{P}(x)\}I(Y < X) + cI(X \le Y)$. As Y increases stochastically, so does Z. Since $f(x)/x$ increases to c as x increases, the loss function is increasing. Hence the expected loss increases. ||

In Example 2.2 both the Hellinger metric and the city-block distance satisfy the conditions of Theorem 2.4. The conditions in Theorem 2.4 are less restrictive than those of Theorem 2.1 in the sense that there is no condition on the distribution of X. Of course the conditions in Theorem 2.4 are more restrictive in the sense that they allow fewer functions f.

3. Measures of Bivariate Dependence. Dependence measures have typically been developed to test for independence between two variables or to measure the degree to which large values of one variable go with large values of the other. Some general notions of dependence are given in the following definition.

DEFINITION 3.1.

1) Positively quadrant dependent (PQD): U and V are positively quadrant dependent if

(11) $$\Pr(U \le u, V \le v) \ge \Pr(U \le u)\Pr(V \le v) \text{ for all } u, v.$$

2) Associated: U and V are associated if

(12) $$\mathrm{Cov}\{\Gamma(U,V), \Delta(U,V)\} \ge 0$$

for all Γ, Δ which are componentwise increasing.

3) Left-Tail Decreasing $(\mathrm{LTD}(V|U))$: V is left-tail decreasing in U if

(13) $$\Pr(V \le v | U \le u) \text{ is decreasing in } u.$$

4) Right-Tail Increasing $(\mathrm{RTI}(V|U))$: V is right-tail increasing in U if

(14) $$\Pr(V > v | U > u) \text{ is increasing in } u.$$

5) Stochastically Increasing $(\mathrm{SI}(V|U))$: V is stochastically increasing in U if

(15) $$\Pr(V > v | U = u) \text{ is increasing in } u.$$

These notions are ordered in strength by:

(16) $$\mathrm{SI}(V|U) \Rightarrow \mathrm{RTI}(V|U) \Rightarrow \text{Association} \Rightarrow \mathrm{PQD}.$$

The sequence of implications is the same when $\mathrm{RTI}(V|U)$ is replaced by $\mathrm{LTD}(V|U)$. For verification of the implications and counterexamples to the reverse implications, see Barlow and Proschan (1981). Most of the above definitions were originally given in Lehmann (1966). The notion of association was introduced in Esary, Proschan, and Walkup (1967).

The inequalities in parts 1–5 of Definition 3.1 are notions of positive dependence for a pair of variables. We now generalize these concepts to compare the levels of dependence of two sets of variables.

DEFINITION 3.2. Given four random variables U_1, U_2, V_1, V_2, we say that:

1) U_1 and V_1 are more PQD than U_2 and V_2 if for all u, v,

(17) $$\begin{aligned} &\Pr(U_1 \le u, V_1 \le v) - \Pr(U_1 \le u)\Pr(V_1 \le v) \\ &\ge \Pr(U_2 \le u, V_2 \le v) - \Pr(U_2 \le u)\Pr(V_2 \le v). \end{aligned}$$

2) U_1 and V_1 are more associated than U_2 and V_2 if

(18) $$\begin{aligned} &\mathrm{Cov}\{\Gamma(U_1,V_1), \Delta(U_1,V_1)\} - \mathrm{Cov}\{\Gamma(U_2,V_2), \Delta(U_2,V_2)\} \ge 0, \\ &\text{for all componentwise increasing functions } \Gamma, \Delta. \end{aligned}$$

3) V_1 is more LTD in U_1 than V_2 is in U_2 if for all $v, u' < u$,

$$\Pr(V_1 \le v | U_1 \le u') - \Pr(V_1 \le v | U_1 \le u) \ \ge \ \Pr(V_2 \le v | U_2 \le u') - \Pr(V_2 \le v | U_2 \le u). \tag{19}$$

4) <u>V_1 is more RTI in U_1 than V_2 is in U_2</u> if for all $v, u' < u$,

$$\Pr(V_1 > v | U_1 > u) - \Pr(V_1 > v | U_1 > u') \ \ge \ \Pr(V_2 > v | U_2 > u) - \Pr(V_2 > v | U_2 > u'). \tag{20}$$

5) <u>V_1 is more SI in U_1 than V_2 is in U_2</u> if for all $v, u' < u$,

$$\Pr(V_1 > v | U_1 = u) - \Pr(V_1 > v | U_1 = u') \ \ge \ \Pr(V_2 > v | U_2 = u) - \Pr(V_2 > v | U_2 = u'). \tag{21}$$

REMARKS. a) With Definition 3.2, comparisons in the censored model are readily made. In our censored data applications we take $U_1 = U_2 = X$ (the survival time random variable), but note that Definition 3.2 does not require that restriction.

b) When U_1 has the same distribution as U_2 and V_1 has the same distribution as V_2, then our notion of "more PQD" given in (17) reduces to Tchen's (1980) notion of the distribution of (U_1, V_1) being "more concordant" than the distribution of (U_2, V_2).

c) Yanagimoto and Okamoto (1969) introduced an ordering which they call monotone regression dependence which is similar but not equivalent to our "more SI" ordering given in (21). They use it to prove monotonicity of some rank correlation statistics with respect to an underlying parameter measuring dependence of the random variables.

d) Schriever (1987) has generalized the ordering of Yanagimoto and Okamato (1969) by introducing an ordering which he terms "more associated." He shows that most well-known rank measures of positive dependence preserve his ordering "more associated" in populations.

e) It is not necessary for the random variables to be positively dependent for any of (17)–(21) to hold.

THEOREM 3.3. *In the censored model the amount of positive quadrant dependence increases as censoring decreases stochastically. That is, if* $Y_1 \overset{st}{\le} Y_2$ *and* $Z_i = \min(X, Y_i)$, $i = 1, 2$, *then* X *and* Z_2 *are more PQD than* X *and* Z_1.

PROOF. Consider $\Pr(X \le x, Z_i \le z) - \Pr(X \le x)\Pr(Z_i \le z)$. There are two cases.

1) If $x \leq z$, then

$$\begin{aligned}
&\Pr(X \leq x, Z_i \leq z) - \Pr(X \leq x)\Pr(Z_i \leq z)\\
&= \Pr(X \leq x) - \Pr(X \leq x)\Pr(Z_i \leq z) = P(x)\{1 - K_i(z)\}\\
&= P(x)\bar{K}_i(z) = P(x)\bar{P}(z)\bar{Q}_i(z),
\end{aligned}$$

where $\bar{K}_i(z) = \bar{P}(x)\bar{Q}_i(z)$, the survival function of Z_i.

2) If $x > z$, then

$$\begin{aligned}
&\Pr(X \leq x, Z_i \leq z) - \Pr(X \leq x)\Pr(Z_i \leq z)\\
&= \Pr\{X \leq x, \min(X, Y_i) \leq z\} - \Pr(X \leq x)\Pr(Z_i \leq z)\\
&= \Pr(X \leq z) + \Pr(z \leq X \leq x, Y_i \leq z) - \Pr(X \leq x)\Pr(Z_i \leq z)\\
&= P(z) + \{P(x) - P(z)\}Q_i(z) - P(x)\{1 - \bar{P}(z)\bar{Q}_i(z)\}\\
&= \bar{Q}_i(z)\{P(z) - P(x) + P(x)\bar{P}(z)\} = \bar{Q}_i(z)P(z)\bar{P}(x). \quad \|
\end{aligned}$$

The following theorem is an easy consequence of Theorem 3.3.

THEOREM 3.4. *For any increasing function ψ, $\int \psi\{\Pr(X \leq x, Z \leq z) - \Pr(X \leq x)\Pr(Z \leq z)\}dxdz$ will increase as censoring decreases stochastically.*

COROLLARY 3.5. *$Cov(X, Z)$ increases as censoring decreases stochastically.*

PROOF. $\mathrm{Cov}(X, Z) = \int\int\{\Pr(X \leq x, Z \leq z) - \Pr(X \leq x)\Pr(Z \leq z)\}dxdz$ and so the result is immediate from Theorem 3.4. $\|$

Covariance is, of course, a well known measure of positive dependence. Many other such measures can also be shown to increase as censoring decreases stochastically. To show this, we state the following theorem.

THEOREM 3.6. *Let $(U_i, V_i^{(1)})$, $i = 1, \ldots, n$, be independent and identically distributed. Let $(U_i, V_i^{(2)})$, $i = 1, \ldots, n$, be independent and identically distributed with $(U_i, V_i^{(1)})$ more PQD than $(U_i, V_i^{(2)})$, $i = 1, \ldots, n$. Let r, s be concordant functions, that is, both r and s monotonic in the same direction in each argument. Then $\{r(U_1, \ldots, U_n), s(V_1^{(1)}, \ldots, V_n^{(1)})\}$ is more PQD than $\{r(U_1, \ldots, U_n), s(V_1^{(2)}, \ldots, V_n^{(2)})\}$.*

The proof is by induction along the lines of Theorems 1 and 2 of Lehmann (1966).

COROLLARY 3.7. *Kendall's τ, Spearman's ρ_s, and Blomqvist's q all increase as censoring decreases stochastically.*

PROOF. Kendall's $\tau = \mathrm{Cov}(\mathrm{sign}(X_2 - X_1), \mathrm{sign}(Z_2 - Z_1))$ and hence is increasing by Theorem 3.6 and Corollary 3.5. Spearman's $\rho_s = 3\mathrm{Cov}(\mathrm{sign}(X_2 - X_1), \mathrm{sign}(Z_3 - Z_1))$ and is increasing by Theorem 3.6 and Corollary 3.5. Blomqvist's $q = 2\{\Pr(X > m_x, Z > m_z) + \Pr(X \leq m_x)\Pr(Z \leq m_z)\} - 1$ where m_x and m_z are the medians of X and Z respectively. This reduces to $2\{\Pr(X > m_x, Z > m_z) - \Pr(X > m_x)\Pr(Z > m_z) + \Pr(X \leq m_x, Z \leq m_z) - \Pr(X \leq m_x)\Pr(Z \leq m_z)\}$, which (from Theorem 3.3) increases as censoring decreases stochastically. ||

In Example 3.8 we show that even though there is less censoring, association may decrease.

EXAMPLE 3.8. Let $\Gamma(X, Z_i) = I(X > x_1, Z_i > z_1), \Delta(X, Z_i) = I(X > x_2, Z_i > z_2)$, $i = 1, 2$, and let $x_1 < x_2 < z_1 < z_2$. Then $\text{Cov}\{\Gamma(X, Z_i), \Delta(X, Z_i)\} = \bar{P}(z_2)\bar{Q}_i(z_2) - \bar{P}(z_1)\bar{Q}_i(z_1)\bar{P}(z_2)\bar{Q}_i(z_2) = \bar{P}(z_2)\bar{Q}_i(z_2)\{1 - \bar{P}(z_1)\bar{Q}_i(z_1)\}$. Choose P, Q_1, Q_2 so that $\bar{P}(z_1) = 1/2$, $\bar{Q}_1(z_1) = 1$, $\bar{Q}_2(z_1) = 1/2$, $\bar{P}(z_2) = 1/4$, $\bar{Q}_1(z_2) = 5/12$, $\bar{Q}_2(z_2) = 1/3$. Note that $\bar{Q}_1(z_i) \geq \bar{Q}_2(z_i), i = 1, 2$. Then $\text{Cov}\{\Gamma(X, Z_1), \Delta(X, Z_1)\} = 5/96$, and $\text{Cov}\{\Gamma(X, Z_2), \Delta(X, Z_2)\} = 6/96$. Thus here $Y_1 \overset{\text{st}}{\geq} Y_2$ but X and Z_2 are more associated than X and Z_1.

Thus a chain of implications similar to (16) using (17)–(21) is not possible. This result is not that surprising as the ordering defined in (18) does not satisfy the properties for a positive dependence ordering as set down in Kimeldorf and Sampson (1989). In particular they show that a bivariate c.d.f. may not be less associated than its Fréchet upper bound.

This leaves the last three notions: LTD, RTI, and SI.

THEOREM 3.9. *If* $Y_1 \overset{st}{\leq} Y_2$ *then*

(i) Z_2 *is more RTI in* X *than* Z_1 *is in* X.

(ii) Z_2 *is more LTD in* X *than* Z_1 *is in* X.

(iii) Z_2 *is more SI in* X *than* Z_1 *is in* X.

PROOF. i) Let $x' < x$. Then

$$\begin{aligned} \Pr(Z > z|X > x) \quad - \quad & \Pr(Z > z|X > x') = \\ & \{\Pr(X > z, Y > z, X > x)/\Pr(X > x)\} - \\ & \{\Pr(X > z, Y > z, X > x')/\Pr(X > x')\}. \end{aligned} \tag{22}$$

There are three cases to consider.

1) Let $x > x' > z$. Then (22) reduces to $\Pr(Y > z) - \Pr(Y > z) = 0$.

2) Let $x > z \geq x'$. Then (22) reduces to $\Pr(Y > z) - \{\Pr(X > z, Y > z)/\Pr(X > x')\} = \bar{Q}(z)[1 - \{\bar{P}(z)/\bar{P}(x')\}]$. This decreases as $\bar{Q}$ decreases.

3) Let $z \geq x > x'$. Then (22) reduces to $\bar{P}(z)\bar{Q}(z)[\{1/\bar{P}(x)\} - \{1/\bar{P}(x')\}] = \bar{P}(z)\bar{Q}(z)\{\bar{P}(x) \cdot \bar{P}(x')\}^{-1}\{\bar{P}(x') - \bar{P}(x)\}$, which decreases as $\bar{Q}$ decreases.

The proofs for LTD and SI follow in an analogous fashion. ||

THEOREM 3.10. *Let ψ be an increasing function. Then*

(1) $\int_z \int_{x<x'} \psi\{\Pr(Z \leq z|X \leq x') - \Pr(Z \leq z|X \leq x)\}dxdx'dz$ *is increasing as censoring decreases stochastically.*

(2) $\int_z \int_{x<x'} \psi\{\Pr(Z > z|X > x) - \Pr(Z > z|X > x')\}dxdx'dz$ *is increasing as censoring decreases stochastically.*

(3) $\int_z \int_{x<x'} \psi\{\Pr(Z > z|X = x) - \Pr(Z > z|X = x')\}dxdx'dz$ *is increasing as censoring decreases stochastically.*

REFERENCES

ALI, S.M. and SILVEY, S.D. (1965a). Association between random variables and the dispersion of a Radon-Nikodym derivative. *J. Roy. Statist. Soc. Ser. B* **27** 100–107.

ALI, S.M. and SILVEY, S.D. (1965b). A further result on the relevance of the dispersion of a Radon-Nikodym derivative to the problem of measuring association. *J. Roy. Statist. Soc. Ser. B* **27** 108–110.

ALI, S.M. and SILVEY, S.D. (1966). A general class of coefficients of divergence of one distribution from another. *J. Roy. Statist. Soc. Ser. B* **28** 131–142.

BARLOW, R. and HSIUNG, J. (1983). Information in a life testing experiment. *Statistician* **32** 35–45.

BARLOW, R. and PROSCHAN, F. (1981). *Statistical Theory of Reliability and Life Testing: Probability Models.* To Begin With, Silver Springs, MD.

BAXTER, L.A. (1989). A note on information and censored absolutely continuous random variables. *Stat. and Dec.* **7** 193–198.

BLACKWELL, D. (1951). Comparison of experiments. *Proc. Second Berkeley Symp. Math. Statist. and Prob.*, University of California Press, 93–102.

CSISZÀR, I. (1963). Eine informations theoretische ungleichung und ihre anwendung auf den beweis der ergodizitat von markoffschen ketten. *Magyar Tud. Akad. Mat. Kutato Int. Közl* **8** 85–108.

CSISZÀR, I. (1967). Information-type measures of difference of probability distributions and indirect observations. *Studia Scientarium Mathmaticarum Hungarica* **2** 299–318.

ESARY, J.D., PROSCHAN, F., and WALKUP, D.W. (1967). Association of random variables with applications. *Ann. Math. Statist.* **38** 1466–1474.

GOEL, P.K. (1986). Comparison of experiments and information in censored data. Technical Report No. 352, Department of Statistics, Ohio State University.

HOLLANDER, M., PROSCHAN, F., and SCONING, J. (1987). Measuring information in right-censored models. *Naval Res. Logist.* **34** 669–681.

HOLLANDER, M., PROSCHAN, F., and SCONING, J. (1985). Measures of dependence for evaluating information in censored models. Technical Report No. M706, Department of Statistics, Florida State University.

KIMELDORF, G. and SAMPSON, A. (1989). A framework for positive dependence. *Ann. Inst. Math. Statist.* **41** 31–45.

KULLBACK, S. (1959). *Information Theory and Statistics.* John Wiley, New York.

LEHMANN, E.L. (1966). Some concepts of dependence. *Ann. Math. Statist.* **37** 1137–1153.

SCHRIEVER, B.F. (1987). An ordering for positive dependence. *Ann. Statist.* **15** 1208–1214.

SCONING, J. (1985). Information in censored models. Florida State University Ph.D. Dissertation.

TCHEN, A.H. (1980). Inequalities for distributions with given marginals. *Ann. Prob.* **8** 814–827.

YANAGIMOTO, T. and OKAMOTO, M. (1969). Monotonicity of rank correlation based on partial orderings. *Ann. Inst. Statist. Math.* **21** 489–506.

ZIV, J. and ZAKAI, M. (1973). On functionals satisfying a data processing theorem. *IEEE Trans. Info. Theory IT-19* 275–283.

DEPARTMENT OF STATISTICS
FLORIDA STATE UNIVERSITY
TALLAHASSEE, FL 32306–3033

DEPARTMENT OF STATISTICS
FLORIDA STATE UNIVERSITY
TALLAHASSEE, FL 32306–3033

DEPARTMENT OF STATISTICS
UNIVERSITY OF IOWA
IOWA CITY, IA 52242

DEPENDENCE IN MULTIVARIATE EXTREME VALUES

By J. Hüsler

University of Bern

We review the limiting behavior of extreme values of sequences of random vectors in R^d by considering mainly the dependence properties of its nondegenerate limit laws. We treat separately the i.i.d. case, the stationary case, the independent non-identically distributed case, and the general nonstationary case. As dependence concepts we discuss total dependence, association, positive lower orthant dependence, and independence.

1. Introduction. Consider a sequence $\{\mathbf{X}_i, i \geq 1\}$ of d-dimensional random vectors with (multivariate) distribution F_i. In this paper, we discuss the behavior of the maximum $\mathbf{M}_n = (M_{n1}, M_{n2}, \ldots, M_{nd})'$ where M_{nj} denotes the maximum up to time n of the j-th components of $\mathbf{X}_i$:

$$M_{nj} = \max(X_{1j}, \ldots, X_{nj}), \;\; j \leq d.$$

Our main interest is the dependence structure of the limiting distribution of properly normalized $\mathbf{M}_n$. More precisely, we deal with the convergence of

$$\begin{aligned} & P\{(\mathbf{M}_n - \mathbf{b}_n)/\mathbf{a}_n \leq \mathbf{z}\} = P\{\mathbf{X}_i \leq \mathbf{a}_n\mathbf{z} + \mathbf{b}_n, \; i \leq n\} \\ (1) \qquad & \xrightarrow{w} G(\mathbf{z}) = P\{\mathbf{Z} \leq \mathbf{z}\} \text{ as } n \to \infty, \end{aligned}$$

and the dependence properties of G. Note that all algebraic operations are componentwise and that the normalization constants satisfy $\mathbf{a}_n > 0$. The univariate case has been treated by many authors; c.f. the textbooks by Leadbetter et al. (1983), Galambos (1987), and Resnick (1987). One additional aspect of extreme value theory in the multivariate case is the dependence properties of the limit law G in (1). This important and interesting question is the primary focus of our review. We investigate the extreme value distribution G for independence, total dependence, association, and positive lower orthant dependence (see Section 2 for definitions).

In Section 2 we quickly review the case of i.i.d. sequences, whose study was initiated by Geffroy (1958/59), Tiago de Oliveira (1958), and Sibuya (1960) for

AMS 1980 subject classifications. Primary 60F05; secondary 62H05.

Key words and phrases. Multivariate extreme values, positive dependence, association, independence, total dependence.

$d = 2$. In the remaining sections we consider more general cases, first beginning with the case of stationary sequences. Most of the classical results remain valid in this situation provided a certain mixing condition is satisfied.

Next we treat the case of independent but non-identically distributed random vectors. This case can only be reasonably treated by introducing a certain uniform asymptotic negligibility condition. Without this restriction, every (multivariate) distribution G can occur as a limit in (1). This condition, however, can be interpreted as a natural extension of the conditions used implicitly in the classical i.i.d. case. Finally, we deal with the general case of non-independent, non-stationary sequences.

2. The i.i.d. Case. For the i.i.d. case ($F_i \equiv F$), equation (1) becomes

$$F^n(\mathbf{a}_n\mathbf{z} + \mathbf{b}_n) \xrightarrow{w} G(\mathbf{z}) \tag{2}$$

or

$$n(1 - F(\mathbf{a}_n\mathbf{z} + \mathbf{b}_n)) \to -\log G(\mathbf{z}).$$

This situation is rather completely discussed in the literature (cf. Galambos (1987) and Resnick (1987) for references). Therefore we only mention the results which are relevant to the following discussion of non i.i.d. random vectors. The limit G is called an extreme value distribution and is characterized by the max stability property, i.e., for every $s > 0$ there exist c_s and d_s such that $G^s(\cdot) = G(c_s \cdot + d_s)$. Note that the univariate marginals G_j of an extreme value distribution G are obviously univariate extreme value distributions. Generally a multivariate distribution on $[0,1]^d$, a so-called dependence function, is used to discuss dependence properties of G (de Haan and Resnick (1977), Deheuvels (1984)). For the purpose of our discussion, however, the following results are more informative and useful in applications. Because of max stability the extreme value distribution G is max i.d. (max infinitely divisible), i.e. G^s is a multivariate distribution for every $s > 0$, (cf. Balkema and Resnick (1977)). Hence G is associated: $\mathrm{Cov}(\phi(\mathbf{Z}), \psi(\mathbf{Z})) \geq 0$ for any (componentwise) nondecreasing functions ϕ and ψ where G is the distribution of $\mathbf{Z}$ (cf. Esary et al. (1967), Resnick (1987)). The result that extreme value distributions are associated, is due to Marshall and Olkin (1983).

THEOREM 2.1. *Assume that (2) holds for the sequence of random vectors* $\{\mathbf{X}_i, i \geq 1\}$ *in* R^d, *with normalization* $\mathbf{a}_n$ *and* $\mathbf{b}_n$. *Then G is associated since it is max stable and max i.d.*

Obviously, it also implies the weaker PLOD (positive lower orthant dependence) property: $G(\mathbf{z}) \geq \Pi_{j=1}^d G_j(z_j)$. If $G(\mathbf{z}) = \Pi_{j=1}^d G_j(z_j)$, we say that $\mathbf{Z}$ has independent components where $\mathbf{Z}$ has distribution G. As an upper bound for any multivariate distribution, we have the inequality $G(\mathbf{z}) \leq G_j(z_j)$ for any $j \leq d$. If this statement holds as an equality, more precisely, if $G(\mathbf{z}) = \min(G_j(z_j), j \leq d)$ for all $\mathbf{z}$, we say that $\mathbf{Z}$ has totally dependent components. Also assuming $G_j \equiv G_1$, this means that $P\{Z_1 = Z_2 = \ldots = Z_d\} = 1$. It defines the strongest possible dependence.

The two questions of independence and of total dependence were treated in the previously mentioned papers by Geffroy (1958/59), Tiago de Oliveira (1958), and Sibuya (1960) for the bivariate case. Since the limit G is max i.d., bivariate independence implies joint multivariate independence (cf. Newman and Wright (1981)). The result on independence was recently improved by Takahashi (1987, 1988). He gave necessary and sufficient conditions such that an extreme value distribution is characterized by its marginal distributions. Combining these facts we obtain the following statement.

THEOREM 2.2. *Let $\{\mathbf{X}_i, i \geq 1\}$ be a sequence of random vectors in R^d. Assume that (2) holds with normalization $\mathbf{a}_n$ and $\mathbf{b}_n$. Then $\mathbf{Z}$ has independent components iff for every $1 \leq j < j' \leq d$*

$$\lim_{n\to\infty} nP\{X_{1j} > a_{nj}z_j + b_{nj},\ X_{1j'} > a_{nj'}z_{j'} + b_{nj'}\} = 0 \tag{3}$$

for some z_j, $z_{j'}$ such that $G_{j,j'}(z_j, z_{j'}) \in (0,1)$, where $G_{j,j'}$ is a bivariate marginal of G.

A similar statement holds for the total dependence in place of independence. The bivariate case was treated by Sibuya (1960). Takahashi's characterization (1988) also improves upon this statement.

THEOREM 2.3. *Let $\{\mathbf{X}_i, i \geq 1\}$ be a sequence of random vectors in R^d. Assume that (2) holds with normalization $\mathbf{a}_n$ and $\mathbf{b}_n$, and that $G_j \equiv G_1$, $j \leq d$. Then $\mathbf{Z}$ has totally dependent components iff for every $1 \leq j \neq j' \leq d$*

$$\lim_{n\to\infty} n(P\{X_{1j} > u_{nj},\ X_{1j'} > u_{nj'}\} - P\{X_{1j} > u_{nj}\}) = 0 \tag{4}$$

for some z, such that $G_1(z) \in (0,1)$, where u_{nj} is defined by $u_{nj} = a_{nj}z + b_{nj}$.

Note that (4) is equivalent to

$$\lim_{n\to\infty} nP\{X_{1j} > a_{nj}z + b_{nj},\ X_{1j'} > a_{nj'}z + b_{nj'}\} = -\log G_1(z)$$

in the i.i.d. case. Equation (4) will be used in the following more general case.

3. The Stationary Case. In this section we consider stationary sequences of random vectors, with $F_i \equiv F$. In this case the extreme value theory is mainly discussed for Gaussian sequences (Lindgren (1974), Amram (1985), Hüsler and Schüpbach (1988)) or for more general sequences in R ($d = 1$) which satisfy a mild mixing condition (cf. Leadbetter et al. (1983)).

In the multivariate situation, if the conditions are such that

$$P\{\mathbf{M}_n \leq \mathbf{a}_n\mathbf{z} + \mathbf{b}_n\} - F^n(\mathbf{a}_n\mathbf{z} + \mathbf{b}_n) \to 0 \text{ as } n \to \infty, \tag{5}$$

then all the results of Section 2 remain valid in this more general situation. However, even under less restrictive conditions, when (5) does not hold, we can still discuss the dependence properties of the limit G, provided it exists.

As mentioned, the univariate problem for stationary sequences is discussed in detail (cf. Leadbetter et al. (1983), Galambos (1987)). The general multivariate stationary problem, however, was only considered in a few papers. An attempt was made by Villasenor (1976), for the case of bivariate exchangeable sequences. Hsing (1987) and Hüsler (1987) independently extended the univariate results, related to Leadbetter's mixing conditions, to the multivariate case. Sbihi (1987) also discusses the multivariate stationary case. In addition, Hüsler (1987) focused more on the dependence properties. These results are reviewed below.

We introduce the following mixing conditions. Given $\mathbf{z}$, we set $\mathbf{u}_n = \mathbf{a}_n\mathbf{z} + \mathbf{b}_n$ and $B_n(I) = \{\mathbf{X}_i \le \mathbf{u}_n, i \in I\}$, where $I \subset \{1, \ldots, n\}$. The set I will also usually depend on n.

Condition $\mathbf{D}_d$ $= \mathbf{D}_d(\{\mathbf{u}_n, n \ge 1\})$ holds for a given $\mathbf{z}$ with normalization $\mathbf{a}_n(> 0)$ and $\mathbf{b}_n$, if there exists an array $\{\alpha_{nm}, n \ge 1, m \le n\}$ such that

i) $|P(B_n(I \cup J)) - P(B_n(I))P(B_n(J))| \le \alpha_{nm}$

for every pair of subsets I and J of $\{1, \ldots, n\}$ which are m-separated (i.e. $\min_{i\in J}(i) - \max_{i\in I}(i) \ge m$ or $\min_{i\in I}(i) - \max_{i\in J}(i) \ge m$) and

ii) $\lim_{n\to\infty} \alpha_{n,m_n^*} = 0$ for some sequence $\{m_n^*, n \ge 1\}$ with $m_n^* \to \infty$ and $m_n^*(1 - F(\mathbf{u}_n)) \to 0$ as $n \to \infty$.

This condition restricts the so-called long-range dependence since it implies that extreme values are asymptotically independent when they occur largely separated in time. Note that this condition is weaker than the usual mixing condition. The following local dependence condition $\mathbf{D}_d'$ excludes the clustering of extreme values in a small time interval.

Condition $\mathbf{D}_d'$ $= \mathbf{D}_d'(\{\mathbf{u}_n, n \ge 1\})$ holds for a given $\mathbf{z}$ with a normalization $\mathbf{a}_n(> 0)$ and $\mathbf{b}_n$, if

$$\lim_{r\to\infty} \limsup_{n\to\infty} n \sum_{1<i\le n/r} P\{\mathbf{X}_1 \not\le \mathbf{u}_n, \mathbf{X}_i \not\le \mathbf{u}_n\} = 0.$$

These two conditions imply that (5) holds. Hence

THEOREM 3.1. *Let $\{\mathbf{X}_i, i \ge 1\}$ be a stationary sequence of random vectors in R^d. Assume that $\mathbf{D}_d$ and $\mathbf{D}_d'$ hold for every $\mathbf{z}$ with $G(\mathbf{z}) > 0$ and $\mathbf{u}_n = \mathbf{a}_n\mathbf{z} + \mathbf{b}_n$, $\mathbf{a}_n(> 0)$, $\mathbf{b}_n$ the normalization. Then (1) is equivalent to (2). Hence*

i) (Association) G is associated, since G is max stable and max i.d.

ii) (Independence) If, in addition, condition (3) holds, then $\mathbf{Z}$ has independent components, and conversely.

iii) (Total Dependence) If, in addition, (4) holds, then $\mathbf{Z}$ *has totally dependent components, and conversely.*

Some of the above statements hold even under weaker conditions.

THEOREM 3.2. *Assume that for some stationary sequence* $\{\mathbf{X}_i, i \geq 1\}$ *in* R^d *the limit* G *in (1) exists and that the condition* $\mathbf{D}_d$ *holds for every* $\mathbf{z}$ *with* $G(\mathbf{z}) > 0$ *and* $\mathbf{u}_n = \mathbf{a}_n\mathbf{z} + \mathbf{b}_n$, $\mathbf{a}_n(> 0)$, $\mathbf{b}_n$ *the normalization. Then* G *is associated since it is max stable. Hence* G *is also PLOD and satisfies the inequality*

$$G(\mathbf{z}) \geq \max(G^*(\mathbf{z}),\ \Pi_{j=1}^d G_j(z_j)),$$

where $G^*(\mathbf{z}) = \lim_n F^n(\mathbf{a}_n\mathbf{z} + \mathbf{b}_n)$.

This statement follows since extreme values which occur in m_n-separated intervals are asymptotically independent by $\mathbf{D}_d$. This implies the max stability in the stationary case.

The case of independence also occurs if the random vectors $\mathbf{X}_i$ are negative dependent in some sense, since the above result shows that under Condition $\mathbf{D}_d$ the limit law G is associated and hence positive dependent. We assume a rather weak form of negative dependence, namely PNQD (pair-wise negative quadrant dependence): F is PNQD if every bivariate marginal $F_{jj'}$ is NQD (negative quadrant dependent), i.e., for every $1 \leq j < j' \leq d$

$$F_{jj'}(x, y) \leq F_j(x)F_{j'}(y) \text{ for all } x, y.$$

THEOREM 3.3. *Let* $\{\mathbf{X}_i, i \geq 1\}$ *be a stationary sequence of random vectors in* R^d *such that* F *is PNQD. Then*

i) Condition (3) holds if $n(1 - F(\mathbf{u}_n)) = O(1)$ *as* $n \to \infty$.

ii) Assume also that (1), $\mathbf{D}_d$ *and* $\mathbf{D}'_d$ *hold for every* $\mathbf{z}$ *with* $G(\mathbf{z}) > 0$ *and* $\mathbf{u}_n = \mathbf{a}_n\mathbf{z} + \mathbf{b}_n$, $\mathbf{a}_n(> 0)$, $\mathbf{b}_n$ *the normalization. Then the limit* $\mathbf{Z}$ *has independent components.*

Independent asymptotic components can also occur in another situation where $G_j \neq G_j^*$. This means that extreme values may occur locally in clusters. G_j is still an extreme value distribution if we assume Condition $\mathbf{D}_d$; more precisely, there exists an extremal index θ_j such that $G_j = (G_j^*)^{\theta_j}$ (cf. Leadbetter (1983)). By Theorem 3.2, G is also an extreme value distribution. By the result of Takahashi, if $G(\mathbf{z}) = \Pi_{j \leq d} G_j(z_j)$ for some $\mathbf{z}$ with $G_j(z_j) \in (0, 1)$ for all $j \leq d$, then $\mathbf{Z}$ has independent components. Therefore the joint behavior of the components has to be restricted in a suitable way to verify the condition of Takahashi. The following result is a slightly extended version of Theorem 3.4 of Hüsler (1987) and follows by similar arguments.

Condition $\mathbf{D}_d''$ holds for $\mathbf{z}$ with normalization $\mathbf{a}_n(>0)$ and $\mathbf{b}_n$ if

$$\lim_{r\to\infty}\limsup_{n\to\infty} r \sum_{1\le i,h\le n/r}\ \sum_{1\le j<l\le d} P\{X_{ij}>u_{nj},\ X_{hl}>u_{nl}\}=0.$$

THEOREM 3.4. *Let $\{\mathbf{X}_i, i\ge 1\}$ be a stationary sequence of random vectors in R^d, such that (1) holds. Assume that Condition $\mathbf{D}_d$ holds for every $\mathbf{z}$ with $G(\mathbf{z})>0$ and $\mathbf{u}_n=\mathbf{a}_n\mathbf{z}+\mathbf{b}_n$, $\mathbf{a}_n(>0)$, $\mathbf{b}_n$ the normalization. If $\mathbf{D}_d''$ holds for some $\mathbf{z}$ with $G_j(z_j)\in(0,1)$, then the limit $\mathbf{Z}$ has independent components.*

Finally we discuss the case of total dependence, without assuming Condition $\mathbf{D}_d$ and $\mathbf{D}_d'$. This result is not stated in Hüsler (1987), but it is an immediate consequence of the more general statement in Section 5, Theorem 5.4.

THEOREM 3.5. *Let $\{\mathbf{X}_i, i\ge 1\}$ be a stationary sequence of random vectors in R^d. Assume that $P\{M_{nj}\le a_{nj}z+b_{nj}\}\stackrel{w}{\to} G_1(z)$ holds for every $j\le d$ with a normalization $a_{nj}(>0)$, b_{nj}. If (4) holds for all z with $G_1(z)\in(0,1)$, then the limit $\mathbf{Z}$ exists and is totally dependent.*

In general, condition (4) is not necessary as is shown in Section 5 by an example. Note also, that if $\mathbf{Z}$ exists with G_1 being an extreme value distribution, it is sufficient that (4) holds only for some z with $G_1(z)\in(0,1)$ by Takahashi's result.

We also mention that Condition $\mathbf{D}_d$, $\mathbf{D}_d'$ and $\mathbf{D}_d''$ can be verified for a Gaussian sequence which satisfies a Berman type condition, i.e.,

$$r_{jj'}(n)\log n\to 0 \text{ as } n\to\infty.$$

Here $r_{jj'}(n)$ is the correlation of X_{1j} and $X_{nj'}$. This verification uses the technique developed in Berman (1964) (cf. Leadbetter et al. (1983)).

In Theorem 3.2–3.4 we did not assume Condition $\mathbf{D}_d'$. These results, however, heavily depend upon Condition $\mathbf{D}_d$. Without Condition $\mathbf{D}_d$ it would not be possible to give such a unified treatment of the behavior of extreme values. Note, for instance, that even negative dependent distributions G could occur as limits in (1) by taking a sequence of random vectors $\mathbf{X}_i\equiv\mathbf{X}_1$ for all $i\ge 1$ with $\mathbf{X}_1$ distributed as G.

Many of the statements can also be formulated for triangular arrays of random vectors. We only mention that, in general, a larger class of limit laws occurs for $\mathbf{M}_n$ (cf. Hüsler and Reiss (1989) for the Gaussian case).

4. The Nonstationary Independent Case. In the independent but nonstationary case, we need to consider the convergence of

$$\Pi_{i\le n}F_i(\mathbf{a}_n\mathbf{z}+\mathbf{b}_n)\stackrel{w}{\to} G(\mathbf{z}) \text{ as } n\to\infty. \tag{6}$$

The following results are contained in Hüsler (1988a). As mentioned in the introduction, we need to impose some restrictions in this case. We assume the following condition $\mathbf{A}_d$; the first part is a uniform asymptotic negligibility (u.a.n.) condition. Without this restriction, any G can occur as a limit in (6).

Condition $\mathbf{A}_d$ holds with normalization $\mathbf{a}_n > 0$ and $\mathbf{b}_n$, if

$$F^*_{\max,n} = \max_{i \le n}\{1 - F_i(\mathbf{a}_n\mathbf{z} + \mathbf{b}_n)\} \to 0 \text{ as } n \to \infty$$

for all $\mathbf{z}$ and if

$$\sum_{i \le n}(1 - F_i(\mathbf{a}_n\mathbf{z} + \mathbf{b}_n)) \xrightarrow{w} w(\mathbf{z}) \ as\ n \to \infty.$$

Assume that $w(\mathbf{z}) < \infty$ for some $\mathbf{z} \in R^d$.

Note that $\mathbf{A}_d$ implies the existence of G. Conversely, if the u.a.n. condition holds, then the existence of G in (6) implies the second part of Condition $\mathbf{A}_d$ with $w(\mathbf{z}) = -\log G(\mathbf{z})$. Note also that in the stationary case, $F^*_{\max,n} = 1 - F(\mathbf{u}_n)$. Thus $n(1 - F(\mathbf{u}_n)) = O(1)$ implies the u.a.n. condition, i.e., the u.a.n. condition is implicitly assumed in the stationary case. In particular, we proved the following result.

THEOREM 4.1. *Let $\{\mathbf{X}_i, i \ge 1\}$ be an independent sequence of random vectors in R^d. Assume that $\mathbf{A}_d$ holds with normalization $\mathbf{a}_n$ and $\mathbf{b}_n$. Then the limit law G in (6) is max i.d., hence associated and PLOD.*

This follows by a result of Balkema and Resnick (1977). By assuming a slightly extended version of Condition $\mathbf{A}_d$, the limits G can be totally characterized. This extended version also implies that the limits of partial maxima $\mathbf{M}_{[nt]}, 0 < t \le 1$ have a max i.d. limit distribution. Since $\mathbf{M}_n = \mathbf{M}_{[nt]} \vee \mathbf{M}_{(nt,n]}$ with $\mathbf{M}_{(m,n]} = \max_{m<i\le n}\{\mathbf{X}_i\}$, a general decomposition of the limit G can be obtained (see Hüsler (1988a)) in an analogous way as for the sup self-decomposable distributions G (see Gerritse (1986)). These arise from the assumption $\mathbf{a}_n \equiv \mathbf{1}$ in (6).

The limit law G in (6) also has a positive dependence structure. If every $\mathbf{X}_i$ has a negative dependence structure, we again expect a limit with independent components as in the former sections. The following condition (7) is weaker and is, in general, the equivalent statement for independence. It follows as in the i.i.d. case and by Theorem 4.1.

THEOREM 4.2. *Let $\{\mathbf{X}_i, i \ge 1\}$ be an independent sequence of random vectors in R^d. Assume that $\mathbf{A}_d$ holds with normalization $\mathbf{a}_n$ and $\mathbf{b}_n$. Then*

$$(7) \qquad S_n^{(2)} = \sum_{i=1}^{n} \sum_{1 \le j < j' \le d} P\{X_{ij} > u_{nj}, X_{ij'} > u_{nj'}\} \to 0 \ as\ n \to \infty$$

for all $\mathbf{z}$ such that $G(\mathbf{z}) > 0$, with $\mathbf{u}_n = \mathbf{a}_n\mathbf{z} + \mathbf{b}_n$, is equivalent to

$$(8) \qquad G(\mathbf{z}) = \Pi_{j=1}^d G_j(z_j),$$

where G_j is the j-th marginal of G.

Corollary 4.3. *Let $\{\mathbf{X}_i, i \geq 1\}$ be an independent sequence of random vectors in R^d. Assume that $\mathbf{A}_d$ holds with normalization $\mathbf{a}_n$ and $\mathbf{b}_n$. If for every $i \geq 1$, F_i is PNQD, then the limit $\mathbf{Z}$ in (6) has independent components.*

These two results on the independence can be slightly improved by only assuming the u.a.n. condition and the existence of all the univariate marginal limits G_j, instead of Condition $\mathbf{A}_d$. This weaker assumption will be used in the last section.

The total dependence case can be treated as before but with an obvious change. Here we assume $\mathbf{A}_d$ which implies the equivalence of (9) and (10).

Theorem 4.4. *Let $\{\mathbf{X}_i, i \geq 1\}$ be an independent sequence of random vectors in R^d. Assume that $\mathbf{A}_d$ holds with normalization $\mathbf{a}_n$ and $\mathbf{b}_n$ and that the existing G satisfies $G_j \equiv G_1$, for every $j \leq d$. Then*

$$G(\mathbf{z}) = G_1(\min_j(z_j)) \tag{9}$$

is equivalent to

$$\lim_{n\to\infty} \left(\sum_{i=1}^{n} P\{X_{ij} > u_{nj}, X_{ij'} > u_{nj'}\} - \sum_{i=1}^{n} P\{X_{ij} > u_{nj}\} \right) = 0 \tag{10}$$

for all $1 \leq j \neq j' \leq d$ and every z with $G_1(z) > 0$ and $u_{nj} = a_{nj}z + b_{nj}$.

In this case, it is generally necessary to assume (7) and (10) for all $\mathbf{z}$ since the results of Takahashi, which were proved for max stable distributions, do not hold for general max i.d. distributions (Hüsler (1989)). Note that because of independence of the random vectors, the second sum in (10) converges to $-\log G_1(z)$. In this case the extreme values M_{nj}, $j \leq d$, occur jointly at the same time point, asymptotically.

In the following situation, a rather restricted but interesting case occurs where we obtain an associated limit law without assuming Condition $\mathbf{D}_d$. This follows by simple properties of association. The PLOD property follows similarly.

Theorem 4.5. *Let $\{\mathbf{X}_i, i \geq 1\}$ be an independent sequence of random vectors in R^d. Assume that $\mathbf{A}_d$ holds with normalization $\mathbf{a}_n$ and $\mathbf{b}_n$. If every F_i is associated (PLOD), then the distributions of $\mathbf{M}_n$ and of $\mathbf{Z}$ are associated (PLOD), respectively.*

5. The General Nonstationary Case. The extension of the results in Section 4 to this more general situation is carried out along the same lines as the extension of the classical i.i.d. case to the stationary situation. If we find conditions such that for every $\mathbf{z}$,

$$P\{\mathbf{X}_i \leq \mathbf{a}_n\mathbf{z} + \mathbf{b}_n\} - \Pi_{i \leq n} F_i(\mathbf{a}_n\mathbf{z} + \mathbf{b}_n) \to 0 \text{ as } n \to \infty, \tag{11}$$

all the results of Section 4 can be reformulated in this general case. But again, we are interested in finding weaker conditions such that the four dependence properties hold for a possible limit G in (1). We obviously assume the u.a.n. condition $\mathbf{A}_d$ in this section again. The following results are discussed and proved in Hüsler (1988b).

The following mixing condition $\mathbf{D}_d$ is an extension of the mixing condition in the stationary case. We use the same notation as before since both mixing conditions are equivalent in the stationary case.

Condition $\mathbf{D}_d = \mathbf{D}_d(\{\mathbf{u}_n, n \geq 1\})$. We assume that there exists an array $\{\alpha_{nm}, n \geq 1, m \leq n\}$ such that

i) $|P(B_n(I \cup J)) - P(B_n(I))P(B_n(J))| \leq \alpha_{nm}$ for every pair of subsets I and J of $\{1, \ldots, n\}$ which are m-separated and

ii) $\lim_{n\to\infty} \alpha_{n,m_n^*} = 0$ for some sequence $\{m_n^*, n \geq 1\}$ with $m_n^* \to \infty$ and $m_n^* F_{\max,n}^* \to 0$ as $n \to \infty$.

Note that in the stationary case, $m_n^* F_{\max,n}^* = m_n^*(1 - F(\mathbf{u}_n)) = O(m_n^*/n)$. To prove (11) we use the following extension of the local mixing condition in the stationary case, which we again denote by $\mathbf{D}_d'$. For any $I \subset \{1, \ldots, n\}$ and $\delta > 0$, define $d_n'(I, \delta)$ by

$$d_n'(I, \delta) = \min_{I^* \subset I} \sum_{i<h \in I^*} P\{\mathbf{X}_i \not\leq \mathbf{u}_n, \mathbf{X}_h \not\leq \mathbf{u}_n\}$$

where $\Sigma_{i\in I\setminus I^*} P\{\mathbf{X}_i \not\leq \mathbf{u}_n\} < \delta$. Let $F_n^*(I) = \Sigma_{i\in I} P\{\mathbf{X}_i \not\leq \mathbf{u}_n\}$ and $F_n^* = F_n^*(\{1, \ldots, n\})$. Note that we define $d_n'(I, \delta)$ as the sum on a suitable subset I^* of I. This idea is very useful in the Gaussian case, where some random vectors may have a heavy weight in the sum $\Sigma_{i<h\in I} P\{\mathbf{X}_i \not\leq \mathbf{u}_n, \mathbf{X}_h \not\leq \mathbf{u}_n\}$, but not in the sum $\Sigma_{i\in I} P\{\mathbf{X}_i \not\leq \mathbf{u}_n\}$ (cf. Hüsler (1983) in the univariate case and Hüsler and Schüpbach (1988) in the multivariate case). Obviously, this idea is also useful in the general non Gaussian case.

Condition $\mathbf{D}_d' = \mathbf{D}_d'(\{\mathbf{u}_n, n \geq 1\})$. We assume that there exist an array $\{\alpha_{nr}', n \geq 1, r \geq 1\}$ and a sequence $\{g_r, r \geq 1\}$ such that $\lim_{r\to\infty} r g_r = 0$, $\lim_{r\to\infty} \limsup_{n\to\infty} r\alpha_{nr}' = 0$ and for every $r \geq 1$ and for all $n \geq n_0(r) : d_n'(I, g_r) \leq \alpha_{nr}'$ for all $I \subset \{1, \ldots, n\}$ such that $F_n^*(I) \leq F_n^*/r$.

Both conditions $\mathbf{D}_d$ and $\mathbf{D}_d'$ together imply that $\mathbf{M}_n$ behaves asymptotically as if $\{\mathbf{X}_i, i \geq 1\}$ would be an independent sequence.

THEOREM 5.1. *Let $\{\mathbf{X}_i, i \geq 1\}$ be a general sequence of random vectors in R^d. Assume that $\mathbf{A}_d$, $\mathbf{D}_d$ and $\mathbf{D}_d'$ hold for every $\mathbf{z}$ with $G(\mathbf{z}) > 0$ (or $w(\mathbf{z}) < \infty$) and with $\mathbf{u}_n = \mathbf{a}_n\mathbf{z} + \mathbf{b}_n$ and normalization $\mathbf{a}_n (> 0)$, $\mathbf{b}_n$. Then (1) is equivalent to (6). Hence*

i) (Association) G is associated, since G is max stable and max i.d.

ii) (Independence) If, in addition, condition (7) holds, then $\mathbf{Z}$ has independent components and conversely.

iii) *(Total Dependence) If, in addition, (10) holds, then* $\mathbf{Z}$ *has totally dependent components and conversely.*

We can still prove that G has a positive dependence structure without assuming the local mixing condition, as in the stationary case.

THEOREM 5.2. *Let* $\{\mathbf{X}_i, i \geq 1\}$ *be a general nonstationary sequence of random vectors in* R^d*. Assume that the conditions* $\mathbf{A}_d$ *and* $\mathbf{D}_d$ *hold for every* $\mathbf{z}$ *such that* $G(\mathbf{z}) > 0$*, with* $\mathbf{u}_n = \mathbf{a}_n\mathbf{z} + \mathbf{b}_n$ *and normalization* $\mathbf{a}_n(> 0)$, $\mathbf{b}_n$*. If*

$$P\{\mathbf{M}_n \leq \mathbf{a}_n\mathbf{z} + \mathbf{b}_n\} \xrightarrow{w} G(\mathbf{z})$$

as $n \to \infty$*, then* G *is max i.d., hence associated.*

The next statement considers the asymptotic independence of the components for the extreme values. If the local mixing condition $\mathbf{D}'_d$ is not assumed, there exists the possibility that the extreme values will cluster in small time intervals. However, the components of $\mathbf{Z}$ can still be independent, if we assume a condition $\mathbf{D}''_d$, similar to the stationary case. To illustrate the possibility of clustering, consider, e.g., the simple case of independent bivariate random vectors with $X_{i1} \equiv X_{i2} - \gamma_i, \gamma_i$ real. Such a sequence satisfies the conditions $\mathbf{D}_d$ and $\mathbf{D}'_d$. Clustering occurs at the same time point, jointly in the two components. An asymptotic result for such an example would still follow from Theorem 5.1. Another interesting case of clustering arises if we consider, in every component, the joint clustering of extreme values in small time intervals. For instance, let Y_{ij} be independent random variables and define $X_{ij} = Y_{[i/\gamma_j]+1,j}$ with γ_j integer. Obviously $\mathbf{M}_n$ has independent components, but $\mathbf{D}'_d$ does not hold if $\gamma_j > 1$. A limit $\mathbf{Z}$ with independent components Z_j is still possible for such a clustering. Hence we define the following restriction.

Condition $\mathbf{D}''_d$ is defined in the same way as Condition $\mathbf{D}'_d$, where the expression $d'_n(I, \delta)$ is replaced by

$$d''_n(I, \delta) = \min_{I^* \subset I} \sum_{i,h \in I^*} \sum_{j < \ell} P\{X_{ij} > u_{nj},\ X_{hl} > u_{nl}\}.$$

Condition $\mathbf{D}''_d = \mathbf{D}''_d(\{\mathbf{u}_n, n \geq 1\})$. We assume that there exist an array $\{\alpha''_{nr}, n \geq 1, r \geq 1\}$ and a sequence $\{g''_r, r \geq 1\}$ such that $\lim_{r\to\infty} rg''_r = 0$, $\lim_{r\to\infty} \limsup_{n\to\infty} r\alpha''_{nr} = 0$ and for every $r \geq 1$ and for all $n \geq n_0(r) : d''_n(I, g_r) \leq \alpha''_{nr}$ for all $I \subset \{1, \ldots, n\}$ such that $F^*_n(I) \leq F^*_n/r$.

Then, analogous to the results in the stationary case, the independence of the components of $\mathbf{Z}$ occurs in the following situation described by Theorem 5.3. Note, however, that the limit G is, in general, not a max stable distribution. Hence we cannot make use of the results of Takahashi to improve upon the general statement of the theorem. For the last two statements we use only the u.a.n. condition $F^*_{\max,n} \to 0$ and the existence of the univariate marginal limits G_j, i.e.,

$$(12) \qquad P\{M_{nj} \leq a_{nj}z + b_{nj}\} \xrightarrow{w} G_j(z) \text{ as } n \to \infty$$

for every $j \leq d$ with normalization a_{nj} and b_{nj}. The convergence in (12) is discussed in Hüsler (1983, 1986), where sufficient conditions such as the univariate versions $\mathbf{D}_1$ and $\mathbf{D}_1'$ are formulated. The results in Section 3 do not follow as a special case of Theorem 5.3 since we are assuming slightly different conditions.

THEOREM 5.3. *Let $\{\mathbf{X}_i, i \geq 1\}$ be a general nonstationary sequence of random vectors in R^d. Assume that the u.a.n. condition, (12) and $\mathbf{D}_d$ hold for every $\mathbf{z}$ such that $G_j(z_j) > 0$, for all $j \leq d$ with $\mathbf{u}_n = \mathbf{a}_n\mathbf{z} + \mathbf{b}_n$ and normalization $\mathbf{a}_n(> 0)$, $\mathbf{b}_n$. Then $G(\mathbf{z}) = \Pi_{j=1}^d G_j(z_j)$ if either, for every $\mathbf{z}$ with $\mathbf{a}_n$ and $\mathbf{b}_n$,*

i) $\mathbf{D}_d'$ holds and every F_i is PNQD, or

ii) $\mathbf{D}_d''$ holds.

Finally we consider the total dependence case. The total dependence result follows rather easily without assuming condition $\mathbf{D}_d$.

THEOREM 5.4. *Let $\{\mathbf{X}_i, i \geq 1\}$ be a sequence of random vectors in R^d. Assume that (12) and (10) hold for every $u_{nj} = a_{nj}z + b_{nj}$ with z such that $G_1(z) > 0$ and $G_j \equiv G_1$ for all $j \leq d$. Then the limit distribution G in (1) exists with*

$$G(\mathbf{z}) = G_1(\min_j(z_j)).$$

Consequently G is totally dependent.

This statement holds for extreme values of random sequences which exhibit a behavior similar to that of independent sequences. More precisely, this means that the extreme values M_{nj}, $j \leq d$, mainly occur jointly at the same time point. It does not include random sequences such as, e.g., $\mathbf{X}_i = (Y_{i+1}, Y_i)$ where $\{Y_i\}$ is an i.i.d. sequence of random variables satisfying (12).

Note also that a general nonstationary Gaussian sequence satisfies Condition $\mathbf{D}_d$, $\mathbf{D}_d'$ and $\mathbf{D}_d''$ for any normalization $\mathbf{u}_n$ satisfying Condition $\mathbf{A}_d$ if a Berman type condition holds (Hüsler and Schüpbach (1988)).

REFERENCES

AMRAM, F. (1985). Multivariate extreme value distributions for stationary Gaussian sequences. *J. Mult. Anal.* **16** 237–240.

BALKEMA, A.A. and RESNICK, S.I. (1977). Max-infinite divisibility. *J. Appl. Prob.* **14** 309–313.

BERMAN, S.M. (1964). Limit theorems for the maximum term in stationary sequences. *Ann. Math. Statist.* **35** 502–516.

DEHEUVELS, P. (1984). Probabilistic aspects of multivariate extremes. In *Statistical Extremes and Applications* (J. Tiago de Oliveira, ed.), Reidel, Dordrecht, 117–130.

ESARY, J.D., PROSCHAN, F., and WALKUP, D.W. (1967). Association of random variables, with applications. *Ann. Math. Statist.* **44** 1466–1474.

GALAMBOS, J. (1987). *The Asymptotic Theory of Extreme Order Statistics.* 2nd. ed. Krieger, Florida.

GEFFROY, J. (1958/59). Contributions à la théorie des valeurs extrémes. *Publ. Inst. Statist.*, Univ. Paris **7/8** 37–185.

GERRITSE, G. (1986). Supremum self-decomposable random vectors. *Probab. Theory and Rel. Fields* **72** 17–34.

HAAN, L. DE and RESNICK, S.I. (1977). Limit theory for multivariate sample extremes. *Z. Wahrscheinlichkeitstheorie verw. Geb.* **40** 317–337.

HSING, T. (1987). Extreme value theory for multivariate stationary sequences. Technical Report, Texas A&M University. *J. Mult. Anal.* **29** 274–291.

HÜSLER, J. (1983). Asymptotic approximation of crossing probabilities of random sequences. *Z. Wahrscheinlichkeitstheorie verw. Geb.* **63** 257–270.

HÜSLER, J. (1986). Extreme values and rare events of non-stationary random sequences. In *Dependence in Probability and Statistics* (E. Eberlein, M.S. Taqqu, eds.), Birkhäuser, Boston, 438–456.

HÜSLER, J. (1987). Multivariate extreme values in stationary random sequences. Technical Report, University of Bern. To appear in *Stoch. Proc. Appl.*

HÜSLER, J. (1988a). Limit properties for multivariate extreme values in sequences of independent non-identically distributed random vectors. *Stoch. Proc. Appl.* **31** 105–116.

HÜSLER, J. (1988b). Limit distribution of multivariate extreme values in nonstationary sequences of random vectors. In *Extreme Value Theory* (J. Hüsler, R.D. Reiss, eds.), *Lecture Notes in Statistics*, Springer, New York, **51** 234–245.

HÜSLER, J. (1989). A note on the independence and total dependence of max i.d. distributions. *Adv. Appl. Probab.* **21** 231–232.

HÜSLER, J. and REISS, R.-D. (1989). Maxima of normal random vectors: Between independence and complete dependence. *Statist. Probab. Letters* **7** 283–286.

HÜSLER, J. and SCHÜPBACH, M. (1988). Limit results for maxima in non-stationary multivariate Gaussian sequences. *Stoch. Proc. Appl.* **28** 91–99.

LEADBETTER, M.R. (1983). Extremes and local dependence in stationary sequences. *Z. Wahrscheinlichkeitstheorie verw. Geb.* **65** 291–306.

LEADBETTER, M.R., LINDGREN, G., and ROOTZÉN, H. (1983). *Extremes and Related Properties of Random Sequences and Processes.* Springer, Series in Statistics, Berlin.

LINDGREN, G. (1974). A note on the asymptotic independence of high level crossings for dependent Gaussian processes. *Ann. Probab.* **2** 535–539.

MARSHALL, A.W. and OLKIN, I. (1983). Domains of attraction of multivariate extreme value distributions. *Ann. Probab.* **11** 168–177.

NEWMAN, C.M. and WRIGHT, A.L. (1981). An invariance principle for certain dependent sequences. *Ann. Probab.* **9** 671–675.

RESNICK, S.I. (1987). *Extreme Values, Regular Variation, and Point Processes.* Springer, Berlin.

SBIHI, A. (1988). Limiting distribution of the maxima of multivariate stationary sequences. Technical Report, McGill University.

SIBUYA, M. (1960). Bivariate extreme statistics. *Ann. Inst. Stat. Math.* **11** 195–210.

TAKAHASHI, R. (1987). Some properties of multivariate extreme value distributions and multivariate tail equivalence. *Ann. Inst. Statist. Math.* **39** A 637–647.

TAKAHASHI, R. (1988). Characterizations of a multivariate extreme value distribution. *Adv. Appl. Probab.* **20** 235–236.

TIAGO DE OLIVEIRA, J. (1958). Extremal distributions. Revista da Fac. Ciencias, Univ. Lisboa A **7** 215–227.

VILLASENOR, J. (1976). On univariate and bivariate extreme value theory. Ph.d. Thesis, Iowa State University.

DEPARTMENT OF MATH. STATISTICS
SIDLERSTR. 5, CH-3012 BERN
SWITZERLAND

INVARIANCE UNDER DEPENDENCE BY MIXING

BY D.R. JENSEN

Virginia Polytechnic Institute and State University

This paper is concerned with arrays of conditionally independent random elements that become dependent by mixing. The principal focus is the preservation of properties known to hold under independence. Findings are reported in the context of limit theory, including laws of large numbers and central limit theory, and topics in statistical inference. Several standard results, ranging from Berry-Esseén bounds in central limit theory to the use of Friedman's (1937) test in the analysis of two-way data, are seen to remain valid under certain models for dependence. The class of limit laws for standardized sums is expanded to include dependent cases, as are bounds on rates of convergence to these limits.

1. Introduction. Independence and conditional independence are central to probability theory and its applications, supporting the theory of Markov chains, Bayesian analysis, limit theory, and the foundations of statistical inference. See Dawid (1979), for example. A systematic study of conditionally independent events dates back at least to de Finetti (1937), who postulated these as models for conditional independence in a random environment. More recently, conditionally independent events have been studied as models for weak dependence (cf. Dykstra et al. (1973), Shaked (1977), and Tong (1980), for example), but there the primary focus centers on inequalities relating unconditional joint probabilities to products of marginal probabilities.

The assumption of independence pervades much of mathematical statistics, including essential portions of parametric and nonparametric statistical inference. That independence is a fundamental mathematical concept is not in doubt. Much less clear is the extent to which it mimics reality. Indeed, apart from highly specialized models, there appear to be no omnibus empirical tests for genuine independence. On physical grounds it even may be argued that observable phenomena at best can be only conditionally independent, owing to the common background energy attributed to the big bang.

AMS 1980 subject classifications. Primary 60E05; secondary 60F05, 62H05.

Key words and phrases. Conditional independence, models for dependence, limit theory, unconditional inference.

In view of these uncertainties, it is essential to regard independence as but one of many model assumptions subject to misspecification. It then is pertinent to examine questions of robustness and even invariance of critical properties to the assumption of independence. That is the focus of this paper, an outline of which follows.

Supporting developments and models for dependence are given in Sections 2 and 3. In Section 4 we consider topics in limit theory. These topics include laws of large numbers, central limit theory, a study of the types of limit laws achieved by conditionally independent sequences, and versions of Berry-Esseén bounds appropriate for these. Section 5 studies topics in statistical inference under dependence. These include the relative sensitivities of experiments, the use of Friedman's (1937) test in the analysis of two-way data under dependence, and the use of Anderson's (1984) classification statistic under dependence by scaling. We infer that numerous properties of these procedures continue to hold exactly under certain models for dependence.

2. Preliminaries. To fix notation $\Re^k$ and $\Re^k_+$ are Euclidean k-space and its positive orthant; $F_{n\times k}$ is the space of real $(n \times k)$ matrices; S_k and S_k^+ consist of symmetric $(k \times k)$ matrices and their positive semidefinite varieties; $\mathbf{X}' = [X_1,\ldots,X_k]$ is the transpose of $\mathbf{X} \in \Re^k$; and $\|\bullet\|$ is the Euclidean norm on $\Re^k$. Special arrays are the $(k \times k)$ identity $\mathbf{I}_k$, the unit vector $\mathbf{1}_k = [1,\ldots,1]' \in \Re^k$, the Kronecker product $\mathbf{A} \times \mathbf{B} = [a_{ij}\mathbf{B}]$, and the block diagonal array Diag $(\mathbf{A}_1,\ldots,\mathbf{A}_r)$. Cumulative distribution, probability density, and characteristic functions are abbreviated as *cdf*, *pdf*, and *chf*, respectively, with $\mathcal{L}(\mathbf{X})$ as the distribution of $\mathbf{X}$. Let $(\Omega, B(\Omega), Q)$ be a probability space. We are concerned with the probability measure P on $(\Re^k, B(\Re^k))$, the *cdf* $F(\mathbf{x})$, and the *chf* $\phi_x(\mathbf{t})$ generated by $\mathbf{X}(\omega) \in \Re^k$, as well as sequences $\{\mathbf{X}_1,\ldots,\mathbf{X}_n\}$ on $(\Re^k)^n$ which, on occasion, are independent and identically distributed (*iid*). Moment arrays for $\mathbf{X} = [X_1,\ldots,X_k]' \in \Re^k$, when defined, are the expected vector value $E(\mathbf{X}) = \theta \in \Re^k$, the dispersion matrix $D(\mathbf{X}) = E(\mathbf{X}-\theta)(\mathbf{X}-\theta)' \in S_k^+$, and the absolute central moments $\{\beta_{\delta j} = E \mid X_j - \theta_j \mid^\delta; 1 \le j \le k\}$ of order $\delta > 0$. Some special distributions on $\Re^k$ are the Gaussian law $N_k(\theta,\Sigma)$ having the mean $\theta \in \Re^k$, the dispersion matrix $\Sigma \in S_k^+$, and the *pdf* $f_N(\mathbf{x};\theta,\Sigma)$, and mixtures of these. In particular, $H_k(\theta,G) = \int_{S_k^+} N_k(\theta,\mathbf{S})dG(\mathbf{S})$ is a Gaussian mixture over S_k^+ with respect to $G(\bullet)$ having the *pdf*

$$f(\mathbf{x};\theta,G) = \int_{S_k^+} f_N(\mathbf{x};\theta,\mathbf{S})dG(\mathbf{S}); \tag{1}$$

and $\mathcal{H}_k = \{H_k(\theta,G); \theta \in \Re^k, G \in M(S_k^+)\}$ is the class of all such mixtures, where $M(S_k^+)$ is the class of all probability measures on S_k^+. Specifically, $\mathcal{H}_{k0} = \{H_{k0}(\mathbf{0},G); G \in M(S_k^+)\} \subset \mathcal{H}_k$ consists of dispersion mixtures of zero-mean Gaussian laws on $\Re^k$. The special case for which $\mathbf{S} = s\Sigma$, with Σ fixed and s random on $\Re^1_+$, yields scale mixtures of Gaussian laws within the class of ellipsoidal distributions on $\Re^k$. The Gaussian law on $F_{n\times k}$ representing an *iid* sample

$\mathbf{X} = [\mathbf{X}_1, \ldots, \mathbf{X}_n]'$ from $N_k(\theta, \Sigma)$ is denoted by $N_{n\times k}(\mathbf{1}_n \times \theta', \mathbf{I}_n \times \Sigma)$.

3. Models for Dependence. We develop models for dependence through mixing conditionally independent arrays. These consist of sequences of random elements arising in limit theory, as well as two-way arrays with applications in inference. It is natural to use *chf*s, for which bounded convergence applies directly.

3.1. Conditionally Independent Sequences. We consider random sequences in $(\Re^k)^n$, on occasion specializing to the case $k = 1$. Let $(\Gamma, B(\Gamma), \mu)$ be a probability space, and consider a collection $\{\phi_i(\mathbf{t}_i;\gamma); 1 \le i \le n\}$ of $B(\Gamma)$-measurable functions such that each $\phi_i(\mathbf{t}_i;\gamma)$ is a *chf* on $\Re^k$ for each $\gamma \in \Gamma$. Let $\{\mathbf{X}_1, \ldots, \mathbf{X}_n\}$ be a sequence on $(\Re^k)^n$ whose joint *chf* is given by

$$\phi_{\mathbf{X}}(\mathbf{t}_1, \ldots, \mathbf{t}_n) = \int_\Gamma \Pi_{i=1}^n \phi_i(\mathbf{t}_i;\gamma) d\mu(\gamma) \tag{2}$$

with $\mathbf{X} = [\mathbf{X}_1, \ldots, \mathbf{X}_n] \in F_{k\times n}$. Clearly $\{\mathbf{X}_1, \ldots, \mathbf{X}_n\}$ are conditionally independent with mixing measure μ. The special case for which $\{\phi_i(\mathbf{t}_i;\gamma) = \phi(\mathbf{t}_i;\gamma); 1 \le i \le n\}$ follows when $\{\mathbf{X}_1, \ldots, \mathbf{X}_n\}$ are conditionally *iid*. For the latter case with $k = 1, \{\{X_1, \ldots, X_n\}; n = 1, 2, \ldots\}$ is a de Finetti sequence on $\Re^\infty$.

Observe that γ may be a scalar, a vector, a matrix, or an arbitrary random element to be denoted by $\boldsymbol{\gamma}$. The concept of dependence by scaling is made precise in the following definition, where γ and the typical element $\mathbf{Z}_i$ conform for multiplication.

DEFINITION 1. Random elements $\{\mathbf{X}_1, \ldots, \mathbf{X}_n\}$ are said to be *dependent by scaling* if there are independent random elements $\{\mathbf{Z}_1, \ldots, \mathbf{Z}_n\}$ and a random element $\boldsymbol{\gamma}$, independent of $\{\mathbf{Z}_1, \ldots, \mathbf{Z}_n\}$, such that $\mathcal{L}(\mathbf{X}_1, \ldots, \mathbf{X}_n) = \mathcal{L}(\boldsymbol{\gamma}\mathbf{Z}_1, \ldots, \boldsymbol{\gamma}\mathbf{Z}_n)$. In particular, random vectors $\{\mathbf{X}_1, \ldots, \mathbf{X}_n\}$ on $(\Re^k)^n$ are called *dependent by coordinate scaling* if $\mathcal{L}(\mathbf{X}_1, \ldots, \mathbf{X}_n) = \mathcal{L}(\boldsymbol{\gamma}\mathbf{Z}_1, \ldots, \boldsymbol{\gamma}\mathbf{Z}_n)$ such that $\boldsymbol{\gamma} = \mathrm{Diag}(\gamma_1, \ldots, \gamma_k)$.

Standard limit theorems depend heavily on moments. It is useful to distinguish orders of dependence under mixing, at issue being the manner in which the conditional moments depend on the mixing variable.

DEFINITION 2. Conditionally independent random variables are said to be *dependent of order r* if their conditional moments of order r depend on the mixing parameter $\gamma \in \Gamma$.

3.2. Conditionally Independent Ensembles. We assemble a collection of random elements in a two-way array having r rows. In particular, let $\{\mathbf{X}_i \in F_{n_i\times k_i}; 1 \le i \le r\}$ be random having the *chf*s $\{\{\phi_i(\mathbf{T}_i;\gamma_i); \gamma_i \in \Gamma_i\}; 1 \le i \le r\}$, and let $\mathbf{X} = [\mathbf{X}_1, \ldots, \mathbf{X}_\mathbf{r}]$. Our model for an ensemble of conditionally independent random components is given by the joint *chf*

$$\phi_{\mathbf{X}}(\mathbf{T}_1, \ldots, \mathbf{T}_r) = \int_\Gamma \Pi_{i=1}^r \phi_i(\mathbf{T}_i;\gamma_i) d\mu(\gamma_1, \ldots, \gamma_r) \tag{3}$$

where $\mu(\bullet, \ldots, \bullet)$ is a mixing measure on $\Gamma = \Gamma_1 \times \cdots \times \Gamma_r$. If $\mu(\bullet)$ is concentrated along the equiangular line $\{\gamma_1 = \cdots = \gamma_r\}$ in Γ, then this model reduces to that of Section 3.1. Different versions of (3) are developed in Section 5.

4. Dependent Limit Theorems. We consider various modes of stochastic convergence, giving unconditional laws of large numbers and results in central limit theory for conditionally independent sequences. The classical limit theorems depend heavily on moments. Here we require conditional moments of given order, whereas unconditional moments need not be defined. Our basic approach uses *chf*s. Together with the Dominated Convergence Theorem, this justifies limits of mixtures as mixtures of limits, so that unconditional versions of the classical limit theorems follow on mixing.

4.1. Laws of Large Numbers. Let $\{\mathbf{X}_n; n = 1, 2, \ldots\}$ be a sequence of conditionally independent random elements on $\Re^k$, and let $\{\theta_n; n = 1, 2, \ldots\}$ be a sequence of parameters. We are concerned with the unconditional convergence of $\{(\mathbf{X}_n - \theta_n); n = 1, 2, \ldots\}$ in various stochastic modes. Under the modes of Section 3.1 our basic tools are the representation $P(A) = \int_\Gamma P_\gamma(A) d\mu(\gamma)$, the variance formula $\mathrm{Var}(X_n) = E_\gamma[\mathrm{Var}(X_n \mid \gamma)] + \mathrm{Var}_\gamma[E(X_n \mid \gamma)]$ on $\Re^1$, and the corresponding dispersion formula $D(\mathbf{X}_n) = E_\gamma[D(\mathbf{X}_n \mid \gamma)] + D[E(\mathbf{X}_n \mid \gamma)]$ on $\Re^k$.

Suppose (i) that $\{\mathbf{X}_n; n = 1, 2, \ldots\}$ are conditionally independent with mixing parameter γ, (ii) that $\{\theta_n; n = 1, 2, \ldots\}$ is a sequence of constant elements not depending on $\gamma \in \Gamma$, and (iii) that $(\mathbf{X}_n - \theta_n) \to \mathbf{0}$ in some mode for each $\gamma \in \Gamma$. Assumption (ii) assures unconditional convergence to a constant rather than to a random variable. Basic relationships among various conditional and unconditional modes of convergence are summarized in Table 1, arranged in pairs as Cases 1–4 in which conditional convergence in the first mode of each pair implies unconditional convergence in the second mode of that pair.

TABLE 1. Unconditional modes of convergence implied by conditional convergence for each $\gamma \in \Gamma$.

Case	Conditional Mode	Unconditional Mode
1	Almost Sure	Almost Sure
2	Mean Square	Mean Square*
3	Mean Square	In Probability
4	In Probability	In Probability

*Assuming unconditional second moments.

The claims in Table 1 are easily verified using standard arguments. For Case 1, to show that almost sure conditional convergence implies almost sure convergence unconditionally, let $A = \{\omega : (\mathbf{X}_n(\omega) - \theta_n) \to \mathbf{0}\}$. Then $\{P_\gamma(A) = 1; \gamma \in \Gamma\}$ by hypothesis, so that $P(A) = \int_\Gamma P_\gamma(A) d\mu(\gamma) = 1$ unconditionally. For Case 2, with $E(X_n \mid \gamma) = \mu_n(\gamma)$ on $\Re^1$, it follows that $E[(X_n - \theta_n)^2] = E_\gamma[\mathrm{Var}(X_n \mid \gamma) + (\mu_n(\gamma) - \theta_n)^2]$, where the expression $[\mathrm{Var}(X_n \mid \gamma) + (\mu_n(\gamma) - \theta_n)^2] \to 0$ for each $\gamma \in \Gamma$ by hypothesis. Parallel arguments apply in the vector case. Case 3 follows from suitable versions of Chebychev inequalities on $\Re^1$ and $\Re^k$ applied conditionally. Case 4 follows on using bounded convergence. Observe that Case 2 applies automatically whenever the conditional moments to second order do not depend on γ.

Various laws of large numbers follow from the foregoing developments unconditionally without difficulty. Details are supplied for the following version of Khintchine's Theorem on $\Re^k$, where moments of the unconditional distribution are not required in order to validate the result.

THEOREM 4.1. *(Khintchine). Let* $\{\mathbf{X}_n; n = 1, 2, \ldots\}$ *be conditionally iid on* $\Re^k$ *having the conditional mean* $E(\mathbf{X}_n \mid \gamma) = \theta$. *Then the sample mean* $\bar{\mathbf{X}}_n = (\mathbf{X}_1 + \cdots + \mathbf{X}_n)/n$ *is weakly consistent for* $\theta \in \Re^k$.

PROOF. Use *chf*s and dominated convergence to write

$$\lim_{n\to\infty} \phi_{\bar{\mathbf{X}}_n}(\mathbf{t}) = \int_\Gamma \lim_{n\to\infty} [\phi(\mathbf{t}/n; \gamma)]^n d\mu(\gamma). \tag{4}$$

Expanding under the integral gives

$$\lim_{n\to\infty} [1 + i\mathbf{t}'\theta/n + o(\| \mathbf{t} \| /n)]^n = e^{i\mathbf{t}'\theta} \tag{5}$$

so that $\lim_{n\to\infty} \phi_{\bar{\mathbf{X}}_n}(\mathbf{t}) = e^{i\mathbf{t}'\theta}$ because θ does not depend on γ. This is the *chf* of the distribution on $\Re^k$ degenerate at θ, thus completing our proof.

It deserves emphasis that moments of the unconditional distribution are not required to exist, yet a weak law of large numbers nonetheless applies unconditionally on $\Re^k$.

4.2. Central Limit Theory. We consider scalar and vector sequences of the types described in Section 3.1. Not only are the limit laws characterized, but dependent versions of local and global Berry-Esseén bounds are developed. In particular, $\{\mathbf{X}_1, \ldots, \mathbf{X}_n\}$ is a conditionally independent sequence on $(\Re^k)^n$ having the typical *cdf* $G_i(\bullet)$ and moments $E(\mathbf{X}_i \mid \gamma) = \theta$ and $D(\mathbf{X}_i \mid \gamma) = \Sigma_i(\gamma)$. Consider $\mathbf{Z}_n = n^{1/2}(\bar{\mathbf{X}}_n - \theta)$; and let $\phi_n(\mathbf{t})$, $F_n(\mathbf{x})$, and $P_n(\bullet)$ respectively be the *chf*, *cdf*, and probability measure induced by $\mathbf{Z}_n$, with $P(\bullet)$ as the weak limit $P(\bullet) = \lim_{n\to\infty} P_n(\bullet)$. Define the Lindeberg function on $\Re^k$ as

$$L_n(z; \gamma) = n^{-1}\Sigma_{i=1}^n \int_{\|\mathbf{x}\|>z} \| \mathbf{x} \|^2 \, dG_i(\mathbf{x}). \tag{6}$$

A principal result is the following.

THEOREM 4.2. *Let* $\{\mathbf{X}_1, \ldots, \mathbf{X}_n\}$ *be conditionally independent on* $(\Re^k)^n$ *having the typical conditional moments* $E(\mathbf{X}_i \mid \gamma) = \mathbf{0}$ *and* $D(\mathbf{X}_i \mid \gamma) = \Sigma_i(\gamma)$. *Suppose that as* $n \to \infty$, $n^{-1}\sum_{i=1}^n \Sigma_i(\gamma) \to \Sigma(\gamma) \neq \mathbf{0}$ *and* $L_n(n^{1/2}\varepsilon; \gamma) \to 0$ *for each* $\varepsilon > 0$

and $\gamma \in \Gamma$. Then the limit distribution of $\mathbf{Z}_n = n^{1/2}\bar{\mathbf{X}}_n$ exists in the class $\mathcal{H}_{k0}$, and its chf is given by

$$\lim_{n\to\infty} \phi_n(\mathbf{t}) = \int_\Gamma e^{-\mathbf{t}'\boldsymbol{\Sigma}(\gamma)\mathbf{t}/2} d\mu(\gamma). \tag{7}$$

PROOF. Write the *chf* $\phi_n(\mathbf{t})$ in its mixture representation and use dominated convergence to get

$$\lim_{n\to\infty} \phi_n(\mathbf{t}) = \int_\Gamma \lim_{n\to\infty} [\Pi_{i=1}^n \phi_i(\mathbf{t}/n^{1/2};\gamma)] d\mu(\gamma) \tag{8}$$

where

$$\lim_{n\to\infty} [\Pi_{i=1}^n \phi_i(\mathbf{t}/n^{1/2};\gamma)] = e^{-\mathbf{t}'\boldsymbol{\Sigma}(\gamma)\mathbf{t}/2} \tag{9}$$

using standard arguments. This completes our proof.

Limit laws for conditionally independent sequences having conditional second moments are seen to be dispersion mixtures of Gaussian laws belonging to the class $\mathcal{H}_{k0}$. This was noted by Taylor et al. (1985) for exchangeable sequences on $\Re^1$. The class of limit distributions is thus larger than for the independent case. A special case of some interest is that $\boldsymbol{\Sigma}(\gamma) = \gamma\boldsymbol{\Sigma}$, with γ a positive scalar, in which case the limit laws are scale mixtures of Gaussian laws. This class contains the ellipsoidal stable laws as a proper subclass.

COROLLARY 4.2.1. *Let $\{\mathbf{X}_1, \ldots, \mathbf{X}_n\}$ be dependent of order r for $r > 2$ but not for $r \leq 2$. Then the limit distribution of $\mathbf{Z}_n$ is Gaussian, as when $\{\mathbf{X}_1, \ldots, \mathbf{X}_n\}$ are independent.*

4.3. Berry-Esseén Bounds. We first consider the scalar case, for which we seek global and local bounds as well as invariance properties of the usual Berry-Esseén bounds. Owing to space constraints, we consider only conditionally *iid* sequences on $\Re^k$; more general results follow along similar lines. The following result gives lower and upper bounds on the rate of convergence for conditionally independent sequences on $\Re^1$.

THEOREM 4.3. *Let $\{X_1, \ldots, X_n\}$ be conditionally independent random variables on $\Re^1$ having the typical cdf $G_i(\bullet;\gamma)$ with conditionally zero means, the variance $\sigma_i^2(\gamma)$, and finite third moments $\beta_{3i}(\gamma)$. Let $F_n(\bullet)$ be the cdf of $Z_n = n^{1/2}\bar{X}_n$, and let $F(\bullet) = \lim_{n\to\infty} F_n(\bullet)$ be its limit. Then there are absolute constants c_1 and c_2 such that the bounds*

$$\begin{aligned} c_1 \int_\Gamma L_n(S_n;\gamma) d\mu(\gamma) &\leq \sup_x \mid F_n(x) - F(x) \mid \\ &\leq c_2 \int_\Gamma \int_0^{S_n(\gamma)} (S_n(\gamma))^{-1} L_n(z;\gamma) dz d\mu(\gamma) \end{aligned} \tag{10}$$

hold whenever the integrals are defined, where $S_n^2(\gamma) = \sigma_1^2(\gamma) + \cdots + \sigma_n^2(\gamma)$ and $L_n(z;\gamma) = [S_n^2(\gamma)]^{-1}\Sigma_{i=1}^n \int_{||x||>z} x^2 dG_i(x;\gamma)$.

PROOF. Apply the bounds of Studnev and Ignat (1967) conditionally.

THEOREM 4.4. *Let the conditionally iid variables* $\{X_1, \ldots, X_n\}$ *on* $\Re^1$ *be dependent by scaling, having the conditional cdf* $G(\bullet;\gamma)$ *with moments as in Theorem 4.3. Then a global bound is given by*

$$\sup_x | F_n(x) - F(x) | \leq \frac{c\beta_3(1)}{n^{1/2}[\sigma^2(1)]^{3/2}} \tag{11}$$

where c is an absolute constant. Moreover, if $G(\bullet;\gamma)$ either has an absolutely continuous component or is of the lattice type, then the local bound

$$| F_n(x) - F(x) | \leq \frac{c\beta_3(1)}{n^{1/2}[\sigma^2(1)]^{3/2}} \int_0^\infty (1 + x/\gamma)^{-1} d\mu(\gamma) \tag{12}$$

holds with c an absolute constant whenever the integral is defined.

PROOF. The first conclusion follows on applying the bounds of Berry (1941) and Esseén (1945) conditionally and noting that the moment ratio is scale-invariant. The second follows on applying a result of Bikjalis (1966) conditionally.

We next turn to Berry-Esseén bounds on $\Re^k$. We consider only conditionally *iid* sequences on $(\Re^k)^n$, noting that more general results follow without difficulty along similar lines. Let $\mathcal{F}$ be the class of all measurable convex subsets of $\Re^k$.

THEOREM 4.5. *Let* $\{\mathbf{X}_1, \ldots, \mathbf{X}_n\}$ *be conditionally iid on* $(\Re^k)^n$ *having conditional moments* $E(\mathbf{X} \mid \gamma) = \theta$, $D(\mathbf{X} \mid \gamma) = \Sigma(\gamma)$, *and* $\{\beta_{3j}(\gamma) = E(| X_j - \theta_j |^3 \mid \gamma); 1 \leq j \leq k\}$, *with* X_j *as the* j*th component of* $\mathbf{X}$. *Then global bounds on the rate of convergence of* P_n *to* P *are given by*

$$\sup_{A\in\mathcal{F}} | P_n(A) - P(A) | \leq \frac{ck^3}{n^{1/2}} \int_\Gamma \sum_{j=1}^k [\xi_{jj}(\gamma)]^{3/2} \beta_{3j}(\gamma) d\mu(\gamma) \tag{13}$$

whenever the integral is defined, where $\Xi(\gamma) = [\xi_{ij}(\gamma)] = [\Sigma(\gamma)]^{-1}$ and c is an absolute constant not depending on k.

PROOF. The proof consists of modifying a result of Bergström (1969) in a form due to Jensen and Mayer (1975), and applying the result conditionally.

COROLLARY 4.5.1. *Let* $\{\mathbf{X}_1, \ldots, \mathbf{X}_n\}$ *be dependent by coordinate scaling with moments as in Theorem 4.5. Then*

$$\sup_{A\in\mathcal{F}} | P_n(A) - P(A) | \leq \frac{ck^3}{n^{1/2}} \sum_{j=1}^k [\xi_{jj}(1)]^{3/2} \beta_{3j}(1). \tag{14}$$

PROOF. The conclusion follows on noting that the expression under the integral on the right of (13) is invariant under scaling the coordinates on $\Re^k$.

COROLLARY 4.5.2. *Let* $\{\mathbf{X}_1, \ldots, \mathbf{X}_n\}$ *be dependent of order* r *for some* $r > 3$ *but not for* $r \leq 3$, *with moments otherwise as in Theorem 4.5. Then*

$$\sup_{A \in \mathcal{F}} | P_n(A) - P(A) | \leq \frac{ck^3}{n^{1/2}} \sum_{j=1}^{k} (\xi_{jj})^{3/2} \beta_{3j}. \tag{15}$$

PROOF. This follows because the expression under the integral on the right of (13) does not depend on γ.

There is a rich literature on central limit theory and bounds of the Berry-Esseén type on $\Re^1$ and $\Re^k$. Many known results carry over to conditionally independent sequences along the lines illustrated here. For example, unconditional Edgeworth series expansions of order s on $\Re^k$ will emerge as mixtures of usual Edgeworth series on $\Re^k$ as given in Chambers (1967). Moreover, on applying results of von Bahr (1967) conditionally, bounds on the errors of these Edgeworth mixtures can be obtained in a manner similar to the unconditional versions of Berry-Esseén bounds given here. If the expression under the integral on the right of (13) is uniformly bounded for all $\gamma \in \Gamma$, then that bound can be used as a nonintegral version of (13). Note, however, that the integral version depends on the particular mixture and thus gives a tighter bound.

In another direction, suppose that second moments are not defined conditionally, but that the conditional distributions are in the domain of attraction of a stable limit on $\Re^1$ or $\Re^k$ with index α. Then limit distributions of standardized sums are conditionally stable with index α, and unconditionally are mixtures of these. Again the class of limit laws is larger than for the independent case. In particular, the *chf*s for these limits can be studied as mixtures of Lévy representations for stable *chf*s on $\Re^1$ or $\Re^k$, as appropriate.

5. Topics In Inference. Many statistical procedures are based on the assumption of independence. A number of these remain valid despite dependencies of certain types. Three examples are given here using variations of the basic model of Section 3.2. In these cases the argument is the same: Conditional distributions of the statistics in question are seen to be free of the conditioning variables and thus are identical to their unconditional forms. In all such cases the classical assumption of independence may be replaced by the much weaker assumption of conditional independence.

EXAMPLE 5.1. *Sensitivities of Experiments.* The sensitivities of alternative experiments in the normal-theory analysis of variance may be studied as follows. Suppose that $\mathcal{L}(\mathbf{X}_1) = N_n(\theta_1, \sigma_1^2 \mathbf{I}_n)$ and $\mathcal{L}(\mathbf{X}_2) = N_n(\theta_2, \sigma_2^2 \mathbf{I}_n)$ are models for two independent experiments pertaining to a parameter θ, and that a linear hypothesis $H : \mathbf{A}\theta = \mathbf{0}$ is to be tested in each experiment. Let F_1 and F_2 be the corresponding variance ratios having noncentrality parameters λ_1 and λ_2. Bradley and Schumann (1957b) studied the ratio $R = F_1/F_2$ both as a gauge of the relative sensitivities of the two experiments, and as a statistic for testing $H : \lambda_1 = \lambda_2$. The distribution of R and various applications are treated in Bradley and Schumann (1957a,b), Schumann and Bradley (1958), Schumann and Bradley (1959); see also Shue and Bain (1982), Subrahmaniam (1979), and Zerbe and Goldgar (1980) for related work.

Now let $\mathbf{X}_1$ and $\mathbf{X}_2$ be conditionally independent as in Section 3.2 such that $\mathcal{L}(\gamma_1\mathbf{X}_1 \mid \gamma_1) = N_n(\theta_1, \sigma_1^2\mathbf{I}_n)$ and $\mathcal{L}(\gamma_2\mathbf{X}_2 \mid \gamma_2) = N_n(\theta_2, \sigma_2^2\mathbf{I}_n)$, where (γ_1, γ_2) have some joint distribution on $\Re_+^2$. Because F_1, F_2 and R are scale-invariant, their conditional and unconditional distributions are identical, so that all the standard properties continue to hold exactly under conditional independence of the type indicated.

EXAMPLE 5.2. *Friedman's Test.* Friedman's (1937) test is used widely in nonparametrics to compare the effectiveness of k treatments using n experimental subjects. Let $\{Y_{ij}; 1 \le j \le k,\ 1 \le i \le n\}$ be outcomes of an experiment such that $\mathbf{Y}_i' = [Y_{i1}, \ldots Y_{ik}]$ has an exchangeable distribution on $\Re^k$ for each $i = 1, \ldots, n$. If R_{ij} denotes the rank of Y_{ij} among $\{Y_{i1}, \ldots Y_{ik}\}$ and if $\{R_j = R_{1j} + \cdots + R_{nj}; 1 \le j \le k\}$, then Friedman's statistic is

$$X_r^2 = \frac{12}{nk(k+1)} \sum_{j=1}^{k} [R_j - n(k+1)/2]^2. \tag{16}$$

The standard assumption is that $\{\mathbf{Y}_1, \ldots, \mathbf{Y}_n\}$ are mutually independent, in which case the exact small-sample null distribution is based on the $(k!)^n$ possible permutations, and the asymptotic distribution is $\chi^2(k-1)$.

Specializing the model of Section 3.2, let $\{\phi_i(t_1, \ldots, t_k); 1 \le i \le n\}$ be *chf*s of exchangeable distributions on $\Re^k$, and let Γ_0 be the class of all monotonic increasing functions $\gamma : \Re^1 \to \Re^1$. For a typical *chf* $\phi(t_1, \ldots, t_k)$ of $\mathbf{Y}' = [Y_1, \ldots Y_k]$, denote by $\phi(t_1, \ldots, t_k; \gamma)$ the joint *chf* of $[\gamma(Y_1), \ldots, \gamma(Y_k)]$. Now choose $[\gamma_1, \ldots, \gamma_n]$ from $\Gamma = \Gamma_0^n$ according to some probability measure $\mu(\bullet)$, and consider the joint distribution of $\{\gamma_i(Y_{ij}); 1 \le j \le k, 1 \le i \le n\}$. This has the form (3). It is well known that $\{[\gamma_i(Y_{i1}), \ldots \gamma_i(Y_{ik})]; 1 \le i \le n\}$ are again exchangeable vectors on $\Re^k$, but now they are dependent. Nonetheless, the conditional null distribution of X_r^2 does not depend on $[\gamma_1, \ldots, \gamma_n]$, and thus its exact small-sample and asymptotic distributions are precisely those occurring when responses from subject to subject are independent. For example, the k responses within each subject may be randomly scaled, with a different scaling for different subjects.

EXAMPLE 5.3. *Classification Rules.* Given samples from two Gaussian populations, $N_k(\theta_1, \Sigma)$ and $N_k(\theta_2, \Sigma)$, and a random observation $\mathbf{X}$ from $N_k(\theta, \Sigma)$ having unknown origins, the problem of classification is to assign $\mathbf{X}$ to one of the two populations. In particular, suppose $\mathcal{L}(\mathbf{X}_1) = N_{n_1 \times k}(\mathbf{1}_{n_1} \times \theta_1', \mathbf{I}_{n_1} \times \Sigma)$, and let $(\bar{\mathbf{X}}_1, \mathbf{S}_1)$ be the corresponding sample mean vector and sample dispersion matrix.

Similarly, consider $\mathcal{L}(\mathbf{X}_2) = N_{n_2\times k}(\mathbf{1}_{n_2} \times \theta_2', \mathbf{I}_{n_2} \times \Sigma)$ and $(\bar{\mathbf{X}}_2, \mathbf{S}_2)$. The standard procedure uses the classification statistic

$$V = [\mathbf{X} - (1/2)(\bar{\mathbf{X}}_1 + \bar{\mathbf{X}}_2)]'\mathbf{S}^{-1}(\bar{\mathbf{X}}_1 - \bar{\mathbf{X}}_2) \tag{17}$$

where $\mathbf{S}$ is the pooled sample estimator for Σ; see Anderson (1984), p. 210. Normal-theory properties of the usual classification rule using V are based on the mutual independence of $\{\mathbf{X}_1, \mathbf{X}_2, \mathbf{X}\}$.

However, we now suppose that $\{\mathbf{X}_1, \mathbf{X}_2, \mathbf{X}\}$ are conditionally independent, given a nonsingular random matrix $\gamma \in F_{k\times k}$, such that $\mathcal{L}(\mathbf{X}_1 \mid \gamma) = N_{n_1\times k}(\mathbf{1}_{n_1} \times \theta_1'\gamma, \mathbf{I}_{n_1} \times \gamma'\Sigma\gamma), \mathcal{L}(\mathbf{X}_2 \mid \gamma) = N_{n_2\times k}(\mathbf{1}_{n_2} \times \theta_2'\gamma, \mathbf{I}_{n_2} \times \gamma'\Sigma\gamma)$, and $\mathcal{L}(\mathbf{X} \mid \gamma) = N_k(\gamma'\theta, \gamma'\Sigma\gamma)$. This is seen to be a special case of model (3) where $r = 3$. Since the statistic V is invariant under nonsingular linear transformations, its conditional distribution $\mathcal{L}(V \mid \gamma)$ is independent of $\gamma \in \Gamma$. It follows that all standard properties of the usual classification rules carry over to scale mixtures of the types indicated.

REFERENCES

ANDERSON, T.W. (1984). *An Introduction to Multivariate Statistical Analysis, 2nd ed.* John Wiley and Sons, New York.

BAHR, B. VON (1967). Multi-dimensional integral limit theorems. *Arkiv för Matematik* **7** 71–88.

BERGSTRÖM, H. (1969). On the central limit theorem in $\Re^k$. *Z. Wahrscheinlichkeitstheorie und Verw. Gebiete* **14** 113–126.

BERRY, A.C. (1941). The accuracy of the Gaussian approximation to the sum of independent variates. *Trans. Amer. Math. Soc.* **49** 122–136.

BIKJALIS, A. (1966). Estimates for the remainder term in the central limit theorem. (In Russian.) *Lit. Matem. Sb.* **6** 321–346.

BRADLEY, R.A. and SCHUMANN, D.E.W. (1957a). The comparison of the sensitivities of similar experiments: Applications. *Biometrics* **13** 496–510.

BRADLEY, R.A. and SCHUMANN, D.E.W. (1957b). The comparison of the sensitivities of similar experiments: Theory. *Ann. Math. Statist.* **28** 902–920.

CHAMBERS, J.M. (1967). On methods of asymptotic approximations for multivariate distributions. *Biometrika* **54** 367–383.

DAWID, A.P. (1979). Conditional independence in statistical theory. *J.R. Statist. Soc.* B **41** 1–31.

DYKSTRA, R.L., HEWETT, J.E. and THOMPSON, W.A. (1973). Events which are almost independent. *Ann. Statist.* **1** 674–681.

ESSEÉN, C.G. (1945). Fourier analysis of distribution functions. A mathematical study of the Laplace-Gaussian law. *Acta Math.* **77** 1–125.

FINETTI, B. DE (1937). La prevision: ses lois logiques, ses sources subjectives. *Ann. Inst. H. Poincaré* **7** 1–68.

FRIEDMAN, M. (1937). The use of ranks to avoid the assumption of normality implicit in the analysis of variance. *J. Amer. Statist. Assoc.* **32** 675–701.

JENSEN, D.R. and MAYER, L.S. (1975). Normal-theory approximations to tests for linear hypotheses. *Ann. Statist.* **3** 429–444.

SCHUMANN, D.E.W. and BRADLEY, R.A. (1958). The comparison of the sensitivities of similar experiments. *Virginia J. Sci.* **9** 303–314.

SCHUMANN, D.E.W. and BRADLEY, R.A. (1959). The comparison of the sensitivities of similar experiments: Model II of the analysis of variance. *Biometrics* **15** 405–416.

SHAKED, M. (1977). A concept of positive dependence for exchangeable random variables. *Ann. Statist.* **5** 505–515.

SHUE, W.K. and BAIN, L.J. (1982). Approximate percentiles for ratios of F-variates. *Comm. Statist. A* **11** 1831–1838.

STUDNEV, Y.P. and IGNAT, Y.I. (1967). Refinement of the central limit theorem and of its global version. *Theory Prob. Applic.* **12** 508–512.

SUBRAHMANIAM, K. (1979). Tests of significance for the odds ratio in a 2×2 table based on the ratio of two independent F variates. *Comm. Statist. B* **8** 245–256.

TAYLOR, R.L., DAFFER, P.Z. and PATTERSON, R.F. (1985). *Limit Theorems for Sums of Exchangeable Random Variables.* Rowman and Allanheld, Totowa, NJ.

TONG, Y.L. (1980). *Probability Inequalities in Multivariate Distributions.* Academic Press, New York.

ZERBE, G.O. and GOLDGAR, D.E. (1980). Comparison of intraclass correlation coefficients with the ratio of two independent F statistics. *Comm. Statist. A* **9** 1641–1645.

DEPARTMENT OF STATISTICS
VIRGINIA POLYTECHNIC INSTITUTE
BLACKSBURG, VA 24061

CONDITIONAL NEGATIVE DEPENDENCE IN STOCHASTIC ORDERING AND INTERCHANGEABLE RANDOM VARIABLES

By Kumar Joag-dev[1]

University of Illinois and Florida State University

A simple proof is provided for the result that when a subset of order statistics is given, the conditional joint distribution of the components of a random sample exhibit a certain form of negative dependence. It is also shown that the assertion remains valid even for the discrete distributions.

1. Introduction. Let $\mathbf{T} = (T_1, \ldots, T_n)$ be a random vector. $\mathbf{T}$ is said to be negatively dependent through stochastic ordering (NDS) if for every coordinatewise nondecreasing function h,

$$E[h(\mathbf{T}(i))|T_i = t_i] \downarrow t_i,\ i = 1 \ldots, n. \tag{1}$$

where $\mathbf{T}(i)$ denotes the $(n-1)$ vector obtained from $\mathbf{T}$ by removing the ith component.

Block, Bueno, Savits, and Shaked (1987) established this negative dependence property for a random sample conditioned by a subset of its order statistics. It was shown that this conditional NDS has interesting applications in the study of the systems formed by *second hand* components. The proof of this result in their paper (see Theorem 3.1 in above) is quite involved and is derived under the assumption that the random variables are independent with a common *continuous distribution* function. Since this conditional NDS may manifest itself in other applications, it may be desirable to have a simpler proof of this property and to seek less restrictive conditions. The present article does both. A simpler proof is provided for this property where random variables are assumed to be independent but their common distribution may be discrete. It is shown that the interchangeability plays an important role and the relevant result is isolated and later used in the main proof.

2. Notation. Let $\mathbf{Z} = (Z_1, Z_2, \ldots, Z_n)$ be a random vector and X be a random variable such that the n components of $\mathbf{Z}$ and X are identically distributed. It will

[1]Research partly supported by Grant DAAGL03-86-K-0094 from the Army Research Office.
AMS 1980 subject classifications. 60E99; 62N05.
Key words and phrases. Negative dependence through stochastic ordering, interchangeable random variables, conditioning by order statistic, systems using second hand components.

be assumed that these $(n+1)$ random variables are interchangeable for deriving a basic result. A stronger condition of independence will be imposed for later results.

The order statistic vector for $(\mathbf{Z}, X)$ will be denoted by $\mathbf{V}$. We write I for the index set $[1, 2, \ldots, (n+1)]$. Since the common distribution of the random variables above could be discrete there is a possibility of ties. In such a case, we interpret $V_i = v_i$ as an event where exactly $(i-1)$ observations are less than v_i and *one or more* at v_i. In case of ties the possible ranks are assigned at random. $R(X)$ will denote the rank of X among the $(n+1)$ components of $(\mathbf{Z}, X)$.

Let $\mathbf{W}$ denote the order statistic corresponding to $\mathbf{Z}$ and $\mathbf{V}(i)$ be the n-vector obtained by removing the ith component of $\mathbf{V}$.

3. Results. Let g be a coordinatewise nondecreasing function defined on R^n. We want to prove the conditional NDS property (1) for $(\mathbf{Z}, X)$, given $\mathbf{V}_B = \mathbf{v}_B$ (see relation (3) below). Here B is a nonempty subset of I and $\mathbf{v}_B$ denotes the set of values v_i, $i \in B$. Note that due to the permutation invariance of $(\mathbf{Z}, X)$, it suffices to consider conditioning on only one component, say X.

Our approach to prove the required monotonicity property is as follows. As a first step it is shown that the conditional expectation decreases as the rank $R(X)$ increases. This is shown under the minimal assumption of interchangeability. For the case where the underlying distribution is continuous, and the random variables are assumed to be independent, it is shown that the conditional expectation depends on the value of X only through its rank $R(X)$. Unfortunately, such a reduction is not possible for the case of discrete distributions and more detailed argument is needed for this case.

LEMMA. *Let the joint distribution of $(n+1)$ components of $(\mathbf{Z}, X)$ be interchangeable. Then*

$$E[g(\mathbf{Z}) | R(X) = i,\ \mathbf{V}_B = \mathbf{v}_B] \downarrow i.$$

PROOF. Due to the assumption of $(\mathbf{Z}, X)$ being interchangeable, it follows that the joint distribution of $(\mathbf{Z}, \mathbf{V}, X)$ is same as that of $(\mathbf{PZ}, \mathbf{V}(\mathbf{P}), X)$, where $\mathbf{P}$ is a permutation of the n components and $\mathbf{V}(\mathbf{P})$ is the order statistic corresponding to $(\mathbf{PZ}, X)$. However, $\mathbf{V}(\mathbf{P}) = \mathbf{V}$ and hence the expectation in the proposition does not change under permutations of $\mathbf{Z}$. Thus, without loss of generality, g may be assumed to be permutation invariant. Whenever convenient, we may replace the argument $\mathbf{Z}$ of g, by $\mathbf{W}$.

Given the order statistic $\mathbf{V}$ it is clear that all $(n+1)!$ permutations of I are equally likely as sets of ranks of $(\mathbf{Z}, X)$. This is clear for the case when the distribution is continuous. For the discrete case this follows from our choice of assigning ranks at random to the tied observations. In either case the rank vector is independent of $\mathbf{V}$. In particular $R(X)$ is independent of $\mathbf{V}$. If the rank $R(X) = i$ then the order statistic $\mathbf{W} = \mathbf{V}(i)$. Due to independence of $\mathbf{V}$ and $R(X)$, the event $\mathbf{V}_B = \mathbf{v}_B$, and $R(X)$ are also independent. Hence the difference

$$E[g(\mathbf{W})|R(X) = j,\ \mathbf{V}_B = \mathbf{v}_B] - E[g(\mathbf{W})|R(X) = i,\ \mathbf{V}_B = \mathbf{v}_B]$$

is the same as

$$E[g(\mathbf{V}(j))|\mathbf{V}_B = \mathbf{v}_B] - E[g(\mathbf{V}(i))|\mathbf{V}_B = \mathbf{v}_B]$$

or

$$E[\{g(\mathbf{V}(j)) - g(\mathbf{V}(i))\}|\mathbf{V}_B = \mathbf{v}_B], \tag{2}$$

where $\mathbf{V}(i)$ is defined above. The components of $\mathbf{V}(j)$ and $\mathbf{V}(i)$ are all the same except for a pair. If $j > i$, the smaller one of that pair belongs to $\mathbf{V}(j)$. Since g is increasing, the difference (2) must be nonpositive. This establishes the assertion of the Lemma.

THEOREM. *Suppose $(n+1)$ random variables $(\mathbf{Z}, X)$ are independent and identically distributed. Let $\mathbf{V}$ be the corresponding order statistic. For every nondecreasing function g and for every nonempty subset B of the index set I,*

$$E[g(\mathbf{Z})|X = x,\ \mathbf{V}_B = \mathbf{v}_B] \downarrow x, \text{ (almost surely)}. \tag{3}$$

PROOF. (a) In this part we assume that the underlying distribution is continuous. The condition $\mathbf{V}_B = \mathbf{v}_B$ creates a partition of the real line with $(b+1)$ open intervals, and b boundary points, where b denotes the cardinality of B. We illustrate our approach by a simple example. Suppose $n = 10$ and $B = \{3, 6, 8\}$ then the condition creates 4 open intervals $(-\infty, v_3)$, (v_3, v_6), (v_6, v_8), and (v_8, ∞). These contain respectively 2,2,1 and 3 observations, while the remaining 3 are at the boundaries. An observation is said to be of the ith *category* if it falls in the ith interval. It is important to note that once the boundaries of these intervals have been given, the observations taking values within these open intervals are independent.

Note that when the distribution is continuous there can be only one observation on the boundary.

Suppose that the conditioned value x of X is in the open interval corresponding to the highest category. In the example, this would imply that it is larger than v_8, or equivalently, $R(X) > 8$. Recalling that $\mathbf{W}$ denotes the order statistic corresponding to $\mathbf{Z}$, it is clear that in this case, $W_i = V_i$ for $i \in B$. Thus one Z value is assigned v_8, while two take values in (v_8, ∞) independently. Due to independence of $\mathbf{Z}$ and X, the expectation in (3) does not depend on the particular value of x but on the indicator of the event $R(X) > 8$. Suppose now that x is decreased to v_8. Then $R(X) = 8$. This changes the conditional distribution of Z. If x now decreases further to the open interval corresponding to the next lower category, in the example this would be the interval (v_6, v_8), then there is a change in the conditional distribution, and again the expectation remains the same as long as x is in this interval. Note that

$$x \in (v_i, v_j) \Longleftrightarrow R(x) \in (i, j).$$

The important observation is that the change in the expectation occurs only when x moves from an open interval to the boundary or from the boundary to the open interval. In either case this change can be expressed entirely in terms of $R(X)$ which is an increasing function of X. Thus the conditional expectation, as a function of x depends only on $R(x)$. The assertion of the Theorem for the continuous case now follows from the Lemma.

(b) Suppose now the common distribution has a discrete part. Then the rank $R(x)$ does not determine whether x is in the open interval or at the boundary. We have to show that in both cases the monotonicity property holds. Proceeding as in (a), suppose that the conditioned value x is in the open interval corresponding to the highest category, or $x > v_8$ in our example. One Z observation has to be at v_8, while two can take values in $[v_8, \infty)$, which includes a three way tie at v_8. When x is decreased to v_8, the conditional distribution has now three Z observations in $[v_8, \infty)$, as opposed to two in the previous case. As in the case of continuous distribution, conditionally, the observations are still independent, however, there is a possibility of a four way tie at the boundary. In any case, one of the observations has become stochastically larger making the conditional distribution of $\mathbf{Z}$ stochastically larger. If x now decreases further to the open interval corresponding to the next lower category, in the example this would be the interval (v_6, v_8), then this would result in promotion of W_7 to v_8. Again this would make the conditional distribution stochastically larger still. Repeating this argument, the Theorem for the discrete case is established.

REFERENCE

Block, H.W., Bueno, V., Savits, T.H., and Shaked, M. (1987). Probability inequalities via negative dependence for random variables conditioned on order statistics. *Nav. Res. Log.* **37** 547–554.

Department of Statistics
University of Illinois
725 S. Wright
Champaign, IL 61820

A LINEAR COMBINATION TEST FOR DETECTING SERIAL CORRELATION IN MULTIVARIATE SAMPLES

By RICHARD A. JOHNSON[1] AND THORE LANGELAND[1]

University of Wisconsin and StatOil–Norway

If multivariate observations taken at adjacent times are correlated the quality of inferences, based on an independence assumption, can be seriously eroded. After illustrating these effects, we propose a new test for detecting dependence among adjacent observations. Our test statistic is the maximum absolute value of the lag 1 correlation obtainable from a linear combination of the observations. We express the statistic in terms of two eigenvalues and then obtain the asymptotic null distribution. Asymptotic power is examined for sequences of local alternatives in a multivariate normal autoregressive process. An explicit expression is obtained for the density of the limit distribution in the bivariate case. We then compare power with the likelihood ratio statistic.

1. Introduction. The presence of even a moderate autocorrelation, among univariate observations, can cause serious difficulties for procedures based on an assumption of independence. To illustrate, suppose normal observations are treated as independent but they actually follow a first order autoregressive (AR) model

$$X_t - \mu = \phi(X_{t-1} - \mu) + \varepsilon_t$$

where the ε_t are independent and identically distributed with mean 0 and variance σ_ε^2 and $|\phi| < 1$. It is well known that $\text{corr}(X_t, X_{t-1}) = \phi$ and $\sqrt{n}(\bar{X} - \mu)/s \xrightarrow{\mathcal{L}} N(0, (1+\phi)(1-\phi)^{-1})$. The coverage of the large sample nominal 95% confidence interval $\bar{X} \pm 1.96s/\sqrt{n}$ depends rather dramatically on ϕ.

Table 1. Coverage Probability of the Interval $\bar{X} \pm 1.96s/\sqrt{n}$

ϕ	-0.3	0	0.3	0.5	0.7
Coverage probability	.992	.950	.849	.742	.590

[1]Sponsored by the Air Force Office of Scientific Research under Grant AFOSR–87–0256.

AMS 1980 subject classifications. 62H20; 62F05.

Key words and phrases. Serial correlation; test of independence; linear combination.

In the context of $\bar{X}$ charts, the value $\bar{X} = \Sigma_{t=1}^{n} X_t/n$ is plotted on a chart with control limits $2s/\sqrt{n}$ where s is based on a large number of observations. The $\bar{X}$ chart will produce an excessive number of false signals under positive correlation.

Table 2. False Signal on $\bar{X}$ Chart

ϕ	-0.3	0	0.3	0.5	0.7
$P[\bar{X}$ outside 2-sigma limits]	.01	.05	.14	.25	.40

Johnson and Bagshaw (1974) have shown a similar deterioration occurs for the distribution of time to signal with *CUSUM* charts.

In the multivariate setting, both inferences about the mean μ, and covariance matrix, Σ, can be severely affected by serial correlation. Let the $k \times 1$ random vectors $\mathbf{X}_t$ follow the multivariate AR(1) model

$$\mathbf{X}_t - \mu = \Phi(\mathbf{X}_{t-1} - \mu) + \varepsilon_t \tag{1}$$

where the ε_t are independent and identically distributed with $E(\varepsilon_t) = \mathbf{0}$ and $\text{Cov}(\varepsilon_t) = \Sigma_\varepsilon$ and all of the eigenvalues of Φ are between -1 and 1. Under this model $\text{Cov}(\mathbf{X}_t, \mathbf{X}_{t-j}) = \Phi^j \Sigma_{\mathbf{X}}$, where

$$\Sigma_{\mathbf{X}} = \text{Cov}(\mathbf{X}_t) = \sum_{j=0}^{\infty} \Phi^j \Sigma_\varepsilon \Phi'^j$$

As a consequence of the ergodic theorem

$$\bar{\mathbf{X}} \overset{a.s.}{\to} \mu \text{ and } \mathbf{S} = \frac{1}{n-1}\sum_{t=1}^{n}(\mathbf{X}_t - \bar{\mathbf{X}})(\mathbf{X}_t - \bar{\mathbf{X}})' \overset{a.s.}{\to} \Sigma_{\mathbf{X}} \tag{2}$$

Also,

$$\text{Cov}(n^{-1/2}\sum_{t=1}^{n}\mathbf{X}_t) \overset{a.s.}{\to} (\mathbf{I} - \Phi)^{-1}\Sigma_{\mathbf{X}} + \Sigma_{\mathbf{X}}(\mathbf{I} - \Phi')^{-1} - \Sigma_{\mathbf{X}}$$

and $\sqrt{n}(\bar{\mathbf{X}} - \mu)$ is asymptotically normal with this limiting covariance matrix. Suppose the underlying process has $\Phi = \phi \mathbf{I}_k$, $|\phi| < 1$, then the nominal 95% large sample confidence ellipsoid

$$\{\mu : n(\bar{\mathbf{X}} - \mu)'\mathbf{S}^{-1}(\bar{\mathbf{X}} - \mu) \leq \chi_k^2(.05)\}$$

has coverage probability $P[\chi_k^2 \leq (1 - \phi)(1 + \phi)^{-1}\chi_k^2(.05)]$.

Table 3. Coverage Probability of the Nominal 95% Confidence Ellipsoid

		ϕ			
		-0.3	0	0.3	0.5
	2	.996	.950	.801	.632
k	5	.999	.950	.690	.405
	10	1.000	.950	.547	.193
	20	1.000	.950	.341	.041

In the context of principal component analysis, suppose we wish to analyze $\boldsymbol{\Sigma}_{\boldsymbol{\varepsilon}}$, which is the covariance matrix for $\mathbf{X}_t$ under independence, but that the AR(1) autocorrelation structure is introduced by selecting a sampling interval that is too short. The first principal component has coefficient vector $\mathbf{e}_1$ where $\boldsymbol{\Sigma}_{\boldsymbol{\varepsilon}}\mathbf{e}_1 = \lambda_1\mathbf{e}_1$ and $\boldsymbol{\Sigma}_{\boldsymbol{\varepsilon}}\mathbf{e}_i = \lambda_i\mathbf{e}_i$ with $\lambda_1 \geq \cdots \geq \lambda_k$. If the underlying process is an AR(1) process with $\Phi = c\boldsymbol{\Sigma}_{\boldsymbol{\varepsilon}}^{-1}$,

$$\boldsymbol{\Sigma}_{\mathbf{X}}\mathbf{e} = \frac{\lambda^3}{\lambda^2 - c^2}\mathbf{e}$$

so, if c is just smaller than λ_k, the ordering of the eigenvalues is reversed. That is, $\mathbf{e}_k$ is incorrectly identified as the coefficient of the first principal component.

The message is clear, a series of observations need to be checked for serial correlation.

Numerous tests have been proposed for the univariate case. The most common tests for independence among a collection of vectors $\{\mathbf{X}_t\}_{t=1}^T$ depend on the sample cross covariance matrix of lag j

(3) $$\mathbf{C}_j = \frac{1}{T}\sum_{t=1}^{T-j}(\mathbf{X}_t - \bar{\mathbf{X}})(\mathbf{X}_{t+j} - \bar{\mathbf{X}})' \text{ for } j = 1, 2, \cdots, T-1$$

where $\bar{\mathbf{X}} = T^{-1}\Sigma_{t=1}^T\mathbf{X}_t$. The likelihood ratio test is derived by considering the multivariate autoregressive process of order p

(4) $$\mathbf{X}_t = \Phi_1\mathbf{X}_{t-1} + \cdots + \boldsymbol{\Phi}_p\mathbf{X}_{t-p} + \boldsymbol{\theta} + \boldsymbol{\varepsilon}_t$$

where the $\boldsymbol{\varepsilon}_t$ are independent and identically distributed normal random vectors with mean $\mathbf{0}$, variance $\boldsymbol{\Sigma}_{\boldsymbol{\varepsilon}}$ and the roots of $|\mathbf{I} - \boldsymbol{\Phi}_1 \cdots - \boldsymbol{\Phi}_p| = 0$ lie outside the unit circle. The likelihood ratio test of $H_0 : [\boldsymbol{\Phi}_1, \cdots, \boldsymbol{\Phi}_p] = \mathbf{0}$ leads to the statistic that is asymptotically equivalent to

(5) $$S_L = -[T - p - 1 - \frac{1}{2}(kp + k + 1)]\log\left(\frac{|\mathbf{C}_0 - \hat{\Phi}\mathbf{C}(p)\hat{\Phi}'|}{|\mathbf{C}_0|}\right)$$

where $\hat{\Phi} = [\Phi_1, \cdots, \Phi_p]$ is the solution of the Yule-Walker equations

$$\mathbf{C}_m = \sum_{j=1}^{p} \hat{\Phi}_j \mathbf{C}_{m-j} \text{ for } m = 1, 2, \cdots, p$$

and

$$\mathbf{C}(p) = \begin{bmatrix} \mathbf{C}_0 & \mathbf{C}_1 & \mathbf{C}_2 & \cdots & \mathbf{C}_{p-1} \\ \mathbf{C}_1' & \mathbf{C}_0 & \mathbf{C}_1 & \cdots & \mathbf{C}_{p-2} \\ \vdots & \vdots & \vdots & & \vdots \\ \mathbf{C}_{p-1}' & \mathbf{C}_{p-2}' & \mathbf{C}_{p-3}' & \cdots & \mathbf{C}_0 \end{bmatrix}.$$

The test based on S_L is seemingly the most popular multivariate test.

Chitturi (1974) proposed testing the same hypothesis using the statistic

$$T \sum_{m=1}^{p} \sum_{u=1}^{k} \sum_{v=1}^{k} \hat{r}_{uv}(m) \hat{r}_{uv}(-m) \tag{6}$$

where $\hat{r}_{uv}(m)$ is the cross autocorrelation of lag m between the u-th and v-th components of $\mathbf{X}_t$.

The extension of Quennouille's test, due to Bartlett and Rajalaksham (1953), is based on the test statistic

$$\sum_{u=1}^{p} \text{tr} (\mathbf{G}_u \mathbf{G}_u')$$

where $\mathbf{G}_u = \mathbf{A}_0^{-1}(\Sigma_{j=0}^{p} \hat{\Phi}_j \mathbf{C}_{u-j}')\hat{\mathbf{B}}_0'$, $\Phi_0 = \mathbf{I}$ and $\hat{\mathbf{A}}_0$ and $\hat{\mathbf{B}}_0$ are given by

$$\hat{\mathbf{A}}_0'\hat{\mathbf{A}}_0 = (\mathbf{C}_0 - \mathbf{C}_1'\mathbf{C}_0^{-1}\mathbf{C}_1 - \cdots - \mathbf{C}_p'\mathbf{C}_0^{-1}\mathbf{C}_p)^{-1}$$
$$\hat{\mathbf{B}}_0'\hat{\mathbf{B}}_0 = (\mathbf{C}_0 - \mathbf{C}_1\mathbf{C}_0^{-1}\mathbf{C}_1' - \cdots - \mathbf{C}_p\mathbf{C}_0^{-1}\mathbf{C}_p')^{-1}.$$

Legget (1977) proposed a multivariate extension of the Bartlett periodogram test.

While this collection of tests, generalized from the univariate case, may be adequate for testing for serial dependence, we found it useful to take an alternative approach. In the next section, we introduce a statistic that concentrates the first order serial correlation into a single linear combination.

2. A Linear Combination Test. Because first order autocorrelation is most common, it is worthwhile to develop a test for first order correlation that is both easy to apply and has a graphic interpretation. We reduce the problem to one dimension by considering linear combinations $\mathbf{a}'\mathbf{X}_t$, $t = 1, 2, \ldots, T$ and selecting $\mathbf{a}$ to maximize the lag 1 correlation

$$r_{\mathbf{a}}(1) = \frac{\Sigma_{t=1}^{T-1} \mathbf{a}'(\mathbf{X}_t - \bar{\mathbf{X}})(\mathbf{X}_{t+1} - \bar{\mathbf{X}})'\mathbf{a}}{\Sigma_{t=1}^{T} [\mathbf{a}'(\mathbf{X}_t - \bar{\mathbf{X}})]^2} = \frac{\mathbf{a}'\mathbf{C}_1\mathbf{a}}{\mathbf{a}'\mathbf{C}_0\mathbf{a}} \tag{7}$$

Our test statistic is then defined as the maximum attainable lag 1 correlation,

$$R_L = \sup_{\mathbf{a} \neq \mathbf{0}} |r_{\mathbf{a}}(1)|.$$

Setting $\mathbf{C}_s = 2^{-1}(\mathbf{C}_1 + \mathbf{C}_1')$, $r_{\mathbf{a}}(1)$ can be expressed in terms of symmetric matrices as

$$(8) \qquad R_L = \sup_{\mathbf{a} \neq \mathbf{o}} \frac{|\mathbf{a}'\mathbf{C}_s\mathbf{a}|}{\mathbf{a}'\mathbf{C}_0\mathbf{a}} = \max\{|\hat{\lambda}_1|, \hat{\lambda}_k\}$$

where $\hat{\lambda}_1 < \hat{\lambda}_2 < \cdots < \hat{\lambda}_k$ are the eigenvalues of $\mathbf{C}_0^{-1/2}\mathbf{C}_s\mathbf{C}_0^{-1/2}$ or $\mathbf{C}_0^{-1}\mathbf{C}_s$. One point of difficulty is that $\mathbf{C}_s$ is not necessarily non-negative definite.

Note that R_L has the properties

(i)

$$R_L > |r_i(1)|, \; r_i(1) = \frac{\Sigma_{t=1}^{T-1}(X_{ti} - \bar{X}_i)(X_{t+1,i} - \bar{X}_i)}{\Sigma_{t=1}^{T}(X_{ti} - \bar{X}_i)^2}.$$

(ii) R_L is invariant under

$$\mathbf{X}_t \rightarrow \mathbf{A}\mathbf{X}_t\mathbf{Q}$$

where $\mathbf{A}$ is non-singular and $\mathbf{Q}$ orthogonal.

A plot of $\hat{\mathbf{a}}'(\mathbf{X}_t - \bar{\mathbf{X}})$ versus $\hat{\mathbf{a}}'(\mathbf{X}_{t+1} - \bar{\mathbf{X}})$ displays the concentrated correlation estimated by R_L.

We now indicate the steps leading to the asymptotic null distribution for R_L leaving the more technical algebraic steps until Section 5. We say that the $k \times k$ matrix $\mathbf{B}$ is $N_{k^2}(0, \boldsymbol{\Sigma} \otimes \boldsymbol{\Sigma}^{-1})$ if $\text{tr}(\mathbf{A}'\mathbf{B})$ is $N(0, \text{tr}\mathbf{A}\boldsymbol{\Sigma}\mathbf{A}'\boldsymbol{\Sigma}^{-1})$ for every $k \times k$ matrix $\mathbf{A}$. Mann and Wald (1943) showed that

$$T^{1/2}\mathbf{C}_0^{-1}\mathbf{C}_1 \xrightarrow{\mathcal{L}} N_{k^2}(0, \boldsymbol{\Sigma}_{\varepsilon} \otimes \boldsymbol{\Sigma}_{\varepsilon}^{-1})$$

so $T^{1/2}\mathbf{C}_0^{-1/2}\mathbf{C}_s\mathbf{C}_0^{\;1/2} \xrightarrow{\mathcal{L}} \mathbf{S}$ where, under the null hypothesis, $\mathbf{S}$ has pdf

$$(9) \qquad f(\mathbf{S}) = \frac{1}{(2\pi)^{k(k+1)/4}} \cdot 2^{k(k-1)/4} \text{etr}(-\frac{1}{2}\mathbf{S}\mathbf{S}'),$$

with respect to $k(k+1)/2$ dimensional Lebesgue measure.

Hsu (1939) encountered the same asymptotic distribution while studying a normal theory one-way MANOVA problem. He established that, if S is distributed as (9), the distribution of its eigenvalues $\lambda_1 < \cdots < \lambda_k$ has pdf

$$(10) \qquad g(\lambda_1, \lambda_2, \ldots, \lambda_k) = [2^{k/2}\Pi_{i=1}^{k}\Gamma(i/2)]^{-1}\Pi_{i<j}^{k}(\lambda_j - \lambda_i) \cdot e^{-\Sigma_{i=1}^{k}\lambda_i^2/2}.$$

Since $T^{1/2}R_L$ is a continuous function of $T^{1/2}\mathbf{C}_0^{-1/2}\mathbf{C}_s\mathbf{C}_0^{-1/2}$,

$$\sqrt{T}R_L \xrightarrow{\mathcal{L}} \max(|\lambda_1|, \hat{\lambda}_k). \tag{11}$$

For $k = 2$, the limit distribution is easy to evaluate

$$P[T^{1/2}R_L < x] \to P[-x < \Lambda_1 < \Lambda_2 < x] = \sqrt{2}\int_{-x}^{x} ue^{-u^2/2}\Phi(u)du = F(x). \tag{12}$$

It is considerably more difficult to present expressions for the general case. Set

$$G_j(t) = \int_{-x}^{t} u^j e^{-u^2/2}du, \quad j = 0, 1, 2, \ldots, k, \tag{13}$$

$$G_{j,\ell}(x) = \int_{-x}^{x} G_j(t)t^\ell e^{-t^2/2}dt, \quad 0 < j,\ \ell < k \tag{14}$$

where it can be shown (see Mehta (1960), p. 399, eqn. (13))

$$G_{j,\ell}(x) = (-1)^{\ell+j}G_{\ell,j}(x). \tag{15}$$

In Section 5, we establish

THEOREM 2.1. *For k even, the asymptotic cdf of the LCT statistic $T^{1/2}R_L$, under the null hypothesis of independence, is*

$$F(x) = (\Pi_{i=1}^k \Gamma(i/2))^{-1}\det(\{G_{j,\ell}(x)\})$$

for $j = 0, 2, 4, \ldots, k-2$ and $\ell = 1, 3, 5, \ldots, k-1$, where $G_{j,\ell}(x)$ is defined in (14).

THEOREM 2.2. *For k odd, the asymptotic cdf of the LCT statistic $T^{1/2}R_L$, under the null hypothesis of independence, is*

$$F(x) = [2^{1/2}\Pi_{i=1}^k\Gamma(i/2)]^{-1}\Sigma_{j=0}^{(k-1)/2}(-1)^{(k-1)/2+j}G_{2j}(x)\det(\mathbf{B}_j)$$

where $G_j(x)$ is defined in (13),

$$\mathbf{B}_j = \begin{bmatrix} G_{0,1}(x) & G_{0,3}(x) & \cdots & G_{0,k-2}(x) \\ G_{2,1}(x) & G_{2,3}(x) & \cdots & G_{2,k-2}(x) \\ \vdots & \vdots & & \vdots \\ G_{2j-2,1}(x) & G_{2j-2,3}(x) & \cdots & G_{2j-2,k-2}(x) \\ G_{2j+2,1}(x) & G_{2j+2,3}(x) & \cdots & G_{2j+2,k-2}(x) \\ \vdots & \vdots & & \vdots \\ G_{k-1,1}(x) & G_{k-1,3}(x) & \cdots & G_{k-1,k-2}(x) \end{bmatrix}$$

for $j = 0, 1, 2, \ldots, (k-1)/2$, and $G_{j,\ell}(x)$ is defined in (14).

A table of 1-st, 5-th, and 10-th percentiles, for $k = 2(1)20$ were calculated using double precision arithmetic (see Langeland (1980)).

3. Some Competing Tests and Power Considerations. Most tests for independence are motivated from consideration of autoregressive alternatives. Let

$$\mathbf{X}_t - \boldsymbol{\mu} = \boldsymbol{\Phi}(\mathbf{X}_{t-1} - \boldsymbol{\mu}) + \boldsymbol{\varepsilon}_t$$

for $t = 1, 2, \ldots, T$. The hypothesis of independence is then

$$H : \boldsymbol{\Phi} = \mathbf{0}. \tag{16}$$

A natural test statistic to use is

$$S_L = -[N - \frac{1}{2}(k + k + 1)] \log[|\mathbf{C}_0 - \hat{\boldsymbol{\Phi}}\mathbf{C}_0\hat{\boldsymbol{\Phi}}'|/|\mathbf{C}_0|) \tag{17}$$

where $N = T - 1 - 1$ and $\hat{\boldsymbol{\Phi}} = \mathbf{C}_1\mathbf{C}_0^{-1}$. If the $\{\varepsilon_t\}_{t=1}^{T}$ are i.i.d. multivariate normal, then the test statistic in (17) has the same asymptotic distribution as the logarithm of the likelihood ratio test statistic. See Hannan (1970, pp. 338–341).

THEOREM 3.1. *Under the null hypothesis of independence (16), the asymptotic distribution of the test statistic (17) is a $\chi^2_{k^2}$-distribution.*

In order to obtain an indication of asymptotic power, we introduce the normal theory AR(1) model (16) where the ε_t are independent $N(\mathbf{0}, \boldsymbol{\Sigma}_\varepsilon)$. Let $\{\boldsymbol{\Phi}_T\}$ be a sequence of alternatives to independence, where $T^{1/2}\boldsymbol{\Phi}_T \to \mathbf{H}$, and let $P_{T,\boldsymbol{\Phi}_T}$ denote the distribution of $\mathbf{X}_1, \ldots, \mathbf{X}_T$. Let P_T be the distribution of $\mathbf{X}_1, \ldots, \mathbf{X}_T$ under independence.

THEOREM 3.2. *Under $\{P_T\}$*

$$\Lambda_T = \ln \frac{dP_{T,\boldsymbol{\Phi}_T}}{dP_T} = tr[\boldsymbol{\Sigma}_\varepsilon^{-1}T^{1/2}\boldsymbol{\Phi}_T T^{1/2}\mathbf{C}_1] - \frac{1}{2}tr[\boldsymbol{\Sigma}_\varepsilon T^{1/2}\boldsymbol{\Phi}_T\mathbf{C}_0 T^{1/2}\boldsymbol{\Phi}_T'] + o_{P_n}(1)$$
$$\xrightarrow{\mathcal{L}} N(-\frac{1}{2}\sigma^2, \sigma^2)$$

so $\{P_T\}$ and $\{P_{T,\boldsymbol{\Phi}_T}\}$ are contiguous.

It can then be shown that $(\Lambda_T, T^{1/2}\mathbf{C}_0{}^{1/2}\mathbf{C}_s\mathbf{C}_0{}^{-1/2})$ is asymptotically normal under P_T so that we can obtain the limiting distribution of the linear combination statistic, R_L, under $\{P_{T,\boldsymbol{\Phi}_T}\}$. Even the bivariate case is complicated. The limit distribution for $T^{1/2}R_L$ is

$$f(x) = 4e^{-(\lambda+\eta)/2}\sum_{j=0}^{\infty}\frac{(\lambda/2)^j}{(j!)^2}\sum_{i=0}^{\infty}\frac{(n/2)^i}{i!}\frac{x^{2(j+i+1)}}{\Gamma(\frac{2i+1}{2})}$$
$$\cdot\int_0^1(1-u)^{2j+1}u^{2i}e^{-x^2[u^2+(1-u)^2]}du$$

for $x > 0$, where $\eta = (\mu_1 + \mu_3)^2/2$, $\lambda = [(\mu_1 - \mu_3)^2 + 4\mu_2^2]/2$ and

$$\boldsymbol{\mu} = \begin{bmatrix} \mu_1 \\ \mu_2 \\ \mu_3 \end{bmatrix} = \begin{bmatrix} 1 & 0 & 0 & 0 \\ 0 & 1/2 & 1/2 & 0 \\ 0 & 0 & 0 & 1 \end{bmatrix} \quad (\boldsymbol{\Sigma}_{\mathcal{E}}^{1/2} \otimes \boldsymbol{\Sigma}_{\mathcal{E}}^{-1/2})\text{vec}(\boldsymbol{\Phi}).$$

It also follows directly that (Λ_T, S_L) are each jointly normal under $\{P_T\}$. From the contiguity, we then obtain

THEOREM 3.3. *Under $\{P_{T,\boldsymbol{\Phi}_T}\}$, the asymptotic distribution of S_L is noncentral $\chi^2_{k^2}$ with noncentrality parameter $tr[\boldsymbol{\Sigma}_{\mathcal{E}}^{-1}\mathbf{H}\boldsymbol{\Sigma}_{\mathcal{E}}\mathbf{H}']$.*

It is well-known that the likelihood ratio test has several large sample optimal properties. However, a calculation of asymptotic power in Table 3.1 with $k = 2$, $\boldsymbol{\Sigma}_{\mathcal{E}} = \mathbf{I}$ shows that the linear combination test has higher power than the others when $T^{1/2}\boldsymbol{\Phi}_T \to \text{diag}(h_{11}, 0)$. In the other cases considered, where $\mathbf{H}$ is of full rank or $\boldsymbol{\Sigma}_{\mathcal{E}}$ is not proportional to $\mathbf{I}$, the likelihood ratio test has higher power. The superiority of the likelihood ratio test prevailed in a number of other cases that are not given in Table 3.1.

4. Example. We consider some data reported by Simon (see Duncan (1959), pp. 626–630) consisting of burning times of 30 fuses as recorded by three observers. Since there is one missing observation for the second observer, we first confine ourselves to the data given by observers one and three. Let $\mathbf{X}_t = (X_{t,1}, X_{t,2})'$, $t = 1, 2, \ldots, 30$ denote the observations. The plot of $X_{t,i}$ versus $X_{t+1,i}$ for $i = 1$ is given below in Figure 4.1. The plot for $i = 2$ is similar. Neither exhibits clear signs of first order serial dependence. The LCT statistic $\sqrt{30}R_L = 2.40$ and it is significant at the 10 percent level. The value of the corresponding eigenvector is $\hat{\mathbf{a}} = (1.0, -.99)'$. The plot of $\hat{\mathbf{a}}'\mathbf{X}_t$ versus $\hat{\mathbf{a}}'\mathbf{X}_{t+1}$ given in Figure 4.2 gives an indication of serial dependence in the two series of data. If the missing observation is estimated, the evidence for dependence with three observers is much stronger. The statistic becomes significant at the 3% level.

5. Derivation of Limiting Null Distribution. The asymptotic cdf of $T^{1/2}R_L$ is given by

$$(18) \quad F(x) = P[-x < \Lambda_1 < \Lambda_k < x] = \int_{Q(-x,x)} \cdots \int g(\lambda_1, \ldots, \lambda_k) d\lambda_1 \ldots d\lambda_k$$

where $g(\cdot)$ is defined in (10) and $Q(a, b) = \{a < \lambda_1 < \lambda_2 \cdots < \lambda_k < b\}$. Since

$$\Pi_{1<i<j<k}(\lambda_j - \lambda_i) = \det \begin{bmatrix} 1 & 1 & \ldots & 1 \\ \lambda_1 & \lambda_2 & \ldots & \lambda_k \\ \vdots & \vdots & & \vdots \\ \lambda_1^{k-1} & \lambda_2^{k-1} & \ldots & \lambda_k^{k-1} \end{bmatrix}$$

Table 3.1
Asymptotic Power

Σ_ε	**H**	R_L	S_L
$\begin{bmatrix} 1 & 0 \\ 0 & 1 \end{bmatrix}$	$\begin{bmatrix} 0.1 & 0 \\ 0 & 0 \end{bmatrix}$	.0513	0.505
$\begin{bmatrix} 1 & 0 \\ 0 & 1 \end{bmatrix}$	$\begin{bmatrix} 0.5 & 0 \\ 0 & 0 \end{bmatrix}$	.0849	.0627
$\begin{bmatrix} 1 & 0 \\ 0 & 1 \end{bmatrix}$	$\begin{bmatrix} 1 & 0 \\ 0 & 0 \end{bmatrix}$	.1769	.1055
$\begin{bmatrix} 1 & 0 \\ 0 & 1 \end{bmatrix}$	$\begin{bmatrix} 2 & 0 \\ 0 & 0 \end{bmatrix}$	.4666	.3201
$\begin{bmatrix} 1 & 0 \\ 0 & 1 \end{bmatrix}$	$\begin{bmatrix} 3 & 0 \\ 0 & 0 \end{bmatrix}$	.7714	.6635
$\begin{bmatrix} 1 & 0 \\ 0 & 1 \end{bmatrix}$	$\begin{bmatrix} 5 & 0 \\ 0 & 0 \end{bmatrix}$	.9952	.9894
$\begin{bmatrix} 1 & 0 \\ 0 & 1 \end{bmatrix}$	$\begin{bmatrix} 0.4 & -0.2 \\ 0.2 & 0.4 \end{bmatrix}$	.0731	.0707
$\begin{bmatrix} 1 & 0 \\ 0 & 1 \end{bmatrix}$	$\begin{bmatrix} 2 & -1 \\ 1 & 2 \end{bmatrix}$	.6338	.7160
$\begin{bmatrix} 1 & 0.5 \\ 0.5 & 1 \end{bmatrix}$	$\begin{bmatrix} 2 & -1 \\ 1 & 2 \end{bmatrix}$	.6890	.7763
$\begin{bmatrix} 1 & 0.5 \\ 0.5 & 1 \end{bmatrix}$	$\begin{bmatrix} 2 & 0 \\ 0 & 0 \end{bmatrix}$	.0834	.0956

Figure 4.1
PLOT OF DATA OF OBSERVER ONE VERSUS
THESE DATA LAGGED ONE UNIT

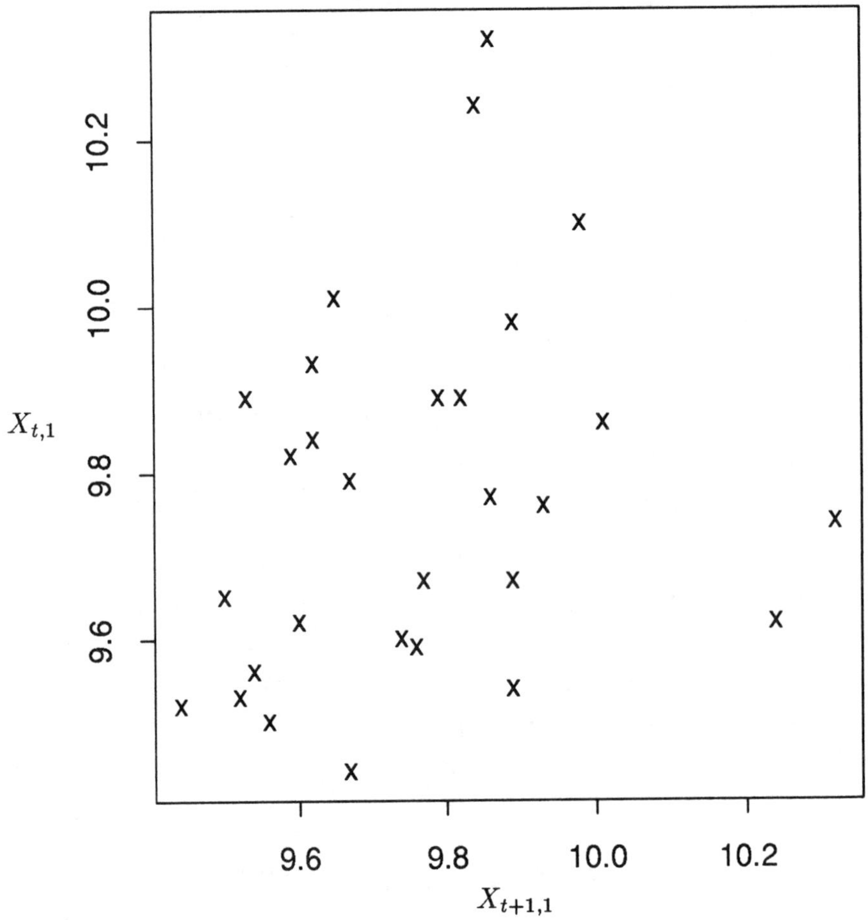

Figure 4.2
PLOT OF $\hat{\mathbf{a}}'\mathbf{X}_t$ VERSUS $\hat{\mathbf{a}}'\mathbf{X}_{t+1}$ FOR DATA FOR OBSERVERS ONE AND THREE ($Y = \hat{\mathbf{a}}'\mathbf{X}$).

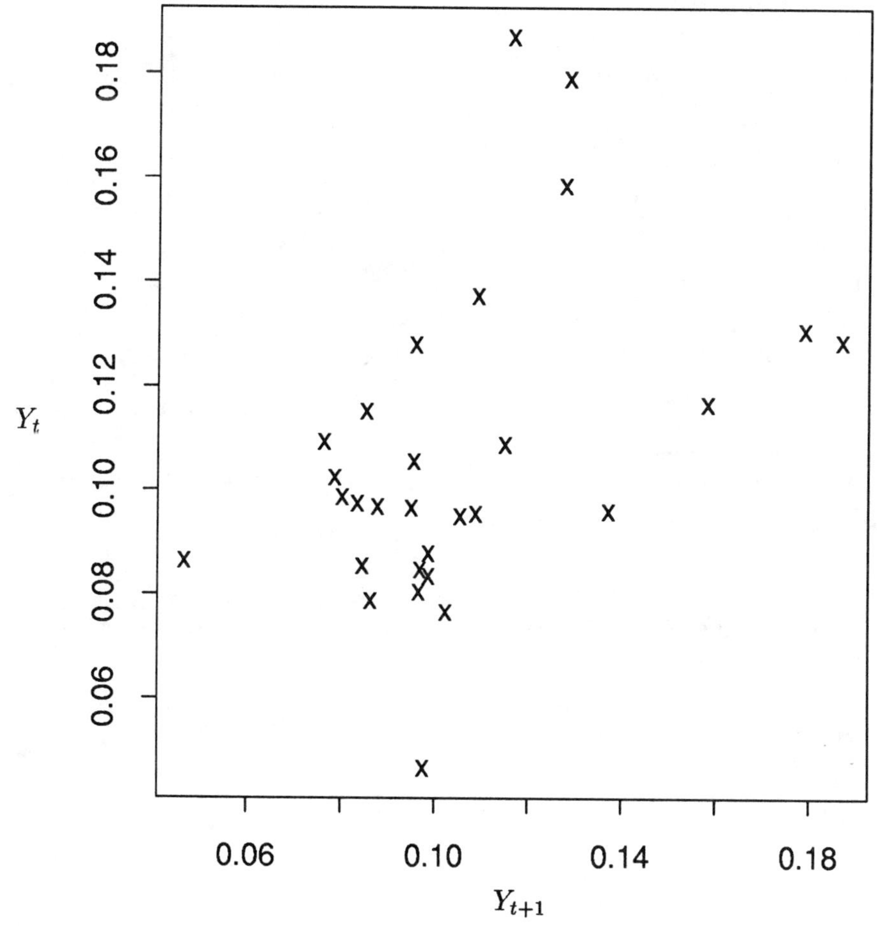

(the Vandermonde determinant), (18) can be rewritten as

$$\int_{Q(-x,x)} \cdots \int c_k \det \begin{bmatrix} e^{-\lambda_1^2/2} & e^{-\lambda_2^2/2} & \cdots & e^{-\lambda_k^2/2} \\ \lambda_1 e^{-\lambda_1^2/2} & \lambda_2 e^{-\lambda_2^2/2} & \cdots & \lambda_k e^{-\lambda_k^2/2} \\ \vdots & \vdots & & \vdots \\ \lambda_1^{k-1} e^{-\lambda_1^2/2} & \lambda_2^{k-1} e^{-\lambda_2^2/2} & \cdots & \lambda_k^{k-1} e^{-\lambda_k^2/2} \end{bmatrix}$$

$$\cdot d\lambda_1 \cdot d\lambda_2 \cdots d\lambda_k$$

where $c_k = [2^{k/2}\Sigma_{j=1}^k \Gamma(j/2)]^{-1}$.

In order to obtain an explicit expression for the densities we need some additional concepts and lemmas (see Aitken (1939), pp. 50 and 111).

The *signature function* $E(x_1, x_2, \ldots, x_k)$ is defined as

$$E(x_1, x_2, \ldots, x_k) = \Pi_{1<i<j<k} \operatorname{sign}(x_j - x_i) \tag{19}$$

for $x = (x_1, x_2, \ldots, x_k)' \in R^k$, $E(x_1, x_2, \ldots, x_k) = 0$ if $x_i = x_j$ for some $i \neq j$, $i, j = 1, 2, \ldots, k$, and $E(x_1) = 1$ for all $x_1 \in R$.

Let $k = 2m$ and $m = 1, 2, \ldots$, and let $\mathbf{A} = \{a_{ij}\}$ be a skew $(k \times k)$ matrix, then the *Pfaffian* of $\mathbf{A}$, $Pf(\mathbf{A})$, is defined as

$$\begin{aligned} Pf(\mathbf{A}) &= (2^m m!)^{-1} \Sigma_{j_1=1}^k \Sigma_{j_2=1}^k \cdots \Sigma_{j_k=1}^k E(j_1, j_2, \ldots, j_k) \\ &\cdot \quad a_{j_1 j_2} \cdot a_{j_3 j_4} \cdots a_{j_{k-1} j_k}. \end{aligned}$$

It is well-known that $[Pf(\mathbf{A})]^2 = \det \mathbf{A}$.

de Bruijn (1955) has established the following expression for k even.

LEMMA 5.1. *Assume* $\det(\{\phi_j(x_j)\}) \in L(R^k)$ *and let* $k = 2m$ *and* $m = 1, 2, \ldots$, *then*

$$\begin{aligned} &\int_{Q(a,b)} \cdots \int \det(\{\phi_j(x_i)\}) dx_1 dx_2 \cdots dx_k \\ &= Pf(\{a_{ij} = \int_a^b \int_a^b \phi_i(x)\phi_j(y) sign(y - x) dx dy\}). \end{aligned} \tag{20}$$

REMARK. de Bruijn (1955) gives a somewhat unusual definition of the Pfaffian and his derivation of the integral on the left-hand side of (20), for k odd, is only valid in a very special case. However, Krishnaiah and Chang (1971, equation 2.6) give a general solution to the odd case. In their notation $\phi_j(x) = x^{r+j-1}\psi(x)$ for $r > 0$ and some function $\psi(x)$ satisfying the integrability conditions. We restate their results as Lemma 5.2 (an alternative proof is given in Langeland (1980)).

LEMMA 5.2. *Assume* $\det(\{\phi_j(x_i)\}) \in L(R^k)$ *and let* k *be odd, then*

$$\int_{Q(a,b)} \cdots \int \det(\{\phi_j(x_i)\}) dx_1 dx_2 \cdots dx_k = \Sigma_{j=1}^{k} (-1)^{j-1} \psi_j(b) Pf(\mathbf{A}_j)$$

where

$$\psi_j(b) = \int_a^b \phi_j(t)dt \text{ for } j = 1, 2, \ldots, k,$$

and

$$\mathbf{A}_j = \begin{bmatrix} 0 & a_{12} & \cdots & a_{1,j-1} & a_{1,j+1} & \cdots & a_{1,k} \\ a_{21} & 0 & \cdots & a_{2,j-1} & a_{2,j+1} & \cdots & a_{2,k} \\ \vdots & \vdots & & \vdots & \vdots & & \vdots \\ a_{j-1,1} & a_{j-1,2} & \cdots & 0 & a_{j-1,j+1} & \cdots & a_{j-1,k} \\ a_{j+1,1} & a_{j+1,2} & \cdots & a_{j+1,j-1} & 0 & \cdots & a_{j+1,k} \\ \vdots & \vdots & & \vdots & \vdots & & \vdots \\ a_{k,1} & a_{k,2} & \cdots & a_{k,j-1} & a_{k,j+1} & \cdots & 0 \end{bmatrix}$$

for $j = 1, 2, \ldots, k$, *and* $Q(a, b)$ *and* a_{ij} *are as in Lemma 5.1.*

We can now establish Theorem 2.1.

PROOF OF THEOREM 2.1. First we notice that

$$\begin{aligned} &\int_{-x}^{x} [\int_{-x}^{x} u^j e^{-u^2/2} t^\ell e^{-t^2/2} \text{sign}(t-u) du] dt \\ &= \int_{-x}^{x} t^\ell e^{-t^2/2} [\int_{-x}^{t} u^j e^{-u^2/2} du - \int_{t}^{x} u^j e^{-u^2/2} du] dt \\ &= G_{j,\ell}(x) - \int_{-x}^{x} t^\ell e^{-t^2/2} [\int_{t}^{x} u^j e^{-u^2/2} du] dt \\ &= G_{j,\ell}(x) - \int_{-x}^{x} u^j e^{-u^2/2} [\int_{-x}^{u} t^\ell e^{-t^2/2} dt] du \\ &= G_{j,\ell}(x) - G_{\ell,j}(x) \text{ for } 0 < j,\ \ell < k - \ell. \end{aligned}$$

By (15), the last quantity equals 0 or $\pm 2G_{j,\ell}(x)$. Lemma 5.1 then gives

$$F(x) = [2^{k/2} \Pi_{j=1}^{k} \Gamma(j/2)]^{-1} \tag{21}$$

$$\cdot Pf \begin{bmatrix} 0 & 2G_{0,1}(x) & 0 & \cdots & 2G_{0,k-1}(x) \\ 2G_{1,0}(x) & 0 & 2G_{1,2}(x) & \cdots & 0 \\ \vdots & \vdots & \vdots & & \vdots \\ 0 & 2G_{k-2,1}(x) & 0 & \cdots & 2G_{k-2,k-1}(x) \\ 2G_{k-1,0}(x) & 0 & 2G_{k-1,2}(x) & \cdots & 0 \end{bmatrix}$$

Let $k = 2m$, then, according to definition of the Pfaffian and the relation for signature functions

$$\begin{aligned} E(x_1, x_2, \ldots, x_k) &= (2^m m!)^{-1} \sum_{j_1=1}^{k} \sum_{j_2=1}^{k} \cdots \sum_{j_k=1}^{k} E(j_1, j_2, \ldots, j_k) \\ &\quad \cdot E(x_{j_1} x_{j_2}) \cdot E(x_{j_3} x_{j_4}) \cdots E(x_{j_{k-1}} x_{j_k}) \end{aligned}$$

established in de Bruijn (1955), the Pfaffian in (21) can be reduced to

$$\begin{aligned} &2^m \sum_{j_1=1}^{m} \sum_{j_2=1}^{m} \cdots \sum_{j_m=1}^{m} E(j_1, j_2, \ldots, j_m) \cdot \\ &\cdot G_{0,2j_1-1}(x)\, G_{2,2j_2-1}(x) \, \ldots \, G_{k-2,2j_m-1}(x). \end{aligned}$$

But this is nothing but 2^m times the determinant in Theorem 2.1. The proof is complete.

PROOF OF THEOREM 2.2.

$$F(x) = [2^{k/2} \Sigma_{i=1}^{k} \Gamma(i/2)]^{-1} \Sigma_{j=0}^{k-1} (-1)^j G_j(x) Pf(\mathbf{A}_j)$$

where $\mathbf{A}_j = \{a_{pq}\}$ is a $(k-1) \times (k-1)$ matrix with entries $a_{pq} = G_{p,q} - G_{q,p}$ for $p, q = 0, 1, \ldots, j-1, j+1, \ldots, k-1$. Next, by (13)

$$G_j(x) = 0$$

for j odd. (It can also be shown that $Pf(\mathbf{A}_j) = 0$ for j odd.) According to (15), for j even, $\mathbf{A}_j$ is

$$\mathbf{A}_j = 2^{(k-1)} \left[\begin{array}{ccccc} 0 & G_{0,1}(x) & 0 & \ldots & G_{0,j+1}(x) \\ G_{1,0}(x) & 0 & G_{1,2}(x) & \ldots & 0 \\ \vdots & \vdots & \vdots & & \vdots \\ G_{j-1,0}(x) & 0 & G_{j-1,2}(x) & \ldots & 0 \\ G_{j+1,0}(x) & 0 & G_{j+1,2}(x) & \ldots & 0 \\ \vdots & \vdots & \vdots & & \vdots \\ G_{k-2,0}(x) & 0 & G_{k-2,2}(x) & \ldots & 0 \\ 0 & G_{k-1,1}(x) & 0 & & G_{k-1,j-1}(x) \end{array} \right.$$

$$\left. \begin{array}{cccc} G_{0,j+1}(x) & \ldots & G_{0,k-2}(x) & 0 \\ 0 & \ldots & 0 & G_{0,k-1}(x) \\ \vdots & & \vdots & \vdots \\ 0 & \ldots & 0 & G_{j-1,k-1}(x) \\ 0 & \ldots & 0 & G_{j+1,k-1}(x) \\ \vdots & & \vdots & \vdots \\ 0 & \ldots & 0 & G_{k-2,k-1}(x) \\ G_{k-1,j+1}(x) & \ldots & G_{k-1,k-2}(x) & 0 \end{array} \right]$$

All entries containing j as a first or as a second index vanish, i.e., all $G_{\ell,j}(x)$ and $G_{j,\ell}(x)$ for $\ell = 1, 3, \ldots, k-2$ vanish. The remaining number of terms $G_{p,q}(x)$, with p even, is exactly $(k-1)/2$. Thus, the Pfaffian of A_j reduces to

$$2^{(k-1)/2} \Sigma_{j_1=1}^{m} \Sigma_{j_2=1}^{m} \cdots \Sigma_{j_m=1}^{m} E(j_1, j_2, \ldots, j_m) G_{0,2j_1-1}(x) \cdot \\ \cdot G_{2,2j_2-1}(x) \cdots G_{j-2,2j_{(j-2)/2}-1}(x) \cdot G_{j+2,2j_{(j+2)/2}-1}(x) \cdots G_{k-1,2j_m-1}(x)$$

where $m = (k-1)/2$. Except for a possible sign change this is nothing but $2^{(k-1)/2}$ times the determinant of the matrix $B_{(j/2)}$ appearing in the statement of Theorem 2.2. By inspection, the sign is given by $(-1)^{(k-1)/2+j}$. The proof is complete.

We remark that nuclear physicists (e.g. Mehta (1967), Wigner (1967)) are interested in distributions of the eigenvalues of **S**.

REFERENCES

AITKEN, A. (1939). *Determinants and Matrices.* Oliver and Boyd, Edinburg.

BARTLETT, S. and RAJALAKSHMAN, D.V. (1953). Goodness-of-fit tests for simultaneous autoregressive series. *J. Roy. Statist. Soc. B* **15** 107–124.

CHITTURI, R.V. (1974). Distribution of residuals autocorrelations in multiple autoregressive schemes. *J. Amer. Statist. Assoc.* **69** 928–934.

DUNCAN, A. (1959). *Quality Control and Industrial Statistics.* Irwin, Homewood, Ill.

DE BRUIJN, N. (1955). On some multiple integrals involving determinants. *J. Indian Math. Soc.* **19** 133–152.

HANNAN, E.J. (1970). *Multiple Time Series.* John Wiley and Sons, New York.

HSU, P.L. (1939). On the distribution of the roots of certain determental equations. *Ann. Eugen.* **9** 250–258.

JOHNSON, R.A. and BAGSHAW, M. (1974). The effect of serial correlation on the performance of cusum tests. *Technometrics* **16** 103–112.

KRISHNAIAH, P. and CHANG, T. (1971). On the exact distributions of the extreme roots of the Wishart and MANOVA matrix. *J. Multiv. Anal.* **1** 108–117.

LANGELAND, T. (1980). Tests for Dependence in Multivariate Observations. Ph.D. Thesis, University of Wisconsin.

LIGGET, W.S. (1977). A test for serial correlation in multivariate data. *Ann. Statist.* **5** 408–413.

MANN, H. and WALD, A. (1943). On the statistical treatment of linear stochastic difference equations. *Econometrika* **11** 173–220.

MEHTA, M. (1967). *Random Matrices and Statistical Theory of Energy Levels.* Academic Press, New York.

SRIVASTAVA, M.S. and KHATRI, C.G. (1979). *Introduction to Multivariate Statistics.* North Holland, New York.

WIGNER, E. (1967). Random matrices in physics. *SIAM Review* **9** 1–23.

DEPARTMENT OF STATISTICS
UNIVERSITY OF WISCONSIN
1210 W. DAYTON STREET
MADISON, WI 53706

STATOIL RESGEO FORUS
STAVANGER 4001
NORWAY

CALCULATING IMPROVED BOUNDS AND APPROXIMATIONS FOR SEQUENTIAL TESTING PROCEDURES

BY JAMES R. KENYON

University of Southern Maine

Recently developed product type bounds are utilized for calculating improvements of the expected stopping time, the variance of the stopping time and the power of various sequential testing procedures. These are presented for testing the mean of a normal distribution, but the results and techniques apply more generally. These bounds have been easily calculated to 5th order by taking advantage of the dependence structure. It should be noted that these improvements are most useful when the distribution for the stopping time has its probability mass shifted towards later stopping times. Testing procedures that are developed with an early stopping time are easily handled with lower order bounds such as 3rd order, which are faster to calculate. This will be demonstrated in the examples.

1. Introduction. Recently, there has been interest in obtaining improvements in bounds for multivariate probabilities (Games (1977), Glaz and Johnson (1984), Miller (1981), and Worsley (1982)). It has been shown that first order bounds such as the usual Bonferroni bound are often not sharp enough to be useful when there are many events, B_i, and the $P(B_i)$ are not "small", or when there is a strong dependence structure in the multivariate distribution (Glaz and Johnson (1984), Miller (1981), Schwager (1984), and Worsley (1982)). Let $Y_1, Y_2, \ldots$ denote independent and identically distributed random variables with mean θ and variance σ^2. Define $\tau = \inf\{k \geq 1 \mid S_k \notin I_k\}$, where $S_k = \sum_{i=1}^{k} Y_i$ and I_k are intervals of the same type: $(-\infty, a_k)$, (b_k, ∞) or $(-c_k, c_k)$. The quantity τ is the stopping time of the sequential testing procedure that is specified by the intervals, I_k. Recall that:

$$\begin{aligned} P(\tau > n) &= P(S_k \in I_k;\ k = 1, \ldots, n), \\ E(\tau) &= \sum_{n=0}^{\infty} P(\tau > n), \end{aligned}$$

AMS 1980 subject classifications. Primary 62L10; secondary 65D30.

Key words and phrases. Probability inequalities, positive dependence, boundary crossing, stopping time, sequential test, normal distribution, Gaussian quadrature.

and

$$\text{Var}(\tau) = E(\tau)(1 - E(\tau)) + 2\sum_{n=1}^{\infty} nP(\tau > n).$$

Additionally, the power function, $\beta(\theta)$, is given by: $\beta(\theta) = P_\theta(S_\tau \in$ Rejection Region) $= \sum_{n=1}^{\infty} P_\theta\{(\cap_{k=1}^{n-1} S_k \in I_k) \cap (S_n \in$ Rejection Region$)\}$.

Power calculations will be performed under the assumption that crossing the boundary results in rejecting the null hypothesis. In the examples of Section 3, this occurs when the upper boundary is crossed.

The difficulty here is in calculating $P_\theta(S_k \in I_k;\ k = 1, \ldots, n)$, or the similar term for $\beta(\theta)$, particularly as n increases. By calculating improved bounds for these probabilities, improvement of bounds for $E(\tau)$, and approximations for $\text{Var}(\tau)$ and $\beta(\theta)$ are obtained.

We consider the following bounds to $\cup_{i=1}^{n} A_i$, where A_i are arbitrary events:

(1) first order upper Bonferroni bound

$$P(\cup_{i=1}^{n} A_i) \leq \sum_{i=1}^{n} P(A_i),$$

(2) second order lower Bonferroni bound

$$P(\cup_{i=1}^{n} A_i) \geq \sum_{i=1}^{n} P(A_i) - \sum_{i<j}^{n} P(A_i \cap A_j),$$

(3) Hunter's upper Bonferroni-type bound (A)

$$P(\cup_{i=1}^{n} A_i) \leq \sum_{i=1}^{n} P(A_i) - 2/n \sum_{i<j}^{n} P(A_i \cap A_j),$$

(4) Hunter's upper Bonferroni-type bound (B)

$$P(\cup_{i=1}^{n} A_i) \leq \sum_{i=1}^{n} P(A_i) - \sum_{i=1}^{n-1} P(A_i \cap A_{i+1}),$$

(5) Kwerel-Galambos lower bound (for $k \geq 2$)

$$P(\cup_{i=1}^{n} A_i) \geq 2/k \sum_{i=1}^{n} P(A_i) - 2/(k(k-1)) \sum_{i<j}^{n} P(A_i \cap A_j),$$

$$(\text{optimal } k = \ \text{INT}(2\sum_{i\leq j}^{n} P(A_i \cap A_j)/\sum_{i=1}^{n} P(A_i)) + 2,$$

where $\text{INT}(x)$ is the greatest integer $\leq x$),

(6) Sidak's first order product-type approximation

$$P(\cup_{i=1}^{n} A_i) \approx (\leq) 1 - \Pi_{i=1}^{n} P(A_i^c),$$

(7) Glaz-Johnson higher order product-type approximations

$$P(\cup_{i=1}^{n} A_i) \approx (\leq) 1 - P(\cap_{h=1}^{k-1} A_h^c) \Pi_{i=k}^{n} P(A_i^c \mid P(\cap_{j=1}^{k-1} A_{i-j}^c), \text{ for } k \in \{1, 2, \ldots, n\}$$

(Note $k = 1$ is Sidak's bound)

We note that the approximations (6) and (7) are bounds in the noted parenthetical directions under certain positive dependence conditions (see Glaz and Johnson (1984)).

Since bounds (1)–(6) did not perform as well as (7), for these and other types of problems, attention has been focused on the bounds of Glaz and Johnson (1984). Also, these bounds exploit the dependence structure of the partial sums. Thus, it is not unexpected that they performed more favorably here. More recently, Glaz (1990) has shown that the product type bounds are superior to the Bonferroni type bounds under positive dependence conditions which are present here.

As observed, the conditional approximations of Glaz and Johnson are not always guaranteed to be bounds. A sufficient, but not necessary condition for these approximations to be bounds, is that the multivariate distribution be multivariate totally positive of order 2, MTP_2, (Karlin and Rinott (1980)).

In some cases, the bounds for $E(\tau)$ and approximations for $\text{Var}(\tau)$ calculated by Glaz and Johnson (1986) for $k = 1, 2$, and 3, were not as good as one might desire. Here further conditioning, $k = 5$, will be employed to improve the bounds and approximations.

2. Calculation Methods. It is expected that evaluation of $P_\theta(S_k \in I_k)$ is available as well as the probability density function of S_k. The conditional probabilities above will be obtained from probability of the joint events and the probability of the conditioning event [e.g., $P(A \mid B) = P(A \cap B)/P(B)$]. For this problem, it should be noted that high dimensional numerical integration is not required. By conditioning on the middle term and using conditional independence we obtain the following formulas, which will be useful in calculating multivariate probabilities.

$$P(S_{k-1} \in I_{k-1}, S_k \in I_k) = \int_{I_{k-1}} P(S_k \in I_k \mid S_{k-1} = s) f_{S_{k-1}}(s) ds$$

$$P(S_{k-2} \in I_{k-2}, S_{k-1} \in I_{k-1}, S_k \in I_k) =$$

$$\int_{I_{k-1}} P(S_{k-2} \in I_{k-2} \mid S_{k-1} = s) f_{S_{k-1}}(s) P(S_k \in I_k \mid S_{k-1} = s) ds$$

$$P(S_{k-3} \in I_{k-3}, S_{k-2} \in I_{k-2}, S_{k-1} \in I_{k-1}, S_k \in I_k)$$

$$= \int_{I_{k-2}} P(S_{k-3} \in I_{k-3} \mid S_{k-2} = s) f_{S_{k-2}}(s) P(S_{k-1} \in I_{k-1}, S_k \in I_k \mid S_{k-2} = s) ds$$

$$= \int_{I_{k-2}} P(S_{k-3} \in I_{k-3} \mid S_{k-2} = s) f_{S_{k-2}}(s)$$

$$\times \int_{I_{k-1}} P(S_k \in I_k \mid S_{k-1} = t, S_{k-2} = s) f_{S_{k-1}|S_{k-2}=s}(t) dt$$

$$P(S_{k-4} \in I_{k-4}, S_{k-3} \in I_{k-3}, S_{k-2} \in I_{k-2}, S_{k-1} \in I_{k-1}, S_k \in I_k)$$

$$= \int_{I_{k-2}} P(S_{k-4} \in I_{k-4}, S_{k-3} \in I_{k-3} \mid S_{k-2} = s)$$

$$\times f_{S_{k-2}}(s) P(S_{k-1} \in I_{k-1}, S_k \in I_k \mid S_{k-2} = s) ds$$

$$= \int_{I_{k-2}} f_{S_{k-2}}(s) \int_{I_{k-4}} P(S_{k-3} \in I_{k-3} \mid S_{k-4} = t, S_{k-2} = s) f_{S_{k-4}|S_{k-2}=s}(t) dt$$

$$\times \int_{I_{k-1}} P(S_k \in I_k \mid S_{k-1} = u, S_{k-2} = s) f_{S_{k-1}|S_{k-2}=s}(u) du\, ds$$

The sequential probability ratio test (SPRT), the triangular boundary test and the asymptotic optimal Bayes sequential test are well known and utilized sequential tests for testing the mean of a normal distribution (e.g., Glaz and Johnson (1986)). These consider the case where the Y_i are independent, normal random variables with mean θ and variance $\sigma^2 = 1$ (or σ^2 known). Consequently, the S_k are normal with mean $k\theta$ and variance k (or $k\sigma^2$), but are not independent. $\mathrm{Cov}(S_k, S_h) = \min(k, h)\sigma^2$. For the multivariate normal distribution, MTP_2 is equivalent to all the partial correlations being ≥ 0 or equivalently $-(\Sigma)^{-1} \geq 0$, where Σ is the variance-covariance matrix (Karlin and Rinott (1980)). This condition is satisfied here (Glaz and Johnson (1986)).

Thus the probability density functions and the conditional distribution functions for the above are all normal with the appropriate mean and variance.

3. Examples for the Normal Distribution. For testing $H_o : \theta = 0$ vs $H_a : \theta = \theta_1 > 0$, consider $X_i = (Y_i - \theta_1/2)$. The transformed hypothesis is $H_o : \theta = -\theta_1/2$ vs $H_a : \theta = \theta_1/2$ and the intervals determining the test boundaries will be symmetric about 0 since it is usually desired that $\alpha = \beta$. The boundaries for the various test procedures are as follows:

Test	Intervals I_k (a, θ_1 such that $\alpha = \beta$)
Wald's SPRT	$(-a/\theta_1, a/\theta_1)$
Triangular	$(-a + k/4\theta_1, a - k/4\theta_1)$ $k \leq M < \infty$
Optimal Bayes	$(-c_k + k/2\theta_1, c_k - k/2\theta_1)$ $c_k = \sqrt{2ak}, a > 0, k \leq M < \infty$.

Now, consider the Gauss-Laguerre formulas:

$$\int_0^\infty e^{-x} f(x) dx \approx \sum_{i=1}^{n} B_i f(x_i)$$

and the Gauss-Legendre formulas:

$$\int_{-1}^{1} f(x)dx \approx \sum_{i=1}^{n} B_i[f(x_i) + f(-x_i)]$$
$$= \sum_{i=1}^{n} 2B_i f(x_i), \text{ when } f(x) \text{ is symmetric about } 0.$$

Since some variables of integration have semi-infinite limits, the Gaussian methods do have a natural advantage. Other methods require either truncation of the integral besides performing an approximation of this truncated integral, or usage of a transformation of the variable of integration which yields finite limits of integration. In addition, Stroud and Secrest (1966) give a comparison of Gaussian quadrature with other methods. For an equal number of points the Gaussian quadrature error is comparable to the other methods, even for cases where it is not believed the "best". They also compare different approaches to calculating some specific integrals with semi-infinite limits, including transformations to integrals with finite limits. All these results appear to indicate this approach for these particular densities. Additionally, the Gaussian methods do have some optimal properties for numerical integration.

These two numerical integration methods can be used for all three cases of I_k. This is accomplished as follows.

For $I_k = (-\infty, a_k)$:

$$\int_{-\infty}^{a_k} f(x)dx = \int_0^{\infty} f(a_k - s)e^{-s+s}ds \approx \sum_{i=1}^{n} B_i f(a_k - s_i)e^{s_i}$$

or

$$\int_{-\infty}^{a_k} f(x)e^{x}dx = \int_0^{\infty} f(a_k - s)e^{-s}e^{a_k}ds \approx e^{a_k}\sum_{i=1}^{n} B_i f(a_k - s_i),$$

or any appropriate transformation to obtain the desired integral in the form, $\int_0^{\infty} e^{-t}g(t)dt$.

For $I_k = (b_k, \infty)$:

$$\int_{b_k}^{\infty} f(x)dx = \int_0^{\infty} f(b_k + s)e^{-s+s}ds \approx \sum_{i=1}^{n} B_i f(b_k + s_i)e^{s_i}$$

or

$$\int_{b_k}^{\infty} f(x)dx = 1 - \int_{-\infty}^{b_k} f(x)dx$$

which then becomes the case of $I_k = (-\infty, a_k)$.

For $I_k = (-c_k, c_k)$:

$$\int_{-c_k}^{c_k} f(x)dx = c_k \int_{-1}^{1} f(s\, c_k)ds \approx c_k \sum_{i=1}^{n} B_i[f(s_i\, c_k) + f(-s_i\, c_k)].$$

We now apply these Gaussian integration formulas, which will be useful in calculating multivariate probabilities. Let $\Phi(\theta, \sigma^2, x)$ be the c.d.f. at x, and $\phi(\theta, \sigma^2, x)$ be the p.d.f. at x for a normal distribution with mean θ and variance σ^2. Then for $I_k = (-c_k, c_k)$:

$$\begin{aligned} P_\theta(S_k \in I_k) &= \Phi(k\theta, k, c_k) - \Phi(k\theta, k, -c_k), \\ P_\theta(S_{k-1} \in I_{k-1}, S_k \in I_k) &\approx c_{k-1} \sum_{i=1}^{n} B_i \\ &\times\ [\Phi(\theta + c_{k-1}s_i, 1, c_k) - \Phi(\theta + c_{k-1}s_i, 1, -c_k)]\phi((k-1)\theta, k-1, c_{k-1}s_i), \end{aligned}$$

$$\begin{aligned} P_\theta(S_{k-2} \in I_{k-2}, S_{k-1} \in I_{k-1}, S_k \in I_k) &\approx c_{k-1} \sum_{i=1}^{n} B_i \\ &\times\ [\Phi((c_{k-1}s_i)(k-2)/(k-1), (k-2)/(k-1), c_{k-2}) \\ &\quad -\Phi((c_{k-1}s_i)(k-2)/(k-1), (k-2)/(k-1), -c_{k-2})] \\ &\times\ \phi((k-1)\theta, k-1, c_{k-1}s_i)[\Phi(\theta + c_{k-1}s_i, 1, c_k) - \Phi(\theta + c_{k-1}s_i, 1, -c_k)], \end{aligned}$$

$$\begin{aligned} P_\theta(S_{k-3} \in I_{k-3}, S_{k-2} \in I_{k-2}, S_{k-1} \in I_{k-1}, S_k \in I_k) &\approx c_{k-2}c_{k-1} \sum_{i=1}^{n} B_i \\ &\times\ [\Phi((c_{k-2}s_i)(k-3)/(k-2), (k-3)/(k-2), c_{k-3}) \\ &\quad -\Phi((c_{k-2}s_i)(k-3)/(k-2), (k-3)/(k-2), -c_{k-3})] \\ &\quad \times\phi((k-2)\theta, k-2, c_{k-2}s_i) \\ &\times\ \sum_{j=1}^{m} B_j\phi(\theta + c_{k-2}s_i, 1, c_{k-1}t_j)[\Phi(\theta + c_{k-1}t_j, 1, c_k) \\ &\quad -\Phi(\theta + c_{k-1}t_j, 1, -c_k)], \end{aligned}$$

$$\begin{aligned} &P_\theta(S_{k-4} \in I_{k-4}, S_{k-3} \in I_{k-3}, S_{k-2} \in I_{k-2}, S_{k-1} \in I_{k-1}, S_k \in I_k) \\ &\approx\ c_{k-3}c_{k-2}c_{k-1} \sum_{i=1}^{n} B_i\phi((k-2)\theta, k-2, c_{k-2}s_i) \\ &\times\ \sum_{j=1}^{n} B_j\phi((c_{k-2}s_i)(k-3)/(k-2), (k-3)/(k-2), c_{k-3}t_j) \\ &\times\ [\Phi((c_{k-3}t_j)(k-4)/(k-3), (k-4)/(k-3), c_{k-4}) \\ &\quad -\Phi((c_{k-3}t_j)(k-4)/(k-3), (k-4)/(k-3), c_{k-4})] \end{aligned}$$

$$\times \sum_{h=1}^{m} B_h \phi(\theta + c_{k-2} s_i, 1, c_{k-1} u_h)[\Phi(\theta + c_{k-1} + u_h, 1, c_k) - \Phi(\theta + c_{k-1} + u_h, 1, -c_k)],$$

where m, and n are the number of points used in the Gauss-Legendre approximation.

For $I_k = (-\infty, a_k)$:

$$P_\theta(S_k \in I_k) = \Phi(k\theta, k, a_k),$$

$$P_\theta(S_{k-1} \in I_{k-1}, S_k \in I_k) \approx \sum_{i=1}^{n} B_i e^{s_i} \Phi(\theta + a_{k-1} - s_i, 1, a_{k-1}) \times \phi((k-1)\theta, k-1, a_{k-1} - s_i),$$

$$P_\theta(S_{k-2} \in I_{k-2}, S_{k-1} \in I_{k-1}, S_k \in I_k) \approx \sum_{i=1}^{n} B_i e^{s_i} \Phi((a_{k-1} - s_i)(k-2)/(k-1), (k-2)/(k-1), a_{k-2}) \times \phi((k-1)\theta, k-1, a_{k-1} - s_i) \Phi(\theta + a_{k-1} - s_i, 1, a_k),$$

$$P(S_{k-3} \in I_{k-3}, S_{k-2} \in I_{k-2}, S_{k-1} \in I_{k-1}, S_k \in I_k) \approx \sum_{i=1}^{n} B_i e^{s_i} \Phi((a_{k-2} - s_i)(k-3)/(k-2), (k-3)/(k-2), a_{k-3}) \times \phi((k-2)\theta, k-2, a_{k-2} - s_i) \times \sum_{j=1}^{m} B_j e^{t_j} \phi(\theta + a_{k-2} - s_i, 1, a_{k-1} - t_j) \Phi(\theta + a_{k-1} + t_j, 1, a_k),$$

$$P(S_{k-4} \in I_{k-4}, S_{k-3} \in I_{k-3}, S_{k-2} \in I_{k-2}, S_{k-1} \in I_{k-1}, S_k \in I_k) \approx \sum_{i=1}^{n} B_i e^{s_i} \phi((k-2)\theta, k-2, a_{k-2} - s_i) \times \sum_{j=1}^{m} B_j e^{t_j} \phi((a_{k-2} - s_i)(k-3)/(k-2), (k-3)/(k-2), a_{k-3} - t_j) \times \Phi((a_{k-3} - t_j)(k-4)/(k-3), (k-4)/(k-3), a_{k-4}) \times \sum_{h=1}^{m} B_h e^{u_h} \phi(\theta + a_{k-2} - s_i, 1, a_{k-1} - u_h) \Phi(\theta + a_{k-1} + u_h, 1, a_k),$$

where m, and n are the number of points used in the Gauss-Laguerre approximation.

The continuation region given by I_n is usually determined by selecting an alternative $\theta_1 > 0$ such that $\beta(0) = \alpha$, and $\beta(\theta_1) = 1 - \alpha$, where α is the desired significance level of the test. Notice that for this situation, P_0(Type I Error) $= P_{\theta_1}$(Type II Error). The power function, $\beta(\theta)$, for the transformed case with symmetric intervals using S_k and I_k is written as follows:

$$\beta(\theta) = \sum_{n=1}^{\infty} P_\theta \left(\cap_{k=1}^{n-1} (S_k \in I_k) \cap (S_n > c_n) \right).$$

This is similar to the evaluation of the $P(\tau > n)$ but, it should be noted that the important difference in this case is that the nth event is not $S_n \in I_n$, but $S_n > c_n$. This requires calculation of an additional term, $P(S_n > c_n \mid \text{prior } S_k^{'}s \in I_k)$. As performed previously, the necessary joint probabilities involving the event $S_n > c_n$ will be calculated and used in calculating an approximation for $\beta(\theta)$. The calculated result is not necessarily a bound because the intervals are not all of the same type (Glaz and Johnson (1986)). The formulas for calculating the approximation to $\beta(\theta)$ are similar to those given earlier.

The following tables illustrate the cases where higher bounds are providing a needed improvement for the tests considered. This is particularly noticeable for $E(\tau) > 30$ (i.e., for the distribution of τ having more mass for larger values of τ). For all examples used here, the approximation for the power of the test did not have much room for improvement above the 3rd order calculations. In each table, the upper value is approximation (7) with $k = 3$ from Glaz and Johnson (1986), the middle value is approximation (7) with $k = 5$, the lower value is from a simulation with 10,000 trials also from Glaz and Johnson (1986), while starred values are exact.

Table I (SPRT)

Lower Bounds and Simulated Values of $E_\theta(\tau)$

	$\theta_0 = 0.25$			$\theta_0 = 0.50$		
α/θ	-0.25	-0.125	0.0	-0.50	-0.25	0.0
0.010	31.39	46.22	59.46	9.05	14.13	19.04
	33.53	51.18	67.39	9.23	14.94	20.59
	36.16	61.31	84.90	9.23	15.25	21.54
0.025	24.70	34.49	41.53	7.13	10.47	13.01
	26.41	38.05	46.61	7.27	10.89	13.71
	28.19	43.32	54.83	7.27	11.03	13.98
0.050	19.51	25.77	29.53	5.62	7.70	8.99
	20.82	28.20	32.71	5.69	7.86	9.22
	22.32	30.58	36.20	5.72	7.99	9.25
0.100	14.12	17.39	19.02	4.02	5.02	5.53
	14.93	18.68	20.57	4.03	5.04	5.56
	15.36	19.49	21.72	4.02	5.10	5.62

Approximated and Simulated Values of $\sigma_\theta(\tau)$

	$\theta_0 = 0.25$			$\theta_0 = 0.50$		
α/θ	-0.25	-0.125	0.0	-0.50	-0.25	0.0
0.010	16.13	26.93	37.31	5.31	9.47	13.86
	18.51	32.27	45.78	5.56	10.65	15.92
	23.02	46.09	68.66	5.70	11.35	17.69
0.025	13.95	21.60	27.37	4.47	7.45	9.79
	16.03	25.73	33.18	4.74	8.13	10.82
	18.83	33.25	44.48	4.77	8.14	11.33
0.050	11.95	17.13	20.35	3.77	5.70	6.94
	13.68	20.17	24.25	3.90	5.99	7.32
	15.86	24.08	29.61	3.97	6.20	7.43
0.100	9.46	12.34	13.80	2.85	3.82	4.32
	10.64	14.12	15.91	2.88	3.87	4.38
	11.55	15.22	17.60	2.86	3.93	4.34

Approximated and Simulated Values of $\beta(\theta)$

	$\theta_0 = 0.25$			$\theta_0 = 0.50$		
α/θ	-0.25	-0.125	0.0	-0.50	-0.25	0.0
0.010	0.0099	0.0906	0.5000	0.0099	0.0905	0.5000
	0.0099	0.0908	0.5000	0.0099	0.0907	0.5000
	0.0097	0.0886	0.5000*	0.0103	0.0891	0.5000*
0.025	0.0244	0.1362	0.5000	0.0245	0.1361	0.5000
	0.0244	0.1364	0.5000	0.0244	0.1363	0.5000
	0.0219	0.1376	0.5000*	0.0230	0.1366	0.5000*
0.050	0.0474	0.1823	0.5000	0.0475	0.1823	0.5000
	0.0476	0.1826	0.5000	0.0476	0.1825	0.5000
	0.0445	0.1798	0.5000*	0.0488	0.1842	0.5000*
0.100	0.0906	0.2399	0.5000	0.0911	0.2401	0.5000
	0.0908	0.2401	0.5000	0.0912	0.2402	0.5000
	0.0884	0.2425	0.5000*	0.0938	0.2391	0.5000*

Table II (Triangular)

Lower Bounds and Simulated Values of $E_\theta(\tau)$

	$\theta_0 = 0.25$			$\theta_0 = 0.50$		
α/θ	-0.25	-0.125	0.0	-0.50	-0.25	0.0
0.010	39.11	51.67	59.82	10.52	14.04	16.19
	40.44	53.89	62.32	10.60	14.19	16.37
	41.67	56.54	64.88	10.61	14.16	16.48
0.025	29.62	37.66	42.11	8.01	10.22	11.39
	30.63	39.17	43.78	8.05	10.29	11.47
	31.46	40.71	45.15	8.03	10.34	11.51
0.050	22.29	27.17	29.53	6.05	7.35	7.97
	23.01	28.13	30.58	6.06	7.37	7.99
	23.26	28.88	31.38	6.09	7.41	8.03
0.100	14.75	16.99	17.94	4.01	4.59	4.83
	15.11	17.42	18.40	4.01	4.59	4.83
	15.29	17.60	18.08	4.01	4.58	4.80

Approximated and Simulated Values of $\sigma_\theta(\tau)$

	$\theta_0 = 0.25$			$\theta_0 = 0.50$		
α/θ	-0.25	-0.125	0.0	-0.50	-0.25	0.0
0.010	13.80	18.22	19.39	3.95	5.13	5.30
	14.97	19.57	20.47	4.05	5.23	5.38
	16.18	21.14	21.43	4.07	5.26	5.41
0.025	11.58	14.32	15.03	3.27	3.96	4.09
	12.48	15.25	15.83	3.32	4.01	4.13
	13.26	16.04	16.36	3.35	4.03	4.11
0.050	9.49	11.11	11.54	2.63	3.02	3.11
	10.12	11.72	12.09	2.65	3.04	3.12
	10.53	12.27	12.51	2.66	3.05	3.16
0.100	6.90	7.64	7.86	1.85	2.03	2.08
	7.21	7.94	8.14	1.85	2.03	2.08
	7.33	8.08	8.14	1.85	2.02	2.09

Approximated and Simulated Values of $\beta(\theta)$

	$\theta_0 = 0.25$			$\theta_0 = 0.50$		
α/θ	-0.25	-0.125	0.0	-0.50	-0.25	0.0
0.010	0.0071	0.1019	0.5000	0.0094	0.1156	0.5000
	0.0084	0.1101	0.5000	0.0099	0.1182	0.5000
	0.0103	0.1177	0.5000*	0.0089	0.1119	0.5000*
0.025	0.0205	0.1470	0.5000	0.0244	0.1583	0.5000
	0.0229	0.1542	0.5000	0.0249	0.1597	0.5000
	0.0238	0.1585	0.5000*	0.0236	0.1616	0.5000*
0.050	0.0450	0.9135	0.5000	0.0496	0.2016	0.5000
	0.0481	0.1991	0.5000	0.0500	0.2023	0.5000
	0.0496	0.2022	0.5000*	0.0507	0.2034	0.5000*
0.100	0.0964	0.2546	0.5000	0.1000	0.2584	0.5000
	0.0991	0.2576	0.5000	0.1001	0.2585	0.5000
	0.0959	0.2573	0.5000*	0.1033	0.2541	0.5000*

Table III (Bayes)

Lower Bounds and Simulated Values of $E_\theta(\tau)$

	$\theta_0 = 0.25$			$\theta_0 = 0.50$		
α/θ	-0.25	-0.125	0.0	-0.50	-0.25	0.0
0.010	33.86	51.48	65.65	9.61	14.48	17.95
	36.49	56.03	71.11	9.86	14.97	18.49
	40.02	62.54	79.90	9.89	15.10	18.78
0.025	26.69	38.75	47.03	7.60	10.82	12.78
	28.84	42.24	51.10	7.76	11.11	13.10
	31.51	47.50	56.30	7.89	11.07	13.20
0.050	21.12	29.13	33.91	5.98	8.02	9.10
	22.82	31.71	36.86	6.07	8.15	9.25
	24.30	34.62	40.36	6.09	8.12	9.33
0.100	15.32	19.70	21.93	4.22	5.21	5.67
	16.45	21.28	23.67	4.24	5.23	5.69
	17.36	22.50	25.13	4.24	5.23	5.69

Approximated and Simulated Values of $\sigma_\theta(\tau)$

	$\theta_0 = 0.25$			$\theta_0 = 0.50$		
α/θ	-0.25	-0.125	0.0	-0.50	-0.25	0.0
0.010	19.02	28.31	33.62	5.40	7.73	8.63
	21.05	31.08	35.99	5.66	8.03	8.85
	24.62	35.68	39.32	5.74	8.15	9.00
0.025	16.61	23.63	27.33	4.63	6.25	6.88
	18.41	25.92	29.42	4.80	6.45	7.03
	21.06	29.30	31.78	4.89	6.51	7.06
0.050	14.39	19.53	22.03	3.87	4.95	5.37
	15.93	21.38	23.80	3.97	5.06	5.46
	17.56	23.63	25.79	3.96	5.01	5.50
0.100	11.62	14.78	16.18	2.87	3.44	3.65
	12.77	16.09	17.49	2.89	3.46	3.68
	13.63	17.09	18.77	2.87	3.50	3.65

Approximated and Simulated Values of $\beta(\theta)$

	$\theta_0 = 0.25$			$\theta_0 = 0.50$		
α/θ	-0.25	-0.125	0.0	-0.50	-0.25	0.0
0.010	0.0083	0.0791	0.5000	0.0086	0.0954	0.5000
	0.0084	0.0841	0.5000	0.0089	0.0991	0.5000
	0.0086	0.0908	0.5000*	0.0079	0.1031	0.5000*
0.025	0.0208	0.1204	0.5000	0.0216	0.1371	0.5000
	0.0211	0.1251	0.5000	0.0222	0.1399	0.5000
	0.0207	0.1279	0.5000*	0.0204	0.1431	0.5000*
0.050	0.0413	0.1645	0.5000	0.0432	0.1805	0.5000
	0.0419	0.1684	0.5000	0.0438	0.1821	0.5000
	0.0471	0.1760	0.5000*	0.0445	0.1822	0.5000*
0.100	0.0816	0.2230	0.5000	0.0854	0.2379	0.5000
	0.0824	0.2287	0.5000	0.0865	0.2382	0.5000
	0.0790	0.2283	0.5000*	0.0854	0.2379	0.5000*

REFERENCES

ANDERSON, T.W. (1960). A modification of the sequential probability ratio test to reduce sample size. *Ann. Statist.* **31** 167–197.

BERK, R.H. (1982). On an asymptotically optimal sequential test. *Scand. J. Statist.* **9** 159–163.

DAVIS, P.J. and RABINOWITZ, P. (1975). *Methods of Numerical Integration.* Academic Press, New York.

GAMES, P.A. (1977). An improved t table for simultaneous control on g contrasts. *J. Amer. Statist. Assoc.* **72** 531–534.

GLAZ, J. and JOHNSON, B. MCK. (1984). Probability inequalities for multivariate distributions with dependence structures. *J. Amer. Statist. Assoc.* **79** 436–440.

GLAZ, J. and JOHNSON, B. MCK. (1986). Approximating boundary crossing probabilities with applications to sequential tests. *Sequential Anal.* **5** 37–42.

GLAZ, J. (1990). A comparison of Bonferroni-type and product-type inequalities in presence of dependence. In *Topics in Statistical Dependence* (H. Block, A.R. Sampson, and T. Savits, eds.), IMS Lecture Notes-Monograph Series, Hayward, CA.

KARLIN, S. and RINOTT, Y. (1980a). Classes of orderings of measures and related correlation inequalities. I: Multivariate total positive distributions. *J. Multivariate Anal.* **10** 467–498.

KARLIN, S. and RINOTT, Y. (1980b). Classes of orderings of measures and related correlation inequalities. II: Multivariate reverse rule distributions. *J. Multivariate Anal.* **10** 499–516.

MILLER, R.G. (1981). *Simultaneous Statistical Inference.* Springer-Verlag, New York.

SCHWAGER, S.J. (1984). Bonferroni sometimes loses. *Amer. Stat.* **38** 192–197.

SIDAK, Z. (1968). On multivariate normal probabilities of rectangles: Their dependence on correlations. *Ann. Statist.* **39** 1425–1434.

STROUD, A.H. (1971). *Approximate Calculation of Multiple Integrals.* Prentice-Hall, Inc., Englewood Cliffs, New Jersey.

STROUD, A.H. and SECREST, D. (1966). *Gaussian Quadrature Formulas.* Prentice-Hall, Inc., Englewood Cliffs, New Jersey.

WALD, A. (1947). *Sequential Analysis.* Dover Publications, Inc., New York, New York.

WHITEHEAD, J. and JONES, D. (1979). The analysis of sequential clinical trials. *Biometrika* **66** 443–452.

WHITEHEAD, J. (1983). *The Design and Analysis of Sequential Clinical Trials.* Ellis Horwood Limited, Chichester, England.

WOODROOFE, M. (1976). Frequentist properties of Bayesian sequential tests. *Biometrika* **63** 101–110.

WOODROOFE, M. (1982). *Nonlinear Renewal Theory in Sequential Analysis.* CBMS-NSF Regional Conference Series in Applied Mathematics, **39** SIAM, Philadelphia, PA.

WORSLEY, K.J. (1982). An improved Bonferroni inequality and applications. *Biometika* **69** 297–302.

DEPARTMENT OF MATHEMATICS AND STATISTICS
UNIVERSITY OF SOUTHERN MAINE
96 FALMOUTH STREET
PORTLAND, ME 04103

INTERRELATIONS AMONG VARIOUS DEFINITIONS OF BIVARIATE POSITIVE DEPENDENCE

BY SAMUEL KOTZ, QIWEN WANG, AND KEN HUNG

University of Maryland, University of Maryland, and Western Washington University

In this paper—based on an extensive computer simulation—a detailed investigation and comparisons of seven types of positive dependence properties appearing in statistical and reliability literature is presented.

A numerical index of the "strength" of positive quadrant dependence (PQD) is proposed and compared with the correlation coefficient. This index can also be adopted to various other definitions of dependence.

1. Introduction. Dependence relations among variables constitute one of the basic topics of applied probability and statistics. This theory goes back to the classical investigations of Pearson in (1900) and (1904). While the concept of independence is mathematically defined by an equality relation, the violation of this equality by definition signifies dependence. Difficulties to provide an adequate measure of dependence are illustrated by the following statement of Cramér (1924): "Every attempt to measure a conception like this by a single number must necessarily contain amount of arbitrariness and suffer from certain inconveniences." In fact, Pearson (1900, 1904), Gini (1914), Fréchet (1951), Cramér (1924), Hoeffding (1940), Rényi (1959), Kolmogorov (1933), Lehmann (1966), and Lai and Robbins (1976) should be mentioned among the leading statisticians and probabilists who have studied this problem. The more recent work of Lehmann (1966) triggered an additional spurt of activity in this area after a certain period of dormancy and, in the last decade, we are witnessing a burgeoning awakening in this field which is closely associated with the renewed interest in statistical and probabilistic reliability methodology pioneered by the works of Barlow and Proschan (and summarized in their monograph (1981)).

A survey of results up to 1975 is presented in the paper by Kotz and Soong (1977) where some 10 properties of positive dependence have been discussed and

---

AMS 1980 subject classification. 62H05.

Key words and phrases. Types of bivariate dependence, simulation, counterexamples, graphical representation, strength of dependence.

The authors express their sincere thanks to Professors H.W. Block, A.R. Sampson, T.H. Savits, and the unknown referees for their comments and suggestions which improved the presentation of this paper.

interrelations among them have been analyzed in some details. See also Kotz (1980) for an updated but a shortened version. Concepts, properties, and measures of dependence are also discussed in Schweizer and Wolff (1976) and Yanagimoto (1972) among other sources.

In spite of substantial advances there still exists a certain confusion in the literature as to various implications of dependence definitions. The main purpose of this paper is to present a clear and self-contained definition of various dependence properties as proposed by various authors, in the last two decades and to provide an empirical study (to the best of our knowledge for the first time in the literature) which will indicate the presence or absence of the particular dependence properties as classified in this paper. These examples indicate the interrelation among these definitions is more delicate than it seems from the first glance. A Monte Carlo simulation was also carried out which may indicate the frequency of presence of particular dependence properties in the so called "typical" or natural models. These simulations are supplemented by appropriate graphical representations. The paper also examines, in some detail, the relationship between various measures of positive dependence and the classical measure of *linear* dependence—the coefficient of correlation.

Finally we propose a new *numerical index* of positive quadrant dependence and compare it with the correlation coefficient. A related extensive investigation for positive quadrant dependence has been conducted by Metry and Sampson (1988) using graph-theoretic methods. A general classification framework for positive dependence was recently developed by Kimeldorf and Sampson (1989).

2. Seven Types of Definitions of Non-negative Dependence Between Two Random Variables X and Y.

1) Non-negativity of the covariance:

(1) $$\mathrm{Cov}(X,Y) = E(XY) - E(X) * E(Y) \geq 0.$$

2) PQD(X,Y) - positive quadrant dependent.

The pair of variables (X,Y) satisfies the PQD(X,Y) property if:

(2) $$P(X \leq x, Y \leq y) \geq P(X \leq x) * P(Y \leq y) \quad \text{for any } x \text{ and } y.$$

3) A(X,Y) - association.

The pair of variables (X,Y) satisfies the $A(X,Y)$ property if:

(3) $$\mathrm{Cov}(f(X,Y), g(X,Y)) \geq 0$$

where f and g are non-decreasing functions of X and Y.

4) LTD($Y \mid X$) - Y is left tail decreasing in X.

The variables X, Y satisfy the LTD($Y \mid X$) property if:

$$(4) \quad P(Y \le y \mid X \le x_1) \ge P(Y \le y \mid X \le x_2) \text{ when } x_1 < x_2 \text{ for all } y.$$

5) RTI($Y \mid X$) - Y is right tail increasing in X.

The variables X, Y satisfy the RTI($Y \mid X$) property if:

$$(5) \quad P(Y > y \mid X > x_1) \le P(Y > y \mid X > x_2) \text{ when } x_1 < x_2 \text{ for all } y.$$

6) a) CR$s(X,Y)$ property.

The pair of variables (X,Y) is said to be column regression dependent of order s (CRs) if for every $t = 1, 2, \ldots, s$, all $x_1 < x_2 < \ldots < x_t$ and all $y_1 < y_2 < \ldots < y_{t+1}$, the $(t+1) * (t+1)$ determinant

$$(6) \quad \begin{vmatrix} P(X \le x_1 \mid Y = y_1) & \ldots & P(X \le x_1 \mid Y = y_{t+1}) \\ \vdots & & \vdots \\ P(X \le x_t \mid Y = y_1) & \ldots & P(X \le x_t \mid Y = y_{t+1}) \\ 1 & \ldots & 1 \end{vmatrix} \ge 0.$$

b) RR$s(X,Y)$ property.

The pair of variables (X,Y) is said to be row regression dependent of order s (RR$s(X,Y)$) if (Y,X) is CR$s(Y,X)$.

c) DR$s(X,Y)$ property.

If (X,Y) is both RRs and CRs dependent of order s, we call (X,Y) to be double regression dependent of order s (DRs).

These definitions are due to Schriever (1985).

RR1(X,Y) is also called PRD($Y \mid X$) $-$ Y is positively regression dependent on X.

The variables X, Y satisfy the PRD($Y \mid X$) property if:

$$(7) \quad P(Y \le y \mid X = x_1) \ge P(Y \le y \mid X = x_2) \text{ when } x_1 < x_2 \text{ for all } y.$$

7) TP$s(X,Y)$ property.

The pair of variables (X,Y) is said to be total positive dependent of order s if their joint density (or the mass function) $f(x,y)$ satisfies the following condition:

for any $t = 1, 2, \ldots, s$, all $x_1 < x_2 < \ldots < x_t$ and $y_1 < y_2 < \ldots < y_t$, the determinant

$$(8)\qquad \begin{vmatrix} f(x_1,y_1) & f(x_1,y_2) & \dots & f(x_1,y_t) \\ \vdots & \vdots & \dots & \vdots \\ f(x_t,y_1) & f(x_t,y_2) & \dots & f(x_t,y_t) \end{vmatrix} \geq 0.$$

The TP2(X,Y) and TP3(X,Y) property will be considered in this paper.

Exchanging X and Y in (4), (5), and (7), the definitions of LTD$(X \mid Y)$, RTI$(X \mid Y)$ and PRD$(X \mid Y)$ are given, respectively.

For brevity below we will use the term positive instead of non-negative.

3. A Network of Relationships Among Seven Types of Positive Dependence Properties. It has been proved by various authors (see citations in References) that the following implications among these 7 types of positive dependence properties between X and Y are valid (provided the appropriate covariances exist).

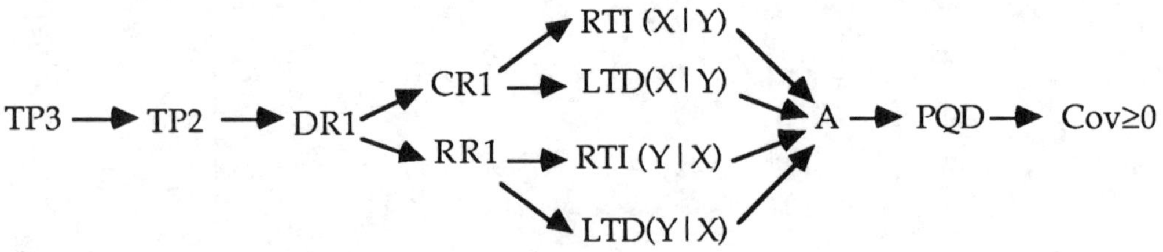

Figure 1

(See e.g. Lehmann (1966); Esary, Proschan, and Walkup (1967); Esary and Proschan (1970); Schriever (1985); and Bilodeau (1989)).

Schriever provides the theorem and a proof that TPs dependence implies DRs−dependence. In our Figure 1 we have the particular case TP2−>DR1.

All of the above implications are strictly held. Some numerical examples will be given in Section 5.

It should be noted that the properties TP3, TP2, DR1, A, PQD, and Cov$\geq$0 (on the middle line of the network) are symmetrical about X and Y, while the remaining ones are non-symmetrical.

4. Computer Simulation. To study the relations among these positive dependence properties, a Monte Carlo simulation was carried out using the following algorithm.

(1) Let g_{ij} be a uniform random number in (0,1), obtained from the BASIC subroutine, for $i = 1, 2, \dots, m$; $j = 1, 2, \dots, n$.

(2) Let $P_{ij} = P(X = j, Y = i) = g_{ij}/[\Sigma_{i=1}^m \Sigma_{j=1}^n \, g_{ij}]$ for $i = 1, 2, \dots, m$; $j = 1, 2, \dots, n$.

It is obvious that $P_{ij} \geq 0$ and $\Sigma_{i=1}^m \Sigma_{j=1}^n P_{ij} = 1$.

Based on this algorithm a computer program was compiled. After running 30 times with different random seeds, 3000 cases with $m = 3$ and $n = 3$ have been generated. Frequencies of occurrences of the seven types of positive dependence properties stated above are presented in the Table 1 and the Figure 2. For brevity we shall denote LTD($Y \mid X$) by LTDx and similarly RTI($Y \mid X$) by RTIx as well as LTD($X \mid Y$) by LTDy and RTI($X \mid Y$) by RTIy.

Several remarks are in order.

1) In the above 3000 cases we did not find a single case which has the PQD property but not the A(X, Y) property. However, one such case was obtained in another simulation run (of *one million* cases related to Table 2).

2) There was not a single case observed such that both LTDx and RTIx were valid but not RR1 or both LTDy and RTIy were valid but not CR1. This is indeed impossible whenever $m = 3$ and $n = 3$. (See e.g. Esary and Proschan (1972).)

3) The frequencies of occurrences of the six one-sided tail properties are almost equal in pairs; a rather substantial decrease in frequencies of two-sided tail properties was observed.

4) Note the sharp decrease of the total number of cases from Cov(X, Y) ≥ 0 to PQD; from LTD and RTI to DR1 and from DR1 to TP2.

5) It is interesting to compare the PQD column of Table 1 with the second row of Metry and Sampson's (1988) Table 6.2 where the exact value 1/6 is obtained by enumeration. This shows that our simulation and their theoretical derivation yield very similar results.

Table 1. Total Numbers (T) and Percentages (P) of Occurrences of Various Dependence Properties in 3000 Simulations

	Cov≥ 0	PQD	A	LTDx	RTIx	LTDy
T	1494	503	503	314	320	308
P(%)	49.8	16.8	16.8	10.5	10.7	10.3
	RTIy	RR1	CR1	DR1	TP2	TP3
T	327	167	163	78	32	22
P(%)	10.9	5.6	5.4	2.6	1.1	0.7

In order to investigate the relation between the coefficent of correlation ρ and the positive dependence properties, we first subdivided the range of ρ from 0 to 0.75 into 15 intervals. Using the same algorithm as for the 3000 cases, additional one million cases were generated. For comparison purposes the first 1000 cases were selected for each of the first 13 intervals of ρ and over 100 cases were obtained for

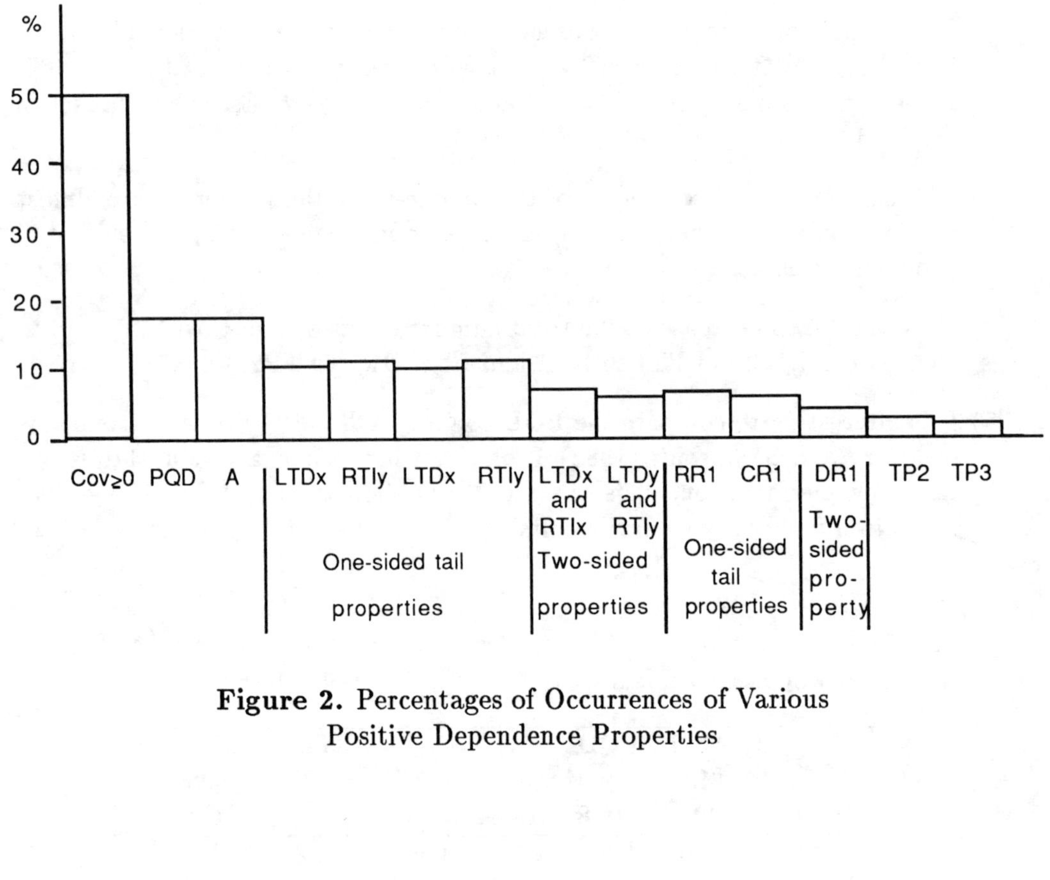

Figure 2. Percentages of Occurrences of Various Positive Dependence Properties

Table 2. Relations Between Frequencies Positive Dependence Properties and the Coefficient of Correlation

$\rho(X,Y)$	Mid-point	No. of CASES	PQD	LTDx	RTIx	RR1	DR1	TP2	TP3
(.00,.05)	.025	1000	5	3	2	1	1	1	1
(.05,.10)	.075	1000	69	32	38	13	3	1	0
(.10,.15)	.125	1000	199	102	99	33	8	4	2
(.15,.20)	.175	1000	368	183	206	69	30	13	9
(.20,.25)	.225	1000	535	316	313	137	65	28	16
(.25,.30)	.275	1000	705	439	440	204	91	50	24
(.30,.35)	.325	1000	810	551	531	305	160	57	32
(.35,.40)	.375	1000	892	636	615	378	210	109	66
(.40,.45)	.425	1000	941	723	692	485	295	152	104
(.45,.50)	.475	1000	962	736	720	505	323	181	136
(.50,.55)	525	1000	980	758	753	544	385	243	187
(.55,.60)	.575	1000	989	806	808	634	453	284	244
(.60,.65)	.625	1000	994	834	825	672	497	310	265
(.65,.70)	.675	407	407	342	338	277	220	141	135
(.70,.75)	.725	134	133	112	118	99	83	55	53
R-square			.838	.924	.946	.980	.968	.937	.881
Intercept			.135	.013	.008	-.085	-.110	-.083	-.088
Slope			1.496	1.352	1.356	1.188	.946	.611	.558

the last two intervals. The results are shown in Table 2. Since the frequency of the property A(X,Y) is almost the same as that of PQD(X,Y) and the frequencies of LTDy, RTIy, and CR1 are very close to that of LTDx, RTIx, and RR1, respectively, only seven properties are included into Table 2. As ρ increases the frequency of virtually every property listed in Table 2 (except TP3) increases, but the effect of the increase of ρ is much more noticeable for the PQD, LTD, and RTI properties and is the least noticeable for TP2 and TP3. As a first approximation, simple linear regression models have been fitted setting ρ as an independent variable (using mid-points of the intervals) and the property as the dependent binary variables (if the property is valid, its value will be 1, and zero otherwise). Table 2 presents the intercept and slope of the regression equations, and the corresponding R-squares. The values of the R-square indicate that there exists a strong linear relationship between the correlation coefficient and the positive regression dependence. Note also a sharp sensitivity in the range (0.075–0.175) and a relative robustness in the range (0.425–0.625) of the values of $\rho(X,Y)$.

5. Ten Numerical Examples. In order to show that all of the implications in

Figure 1 are strict, we present ten numerical examples. Eight of them are obtained from the above 3000 cases and the remaining two involve additional computations. These examples seem to be the simplest for our purposes.

EXAMPLE 1.

$Y \setminus X$	1	2	3	
1	0.25	0.10	0.01	0.36
2	0.01	0.03	0.22	0.26
3	0.02	0.26	0.10	0.38
	0.28	0.39	0.33	

The table gives the probabilities $P(X = j, Y = i)$ and the marginal distributions $P(X = j)$ and $P(Y = i)$ for $j = 1,2,3$ and $i = 1,2,3$. For example, $P(X = 2, Y = 3) = 0.26$, $P(X = 1) = 0.28$, $P(Y = 2) = 0.26$. In this case, we have the coefficient of correlation $= 0.476$. However, the distribution does not possess PQD property since

$$P(X \leq 2, Y \leq 2) = 0.39 \text{ while } P(X \leq 2) * P(Y \leq 2) = 0.4154.$$

EXAMPLE 2.

$Y \setminus X$	1	2	3	
1	0.23	0.01	0.11	0.35
2	0.01	0.28	0.01	0.30
3	0.12	0.01	0.22	0.35
	0.36	0.30	0.34	

To show that the distribution possesses PQD(X, Y) property we form the table:

(i, j)	(1,1)	(1,2)	(2,1)	(2,2)
$P(X \leq j, Y \leq i)$	0.23	0.24	0.24	0.53
$P(X \leq j) * P(Y \leq i)$	0.126	0.231	0.234	0.429

For $i = 1,2$ and $j = 1,2$ we have strict inequality in (2) whereas equality holds for $i = 3$ or $j = 3$.

Hence the PQD property is valid. But there is no A(X, Y) property.

Indeed, let

$$f(X,Y) = \begin{cases} 0 & \text{if } X < 3 \text{ and } Y < 3 \\ 1 & \text{if } X = 3 \text{ or } Y = 3 \end{cases}$$

$$g(X,Y) = \begin{cases} 0 & \text{if } X = 1 \text{ or } Y = 1 \\ 1 & \text{if } X > 1 \text{ and } Y > 1 \end{cases}$$

f and g are non-decreasing functions of X or Y.

The distribution of $f(X,Y)$ and $g(X,Y)$ is then

$g \setminus f$	0	1
0	0.25	0.23
1	0.28	0.24

Since $\mathrm{Cov}(f(X,Y), g(X,Y)) = -0.0044$, the property $\mathrm{A}(X,Y)$ is not valid.

EXAMPLE 3.

$Y \setminus X$	1	2	3	
1	0.07	0.13	0.16	0.36
2	0.06	0.15	0.11	0.32
3	0.05	0.09	0.18	0.32
	0.18	0.37	0.45	

To show that $\mathrm{A}(X,Y)$ is valid we use the following theorem given in Esary, Proschan, and Walkup (1967).

Theorem. If $\mathrm{Cov}(f(X,Y), g(X,Y)) \geq 0$ *for all* binary *non-decreasing functions* f *and* g*, then the pair of variables* (X,Y) *are associated.*

There are 20 different binary non-decreasing functions for this example (the case $m = n = 3$) which result in 400 different covariances. A computer program has been run and it was verified that all of the 400 covariances are non-negative.

However, neither $\mathrm{LTD}(Y \mid X)$ nor $\mathrm{RTI}(Y \mid X)$ are true in this case. Indeed: $P(Y \leq 2 \mid X \leq 1) = 0.722$, $P(Y \leq 2 \mid X \leq 2) = 0.745$ (so it is not LTDx) and $P(Y > 1 \mid X > 1) > P(Y > 1 \mid X > 2)$ (so it is not RTIx.)

Note that among our 3000 cases only 36 possess the properties of Example 3.

EXAMPLE 4.

$Y \setminus X$	1	2	3	
1	0.10	0.13	0.15	0.38
2	0.04	0.12	0.10	0.26
3	0.07	0.13	0.16	0.36
	0.21	0.38	0.41	

The corresponding conditional probabilities $P(Y \leq i \mid X \leq j)$, $i = 1, 2, 3$ and $j = 1, 2, 3$, indicate that LTDx property is valid. However, since $P(Y > 1 \mid X > 1) = 0.646$ and $P(Y > 1 \mid X > 2) = 0.634$ the RTIx does not hold.

EXAMPLE 5.

$Y \setminus X$	1	2	3	
1	0.37	0.11	0.02	0.50
2	0.04	0.12	0.16	0.32
3	0.05	0.02	0.11	0.18
	0.46	0.25	0.29	

The corresponding conditional probabilities show that RTIx property is valid. However, LTDx is not valid, since $P(Y \leq 2 \mid X \leq 1) = 0.89$ and $P(Y \leq 2 \mid X \leq 2) = 0.90$.

EXAMPLE 6.

$Y \setminus X$	1	2	3	4	
1	0.06	0.17	0.19	0.05	0.47
2	0.03	0.19	0.19	0.12	0.53
	0.09	0.36	0.38	0.17	

For this example, the values of $P(Y < 1 \mid X > j)$ and $P(Y > 1 \mid X > j)$, $j = 1, 2, 3, 4$, indicate that both LTDx and RTIx are true.

However, PRD$(Y \mid X)$ is not valid since $P(Y \leq 1 \mid X = 2) = 0.47$ and $P(Y \leq 1 \mid X = 3) = 0.5$.

To avoid confusion we remind the reader that PRD$(Y \mid X)$ property is the same as RR1(X, Y) and below we shall use the later terminology.

EXAMPLE 7.

$Y \setminus X$	1	2	3	
1	0.19	0.15	0.09	0.43
2	0.02	0.02	0.01	0.05
3	0.21	0.19	0.12	0.52
	0.42	0.36	0.22	

The values of $P(Y \leq i \mid X = j)$, $i = 1, 2, 3$, $j = 1, 2, 3$, show that RR1 pattern exists.

However, the values of $P(X \leq j \mid Y = i)$, $(i = 1,2,3, j = 1,2,3)$, indicate that CR1(X,Y) property is not valid in this case.

Notice that neither LTD$(X \mid Y)$ nor RTI$(X \mid Y)$ are valid in this case.

EXAMPLE 8.

$Y \setminus X$	1	2	3	
1	0.10	0.04	0.06	0.20
2	0.25	0.01	0.24	0.50
3	0.04	0.10	0.16	0.30
	0.39	0.15	0.46	

The values of $P(X \leq j \mid Y = i)$ and $P(Y \leq i \mid X = j)$ indicate that this distribution is CR1, but not RR1.

EXAMPLE 9.

$Y \setminus X$	1	2	3	
1	0.14	0.13	0.09	0.36
2	0.11	0.13	0.10	0.34
3	0.09	0.12	0.09	0.30
	0.34	0.38	0.28	

The values of $P(Y \leq i \mid X = j)$ indicate that CR1(X,Y) is valid. The values of $P(Y \leq i \mid X = j)$ also indicate the existence of the RR1(X,Y) pattern. Therefore in this case the DR1(X,Y) property is valid. However, since

$$\begin{vmatrix} P(X=2,Y=2) & P(X=2,Y=3) \\ P(X=3,Y=2) & P(X=3,Y=3) \end{vmatrix} = -0.0003 < 0$$

the TP2 property is not valid.

EXAMPLE 10.

$Y \setminus X$	1	2	3	
1	0.09	0.16	0.11	0.36
2	0.04	0.08	0.08	0.20
3	0.08	0.17	0.19	0.44
	0.21	0.41	0.38	

The TP2(X,Y) property is valid since nine 2×2 related determinants are positive. However, the determinant

$$\begin{vmatrix} 0.09 & 0.16 & 0.11 \\ 0.04 & 0.08 & 0.08 \\ 0.08 & 0.17 & 0.19 \end{vmatrix} = -0.000004 < 0$$

indicates that TP3(X,Y) is not valid.

6. Pitfalls in the Relationship Between the Coefficient of Correlation and Other Six Types of Positive Dependence Properties. The implications of the seven types of positive dependence properties as presented in Figure 1 presumably indicates the degree of dependence between variables. However this relationship is somewhat more complex than it was originally envisioned.

In Example 1, Cov$(X,Y) = 0.319$ and the coefficient of correlation is the second largest among 10 examples ($\rho = 0.476$); however the ("weak") PQD property is not valid. At the same time in Example 7, $\rho = 0.037$ (less than 8% of the value obtained in Example 1), while here PRD$(Y \mid X)$ property holds. Example 10 provides the second strongest positive dependence as indicated by Figure 1, however the value of $\rho(0.114)$ is very small. If we have used ρ as an indicator of dependence, we would have concluded that X and Y are more "independent" in the situation of the Example 7 and 10 than in Example 1 and some other examples.

Our last example chosen from the 3000 generated cases illustrates a situation in which TP3(X,Y) property is valid.

EXAMPLE 11.

$Y \setminus X$	1	2	3	
1	0.07	0.06	0.04	0.17
2	0.14	0.14	0.11	0.39
3	0.13	0.16	0.15	0.44
	0.34	0.36	0.30	

The TP3(X,Y) property is valid. The verification is left to readers.

Again, we have a very low value of the correlation coefficient ($\rho = 0.104$) coupled with the presence of the strongest positive dependence TP3 property.

In conclusion we present two extreme examples that are not based on a computer simulation.

Let X and Y be independent and uniform:

$$P(X = x_j, Y = y_i) = 1/(mn) \text{ for } i = 1, 2, \ldots, m; \; j = 1, 2, \ldots, n.$$

Here we have $\rho = 0$, and moreover TP3(X,Y) and all of the seven types of positive dependence properties hold with the equality sign.

At the other extreme consider the family distribution parameterized by t:

$Y \setminus X$	1	2	3	
1	0	t	0	t
2	t	0.5-t	0	0.5
3	0	0	0.5-t	0.5-t
	t	0.5	0.5-t	

where t takes values in the interval (0,0.5).

Here $\text{Cov}(X,Y) = 0.25 + t - 4t^2$, $\text{Var}(X) = \text{Var}(Y) = 0.25 + 2t - 4t^2$, and $\rho = (0.25 + t - 4t^2)/(0.25 + 2t - 4t^2)$. As $t \to 0$, the probability mass concentrates at $(X = 2, Y = 2)$ and $(X = 3, Y = 3)$ and $\rho \to 1$, hence X and Y possess a very high linear relation. However, in this case even the PQD property is not valid, since

$$P(X < 1, Y < 1) = 0 < t^2 = P(X < 1)P(Y < 1).$$

7. Surface Representation of Bivariate Distributions With Selected Types of Positive Dependence. In this section we present several graphs which depict characteristic structures of the joint probability mass distributions for selected types of bivariate dependence.

1) Characteristic structure of distributions with a high positive correlation lacking most of the other positive dependence properties is presented in Graph 1 (corresponding to Example 1). The largest values of P_{ij} are on the main diagonal ($P_{11} = 0.25$) or near the main diagonal ($P_{23} = 0.22$ and $P_{32} = 0.26$), while P_{ij}'s that are far away from the main diagonal (at the corners) have low values ($P_{13} = 0.01$ and $P_{31} = 0.02$). The correlation coefficient $\rho = 0.476$ is "moderately high" (the second largest among all the examples). However, the probability P_{22} on the main diagonal at the center is only 0.03, which is too small to possess PQD and other positive dependence properties.

2) Graph 2 possesses the strongest positive dependence property TP3, while the correlation coefficient is very small ($\rho = 0.104$). The graph indicates that this distribution is closer to the uniform distribution than those in other examples.

3) The comparison between RR1 (but not CR1 or LTDy and RTIy) and CR1 (but not RR1 or LTDx and RTIx) is presented in Graphs 3 and 4 respectively. Note that the "horizontal" edges are prominent in Graph 3 while the "vertical" ones contain a substantial amount of probability mass in Graph 4.

8. A Proposed Modification of Definitions of Dependence Between Random Variables. The main difference between covariance (or the coefficient of correlation) and the other six properties of positive dependence discussed in this paper is that the latter simply provide a "Yes" or "No" answer to a certain probabilistic relationship, while the former presents a numerical measure of a linear dependence. Basically, the coefficient of correlation deals with an overall appraisal of "whole forest," while the other dependence properties attempt to check "the trees" individually.

It would seem appropriate to generalize the definitions of dependence between two variables by introducing a probabilistic component within the specified relationship. As an example, we propose an index of positive quadrant dependence defined below.

DEFINITION. An index of positive quadrant dependence of two random variables X and Y is given by:

$$\mathrm{IPQD}\phi(X \mid Y) = \Sigma_{(x,y)\epsilon R\phi} P(X = x, Y = y)/[1 - \Sigma_{(x,y)\epsilon R_b} P(X = x, Y = y)]$$

where $x = 1, 2, \ldots, n$ and $y = 1, 2, \ldots m$; $R\phi = \{(x, y) \mid P(X \leq x, Y \leq y) > P(X \leq x)P(Y \leq y) + \phi\}$, ϕ is a non negative parameter; and the boundary $R_b = \{(x, y) \mid x = n \text{ or } y = m)\}$, in this case.

Table 3 provides the index IPQDϕ for the examples presented in this paper for $\phi = 0, 0.005, 0.01, 0.015, 0.02$ and 0.05.

Note that in the column corresponding to $\phi = 0$ we obtain essentially "Yes" (IPQD$\phi = 1$) or "No" (IPQD$\phi < 1$) answers to the question "is the PQD property valid?". The other columns present values which successively indicate the strength of validity of the PQD property when ϕ ranges from 0.005 to 0.05.

An alternative—perhaps even more important—justification for the introduction of parameter ϕ is to assess the robustness of the properties as far as possible random errors are concerned. If our data is subject to random errors, the situation as presented in Examples 7, 9, and 11 may not indicate the presence of the PQD property while the values appearing in Examples 5, 8, 2, and 1 will very likely provide a "significant" PQD distribution.

Evidently the other definitions of positive dependence properties can be extended along these lines.

REFERENCES

BARLOW, R. and PROSCHAN, F. (1981). *Statistical Theory of Reliability and Life Testing.* To Begin With, Silver Spring, Md.

BILODEAU, M. (1989). On the monotone regression dependence for Archimedian bivariate uniform. *Commun. Statist.-Theor. Meth.* **18(3)** 981–988.

CRAMÉR, H. (1924). Remarks on correlation. *Skand. Aktuarietidskr.* **7** 220–240.

ESARY, J.D. and PROSCHAN, F. (1972). Relationships among some concepts of bivariate dependence. *Ann. Math. Statist.* **43** 651–655.

ESARY, J.D., PROSCHAN, F. and WALKUP, D.W. (1967). Association of random variables with applications. *Ann. Math. Statist.* **38** 1466–1474.

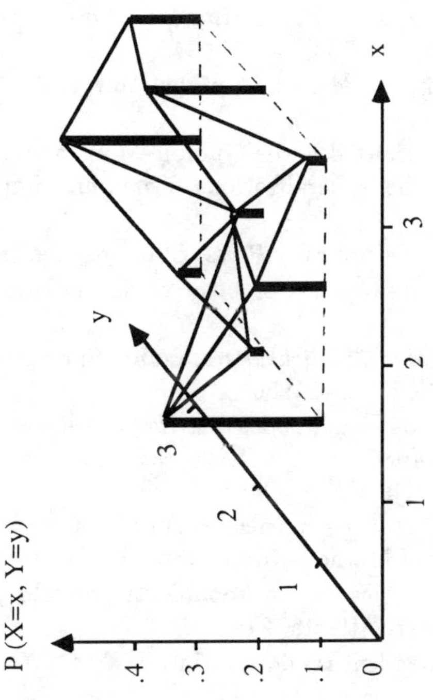

Graph 1 (Example 1) ρ = 0.478, PQD=0

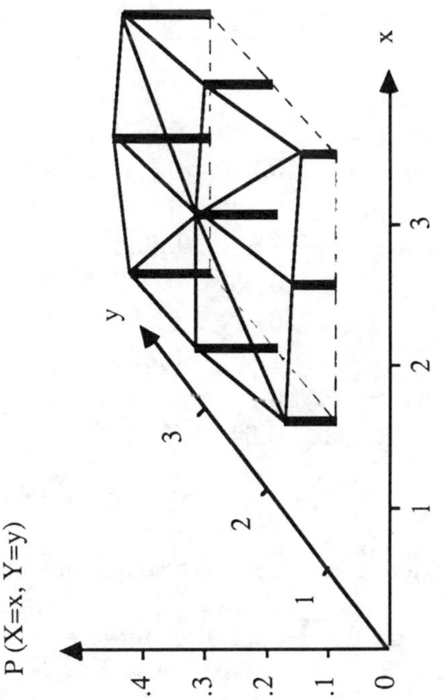

Graph 2 (Example 11) ρ = 0.104, TP3=1

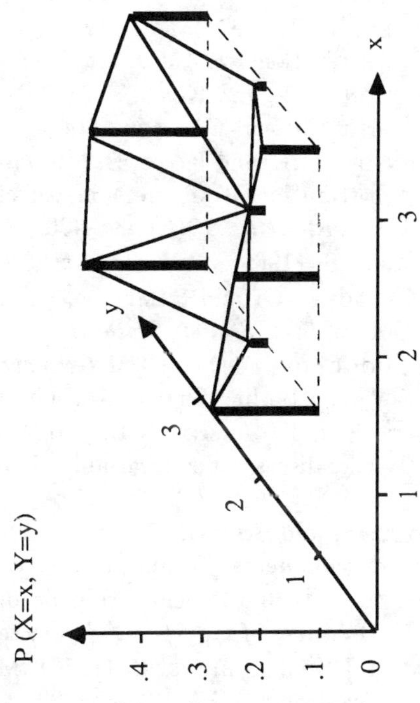

Graph 3 (Example 7) ρ = 0.037, RR1=1, CR1=0

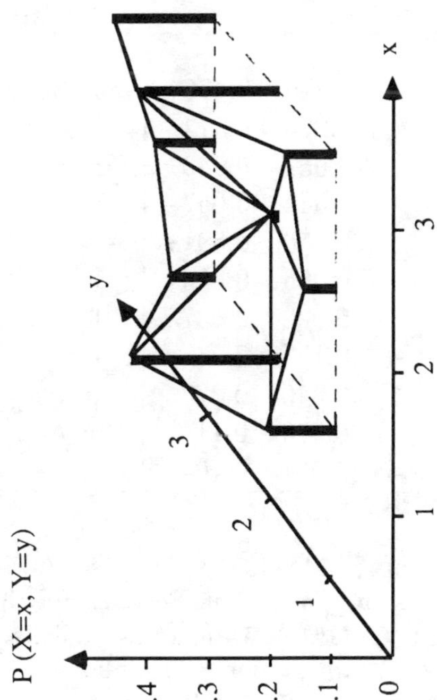

Graph 4 (Example 8) ρ = 0.238, CR1=1, RR1=0

Table 3. Values of IPQDϕ for selected values of

Exmp.	Cov	ρ	$\phi = 0$	0.005	0.010	0.015	0.020	0.050
1	0.319	0.476	0.923	0.923	0.923	0.923	0.923	0.923
2	0.220	0.314	1	1	0.962	0.962	0.962	0.962
3	0.051	0.083	1	0.683	0.366	0.366	0.366	0
4	0.044	0.067	1	1	0.564	0.256	0.256	0
5	0.356	0.551	1	1	1	1	1	0.938
6	0.066	0.153	1	1	1	1	0.452	0
7	0.028	0.037	1	0.605	0	0	0	0
8	0.153	0.238	1	1	1	1	1	0.625
9	0.046	0.073	1	1	0.745	0.275	0	0
10	0.076	0.114	1	1	1	0.649	0.649	0
11	0.061	0.104	1	1	1	0.683	0	0

FRÉCHET, M. (1951). Sur les tableaux de correlation dont les marges sont donnees. *Ann. Univ. Lyon*, Sect. A, Ser. 3, **14** 53–57.

GINI, C. (1914). Di una misura della dissomiglianza tra due gruppi di quantita e delle due applicazioni allo studio delle relazioni statistiche. *Atti R. Ist. Veneto Sci. Lett. Arti.* (8) **74** 185–213.

HOEFFDING, W. (1940). Masstabinvariante korrelations-theorie. *Schriften Math. Inst. Univ. Berlin* **5** 181–233.

JENSEN, D.R. (1971). A note on positive dependence and the structure of bivariate distributions. *SIAM J. Appl. Math.* **20**, No. 4, 749–753.

JOHNSON, N.L. and KOTZ, S. (1975). On some generalized Farlie-Gumbel-Morgenstern distributions. *Commun. Statist.* **4** 415–427.

JOHNSON, N.L. and KOTZ, S. (1977). On some generalized Farlie-Gumble-Morgenstern distributions–II. Regression, correlation and further generalizations. *Commun. Statist. Theor. Meth.* **A6(6)** 485–496.

KARLIN, S. (1968). *Total Positivity, Vol. I.* Stanford University Press, Stanford, Calif.

KIMELDORF, G. and SAMPSON, A.R. (1989). A framework for positive dependence. *Ann. Inst. Statist. Math.* **41** 31–45.

KOLMOGOROV, A.N. (1933). *Grundbegriffe der Wahrscheinlichkeitsrechnung.* Springer-Verlag, Berlin (English translation: Chelsea, New York, 1950).

KOTZ, S. and JOHNSON, N.L. (1977). Propriétés de dépendance des distributions itérées, généralisées â deux variables Farlie-Gumbel-Morgenstern. *C.R. Acad. Sc. Paris* **285**, Ser. A, 277–280.

KOTZ, S. and SOONG, C. (1975). On the measures of dependence, a survey of recent developments. Technical Report, Department of Mathematics, Temple University.

KOTZ, S. (1980). Dependence concepts and their application in probabilistic modeling. *Transact. of the 34-th ASQC Techn. Conference*, Atlanta, 245–251.

LAI, T.L. and ROBBINS, H. (1976). Maximally dependent random variables. *Proc. Nat. Acad. Sci.* U.S.A. **73** 286–288.

LEHMANN, E.L. (1966). Some concepts of dependence. *Ann. Math. Statist.* **37** 1137–1153.

METRY, M.H. and SAMPSON, A.R. (1988). Characterizing and generating bivariate empirical rank distributions satisfying certain positive dependence concepts. Technical Report 88–04, Department of Mathematics and Statistics, University of Pittsburgh, Pittsburgh, PA.

PEARSON, K. (1900). *Mathematical Contributions to the Theory of Evolution.* VII: On the correlation of characters not quantitatively measurable. *Philos. Trans. Roy. Soc.* A **195** 1–47.

PEARSON, K. (1904). *Mathematical Contributions to the Theory of Evolution.* XIII: On the theory of contingency and its relation to association and normal correlation. Drapers' Company Research Memoirs, *Biometric Series 1.*

RÉNYI, A. (1959). On measures of dependence. *Acta Math. Acad. Sci. Hungar.* **10** 441–451.

SCHRIEVER, B.F. (1985). *Order Dependence.* Vrije Universiteit te Amsterdam, Centrum voor Wiskunde en Informatica, Amsterdam, Holland.

SCHWEIZER, B. and WOLFF, E.F. (1976). Sur une measure dedependance pour les variables aleatoire. *C.R. Acad. Sci. Paris,* Ser. A, **283** 609–611.

SEEGER, J. (1978). Dependence with fixed marginals from a density viewpoint. Ph.D Dissertation, Department of Mathematics, Temple University, Philadelphia, Pa.

YANAGIMOTO, T. (1972). Families of positively dependent random variables. *Annals Inst. Statist. Math.* **24** 559–573.

DEPARTMENT OF MANAGEMENT SCIENCE AND STATISTICS
COLLEGE OF BUSINESS AND MANAGEMENT
UNIVERSITY OF MARYLAND
COLLEGE PARK, MD 20742

DEPARTMENT OF MANAGEMENT SCIENCE AND STATISTICS
COLLEGE OF BUSINESS AND MANAGEMENT
UNIVERSITY OF MARYLAND
COLLEGE PARK, MD 20742

DEPARTMENT OF FINANCE, MARKETING & DECISION SCIENCES
COLLEGE OF BUSINESS AND ECONOMICS
WESTERN WASHINGTON UNIVERSITY
BELLINGHAM, WA 98225

TESTS OF INDEPENDENCE AGAINST LIKELIHOOD RATIO DEPENDENCE IN ORDERED CONTINGENCY TABLES

By Mei-Ling Ting Lee

Boston University

Tests of independence against positive likelihood ratio dependence in ordered contingency tables are reviewed. Some exact tests and an intuitive sign test are discussed.

1. Introduction. In the analysis of contingency tables, a classical test of independence is the chi-square test. However, if the chi-square test strongly rejects the null hypothesis of independence, the statistician receives little information as to what kind of dependency may exist in the data. When both categorical variables are ordinal, it might be interesting to check whether there is a monotonic relationship between the variables. Grove (1984) and Agresti (1984) explored a variety of definitions of positive dependence for contingency tables.

On testing the independence in contingency tables, various types of alternatives have been investigated. Armitage (1955) considered tests for linear trends in proportions. Goodman (1985) discussed association models, correlation models, and asymmetry models for contingency tables. Grove (1980, 1984) considered positive association and tests in a two-way contingency table. Nguyen and Sampson (1987) considered the alternative hypothesis of positive quadrant dependence. Agresti et al. (1979) and Patefield (1982) considered several exact tests of independence against positively likelihood ratio dependence. Lee (1988) investigated an intuitive sign test to test against likelihood ratio dependence.

In this review article, tests of independence against likelihood ratio dependence in ordered contingency tables are discussed.

2. Likelihood Ratio Dependence in Contingency Tables. Lehmann (1966) defined that random variables X and Y are said to be positively likelihood ratio dependent if their joint density f satisfies the following properties:

$$f(x,y)f(x',y') \geq f(x',y)f(x,y') \text{ for all } x \leq x',\ y \leq y'.$$

In this section, we shall discuss the concepts of positive likelihood ratio dependence in ordered contingency tables.

AMS 1980 subject classifications. Primary 62G10, 62H20; secondary 62P10.

Key words and phrases. Cross product difference statistics, exact test, likelihood ratio dependence, ordered contingency tables, U-statistics, sign test.

I thank the editors and a referee for helpful comments.

Let X and Y be ordinal categorical random variables such that X has values $x_1 < x_2 < \ldots < x_r$ and Y has values $y_1 < y_2 < \ldots < y_c$. Consider a random sample (X_k, Y_k), $1 \leq k \leq N$ of size N. Collect the sample data into an $r \times c$ contingency table (n_{ij}) where n_{ij} denotes the number of observations (X_k, Y_k) such that $X_k = x_i$, $Y_k = y_j$, $k = 1, \ldots N$. Let $p_{ij} = P(X = x_i, Y = y_j)$ for $1 \leq i \leq r, 1 \leq j \leq c$. For any two rows i and k and any two columns j and l, the corresponding odds ratio parameter is given by $(p_{i,j}p_{k,l})/(p_{i,l}p_{k,j})$. There are a total of $rc(r-1)(c-1)/4$ odds ratios formed this way. However, using all the odds ratios is redundant. It can be shown that local odds ratios, formed by using cells in adjacent rows and adjacent columns, determine all possible odds ratios that can be formed from any pairs of rows and any pairs of columns. Therefore, it suffices to work on the $(r-1)(c-1)$ local odds ratios defined on adjacent rows and columns, namely

$$\theta^{(i,j)} = (p_{i,j}p_{i+1,j+1})/(p_{i+1,j}p_{i,j+1}) \text{ for } 1 \leq i \leq r-1, \ 1 \leq j \leq c-1.$$

For cross-classifications of ordinal variables, positive likelihood ratio dependence corresponds to the property that all local odds ratios are greater or equal than 1. A lot of research has been done based upon the estimated odds-ratio

$$\hat{\theta}^{(i,j)} = (\hat{p}_{i,j}\hat{p}_{i+1,j+1})/(\hat{p}_{i+1,j}\hat{p}_{i,j+1}),$$

where $\hat{p}_{i,j} = n_{ij}/N$ is an unbiased estimator for $p_{i,j}$. See Bishop, Fienberg, and Holland (1975) and Agresti (1984) for a review of theories and methods based on odds ratios statistics.

However, $\hat{\theta}^{(i,j)}$ may equal to 0 or ∞ if any of the $n_{ij} = 0$. One way to get around this is to collapse categories and therefore increase the cell values. But this kind of procedures would sometimes result in wasting information collected from the observed data. To avoid these situations, Lee (1988) investigated the cross-product difference parameters. Similarly as in the case of odds-ratios, it can be shown that it suffices to consider only the local cross-product difference parameters $u^{(i,j)}$, where

$$u^{(i,j)} = p_{i,j}p_{i+1,j+1} - p_{i,j+1}p_{i+1,j} \text{ for } 1 \leq i \leq r-1, \ 1 \leq j \leq c-1.$$

For cross-classification tables, positive likelihood ratio dependence corresponds to the property that all local cross-product differences are greater or equal to 0.

3. Some Exact Tests. Patefield (1982) develops exact tests of independence against trends of positive likelihood ratio dependence in $r \times c$ contingency tables. The exact test was constructed according to procedures outlined by Agresti, Wackerly, and Boyett (1979). This procedure is developed to construct four tests based on differing criteria for measuring departures from H_0 in favor of H_1, where the null hypothesis is $H_0 : \theta^{(i,j)} = 1$ for all $1 \leq i \leq r-1$, $1 \leq j \leq c-1$, and the alternative hypothesis is $H_1 : \theta^{(i,j)} \geq 1$ for all $1 \leq i \leq r-1$, $1 \leq j \leq c-1$, with strict inequality holding for at least one pair of (i,j). The tests Patefield considered are based on: (1) the likelihood ratio, (2) the maximized score correlation, (3) the natural score correlation, and (4) the measure of association proposed by Goodman and Kruskal (1954). For each of the four criteria, the test procedure requires calculation of the appropriate test statistic for each element of the conditional sample space, i.e. for each table having the same marginal totals as the observed table. Of the four tests Patefield considered, it was concluded that test 2, the maximized score correlation, should be preferred to tests 3 and 4 as it is more flexible and appropriate in practical applications. The statistic of Patefield's test 2 considers the correlation between row and column scores maximized over ordered values of those row and column scores, i.e.

$$\lambda_2 = \sup_R \left\{ n_{..}^{-1} \sum_{i=1}^{r} \sum_{j=1}^{c} n_{ij} w_i s_j \right\}$$

where

$$R = \left\{ w, s : \sum_i n_{i.} w_i = 0,\ \sum_j n_{.j} s_j = 0,\ \sum_i n_{i.} w_i^2 = n_{..},\right.$$
$$\left. \sum_j n_{.j} s_j^2 = n_{..},\ w_1 \le w_2 \le w_r,\ s_1 \le s_2 \le \ldots \le s_c \right\}$$

and

$$n_{i.} = \sum_j n_{ij},\quad n_{.j} = \sum_i n_{ij},\quad n_{..} = \sum_i \sum_j n_{ij}.$$

Through simulation and Monte Carlo power study, it was shown that test 2 and test 1, the likelihood ratio test, have similar power. Therefore, test 2 should be preferred to test 1, the likelihood ratio test, on the ground of computation feasibility. However, when the sample size is large or when tables have higher dimensions, full enumeration of the conditional sample space is impossible and the random sampling technique is used.

4. A Sign Test. In this section, an intuitive sign test is considered to test the hypothesis of $H_0 : u^{(i,j)} = 0$ for all i,j against the alternative hypothesis that $H_1 : u^{(i,j)} \ge 0$ for all i,j and $u^{(i,j)} > 0$ for at least one pair of (i,j).

Considering

$$\hat{u}_N^{(i,j)} = \frac{1}{N(N-1)}(n_{i,j}n_{i+1,j+1} - n_{i+1,j}n_{i,j+1})$$

as an unbiased estimator of the cross-product difference $u^{(i,j)}$, Lee (1988) introduced a test statistic as follows. Let

$$S = \sum_{\substack{1 \le i \le r-1 \\ 1 \le j \le c-1}} sgn\ \hat{u}_N^{(i,j)},$$

where $sgn\ x = 1\{x > 0\}$ denote the indicator function for positive values of x. That is, the test statistic S is simply the number of 2×2 subtables, formed by adjacent rows and adjacent columns, such that the corresponding cross product difference statistic $\hat{u}_N^{(i,j)}$ is positive. Since positive association in many 2×2 subtables (i.e. many $\hat{u}_N^{(i,j)} > 0$) will provide strong evidence for the alternative H_1, one rejects H_0, if the value of S is sufficiently large.

Instead of using the delta method which would lead to lengthy calculations, the method of U statistics can be used to derive the asymptotic distributions of the $\hat{u}_N^{(i,j)}s$. It can be shown that the joint distribution of

$$\sqrt{N}\{\hat{u}_N^{(i,j)} - u^{(i,j)},\ 1 \le i \le c-1,\ 1 \le j \le r-1\}$$

is asymptotically normal with mean vector zero and variance-covariance matrix 4Σ (see Serfling (1980)), where $\Sigma = (\sigma_{(i,j)(i',j')})$ is given by

$$\sigma_{(i,j)(i',j')} = cov(g^{(i,j)}(X_1, Y_1); g^{(i',j')}(X_1, Y_1)), \text{ with}$$

$$var(g^{(i,j)}) = (p_{i,j}p_{i+1,j+1}^2 + p_{i+1,j+1}p_{i,j}^2 + p_{i,j+1}p_{i+1,j}^2 + p_{i+1,j}p_{i,j+1}^2)/4 - (u^{(i,j)})^2$$
$$cov(g^{(i,j)}, g^{(i-1,j-1)}) = (p_{i,j}p_{i-1,j-1}p_{i+1,j+1})/4 - u^{(i,j)}u^{(i-1,j-1)}$$
$$cov(g^{(i,j)}, g^{(i-1,j+1)}) = (p_{i,j+1}p_{i+1,j}p_{i-1,j+2})/4 - u^{(i,j)}u^{(i-1,j+1)}$$
$$cov(g^{(i,j)}, g^{(i-1,j)}) = -\left[p_{i,j+1}p_{i-1,j}p_{i+1,j} + p_{i,j}p_{i+1,j+1}p_{i-1,j+1}\right]/4 - u^{(i,j)}u^{(i-1,j)}$$
$$cov(g^{(i,j)}, g^{(i,j-1)}) = -\left[p_{i,j}p_{i+1,j+1}p_{i+1,j-1} + p_{i+1,j}p_{i,j-1}p_{i,j+1}\right]/4 - u^{(i,j)}u^{(i,j-1)}$$
$$cov(g^{(i,j)}, g^{(s,t)}) = 0 \text{ for all other } (s,t).$$

To derive the asymptotic null distribution of the test statistic S, it suffices to compute the probability $P(S = m)$ under H_0 for any $m = 1, 2, \ldots (r-1)(c-1)$.

$$\begin{aligned} P(S &= m) \\ &= \sum_{C_m} P(\hat{u}_N^{(i_1,j_1)} > 0, \ldots, \hat{u}_N^{(i_m,j_m)} > 0,\ \hat{u}_N^{(i_{m+1},j_{m+1})} \le 0, \ldots, \\ &\qquad \hat{u}_N^{(i_{(r-1)(c-1)},j_{(r-1)(c-1)})} \le 0) \end{aligned}$$

where $\sum_{C_m}$ denotes summation over all possible combinations of m distinct elements $\{(i_1, j_1), \ldots, (i_m, j_m)\}$ from $\{(1,1), \ldots, (r-1, c-1)\}$ and where $\{(i_{m+1}, j_{m+1}),$

$\dots, (i_{(r-1)(c-1)}, j_{(r-1)(c-1)})\}\}$ is its complementary set. It can be shown that $P(S = m)$ converges to a sum of integrals of a multivariate normal over certain quadrants.

Since this is essentially a test based on signs, it is not very powerful for 2×2 tables. The proposed sign test is not recommended for 2×2 tables. For the case of $(r-1)(c-1) = 2$, that is, for 3×2 (or 2×3) tables, explicit formulas for $P(S = m)$ can be derived.

EXAMPLE 4.1: 3×2 tables. In this case, the U-statistic considered has values in a two-dimensional space. It is known that the mass attributed to the positive quadrant by a standard bivariate normal random vector with correlation coefficient ρ is given by $1/4 + (\arcsin \rho)/2\pi$, (see Johnson and Kotz (1976)), and this probability is invariant under the scale transformation.

Therefore,

$$\begin{aligned} P(S = 2) &= P(\hat{u}_N^{(1,1)} > 0,\ \hat{u}_N^{(2,1)} > 0) \\ &= \frac{1}{4} + \frac{1}{2\pi} \arcsin\left(-1(1 + \frac{p_{2.}}{p_{1.}p_{3.}})^{-\frac{1}{2}}\right), \\ P(S = 1) &= 1 - 2P(S = 2), \\ P(S = 0) &= P(\hat{u}_N^{(1,1)} \le 0,\ \hat{u}_N^{(2,1)} \le 0) = P(S = 2). \end{aligned}$$

Consider the following 3×2 table from the sample.

	y_1	y_2	total
x_1	23	13	36
x_2	1	1	2
x_3	20	31	51
total	44	45	89

The test statistic S is equal to 2. The proposed test of independence has a significance level of 0.048. In this case a chi-square test has a significance level of 0.077.

For higher dimensional tables, one may use Monte Carlo methods to evaluate multivariate normal orthant probabilities and then compute the asymptotic null distribution if the table dimensions satisfy the condition that $(r - 1)(c - 1) \le 20$. An efficient evaluation method is given by Evans and Schwartz (1986). Note that unlike the exact test procedures, the proposed sign test has the advantage that increasing sample sizes will not add difficulties to the computation of the null distribution. Hence the proposed sign test procedure is more appropriate for tables with large sample sizes.

EXAMPLE 4.2: Monte Carlo Simulations for Higher Dimensional Tables. Consider the following data set referred to by Kasser and Bruce (1969), BMDP (1979), and Nguyen and Sampson (1987).

Coronary Function and Activity for Patients Under Age 51

	Functional Class			
Active	None or Minimal	Moderate	Severe	Total
Very	4	2	0	6
Normal	8	14	2	24
Limited	1	2	4	7
Total	13	18	6	37

In this example the test statistic S is equal to 4. Using the Monte Carlo simulation method derived by Evans and Schwartz (1986), based on a sample of 10,000, the estimated level of significance is equal to 0.028492 with a standard error of 0.0000534. A chi-square test has a significance level of 0.011.

5. Discussion. For each exact test reviewed in Section 3, the test procedure requires calculation of the corresponding test statistic for each element of the conditional sample space, i.e. for each table having the same marginal totals as the observed tables. This kind of approach is feasible for small tables with small sample sizes, but when the sample size is large or when the table has much higher dimensions, full enumeration of the conditional sample space is impossible. See Gail and Mantel (1977) for an approximation of the number of $r \times c$ contingency tables with fixed marginals.

The sign test discussed in Section 4 does not have this restriction as it does not depend on the conditional sample space of tables having fixed marginals. The simulation method used for evaluation of the null distribution of the sign test for high dimensional tables is relatively efficient. See Evans and Schwartz (1986) for a discussion of the efficiency of the simulation method for evaluating the orthant probability of a multivariate normal distribution. Comparisons of the sign test with likelihood ratio tests using Goodman's association models are being considered. The comparison results will be discussed in another paper.

REFERENCES

AGRESTI, A. (1984). *Analysis of Ordinal Categorical Data.* John Wiley, New York.

AGRESTI, A., WACKERLY, D., and BOYETT, J. (1979). Exact conditional tests for cross classifications, approximation of attained significance levels. *Psychometrika* **44** 75–83.

ARMITAGE, P. (1955). Tests for linear trends in proportions and frequencies. *Biometrics* **11** 375–386.

BISHOP, Y.M.M., FIENBERG, S.E., and HOLLAND, P.W. (1975). *Discrete Multivariate Analysis.* The MIT Press, Cambridge.

EVANS, M. and SWARTZ, T. (1986). Monte Carlo computation of some multivariate normal probabilities. Technical Report No. 7, University of Toronto, Toronto, Canada.

GAIL, M. and MANTEL, N. (1977). Counting the number of $r \times c$ tables with fixed marginals. *J. Amer. Statist. Assoc.* **72** 859–862.

GOODMAN, L.A. (1985). The analysis of cross-classified data having ordered and/or unordered categories. *Ann. Statist.* **13** No. 1, 10–69.

GOODMAN, L.A. and KRUSKAL, W.H. (1954). Measures of association for cross classifications. *J. Amer. Statist. Assoc.* **49** 732–764.

GROVE, D.M. (1980). A test of independence against a class of ordered alternatives in a $2 \times c$ contingency table. *J. Amer. Statist. Assoc.* **75** 454–459.

GROVE, D.M. (1984). Positive association in a two-way contingency table: Likelihood ratio tests. *Comm. Stat.- Theor. Math.* **13** 931–945.

JOHNSON, N.L. and KOTZ, S. (1976). *Distributions in Statistics, Continuous Multivariate Distributions.* John Wiley, New york.

LEE, M.-L.T. (1988). Some cross-product difference statistics and a test for trends in ordered contingency tables. *Statist. and Prob. Letters* **7** 41–46.

LEHMANN, E.L. (1966). Some concepts of dependence. *Ann. Math. Statist.* **37** 1137–1153.

NGUYEN, T.T. and SAMPSON, A.R. (1987). Testing for positive quadrant dependence in ordinal contingency tables. *Naval Res. Logist. Quar.* **34** 859–877.

KASSER, I. and BRUCE, R.A. (1969). Comparative effects of aging and coronary heart diseases and submaximal and maximal exercise. *Circulation* **39** 759–774.

PATEFIELD, W.M. (1982). Exact tests for trends in ordered contingency tables. *Appl. Statist.* **31** No. 1, 32–43.

SERFLING, R.J. (1980). *Approximation Theorems of Mathematical Statistics.* John Wiley, New York.

DEPARTMENT OF MATHEMATICS
BOSTON UNIVERSITY
111 CUMMINGTON STREET
BOSTON, MA 02215

CLASSES OF MULTIVARIATE EXPONENTIAL AND MULTIVARIATE GEOMETRIC DISTRIBUTIONS DERIVED FROM MARKOV PROCESSES

By N.T. Longford

Educational Testing Service

We define a class of multivariate exponential distributions as the distributions of occupancy times in upwards skip-free Markov processes in continuous time. These distributions are infinitely divisible, and the multivariate gamma class defined by convolutions and fractions is a substantial generalization of the class defined by Johnson and Kotz (1972). Parallel classes of multivariate geometric and multivariate negative binomial distributions are constructed from occupancy times in "instant" upwards skip-free Markov chains. Maximum likelihood estimation and times series applications are discussed.

1. Introduction. The exponential distribution plays a central role in several fields of probability and statistics, and ranks in overall importance next to the normal distribution. While for the normal case we have a well established multivariate normal distribution, in the exponential case the situation is far from clear-cut. A variety of bivariate exponential distributions (**BVE**) have been defined in the past, some of them extendable to higher dimensions and to gamma distributions.

The univariate exponential distribution has a number of important characterizations. Multivariate extensions of some of these characterizations were used for construction of multivariate exponential distributions (**MVE**) by Marshall and Olkin (1967) (lack of memory property) and Paulson (1973) (a stochastic difference equation). Standard transformation techniques were exploited by Kibble (1941) using a χ^2-type derivation, and Moran (1969) using the distribution-function transformation and then the '**-log**' transformation to obtain a **MVE** from the multivariate normal via a multivariate uniform distribution. Gumbel (1960) explored possibilities of defining **BVE** and **MVE** classes based on the form of the joint distribution and density functions. Arnold (1975) constructed nested classes of **BVE**'s by repeated application of geometric compounding. His constructions are equally applicable to multivariate geometric distributions (**MVG**). Wang Zi Kun (1980) derived the distributions of occupancy times (sojourn times) in birth-death

AMS 1980 subject classifications. Primary 60E05; secondary 60G40.

Key words and phrases. Multivariate exponential, multivariate geometric, skip-free Markov chain.

processes; these distributions are **MVE**. This brief review of related research is not exhaustive; our research was motivated by these references. We extend the results of Wang Zi Kun to all upwards skip-free Markov processes on $\{0, 1, 2, \dots\}$, and define a parallel **MVG** class.

In Section 2 we give the definition of occupancy times and related notation, and construct a new class of **MVE** distributions. We will work mainly with moment generating functions (**mgf**), and we derive for them recursive formulae which involve the matrix of transition intensities of the underlying Markov process. Owing to infinite divisibility we have also a class of multivariate gamma distributions (**MVΓ**).

In Section 3 we construct a new class of **MVG** distributions from occupancy times in '**instant**' Markov chains. The constructions in Sections 2 and 3 are completely analogous, and there is a one-to-one correspondence between our **MVE** and **MVG** classes, which is also a one-to-one correspondence between the underlying stochastic processes. This one-to-one correspondence is an extension of the well-known relationship between the **mgf**'s of univariate exponentials and probability generating functions (**pgf**) of geometric distributions:

$$\varphi_\Theta(s) = P_p(s+1) \ \text{ for } \ \Theta = p^{-1} - 1,$$

where φ_Θ is the **mgf** of the exponential with parameter Θ, $\Theta/(\Theta - s)$, and P_p is the **pgf** of the geometric distribution with parameter p, $(1-p)/(1-pz)$.

In Section 4 we discuss maximum likelihood estimation with our **MVE** class and indicate some time series applications. We propose estimation methods based on the **mgf** because it has a much more tractable form than the density function. Our comments in the Section are equally applicable to the **MVG** class, although there the scope of applications is probably limited.

In Section 5 we discuss an alternative definition of an **MVE** class based on the generalization of the χ_2^2 distribution, and state a conjecture that the occupancy times and this generalization define the same class of **MVE**. The support for this conjecture is the equivalence between a pair of subclasses of these distributions, proved by Kent (1982).

2. Occupancy Times. Let $\{Z_t\}_{t \geq 0}$ be an upwards skip-free Markov process on the state space of the non-negative integers $\{0, 1, 2, \dots\}$ in continuous time $t \geq 0$, given by the matrix of transition intensities

$$Q = \begin{pmatrix} -\nu_0 & \lambda_0 & \cdot & \cdot & \cdot & & \\ \mu_1 & -\nu_1 & \lambda_1 & 0 & \cdot & \cdot & \cdot \\ \xi_{2,0} & \mu_2 & -\nu_2 & \lambda_2 & 0 & \cdot & \cdot \\ \xi_{3,0} & \xi_{3,1} & \mu_3 & -\nu_3 & \lambda_3 & 0 & \cdot \\ \cdot & \cdot & \cdot & \cdot & \cdot & \cdot & \cdot \\ \cdot & \cdot & \cdot & \cdot & \cdot & \cdot & \cdot \\ \cdot & \cdot & & \cdot & \cdot & \cdot & \cdot \end{pmatrix}, \tag{1}$$

where $\lambda_i > 0$ $(i \geq 0)$, $\mu_i \geq 0$ $(i \geq 1)$, $\xi_{i,j} \geq 0$ $(i \geq j+2 \geq 2)$, and all the row-totals of Q are equal to zero. We denote by Q_n the $n \times n$ upper left-hand corner submatrix of Q, corresponding to the states $0, 1, \ldots, n-1$. Let $S_n = \mathbf{diag}\ \{s_0, s_1, \ldots, s_{n-1}\}$ be the diagonal matrix with real numbers s_i on the diagonal; they will be subsequently used as the arguments of a joint **mgf**

$$\varphi(\mathbf{s}_n) = \int \ldots \int \mathbf{exp}\ (\mathbf{x}'\mathbf{s}_n) f(\mathbf{x}) d\mathbf{x}, \tag{2}$$

where $\mathbf{s}_n = (s_0, s_1, \ldots, s_{n-1})$. The index n will be omitted whenever its value is obvious from the context.

The first hitting time from a state k to a state $n \geq k$ is formally defined as

$$\tau_{k,n} = \mathbf{min}\ \{t; Z_t = n \mid Z_0 = k\}, \tag{3}$$

and the vector of occupancy times during passage from k to $n \geq k$ (denoted by $T_{k,n}$) is the decomposition of the first hitting time $\tau_{k,n}$ into the sum of the times that the Markov process Z has spent in each of the states $0, 1, \ldots, n-1$:

$$T_{k,n} = (\tau_{k,n}^{(0)}, \tau_{k,n}^{(1)}, \ldots, \tau_{k,n}^{(n-1)}),$$

where

$$\tau_{k,n}^{(h)} = \int I\{Z_t = h \mid Z_0 = k\} dt$$

($\mathbf{I}\{A\}$ is the indicator function for the event A, and the integral is over the interval $[0, \tau_{k,n}]$).

The joint **mgf** $\varphi_{k,n}(\mathbf{s})$ for the vector of occupancy times during the passage from k to $n \geq k$ can be derived using a backwards equations argument; the derivations below are similar to those for the first hitting times (**fht**) given by Rosenlund (1977). Firstly, owing to the strong Markov property of the process Z we have

$$T_{k,n} = T_{k,k+1} + T_{k+1,k+2} + \ldots + T_{n-1,n}, \tag{4}$$

with mutually independent summands, and trivially $T_{k,k} = (0, 0, \ldots, 0)$. Correspondingly for the **mgf**'s we have

$$\varphi_{k,n}(\mathbf{s}) = \varphi_{k,k+1}(\mathbf{s}) \varphi_{k+1,k+2}(\mathbf{s}) \ldots \varphi_{n-1,n}(\mathbf{s}), \tag{5}$$

and $\varphi_{k,k}(\mathbf{s}) \equiv 1$. The backwards equations for the 'one-step' **mgf** $\varphi_k^+(\mathbf{s}) = \varphi_{k,k+1}(\mathbf{s})$ can be expressed in the form

$$\begin{aligned} \varphi_k^+(\mathbf{s}) = (\nu_k - s_k)^{-1}[\lambda_k + \mu_k \varphi_{k-1,k+1}(\mathbf{s}) + \xi_{k,k-2} \varphi_{k-2,k+1}(\mathbf{s}) + \\ \ldots + \xi_{k,0} \varphi_{0,k+1}(\mathbf{s})], \end{aligned} \tag{6}$$

where $\varphi_0^+(\mathbf{s}) = \nu_0/(\nu_0 - s_0)$, and $\xi_{1,0} = 0$. The equation (6) can be reexpressed as

$$\varphi_k^+(\mathbf{s}) = \lambda_k/(\nu_k - s_k)[1 - \mu_k/(\nu_k - s_k)\varphi_{k-1}^+(\mathbf{s})$$
$$-\xi_{k,k-2}/(\nu_k - s_k)\varphi_{k-2,k}(\mathbf{s}) - \ldots - \xi_{k,0}/(\nu_k - s_k)\varphi_{0,k}(\mathbf{s})]^{-1}, \quad (7)$$

which implies that the occupancy times vector $T_k^+ = T_{k,k+1}$ is a convolution of a univariate exponential distribution and a geometric compound distribution. Hence $T_k^+(\varphi_k^+)$ is infinitely divisible, and owing to (5) so are all the occupancy times.

It is easy to show by induction, using (5) and (7), that $\varphi_k^+(\mathbf{s})$ is a ratio of polynomials

$$\varphi_k^+(\mathbf{s}) = \lambda_k R_k(\mathbf{s})/R_{k+1}(\mathbf{s}), \quad (8)$$

where

$$\begin{aligned} R_0(\mathbf{s}) &\equiv 1, \\ R_1(\mathbf{s}) &= \nu_0 - s_0, \\ R_2(\mathbf{s}) &= (\nu_0 - s_0)(\nu_1 - s_1) - \lambda_0\mu_1, \end{aligned}$$

and generally,

$$\begin{aligned} R_{k+1}(\mathbf{s}) &= (\nu_k - s_k)R_k(\mathbf{s}) - \mu_k\lambda_{k-1}R_{k-1}(\mathbf{s}) - \xi_{k,k-2}\lambda_{k-1}\lambda_{k-2}R_{k-2}(\mathbf{s}) \\ &- \ldots - \xi_{k,1}\lambda_{k-1}\ldots\lambda_2\lambda_1R_1(\mathbf{s}) - \xi_{k,0}\lambda_{k-1}\ldots\lambda_1\lambda_0, \end{aligned} \quad (9)$$

which is exactly the expansion for **det** $(Q_{k+1} - S_{k+1})$ with respect to the bottom row. Hence

$$R_n(\mathbf{s}) = \mathbf{det}\ (Q_n - S_n) \quad (10)$$

for $n \geq 1$. As a by-product we have the identity $\mathbf{det}(-Q_n) = \lambda_0\lambda_1\ldots\lambda_{n-1}$.

The vectors of occupancy times during passage from 0 to n define our class of n-variate exponential distributions. Their **mgf**'s have the form

$$\varphi_{0,n}(\mathbf{s}) = \lambda_0\lambda_1\ldots\lambda_{n-1}/R_n(\mathbf{s}), \quad (11)$$

where the polynomials R_n, linear in the variables $s_0, s_1, \ldots, s_{n-1}$, are generated recursively by (9).

The versions of the identities (8) and (9) for the birth-death process (all $\xi_{i,j}$ equal to 0) were obtained by Wang Zi Kun (1980).

The bivariate exponential distribution generated by occupancy times during passage from 0 to 2 has the **mgf**

$$\lambda_0\lambda_1/R_2(\mathbf{s}) = \lambda_0\lambda_1/[(\nu_0 - s_0)(\nu_1 - s_1) - \lambda_0\mu_1],$$

and the joint density

$$\lambda_0\lambda_1 \exp(-\nu_0 x_0 - \nu_1 x_1) L_0(\lambda_0\mu_1 x_0 x_1) \tag{12}$$

where

$$L_0(x) = \Sigma_k x^k/(k!)^2.$$

The distribution (12) has been previously defined by Downton (1970), and in a more general context by Kibble (1941).

We define for $h > 0$

$$L_h(x) = \Sigma_k x^k/k!\Gamma(k+h+1), \tag{13}$$

which is an analytic version of the Besel function of order h. The bivariate gamma distribution (**BVΓ**) with scale $\sigma > 0$ corresponding to (11) has the density

$$(\lambda_0\lambda_1)^\sigma (x_0 x_1)^{\sigma-1} \exp(-\nu_0 x_0 - \nu_1 x_1) L_\sigma(\lambda_0\mu_1 x_0 x_1). \tag{14}$$

The **BVE** distribution (11) has the mean $\{(1+\mu_1/\lambda_1)/\lambda_0, 1/\lambda_1\}$ and correlation $\mu_1/\nu_1 \in [0,1)$. No correlation corresponds to independence. The conditional exponential distributions defined from (11) by conditioning on x_0 or x_1 have linear regressions:

$$\mathbf{E}(X_0 \mid X_1 = x_1) = \lambda_0^{-1}(1 + \mu_1 x_1)$$

and

$$\mathbf{E}(X_1 \mid X_0 = x_0) = \nu_1^{-1}(1 + \lambda_0\mu_1 x_1/\nu_1).$$

The occupancy times during passage from 0 to $n > 2$ in birth-death processes form a conditionally independent sequence and their joint density can be partitioned into a product of conditional exponential densities, see Johnson and Kotz (1972) or Longford (1982). Such a sequence can be used to model an **AR**(1) times series, although the innovation distribution is difficult to describe.

The general trivariate exponential density has the form

$$\begin{aligned} &\lambda_0\lambda_1\lambda_2 \exp(-\nu_0 x_0 - \nu_1 x_1 - \nu_2 x_2) \\ &\times \Sigma_k L_k(\lambda_0\mu_1 x_0 x_1) L_k(\lambda_1\mu_2 x_1 x_1)(\lambda_0\lambda_1\xi_{2,0} x_0 x_1 x_2)^k/k!. \end{aligned} \tag{15}$$

The proof is given in the Appendix. Clearly this and densities for higher dimensions are not suitable for direct maximum likelihood estimation. An alternative approach to **MLE** is discussed in Section 4.

All bivariate marginals of the trivariate exponential (13) belong to the **BVE** class. The correlation matrix for the trivariate exponential is

$$\begin{pmatrix} 1 & & \\ 1-\lambda_1\lambda_2\tau & 1 & \\ 1-\nu_1\lambda_2\tau & 1-\lambda_1/\nu_2 & 1 \end{pmatrix}$$

and the means are $\{1/(\lambda_0\lambda_1\lambda_2\tau), \nu_2/\lambda_1\lambda_2, 1/\lambda_2\}$, where $\tau = 1/(\nu_1\nu_2 - \lambda_1\mu_2)$.

3. Multivariate Geometric Class. The constructions of the **MVE** class from occupancy times in Markov processes have their obvious analogues for discrete distributions in occupancy times in Markov chains. For example, the birth-death process has its analogue in the Markov chain which allows jumps only one step up or down (discrete random walk, skip-free in both directions). However, the distributions of the occupancy times in such Markov chains have a more complex structure than their birth-death analogues. Rather than give an example we return to this point in the conclusion of this Section.

Let

$$A = \begin{pmatrix} a_0 & u_0 & 0 & . & . & . & \\ d_1 & a_1 & u_1 & 0 & . & . & . \\ r_{2,0} & d_2 & a_2 & u_2 & 0 & . & . & . \\ r_{3,0} & r_{3,1} & d_3 & a_3 & u_3 & 0 & . & . \\ . & . & . & . & . & . & . \\ . & . & & . & . & . & . & . \\ . & . & & . & . & . & . & . \end{pmatrix}$$

be a matrix of transition probabilities of an upwards skip-free Markov chain (all entries non-negative, $u_i > 0$ for all i, row totals equal to 1). In analogy with Section 2 we denote by A_n the $n \times n$ upper left-hand corner submatrix of A, and define the first hitting times and occupancy times vectors.

The occupancy time in a state k during passage from 0 to n $(0 \le k \le n)$ is a compound distribution of the individual waiting times in the state which are geometric starting at 1. To obtain a geometric distribution we could either subtract the constant 1 from the occupancy time, or subtract 1 from every waiting time. We choose the latter option, and refer to the underlying Markov process as the '**instant**' Markov chain. In this Markov chain zero waiting times have positive probability, and so a sequence of states can be visited within the same time-instant. Our main motivation for this definition of a Markov process is to construct a class of multivariate geometric distributions (**MVG**) with analogous structure to the **MVE** class defined in Section 2.

For the first hitting times and the occupancy times in instant Markov chains given by the probability transition matrix A we use the notation identical to that introduced in Section 2, $\tau_{k,n}$ or τ_k^+, and $T_{k,n}$ or T_k^+, respectively. The vector $\mathbf{z} = (z_0, z_1, z_2, \ldots)$ will be used as the argument in the probability generating functions (**pgf**) $P_{k,n}$ for the occupancy times vectors:

$$P_{k,n}(\mathbf{z}) = \mathbf{E}(\mathbf{z}^{T_{k,n}}).$$

The formula (5) has a direct analogue in

$$P_{k,n}(\mathbf{z}) = P_k^+(\mathbf{z})P_{k+1}^+(\mathbf{z})\ldots P_{n-1}^+(\mathbf{z}), \tag{16}$$

where $P_h^+(\mathbf{z}) = P_{h,h+1}(\mathbf{z})$, and the backwards equations yield

$$P_k^+(\mathbf{z}) =$$
$$(1 - a_k z_k)^{-1}[u_k + d_k P_{k-1,k+1}(\mathbf{z}) + r_{k,k-2}P_{k-2,k+1}(\mathbf{z}) + \ldots + r_{k,0}P_{0,k+1}(\mathbf{z})] \tag{17}$$

with the solution in a recursive form

$$\begin{aligned} P_k^+(\mathbf{z}) = \\ u_k/(1 - a_k z_k)[1 - d_k/(1 - a_k z_k)P_{k-1}^+(\mathbf{z}) - r_{k,k-2}/(1 - a_k z_k)P_{k-2,k}(\mathbf{z}) + \\ \ldots + r_{k,0}/(1 - a_k z_k)P_{0,k}(\mathbf{z})]^{-1}, \end{aligned} \tag{18}$$

$$P_0^+(\mathbf{z}) = (1 - a_0)/(1 - a_0 z_0).$$

The formula (18) is a convolution of a univariate geometric and a geometric compound distribution. This, together with (16), implies infinite divisibility of all occupancy times distributions. We declare the class of all distributions generated by occupancy times during passage from 0 to n as our **MVG** class. This definition can be extended to the class of multivariate negative binomial distributions (**MVNb**) in the obvious way.

The identities (17) and (18), compared with (6) and (7) indicate a one-to-one correspondence between occupancy times in continuous and discrete processes. Moreover, we have a one-to-one correspondence between the underlying processes: If

$$\begin{aligned} u_k &= \lambda_k/\nu_k & (k \geq 0) \\ d_k &= \mu_k/\nu_k & (k \geq 1) \\ r_{k,h} &= \xi_{k,h}/\nu_k & (k \geq h + 2 \geq 2) \end{aligned}$$

then

$$\varphi_{k,n}(\mathbf{s}) = P_{k,n}(\mathbf{s} + \mathbf{1}), \tag{19}$$

where $\mathbf{1} = (1, 1, \ldots)$. This one-to-one correspondence between the **MVE** and **MVG** classes is the natural extension of the one-to-one correspondence for the univariate exponential and geometric distributions.

In complete analogy with the continuous case we obtain the identity

$$P_{0,n}(\mathbf{z}) = u_0 u_1 \ldots u_{n-1}/T_n(\mathbf{z}), \tag{20}$$

where T_n are polynomials linear in $z_0, z_1, \ldots, z_{n-1}$, generated by the recursive formula

$$\begin{aligned} &T_{n+1}(\mathbf{z}) = (1 - a_n z_n)T_n(\mathbf{z}) - d_n u_{n-1} T_{n-1}(\mathbf{z}) - r_{n,n-2} u_{n-2} u_{n-1} T_{n-2}(\mathbf{z}) - \\ &\ldots - r_{n,0} u_0 u_1 \ldots u_{n-1}, \end{aligned} \tag{21}$$

with $T_0(\mathbf{z}) = 1$ and $T_1(\mathbf{z}) = 1 - a_0 z_0$. It is easy to show by induction that

$$T_n(\mathbf{z}) = \mathbf{det}\,\{I_n - A_n(\mathbf{z})\}, \tag{22}$$

where $A_n(\mathbf{z})$ is formed from the matrix A_n by replacing its diagonal elements a_k by $a_k z_k$ $(0 \leq k \leq n)$, and is I_n the $n \times n$ unit matrix.

For the bivariate and trivariate geometric distributions the joint probabilities and moments (correlations) can be obtained by standard methods, in complete analogy with the exponential case. For higher dimensions the formulae are not tractable.

The backwards equations for the occupancy times in classical Markov chains also define a class of **MVG** distributions (and are infinitely divisible), but the one-to-one correspondence with our **MVE** has a substantially more complex and less natural form. Even the distributions of the first hitting times in Markov chains have a substantially more complex structure than the **fht**'s in continuous time; for details see Kent and Longford (1983).

4. Maximum Likelihood Estimation and Time Series Applications. Maximum likelihood estimation for the **BVE** and **BVΓ** given by the densities (12) and (14), respectively, can be efficiently carried out by application of standard numerical methods using some well-known recursive formulae for computation of Bessel functions and ratios of Bessel functions.

Since $dL_k(x)/dx = L_{k+1}(x)$, the derivatives of the log-likelihood involving the bivariate densities of the form (14) involve ratios of Bessel functions, $L_{k+1}(x)/L_k(x)$. Efficient recursive algorithms for calculation of such ratios were derived by Amos (1974); other useful identities are given in Abramowicz and Stegun (1972). The natural parameter space for the **BVE** is not an open space because of the boundary $\mu_1 = 0$. It is easy to show that the maximum likelihood estimate of μ_1 is positive if and only if the sample covariance $N^{-1}\Sigma_i x_{0i} x_{1i} - \bar{x}_0 \bar{x}_1$ is positive; see Longford (1982) where other numerical details are discussed.

The **MVE** class generated as the occupancy times vectors from birth-death processes have conditionally independent components, and they can be used for modelling of exponential **AR**(1) time series. Since the likelihood for such a time series factors into univariate conditional exponential densities, direct maximum likelihood is feasible.

The form of the density of the general trivariate exponential distribution renders standard maximum likelihood methods impossible, even though the corresponding **mgf**'s have a very simple structure. Feuerverger and McDonough (1981) have developed procedures for maximum likelihood estimation based on the empirical **mgf** and proved that these procedures can be 'fine-tuned' to arbitrarily high relative efficiency, given some information about the estimated parameters. Their methods appear to be tailor-made for our classes of multivariate distributions (**MVE**, **MVΓ**, **MVG**, and **MVNb**) because they offer a unified approach to estimation in all these classes with generating functions of similar functional form. The main practical point in application of the methods of Feuerverger and McDonough is in determining the number and location of the points in which the **mgf/pgf** would be approximated. These issues could be explored in the special case of **BVE** where direct maximum likelihood estimation is available. It is not clear though to what extent these results could be generalized to **MVE**. Of course, the moment method of estimation is another tractable option, owing to the simple form of the **mgf/pgf**.

The **MVE** class of n-variate distributions ($n > 2$) has the subclass of n independent univariate exponentials and the larger subclass of the distributions with conditionally independent components (**AR**(1), generated from birth-death processes). It appears natural to define a whole set of nested classes of **MVE** distributions by allowing the generating Markov process to have the first $2, 3, \ldots, n-1$ non-zero subdiagonals in the transition intensities matrix Q. If Q has only the first subdiagonal non-zero, we have an **AR**(1) time series. We conjecture that if the first k subdiagonals are non-zero then the resulting **MVE** has an **AR**(k) structure, i.e., it forms a k-step conditionally independent sequence: Z_h and Z_{k+h+1} are conditionally independent, given $Z_{h+1}, Z_{h+2}, \ldots, Z_{k+h}$. Definition of these subclasses of **MVE** imposes a structure upon the entire **MVE** class that could be used for description of the complexity of the correlation structure of an exponential time series or a multivariate sample from **MVE**.

5. MVE As a Generalized Chi-square Distribution. Let $\mathbf{X}_1 = (X_{11}, X_{12}, \ldots, X_{1n})$ and $\mathbf{X}_2 = (X_{21}, X_{22}, \ldots, X_{2n})$ be a pair of independent and identically distributed normal random vectors with mean $\mathbf{0}$ and variance matrix Ω. Then the random vector $\mathbf{Y} = (Y_1, Y_2, \ldots, Y_n)$ given by $Y_k = X_{1k}^2 + X_{2k}^2$ defines an n-variate exponential distribution. The original idea for this definition is due to Kibble (1941). We will refer to this derivation as the generalized χ_2^2. It is easy to show that the **mgf** for **Y** is

$$\mathbf{det}\,(1/2\,\Omega^{-1}) / \,\mathbf{det}\,(1/2\,\Omega^{-1} - S_n),$$

which closely resembles the functional form of our **MVE**, see (11) and (10). Kent (1982) has in fact proved that the subclass of our **MVE** class arising from birth-death processes coincides with the subclass of the generalized χ_2^2 distributions derived from variance matrices Ω for which Ω^{-1} is tridiagonal.

An obvious extension of this identity is the following conjecture: The distributions of the **MVE** with **AR**(k) structure (as defined in Section 4) coincide with the generalized χ_2^2 distributions derived from variance matrices Ω such that Ω^{-1} have k non-zero rows below and above the main diagonal. The proof of Kent (1982) cannot be extended for this general proposition, and we do not have an alternative method of proof.

REFERENCES

ABRAMOWICZ, M. and STEGUN, I.A. (1972). *Handbook of Mathematical Functions.* Dover, New York.

AMOS, D.E. (1974). Computation of modified Bessel functions and their ratios. *Math. Comp.* **28** 239–251.

ARNOLD, B.C. (1975). Multivariate geometric compounding. *Sankhya A* **37** 164–173.

BLOCK, H.W. (1977). A characterization of a bivariate exponential distribution. *Ann. Statist.* **5** 808–812.

DOWNTON, F. (1970). Bivariate exponential distributions in reliability theory. *J. Roy. Statist. Soc.* B **34** 129–131.

GUMBEL, E.J. (1960). Bivariate exponential distributions. *Biometrika* **44** 265–268.

JOHNSON, N.L. and KOTZ, S. (1972). Distributions in statistics: *Continuous Multivariate Distributions.* Wiley Series in Probability and Mathematical Statistics–Applied. John Wiley and Sons, Inc., New York.

KENT, J.T. (1982). The appearance of a multivariate exponential distribution in sojourn times for birth-death and diffusion processes. In *Probability, Statistics and Analysis* (J.F.C. Kingman and G.E.H. Reuter, eds.), Cambridge University Press, Cambridge.

KENT, J.T. and LONGFORD, N.T. (1983). An eigenvalue decomposition for first hitting times in random walks. *Z. Wahrsch. verw. Geb.* **63** 71–84.

KIBBLE, W.F. (1941). A two-variate gamma-type distribution. *Sankhya* **5** 137–150.

LAWRANCE, A.J. and LEWIS, P.A.W. (1985). Modelling and residual analysis of nonlinear autoregressive time series in exponential variables. *J. Roy. Statist. Soc.* B **47** 165–202.

LONGFORD, N.T. (1982). Hitting times in stochastic processes. Ph.D. Thesis, Leeds University, U.K.

MARSHALL, A.W. and OLKIN, I. (1967). A multivariate exponential distribution. *J. Amer. Statist. Assoc.* **62** 30–44.

MORAN, P.A.P. (1969). Statistical inference with bivariate gamma distributions. *Biometrika* **56** 627–634.

PAULSON, A.S. (1973). A characterization of the exponential distribution and a bivariate exponential distribution. *Sankhya A* **35** 69–78.

ROSENLUND, S.I. (1977). Upwards passage times in the non-negative birth-death process. *Scand. J. Statist.* **4** 90–92.

WANG, Z.K. (1980). Sojourn times and first passage times for birth and death processes. *Sci. Sinica* **23** 269–279.

APPENDIX

The density of trivariate exponential distribution

The mgf of the trivariate exponential distibution is

$$\frac{\lambda_0\lambda_1\lambda_2}{(\nu_0-s_0)(\nu_1-s_1)(\nu_2-s_2)-\lambda_0\mu_1(\nu_2-s_2)-\lambda_1\mu_2(\nu_0-s_0)-\xi_{2,0}\lambda_0\lambda_1}$$

$$= \frac{\lambda_0\lambda_1\lambda_2}{(\nu_0-s_0)(\nu_1-s_1)(\nu_2-s_2)}\cdot\sum_{k=0}^{\infty}\frac{1}{(\nu_1-s_1)^k}\left[\frac{\lambda_0\mu_1}{\nu_0-s_0}+\frac{\lambda_1\mu_2}{\nu_2-s_2}+\frac{\xi_{2,0}\lambda_0\lambda_1}{(\nu_0-s_0)(\nu_2-s_2)}\right]^k$$

$$= \lambda_0\lambda_1\lambda_2\sum_{k_1=0}^{\infty}\sum_{k_2=0}^{\infty}\sum_{k_3=0}^{\infty}\frac{(k_1+k_2+k_3)!}{k_1!k_2!k_3!}\frac{1}{(\nu_1-s_1)^{k_1+k_2+k_3+1}}\cdot\frac{1}{(\nu_0-s_0)^{k_1+k_3+1}}$$
$$\cdot\frac{1}{(\nu_2-s_2)^{k_2+k_3+1}}\cdot(\lambda_0\mu_1)^{k_1}(\lambda_1\mu_2)^{k_2}(\xi_{2,0}\lambda_0\lambda_1)^{k_3}$$

The summands above are independent gammas, and the corresponding density is

$$\lambda_0\lambda_1\lambda_2\ \exp\left(-\nu_0x_0-\nu_1x_1-\nu_2x_2\right)\cdot\sum_{k_1}\sum_{k_2}\sum_{k_3}\frac{1}{k_1!k_2!k_3!(k_1+k_3)!(k_2+k_3)!}$$
$$\cdot(\lambda_0\mu_1x_0x_1)^{k_1}(\lambda_1\mu_2x_1x_2)^{k_2}(\lambda_0\lambda_1\xi_{2,0}x_0x_1x_2)^{k_3}$$
$$= \lambda_0\lambda_1\lambda_2\ \exp\left(-\nu_0x_0-\nu_1x_1-\nu_2x_2\right)$$
$$\sum_{k=0}^{\infty}\frac{(\lambda_0\lambda_1\xi_{2,0}x_0x_1x_2)^k}{k!}L_k(\lambda_0\mu_1x_0x_1)L_k(\lambda_1\mu_2x_1x_2)$$

Note that if $\xi_{2,0}=0$ (conditional independence), this density collapses to

$$\lambda_0\lambda_1\lambda_2\ \exp\left(-\nu_0x_0-\nu_1x_1-\nu_2x_2\right)\cdot L_0(\lambda_0\mu_1x_0x_1)L_0(\lambda_1\mu_2x_1x_2).$$

Educational Testing Service
Research Statistics Group
21-T Rosedale Road
Princeton, NJ 08541

MULTIVARIATE DISTRIBUTIONS GENERATED FROM MIXTURES OF CONVOLUTION AND PRODUCT FAMILIES

BY ALBERT W. MARSHALL[1,2] AND INGRAM OLKIN[1]

University of British Columbia and Western Washington University, and Stanford University

A number of standard univariate distributions can be represented as mixtures of other standard distributions. In this paper such mixture representations are exploited to generate families of multivariate distributions with given marginals. Attention is confined to mixtures of parametric families where the parameter appears as the order of a convolution or as a power of the distribution or survival function. The mixture structure yields properties of the generated multivariate distributions such as total positivity, association and infinite divisibility. Examples obtained include the bivariate Poisson, binomial, negative binomial, normal, chi-square, logistic and Pareto distributions.

1. Introduction. For any given parametric family of distributions $F(\cdot \mid \theta)$, it is possible to regard the parameter θ as the value of a random variable Θ with distribution G, say. Then $F(\cdot \mid \theta)$ is a conditional distribution given $\Theta = \theta$ and the corresponding unconditional distribution

$$H(x) = \int F(x \mid \theta) dG(\theta) \tag{1}$$

is a mixture.

Here, both x and θ can be vectors, often of different dimensions. Many examples arise in which θ is a scalar and $F(\cdot \mid \theta)$ is the product of its marginals. Then (1) takes the form

$$H(\mathbf{x}) = \int \Pi F_i(x_i \mid \theta) dG(\theta), \tag{2}$$

[1]Supported in part by the National Science Foundation, and

[2] The Natural Sciences and Engineering Research Council of Canada.

AMS Classification. 60E05, 62E10.

Key words and phrases. Associated random variables, total positivity, multivariate distributions with given marginals, infinite divisibility, bivariate Poisson distribution, bivariate binomial distribution, bivariate negative binomial distribution, bivariate Pareto distribution.

where G and each F_i is a univariate distribution function. Clearly, multivariate distributions H of the form (1) or (2) have univariate marginals of the form (2). Of course many univariate examples are well known.

This paper is concerned with properties and examples of mixtures of the form (1) or (2) for two kinds of parametric families $\{F(\cdot \mid \theta) : \theta \in A\}$ which we call "convolution families" and "product families." These families arise in a natural fashion which is here described in a univariate setting.

Suppose that $X_1, \cdots, X_n$ are independent random variables with common distribution F, and suppose that $\Theta \geq 0$ is a random nonnegative integer having distribution G. Denote the θ-th convolution of F with itself by $F^{\theta *}$. With the conventions that $F^{\theta *}$ is degenerate at 0 and that an empty sum is 0, the random variable $U = X_1 + \cdots + X_\Theta$ has the distribution function H given by

$$H(x) = \int_0^\infty F^{\theta *}(x) dG(\theta). \tag{3}$$

In case F is infinitely divisible, (3) has meaning and H is a distribution function whenever G satisfies $G(0-) = 0$. Distributions of the form (3) are often called *compound distributions* (see, e.g., Feller, 1968, p. 286).

Again, suppose that $X_1, X_2, \cdots$ are independent random variables with common distribution F and suppose this time that $\Theta \geq 1$ is a random positive integer having distribution G. Then the random variables

$$V = \min(X_1, \cdots, X_\Theta), \quad W = \max(X_1, \cdots, X_\Theta)$$

have respective distributions H and K given by

$$\bar{H}(x) = \int \bar{F}^\theta(x) dG(\theta), \tag{4}$$

$$K(x) = \int F^\theta(x) dG(\theta), \tag{5}$$

where for any distribution function L, $\bar{L}$ is the corresponding survival function. The mixtures (3)–(5) give rise to the following definition.

DEFINITION 1.1. Let $\mathcal{F} = \{F(\cdot \mid \theta) : \theta \in A\}$ be an indexed family of n-dimensional distributions with index set $A \subset \mathcal{R}^k$ satisfying

$$\alpha \in A,\ \beta \in A \Rightarrow \alpha + \beta \in A. \tag{6}$$

$\mathcal{F}$ is said to be a *convolution family* if

$$F(\cdot \mid \alpha) * F(\cdot \mid \beta) = F(\cdot \mid \alpha + \beta), \quad \alpha, \beta \in A; \tag{7}$$

$\mathcal{F}$ is said to be a *survival product family* if

$$\bar{F}(\cdot \mid \alpha)\, \bar{F}(\cdot \mid \beta) = \bar{F}(\cdot \mid \alpha + \beta), \quad \alpha, \beta \in A; \tag{8}$$

finally $\mathcal{F}$ is said to be a *distribution product family* if

$$F(\cdot \mid \alpha)\, F(\cdot \mid \beta) = F(\cdot \mid \alpha + \beta), \quad \alpha, \beta \in A. \tag{9}$$

Convolution, survival product and distribution product families can be defined as semi-groups under the appropriate operation without the aid of an index set, but in this paper the index plays an important role.

EXAMPLE 1.2. For any distribution F, $\{F^{k*} : k = 0, 1, \ldots\}$ is a convolution family. In case F is a Bernoulli distribution with parameter p, F^{k*} is a binomial distribution with parameters k and p, $k = 0, 1, \ldots$.

More generally, if $F_1, \ldots, F_\ell$ is a finite collection of distributions each having a support in $\mathcal{R}^n$, then the set of all distributions of the form

$$F_1^{k_1 *} * \cdots * F_\ell^{k_\ell *}$$

is a convolution family.

EXAMPLE 1.3. The prototype survival product family is the family of univariate exponential distributions. For any univariate distribution function F, $\{\bar{F}^\theta \mid \theta > 0\}$ is a survival product family of distributions with proportional hazards. Some, but not all, bivariate distributions can be used in the same way to generate a survival product family of bivariate distributions (see Theorem 3.4).

In the study of both convolution families (Section 2) and product families (Section 3), the notions of total positivity and association play an important role. Some results concerning these notions are reviewed in an Appendix (Section 4).

2. Convolution Families.

Convolution families involve infinite divisibility as well as the dependency property of total positivity (see Section 4). These properties sometimes carry over to mixtures and sometimes can be easily obtained from mixture representations.

Convolution families combine in obvious ways to give new convolution families.

OBSERVATION 2.1. If $\{F_{(i)}^{\theta *} : \theta \in A_i\}$ is a convolution family of k_i-variate distributions, $i = 1, 2$, then the distributions of the form

$$F(\mathbf{x}_1, \mathbf{x}_2 \mid \theta_1, \theta_2) = F_{(1)}^{\theta_1 *}(\mathbf{x}_1)\, F_{(2)}^{\theta_2 *}(\mathbf{x}_2), \quad \theta_1 \in A_1, \ \theta_2 \in A_2,$$

constitute a convolution family of $(k_1 + k_2)$-variate distributions.

OBSERVATION 2.2. If $\{F_{(i)}^{\theta *} : \theta \in A_i\}$ is a convolution family of n-variate distributions, $i = 1, 2$, then the distributions of the form

$$F(\mathbf{x} \mid \theta_1, \theta_2) = (F_{(1)}^{\theta_1 *} * F_{(2)}^{\theta_2 *})(\mathbf{x}), \quad \theta_1 \in A_1, \ \theta_2 \in A_2$$

form a convolution family of n-variate distributions.

For the study of mixtures of convolution families, a basic fact is that a convolution of mixtures is a mixture of convolutions.

LEMMA 2.3. *If* $H_{(i)}(\mathbf{x}) = \int F_{(i)}(\mathbf{x} \mid \theta) dG_i(\theta), \quad i = 1, 2$ *and*

$$F(\mathbf{x} \mid \theta, \eta) = \int F_{(1)}(\mathbf{x} - \mathbf{t} \mid \theta) dF_{(2)}(\mathbf{t} \mid \eta),$$

then

$$(H_{(1)} * H_{(2)})(\mathbf{x}) = \int\int F(\mathbf{x} \mid \theta, \eta)\, dG_{(1)}(\theta) dG_{(2)}(\eta).$$

PROOF.

$$\begin{aligned}
&(H_{(1)} * H_{(2)})(\mathbf{x}) \\
= &\int H_{(1)}(\mathbf{x} - \mathbf{z})\, dH_{(2)}(\mathbf{z}) = \int_{\mathbf{z}} \int_{\theta} F_{(1)}(\mathbf{x} - \mathbf{z} \mid \theta)\, dG_{(1)}(\theta)\, dH_{(2)}(\mathbf{z}) \\
= &\int_{\theta} \int_{\mathbf{z}} F_{(1)}(\mathbf{x} - \mathbf{z} \mid \theta)\, dH_{(2)}(\mathbf{z})\, dG_{(1)}(\theta) = \int_{\theta} \int_{\mathbf{z}} H_{(2)}(\mathbf{x} - \mathbf{z})\, dF_{(1)}(\mathbf{z} \mid \theta)\, dG_{(1)}(\theta) \\
= &\int_{\theta} \int_{\mathbf{z}} \int_{\eta} F_{(2)}(\mathbf{x} - \mathbf{z} \mid \eta)\, dG_{(2)}(\eta)\, dF_{(1)}(\mathbf{z} \mid \theta)\, dG_{(1)}(\theta) \\
= &\int_{\theta} \int_{\eta} F(\mathbf{x} \mid \theta, \eta)\, dG_{(2)}(\eta)\, dG_{(1)}(\theta). \quad \|
\end{aligned}$$

Infinite Divisibility in Convolution Families.

LEMMA 2.4. *If* $\{F(\cdot \mid \theta) : \theta > 0\}$ *is a convolution family then* $F(\cdot \mid 1)$ *is infinitely divisible and*

$$F(\cdot \mid \theta) = F^{\theta *}(\cdot \mid 1).$$

PROOF. Let ϕ_θ be the characteristic function of $F(\cdot \mid \theta), \theta > 0$. From (7) it follows that

$$\phi_\theta\, \phi_\eta = \phi_{\theta+\eta}, \quad \theta, \eta > 0.$$

This functional equation has the solution $\phi_\theta = \phi_1^\theta$ (Aczél, 1966, p. 36). ||

In the following theorem, the assumption is made that the index set A is a convex cone. A subset $\mathcal{T}$ of a convex cone is said to be a *frame* for the cone if $\mathcal{T}$, but no proper subset of $\mathcal{T}$, spans the cone positively.

THEOREM 2.5. *If* $\{F(\cdot \mid \theta) : \theta \in A\}$ *is a convolution family indexed by a convex cone* A, *then for all* $\theta \in A$, $F(\cdot \mid \theta)$ *is infinitely divisible. If the convex cone* A *has a finite frame* $\mathcal{T} = \{t_1, \ldots, t_\ell\}$, *then* F *is of the form*

$$F(\cdot \mid \theta) = F^{\theta_1 *}(\cdot \mid t_1) * \cdots * F^{\theta_\ell *}(\cdot \mid t_\ell),$$

where $\theta = \sum_1^\ell \theta_i t_i$.

PROOF. This result follows from (7) and Lemma 2.4. ||

Various versions of the following results can be found in the literature. Lemma 2.6 is given for the univariate case by Keilson and Steutel (1974, p. 116). Theorem 2.8 in one dimension is due to Feller (1971, p. 538). With the assumption that $F(\cdot \mid \theta)$ is infinitely divisible, Theorem 2.8 is given in a very general setting by Kent (1981), who also lists some additional relevant references.

LEMMA 2.6. *If* $\{F(\cdot \mid \theta) : \theta \in A\}$ *is a convolution family of multivariate distributions and*

$$H_{(i)}(\mathbf{x}) = \int F(\mathbf{x} \mid \theta)\, dG_{(i)}(\theta), \quad i = 1, 2,$$

then

$$(H_1 * H_2)(\mathbf{x}) = \int F(\mathbf{x} \mid \theta)\, d(G_1 * G_2)(\theta).$$

PROOF. In Lemma 2.3, take $F_{(1)} = F_{(2)} = F$. Since $\{F(\cdot \mid \theta) : \theta \in A\}$ is a convolution family, the distribution $F(\cdot \mid \theta, \eta)$ of Lemma 2.3 is just $F(\cdot \mid \theta + \eta)$ and consequently the result follows from Lemma 2.3. ||

THEOREM 2.7. *If* $\{F(\cdot \mid \theta) : \theta \in A\}$ *and* $\{G(\cdot \mid \alpha) : \alpha \in B\}$ *are convolution families, and if*

$$H(\mathbf{x} \mid \alpha) = \int F(\mathbf{x} \mid \theta)\, dG(\theta \mid \alpha), \quad \alpha \in B, \ \mathbf{x} \in \mathcal{R}^n, \tag{10}$$

then $\{H(\cdot \mid \alpha), \alpha \in B\}$ *is a convolution family.*

PROOF. This is immediate from Lemma 2.6. ||

THEOREM 2.8. *If* $\{F(\cdot \mid \theta) : \theta \in A\}$ *is a convolution family and* G *is an infinitely divisible distribution, then* H *given by (10) is infinitely divisible. Moreover,*

$$H^{\alpha *}(\mathbf{x}) = \int F(\mathbf{x} \mid \theta)\, dG^{\alpha *}(\theta), \quad \alpha > 0, \ \mathbf{x} \in \mathcal{R}^n. \tag{11}$$

PROOF. Suppose that for some positive integer m, $\alpha = 1/m$ so that $(G^{\alpha *})^{m *} = G$. If $H^{\alpha *}$ is defined by (11) then by Lemma 2.6 $(H^{\alpha *})^{m *} = H$. This proves that H is infinitely divisible and (11) is satisfied for $\alpha = 1/m$, $m = 1, 2, \ldots$. Again from Lemma 2.6, it follows immediately that (11) holds for rational α and the proof is completed by a limiting argument. ||

Total Positivity in Convolution Families.

The following theorem shows how the positive dependency notion of total positivity (see Section 4) arises in the context of convolution families. There, $\tilde{F}$ can be taken to be either F or $\bar{F}$. For typographical simplicity, $\overline{F^{\theta *}}$ is written $\bar{F}^{\theta *}$.

THEOREM 2.9.

(i) If $\tilde{F}(x \mid \theta) = \tilde{F}^{\theta *}(x)$, $\theta = 0, 1, 2, \ldots$ *where* F *is a univariate distribution function such that* $F(0-) = 0$ *and* $\tilde{F}(x-y)$ *is* TP_2 *in* x *and* y*, then* $\tilde{F}(x \mid \theta)$ *is* TP_2 *in* x *and* θ.

(ii) If $\tilde{F}(x \mid \theta) = \tilde{F}^{\theta *}(x)$, $\theta \geq 0$, *where* F *is a univariate infinitely divisible distribution function such that* $F(0-) = 0$ *and* $\tilde{F}^{\theta *}(x-y)$ *is* TP_2 *in* x *and* y *for all* θ*, then* $\tilde{F}(x \mid \theta)$ *is* TP_2 *in* x *and* θ.

(iii) If $F(x \mid \theta) = F^{\theta *}(x)$, $\theta = 0, 1, 2, \ldots$ *where* F *is a univariate distribution function with a density* f *such that* $f(x-y)$ *is* TP_2 *in* x *and* y*, then* $F(\cdot \mid \theta)$ *has a density* $f(\cdot \mid \theta)$ *which is* TP_2 *in* x *and* $\theta = 0, 1, \ldots$.

PROOF. Let $x_1 < x_2$ and $\theta_1 < \theta_2$, and suppose that $\tilde{F}$ is $\bar{F}$. Then

$$\begin{vmatrix} \bar{F}(x_1 \mid \theta_1) & \bar{F}(x_1 \mid \theta_2) \\ \bar{F}(x_2 \mid \theta_1) & \bar{F}(x_2 \mid \theta_2) \end{vmatrix} = \begin{vmatrix} \bar{F}^{\theta_1 *}(x_1) & \bar{F}^{\theta_2 *}(x_1) \\ \bar{F}^{\theta_1 *}(x_2) & \bar{F}^{\theta_2 *}(x_2) \end{vmatrix}$$

$$= \int_0^\infty \begin{vmatrix} \bar{F}^{\theta_1 *}(x_1) & \bar{F}^{\theta_1 *}(x_1 - u) \\ \bar{F}^{\theta_1 *}(x_2) & \bar{F}^{\theta_1 *}(x_2 - u) \end{vmatrix} df^{(\theta_2 - \theta_1)*}(u) \geq 0$$

because $\bar{F}(x-y)$ is TP_2 in x and y implies that the same is true for $\bar{F}^{\theta *}$, $\theta = 0, 1, \ldots$ (Barlow and Proschan, 1975, p. 100), and this means the integrand is nonnegative. The proofs for other cases are similar. For a proof of (iii), see Karlin (1968, p. 150). ||

THEOREM 2.10. *Let*

$$\tilde{H}(\mathrm{x}) = \int \Pi \tilde{F}^{\theta *}(x_i) \, dG(\theta) \tag{12}$$

where each F_i *is a univariate distribution function such that* $F_i(0-) = 0$.

(i) If for each i, $\tilde{F}_i(x-y)$ *is* TP_2 *in* x *and* y*, then* $\tilde{H}$ *is* MTP_2.

(ii) If for each i, $f_i(x-y)$ *is* TP_2 *in* x *and* y*, then* H *has a density* h *that is* MTP_2.

PROOF. This is immediate from Theorem 2.9 and Theorem 4.15. ||

Examples

EXAMPLE 2.11. BIVARIATE POISSON DISTRIBUTION. In the mixture (2), let F_i be a binomial distribution with parameters (θ, p_i), where p_i is fixed, and suppose that G is a Poisson distribution with parameter η. It is well known that this mixture has Poisson marginals. With $q_i = 1 - p_i, i = 1, 2$, the probability mass function of this mixture is given by

$$h(k,\ell) = \sum_{\theta=0}^{\infty} \binom{\theta}{k} p_1^k q_1^{\theta-k} \binom{\theta}{\ell} p_2^\ell q_2^{\theta-\ell} e^{-\eta} \frac{\eta^\theta}{\theta!}, \quad k, \ell = 0, 1, \ldots. \tag{13}$$

Since $\binom{x}{y}$ is TP_∞ in x and $y = 0, 1, \ldots$ (Karlin, 1968, p. 137), it follows from the basic composition theorem for totally positive functions that $h(k, \ell)$ is TP_∞ in k and $\ell = 0, 1, \ldots$. Consequently, by Corollary 4.8 it follows that variables having the probability mass function (13) are associated. From Theorem 2.8, it follows that the bivariate Poisson distribution of (13) is also infinitely divisible.

A different construction of a bivariate Poisson distribution starts with independent random variables U_1, U_2 and Θ having Poisson distributions with respective parameters λ_1, λ_2 and λ_{12}. If $X_i = U_i + \Theta$, $i = 1, 2$, then (X_1, X_2) has the bivariate Poisson distribution of M'Kendrick (see Marshall and Olkin, 1985). The joint probability mass function of (X_1, X_2) is

$$\begin{aligned} h(k,\ell) &= \sum_{\theta=0}^{\infty} P(U_1 + \theta = k)\, P(U_2 + \theta = \ell)\, e^{-\lambda_{12}} \frac{\lambda_{12}^\theta}{\theta!} \\ &= \sum_{\theta} \frac{\lambda_{12}^\theta \lambda_1^{k-\theta} \lambda_2^{\ell-\theta}}{\theta!(k-\theta)!(\ell-\theta)!} e^{-(\lambda_1+\lambda_2+\lambda_{12})}, \quad k, \ell = 0, 1, \ldots. \end{aligned} \tag{14}$$

Dwass and Teicher (1957) show that this is the only infinitely divisible bivariate Poisson distribution, so it is reassuring to note by comparing Laplace transforms that when $\lambda_1 = \eta p_1 q_2$, $\lambda_2 = \eta p_2 q_1$ and $\lambda_{12} = \eta p_1 p_2$, (13) and (14) define the same distribution.

EXAMPLE 2.12. BIVARIATE NEGATIVE BINOMIAL DISTRIBUTION. It is well known that if F_1 and F_2 are Poisson distributions with respective parameters $\alpha\theta$ and $\beta\theta$, and if G is a gamma distribution with shape parameter r and scale parameter $\lambda = 1$, then the mixture (2) has negative binomial marginals and probability mass function

$$\begin{aligned} h(k,\ell \mid r) &= \int_0^\infty \frac{(\alpha\theta)^k e^{-\alpha\theta}}{k!} \frac{(\beta\theta)^\ell e^{-\beta\theta}}{\ell!} \frac{\theta^{r-1}e^{-\theta}}{\Gamma(r)} d\theta \\ &= \frac{\Gamma(k+\ell+r)}{k!\,\ell!\,\Gamma(r)} \left(\frac{\alpha}{\alpha+\beta+1}\right)^k \left(\frac{\beta}{\alpha+\beta+1}\right)^\ell \left(\frac{1}{\alpha+\beta+1}\right)^r, \\ & \qquad k,\ell = 0,1,\ldots\,. \end{aligned}$$

The above derivation of this bivariate negative binomial distribution is due to Arbous and Kerrich (1951). It follows from Theorem 2.10 that $h(k,\ell \mid r)$ is MTP_2; by Corollary 4.8, this means the distribution is associated, so that the correlation is non-negative. By Theorem 2.8, $h(k,\ell \mid r)$ is infinitely divisible and moreover

$$h(\cdot \mid r_1) * h(\cdot \mid r_2) = h(\cdot \mid r_1 + r_2), \quad r_1, r_2 > 0.$$

EXAMPLE 2.13. BIVARIATE NORMAL DISTRIBUTION. If F_1 and F_2 are normal distributions with means μ and $\alpha\mu$, $(\alpha = \pm 1)$ and variances σ_1^2 and σ_2^2, respectively, and if μ has a $\mathcal{N}(0,\sigma_0^2)$ distribution, then the mixture (2) has density function

$$\begin{aligned} h(x,y) &= \frac{1}{(2\pi)^{3/2}\sigma_0\sigma_1\sigma_2} \int_{-\infty}^{\infty} [\exp -\frac{1}{2}\{\frac{(x-\mu)^2}{\sigma_1^2} + \frac{(y-\alpha\mu)^2}{\sigma_2^2} + \frac{\mu^2}{\sigma_0^2}\}]d\mu \\ &= \frac{|\Sigma|^{-\frac{1}{2}}}{(2\pi)} \exp[-\frac{1}{2}(x,y)\Sigma^{-1}(x,y)'], \end{aligned}$$

where

$$\Sigma = \frac{1}{d}\begin{pmatrix} \sigma_0^2 + \sigma_1^2 & \alpha\sigma_0^2 \\ \alpha\sigma_0^2 & \sigma_0^2 + \sigma_2^2 \end{pmatrix}$$

and $d = (\sigma_1^2\sigma_2^2 + \sigma_0^2\sigma_1^2 + \sigma_0^2\sigma_2^2)/[(\sigma_0^2+\sigma_1^2)(\sigma_0^2+\sigma_2^2) - \alpha^2\sigma_0^4]$.

The choice $\alpha = +1$ yields a positive correlation and $\alpha = -1$ yields a negative correlation.

EXAMPLE 2.14. A BIVARIATE CHI-SQUARE DISTRIBUTION. If F_1 and F_2 are gamma distributions with common shape parameter $m+2\theta$ and with respective scale parameters $\frac{1}{2}\lambda_1$ and $\frac{1}{2}\lambda_2$ and G is a negative binomial distribution, then the mixture (2) has density function

$$\begin{aligned} & h(x,y \mid m) \\ &= \sum_{\ell=0}^{\infty} \left(\frac{\lambda_1^{m+\ell} x^{m+\ell-1} e^{-\frac{1}{2}\lambda_1 x}}{2^{m+\ell}\Gamma(m+\ell)}\right) \left(\frac{\lambda_2^{m+\ell} y^{m+\ell-1} e^{-\frac{1}{2}\lambda_2 y}}{2^{m+\ell}\Gamma(m+\ell)}\right) \frac{\Gamma(m+\ell)(\rho^2)^\ell(1-\rho^2)^m}{\Gamma(m)\ell!}. \end{aligned}$$

This is the joint density of the sample variances s_{11} and s_{12} from a sample of size $n = m/2$ from a bivariate normal distribution with inverse covariance matrix $\Sigma^{-1} = (\sigma^{ij})$, $i, j = 1, 2$. Here $\lambda_1 = \sigma^{11}, \lambda_2 = \sigma^{22}$ and $\rho = \sigma^{12}/\sqrt{\sigma^{11}\sigma^{22}}$ is the correlation.

Since the gamma density is TP_2, it follows from Theorem 2.10 that $h(x, y \mid m)$ is MTP_2, and hence, by Corollary 4.8, is associated. Because the negative binomial distribution is infinitely divisible, by Theorem 2.8 $h(x, y \mid m)$ is infinitely divisible.

EXAMPLE 2.15. BIVARIATE NON-CENTRAL CHI-SQUARE DISTRIBUTIONS. Because a non-central chi-square distribution is a Poisson mixture of central chi-square distributions, a bivariate non-central chi-square distribution can formally be obtained using (2). The corresponding density is

$$h(x, y) = \sum_{\theta=0}^{\infty} \frac{x^{\frac{n}{2}+\theta-1} e^{-\frac{x}{2}}}{\Gamma(\frac{n}{2}+\theta) 2^{\frac{n}{2}+\theta}} \frac{y^{\frac{m}{2}+\theta-1} e^{-\frac{y}{2}}}{\Gamma(\frac{m}{2}+\theta) 2^{\frac{m}{2}+\theta}} \frac{\alpha^{\theta} e^{-\alpha}}{\theta!}, \qquad x, y \geq 0.$$

It follows from Theorem 2.10 that $h(x, y)$ is MTP_2 and hence (by Corollary 4.8) is associated.

A possibly more meaningful bivariate non-central chi-square distribution is obtained from the representation

$$X = U_1^2 + Z_1, \quad Y = U_2^2 + Z_2,$$

where U_1, U_2 and (Z_1, Z_2) are independently distributed with $U_i \sim \mathcal{N}(\mu_i, 1/\lambda_{ii}), i = 1, 2$ and (Z_1, Z_2) has the bivariate chi-square distribution of Example 2.14. The corresponding bivariate chi-square density function is

$$h(x, y \mid n) = \sum_{\ell=0}^{\infty} \sum_{\theta_1=0}^{\infty} \sum_{\theta_2=0}^{\infty} f_1(x \mid \ell, \theta_1)\, f_2(y \mid \ell, \theta_2)\, g(\ell \mid \rho^2) \frac{e^{-\alpha_1} \alpha_1^{\theta_1}}{\theta_1!} \frac{e^{-\alpha_2} \alpha_2^{\theta_2}}{\theta_2!},$$

where $\alpha_i = \lambda_{ii}\mu_i^2/2$, $i = 1, 2, g(\ell \mid \rho^2)$ is negative binomial distribution of 2.14, and

$$f_i(t \mid \ell, \theta_i) = \frac{\lambda_{ii}^{(n+2\ell+2\theta_i+1)/2} t^{(n+2\ell+2\theta_i+1)/2} e^{-\lambda_{ii} t/2}}{2^{(n+2\ell+2\theta_i)/2} \Gamma((n + 2\ell + 2\theta_i)/2)},$$

$i = 1, 2$. When $\rho = 0$ we obtain the independent case with X and Y each having a non-central chi-square distribution.

The association of $h(x, y)$ follows from the representation of X and Y given above, or from Theorem 2.10. By Theorem 2.8, $h(x, y \mid n)$ is infinitely divisible.

Multivariate Extension of Convolution Families.

Let $\mathcal{F}$ be an indexed family of distributions and let $\mathcal{S}$ be the class of nonempty subsets of $\{1, \ldots, n\}$. For each $S \in \mathcal{S}$, let U_S have a distribution $F(\cdot \mid \theta_S)$ in $\mathcal{F}$. Suppose that the random variables $U_S, S \in \mathcal{S}$ are independent, and let

(15) $$X_i = \sum_{S:i\in S} U_S, \quad i = 1, \ldots, n.$$

Denote by $\mathcal{F}^{n*}$ the family of n-dimensional distributions for random vectors of the form $(X_1, \ldots, X_n)$. Clearly distributions in $\mathcal{F}^{n*}$ have $n-1$ dimensional marginals in $\mathcal{F}^{(n-1)*}$. If $\mathcal{F}$ is a convolution family, then distributions in $\mathcal{F}^{n*}$ have one dimensional marginals in $\mathcal{F}$; in particular, X_i has distribution $F(\cdot \mid \sum_{S:i\in S} \theta_S)$, $i = 1, \ldots, n$. From (15) it is clear that distributions in $\mathcal{F}^{n*}$ are associated. The family $\mathcal{F}^{n*}$ has various other desirable properties (see Marshall and Shaked, 1986).

THEOREM 2.16. *If $\mathcal{F} = \{F(\cdot \mid \theta) : \theta \in A\}$ is a convolution family of univariate distributions, then $\mathcal{F}^{n*}$ is a convolution family (indexed by A^{2^n-1}).*

PROOF. Clearly A^{2^n-1} satisfies (6). Suppose X and Y are independent random vectors with respective distributions $F(\cdot \mid \alpha)$ and $F(\cdot \mid \beta)$ in $\mathcal{F}^{n*}$. Then there exist independent random variables $U_S, V_S, S \in \mathcal{S}$, such that

$$X_i = \sum_{S:i\in S} U_S, \quad Y_i = \sum_{S:i\in S} V_S, \quad X_i + Y_i = \sum_{S:i\in S} (U_S + V_S), \quad i = 1, \ldots, n.$$

If U_S and V_S have respective distributions $F(\cdot \mid \alpha_S)$ and $F(\cdot \mid \beta_S)$ in $\mathcal{F}$, then by (7) $U_S + V_S$ has the distribution $F(\cdot \mid \alpha_S + \beta_S)$. Thus

$$F(\cdot \mid \alpha) * F(\cdot \mid \beta) = F(\cdot \mid \alpha + \beta). \quad \|$$

THEOREM 2.17. *If F is infinitely divisible and $\mathcal{F} = \{F^{\theta *} : \theta \geq 0\}$, then distributions in $\mathcal{F}^{n*}$ are infinitely divisible.*

The proof of this result is similar to the proof of Theorem 2.16.

EXAMPLE 2.18. BIVARIATE BINOMIAL AND POISSON DISTRIBUTIONS. Let $X_i = U_i + U_{12}, i = 1, 2$, where U_1, U_2 and U_{12} are independent random variables with distributions in $\mathcal{F}$. Then the joint distribution of X_1 and X_2 is in $\mathcal{F}^{2*}$. When $\mathcal{F}$ consists of binomial distributions, (X_1, X_2) has the bivariate binomial distribution of Wicksell (see Marshall and Olkin, 1985); it follows from Theorem 4.13 that such distributions form a convolution family. When $\mathcal{F}$ consists of the Poisson distributions then (X_1, X_2) has the bivariate Poisson distribution of Example 2.11; it follows from Theorem 2.17 that such distributions are infinitely divisible. See also Example 2.20.

Mixtures of distributions in $\mathcal{F}^{n*}$ can take various forms, but discussion here is confined to the case that the parameters $\theta_S, S \in \mathcal{S}$ are independent random variables with respective distributions $G(\cdot \mid \alpha_S)$, $S \in \mathcal{S}$. Denote the joint distribution of $X_1, \ldots X_n$ given $\theta_S, S \in \mathcal{S}$, by $F^*(\cdot \mid \theta_S,\, S \in \mathcal{S})$.

The next observation says that the operations of mixing and of extending $\mathcal{F}$ to $\mathcal{F}^{n*}$ commute. Alternatively, it can be viewed as saying that the structure exhibited in (15) is preserved under mixing.

PROPOSITION 2.19. *Let $\mathcal{F} = \{F(\cdot \mid \theta) : \theta \in A\}$, and let $\mathcal{H} = \{H(\cdot \mid \alpha) : \alpha \in B\}$ be the family of distributions of the form*

$$H(x \mid \alpha) = \int F(x \mid \theta) dG(\theta \mid \alpha),$$

where $F \in \mathcal{F}$ and $\mathcal{G} = \{G(\cdot \mid \alpha) : \alpha \in B\}$ is a family of distributions having support contained in A. Then $\mathcal{H}^{n*}$ consists of distributions having the form

$$(16) \qquad H^{n*}(\mathbf{x} \mid \alpha_T, T \in \mathcal{S}) = \int F^*(\mathbf{x} \mid \theta_S,\ S \in \mathcal{S}) \prod_{S \in \mathcal{S}} dG(\theta_S \mid \alpha_T).$$

PROOF. Let $U_S, S \in \mathcal{S}$ be independent random variables with distributions in $\mathcal{F}$ such that

$$X_\ell = \sum_{\ell \in S} U_S, \quad \ell = 1, \ldots, n$$

have joint distribution F^*. Denote the characteristic function of U_S by ϕ_S. Then $X_1, \ldots, X_n$ have joint characteristic function

$$Ee^{i\Sigma t_\ell X_\ell} = Ee^{i\Sigma_{S\in\mathcal{S}} \tau_S U_S} = \Pi_{S \in \mathcal{S}} \phi_S(\tau_S),$$

where $\tau_S = \Sigma_{\ell : \ell \in S} t_\ell$. Consequently, the characteristic function of H^{n*} is

$$(17) \qquad \int \prod_{S \in \mathcal{S}} \phi_S(\tau_S) \prod_{S \in \mathcal{S}} dG(\theta_S \mid \alpha_T) = \prod_{S \in \mathcal{S}} \int \phi_S(\tau_S)\, dG(\theta_S \mid \alpha_T).$$

Now let $V_T, T \in \mathcal{S}$ be independent random variables such that V_T has the distribution $H(\cdot \mid \alpha_T)$ and let

$$Y_j = \sum_{T : j \in T} V_T, \ j = 1, \ldots, n.$$

Then $Y_1, \ldots, Y_n$ have a joint distribution in $\mathcal{H}^{n*}$ and joint characteristic function given by (17). ||

EXAMPLE 2.20. A BIVARIATE NEGATIVE BINOMIAL DISTRIBUTION. Let $X_1 = U_1 + U_{12}, X_2 = U_2 + U_{12}$, where U_1, U_2 and U_{12} are independent random variables having Poisson distributions with respective parameters θ_{10}, θ_{01} and θ_{11} (i.e., X_1 and X_2 have the bivariate Poisson distribution of Example 2.11). Suppose that θ_{10}, θ_{01} and θ_{11} are independent random variables having gamma distributions with respective parameters $(\alpha_{10}, \lambda), (\alpha_{01}, \lambda)$ and (α_{11}, λ). Then for $k, \ell = 0, 1, \ldots,$

$$h(k,\ell) = \int\int\int \sum_j \frac{\theta_{11}^j \theta_{10}^{k-j} \theta_{01}^{\ell-j}}{j!(k-j)!(\ell-j)!} e^{-(\theta_{10}+\theta_{01}+\theta_{11})}$$
$$\frac{\lambda^{\alpha_{10}} \theta_{10}^{\alpha_{10}-1} e^{-\lambda\theta_{10}}}{\Gamma(\alpha_{10})} \frac{\lambda^{\alpha_{01}} \theta_{01}^{\alpha_{01}-1} e^{-\lambda\theta_{01}}}{\Gamma(\alpha_{01})} \frac{\lambda^{\alpha_{11}} \theta_{11}^{\alpha_{11}-1} e^{-\lambda\theta_{11}}}{\Gamma(\alpha_{11})} d\theta_{01} d\theta_{10} d\theta_{11}$$

$$= \sum_j \frac{\lambda^\alpha}{\Gamma(\alpha_{10})\Gamma(\alpha_{01})\Gamma(\alpha_{11})} \frac{\Gamma(\alpha_{10}+k-j-1)\Gamma(\alpha_{01}+\ell-j-1)\Gamma(\alpha_{11}+j-1)}{j!(k-j)!(\ell-j)!(\lambda+1)^{\alpha+k+\ell-j}}$$
$$= \sum_j \frac{\Gamma(\alpha_{10}+k-j-1)}{\Gamma(\alpha_{10})(k-j)!} \frac{\Gamma(\alpha_{01}+\ell-j-1)}{\Gamma(\alpha_{01})(\ell-j)!} \frac{\Gamma(\alpha_{11}+j-1)}{\Gamma(\alpha_{11})j!} p^\alpha (1-p)^{k+\ell-j},$$

where $\alpha = \alpha_{11} + \alpha_{10} + \alpha_{01}$, $p = \lambda/(\lambda+1)$.

This distribution has negative binomial marginals; with $\alpha_{10} = \alpha_{01} = 1 - \alpha_{11}$, this is a bivariate geometric distribution.

From Proposition 2.19, it follows that this negative binomial distribution is in $\mathcal{F}^{2*}$ when $\mathcal{F}$ consists of negative binomial distributions with fixed parameter p.

3. Product Families. Results which follow are stated for survival product families, but it should be understood that parallel results hold for distribution product families.

OBSERVATION 3.1. If $\{F_{(i)}(\cdot \mid \theta) : \theta \in A_i\}$ is a survival product family, $i = 1, 2$, then distributions of the form

$$\bar{F}(\mathbf{x}_1, \mathbf{x}_2, \mathbf{x}_3 \mid \theta_1, \theta_2) = \bar{F}_1(\mathbf{x}_1, \mathbf{x}_2 \mid \theta_1)\, \bar{F}_2(\mathbf{x}_2, \mathbf{x}_3 \mid \theta_2), \quad \theta_1 \in A_1, \ \theta_2 \in A_2$$

constitute a survival product family.

LEMMA 3.2. *A survival product family of distributions $F(\cdot \mid \theta)$ indexed by $A = (0, \infty)$ must be a proportional hazard family, that is,*

$$\bar{F}(\cdot \mid \theta) = \bar{F}^\theta(\cdot \mid 1), \quad \theta > 0. \tag{18}$$

PROOF. Fix $\mathbf{x}$ and let $\phi(\theta) = \bar{F}(\mathbf{x} \mid \theta)$. From (8) it follows that

$$\phi(\alpha + \beta) = \phi(\alpha)\, \phi(\beta), \quad \alpha, \beta > 0.$$

Since ϕ is bounded, it follows that for some real number $\gamma, \phi(\alpha) = e^{-\gamma\alpha}$, that is $\phi(\alpha) = [\phi(1)]^\alpha$. But this is (18). ||

THEOREM 3.3. *If $\{F(\cdot \mid \theta) : \theta \in A\}$ is a survival product family indexed by a convex cone $A \subset \mathcal{R}^k$ with finite frame $\mathcal{T} = \{t_1, \ldots, t_\ell\}$ then*

$$\bar{F}(\mathbf{x} \mid \theta) = \prod_{i=1}^{\ell} \bar{F}^{\theta_i}(\mathbf{x} \mid t_i),$$

where $\theta = \sum_1^\ell \theta_i t_i$.

PROOF. This is immediate from (8) and Lemma 3.2. ||

Proportional Hazard Families in Higher Dimensions.

If $\bar{F}$ is a univariate survival function, then for all $\theta > 0$, $\bar{F}^\theta$ is also a univariate survival function. In the multivariate case, $\bar{F}^k$, $k = 1, 2, \ldots$ is always a survival function, but $\bar{F}^\theta$ is a survival function for all $\theta > 0$ only in special circumstances.

THEOREM 3.4. *Let* $\bar{F}$ *be a bivariate survival function. Then* $\bar{F}^\theta$ *is a bivariate survival function for all* $\theta > 0$ *if and only if* $\bar{F}(x, y)$ *is* TP_2 *in* (x, y).

PROOF. Suppose first that $\bar{F}^\theta$ is a survival function for all $\theta > 0$. Then for all $\epsilon, \delta \geq 0$,

$$\xi(\theta) = \bar{F}^\theta(x, y) - \bar{F}^\theta(x, y+\epsilon) - \bar{F}^\theta(x+\delta, y) + \bar{F}^\theta(x+\delta, y+\epsilon) \geq 0.$$

Since $\xi(\theta) \geq 0$ for all $\theta > 0$ and $\xi(0) = 0$, it follows that the derivative $\xi'(0) \geq 0$; but this is just the condition

$$\frac{\bar{F}(x, y)\bar{F}(x+\delta, y+\epsilon)}{\bar{F}(x, y+\epsilon)\bar{F}(x+\delta, y)} \geq 1$$

that $\bar{F}$ is TP_2.

Next, suppose that $\bar{F}$ is TP_2. With $R(x, y) = -\log \bar{F}(x, y)$, this condition can be written in the form

$$(19) \qquad R(x+\delta, y+\epsilon) \leq R(x, y+\epsilon) + R(x+\delta, y) - R(x, y) \quad \text{for all } \epsilon, \delta \geq 0 \text{ and all } x, y.$$

Note that $R(x+\delta, y) - R(x, y) \geq 0$ and write

$$R(x, y+\epsilon) - R(x, y) = [R(x, y+\epsilon) + R(x+\delta, y) - R(x, y)] - R(x+\delta, y).$$

Since $\psi(x) = e^{-x}$ is decreasing and convex, it follows that for all $\theta > 0$,

$$\begin{aligned} e^{-\theta R(x,y)} - e^{-\theta R(x,y+\epsilon)} &\geq e^{-\theta R(x+\delta,y)} - e^{-\theta[R(x,y+\epsilon)+R(x+\delta,y)-R(x,y)]} \\ &\geq e^{-\theta R(x+\delta,y)} - e^{-\theta R(x+\delta,y+\epsilon)}; \end{aligned}$$

the last inequality because ψ is decreasing and (19) holds. But this says that $\bar{F}^\theta$ is a survival function. ||

As noted in Section 4, the condition that $\bar{F}$ is TP_2 is a positive dependency property which implies that the correlation is non-negative (when it exists).

REMARK 3.5. If F is a bivariate distribution with density f such that $f(x,y)$ is TP_2 in x and y, then it follows from Theorem 4.9 that $\bar{F}(x,y)$ is also TP_2 in x and y. Several examples from Section 2 have this property, including 2.11, 2.12. See also Examples 3.8 and 3.9.

Mixtures of Product Families.

LEMMA 3.6. *If $\bar{F}_i^\theta$ is a survival function for all θ in the support of G_i and $\bar{H}_i(\mathbf{x}) = \int \bar{F}_i^\theta(\mathbf{x}) dG_i(\theta)$, $i = 1, 2$, then*

$$\bar{H}_1(\mathbf{x})\,\bar{H}_2(\mathbf{y}) = \int \bar{F}^\theta(\mathbf{x})\, d(G_1 * G_2)(\theta).$$

PROOF. Write $F^\theta(\mathbf{x}) = e^{-\theta R(\mathbf{x})}$, where $R(\mathbf{x}) = -\log \bar{F}(\mathbf{x})$. Then the result is easily seen to be a reflection of the fact that the Laplace transform of a convolution is the product of Laplace transforms. ||

THEOREM 3.7. *Let $\{G(\cdot \mid \alpha) : \alpha \in B\}$ be a convolution family of distributions such that for each α, $G(\cdot \mid \alpha)$ has support contained in A. Let $\{F(\cdot \mid \theta) : \theta \in A\}$ be a survival product family of distributions such that $F(x \mid \theta)$ is measurable in θ for each fixed x.*

If

$$H(x \mid \alpha) = \int F(x \mid \theta)\, dG(\theta \mid \alpha), \quad \alpha \in B,$$

then

$$\bar{H}(x \mid \alpha + \beta) = \bar{H}(x \mid \alpha)\, \bar{H}(x \mid \beta) \text{ for all } \alpha, \beta \in B, \quad -\infty < x < \infty. \tag{20}$$

PROOF. This is immediate from Lemma 3.6. ||

EXAMPLE 3.8. MULTIVARIATE LOGISTIC DISTRIBUTION. If F_i are iterated exponential extreme value distributions for minima, that is, $\bar{F}_i(x_i \mid \theta) = \exp\{-\theta e^{x_i}\}, -\infty < x_i < \infty$, $\theta > 0$, $i = 1, \ldots, n$, and if G is a gamma distribution with shape parameter r and scale parameter λ, then the distribution H of (2) takes the form

$$\bar{H}(\mathbf{x} \mid r) = \lambda^r / (\lambda + \sum_1^n e^{x_i})^r, \quad \lambda, r > 0.$$

EXAMPLE 3.9. MULTIVARIATE PARETO DISTRIBUTIONS. If

$$\bar{F}_i(x_i) = \exp\left\{-\theta\left(\frac{x_i - \mu_i}{\sigma_i}\right)^{1/\gamma_i}\right\}, \quad x_i \geq \mu_i, \quad i = 1, \dots, n$$

are Weibull survival functions and if G is a $\mathrm{Gam}(\alpha, 1)$ distribution then the mixture (2) is a Type IV multivariate Pareto distribution as defined by Arnold (1983). This mixture has survival function

$$\text{(21)} \qquad \bar{H}(\mathbf{x} \mid \alpha) = \left[1 + \sum_{i=1}^{n}\left(\frac{x_i - \mu_i}{\sigma_i}\right)^{1/\gamma_i}\right]^{-\alpha}, \quad x_i \geq \mu_i, \quad i = 1, \dots, n.$$

The representation of (21) as a mixture was used with a minor variation by Takahasi (1965) to define a multivariate Burr distribution.

It follows from Theorem 4.15 that the density corresponding to H is TP_∞ in each pair of arguments, the other arguments being fixed. Consequently the distribution is associated.

Lemma 3.6 and Theorem 3.7 can be generalized to allow the F_i to involve different sets of variables.

LEMMA 3.10. *Let $\mathbf{x}^{(i)}$ be a subvector of $\mathbf{x}$ of dimension n_i, and let F_i be a distribution fuction of dimension n_i such that $\bar{F}_i^\theta$ is a survival function for all $\theta \in A_i$, $i = 1, \dots, k$. Let G_j be a distribution of dimension k such that $G_j(A_1 \times \cdots \times A_k) = 1$, $j = 1, 2$. If*

$$\bar{H}_j(\mathbf{x}) = \int \prod_{i=1}^{k} \bar{F}_i^{\theta_i}(\mathbf{x}^{(i)})\, dG_j(\theta), \quad j = 1, 2,$$

then

$$\bar{H}_1(\mathbf{x})\, \bar{H}_2(\mathbf{x}) = \int \prod_{i=1}^{k} \bar{F}_i^{\theta_i}(\mathbf{x}^{(i)})\, d(G_1 * G_2)(\theta).$$

PROOF. Let $R_i(\mathbf{x}^{(i)}) = -\log \bar{F}_i(\mathbf{x}^{(i)})$ so that

$$\bar{H}_j(\mathbf{x}) = \int \exp\{-\Sigma_{i=1}^k \theta_i R_i(\mathbf{x}^{(i)})\}\, dG_j(\theta).$$

Then

$$\begin{aligned} \bar{H}_1(\mathbf{x})\, \bar{H}_2(\mathbf{x}) &= \int\int \exp\{-\sum_{i=1}^{k}(\theta_i + \eta_i) R_i(\mathbf{x}^{(i)})\} dG_1(\theta) dG_2(\eta) \\ &= \int \exp\{-\sum_{i=1}^{k} \theta_i R_i(\mathbf{x}^{(i)})\} d(G_1 * G_2)(\theta). \quad \| \end{aligned}$$

Special cases of particular interest include

(i) $k = 1$ and $\mathbf{x}^{(1)} = \mathbf{x}$,

(ii) $k = n$ and $\mathbf{x}^{(i)} = x_i$ where $\mathbf{x} = (x_1, \ldots, x_n)$.

(iii) $k = 2^n - 1$ and $\mathbf{x}^{(i)}$ ranges through the nonempty subvectors of $\mathbf{x}$.

THEOREM 3.11. *Let $\mathcal{G} = \{G(\cdot \mid \alpha),\ \alpha \in A\}$ be a convolution family of k-dimensional distribution functions such that*

$$\bar{G}(0 \mid \alpha) = 1 \quad \textit{for all} \quad \alpha \in A. \tag{22}$$

With the notation of Lemma 3.10, let

$$\bar{H}(\mathbf{x} \mid \alpha) = \int \prod_{i=1}^{k} \bar{F}_i^{\theta_i}(\mathbf{x}^{(i)})\, dG(\theta \mid \alpha), \quad \mathbf{x} \in \mathcal{R}^n,\ \alpha \in A.$$

Then $\mathcal{H} = \{H(\cdot \mid \alpha),\ \alpha \in A\}$ is a survival product family.

PROOF. This follows from Lemma 3.10. ||

Multivariate Extensions of Survival Product Families.

Let $\mathcal{F}$ be an indexed family of distributions and for each $S \in \mathcal{S}$, the non-empty subsets of $\{1, 2, \ldots, n\}$, let U_S have distribution $F(\cdot \mid U_S) \in \mathcal{F}$. Suppose that the random variables U_S, $S \in \mathcal{S}$ are independent, and let

$$X_i = \min_{S: i \in S} U_S, \quad i = 1, \ldots, n. \tag{23}$$

Then the joint survival function of the X_i's is given by

$$\bar{F}_{(n)}(\mathbf{x} \mid \theta_S,\ S \in \mathcal{S}) = \prod_{S \in \mathcal{S}} \bar{F}(\max_{j \in S} x_j \mid \theta_S). \tag{24}$$

The family $\bar{\mathcal{F}}_{(n)}$ of such distributions has various desirable properties (see Marshall and Shaked, 1986); in particular if $\mathcal{F}$ is a survival product family, then distributions in $\bar{\mathcal{F}}_{(n)}$ have univariate marginals in $\mathcal{F}$. If $\mathcal{F}$ is a survival product family indexed by $(0, \infty)$, then Lemma 3.2 applies and

$$\bar{F}_{(n)}(\mathbf{x} \mid \theta_S,\ S \in \mathcal{S}) = \prod_{S \in \mathcal{S}} \bar{F}^{\theta_S}(\max_{j \in S} x_j \mid 1). \tag{25}$$

EXAMPLE 3.12. If $X_1 = \min(U_1, Z)$ and $X_2 = \min(U_2, Z)$, where U_1, U_2 and Z are independent random variables with distributions in the family $\mathcal{F}$ then (X_1, X_2) has a distribution in $\bar{\mathcal{F}}_{(2)}$. When $\mathcal{F}$ consists of the exponential distributions, $\bar{\mathcal{F}}_{(2)}$ consists of the bivariate exponential distributions of Marshall and Olkin (1967).

For distribution product families, "min" in (23) is replaced by "max" and (24) is replaced by

$$F(\mathbf{x} \mid \theta) \equiv F_{(n)}(\mathbf{x} \mid \theta) = \prod_{S \in \mathcal{S}} F(\min_{j \in S} x_j \mid \theta_S), \tag{26}$$

and the family of such distributions is denoted by $\mathcal{F}_{(n)}$.

THEOREM 3.13. *If $\mathcal{F} = \{F(\cdot \mid \theta) : \theta \in A\}$ is a survival product family then $\bar{\mathcal{F}}_{(n)}$ is a survival product family; if $\mathcal{F}$ is a distribution product family, then $\mathcal{F}_{(n)}$ is a distribution product family.*

The next observation says that the operations of mixing and of extending $\mathcal{F}$ to $\bar{\mathcal{F}}_{(n)}$ commute (cf. Proposition 2.19).

PROPOSITION 3.14. *Let $\mathcal{F} = \{F(\cdot \mid \theta) : \theta \in A\}$, and let $\mathcal{H} = \{H(\cdot \mid \alpha) : \alpha \in B\}$ be the family of distributions of the form*

$$H(x \mid \alpha) = \int F(x \mid \theta)\, dG(\theta \mid \alpha),$$

where $F \in \mathcal{F}$ and $\mathcal{G} = \{G(\cdot \mid \alpha) : \alpha \in B\}$ is a family of distributions having support contained in A. Then $\bar{\mathcal{H}}_{(n)}$ consists of distributions having the form

$$\bar{H}_{(n)}(\mathbf{x} \mid \alpha_S, S \in \mathcal{S}) = \int \bar{F}_{(n)}(\mathbf{x} \mid \theta_S, S \in \mathcal{S}) \prod_{S \in \mathcal{S}} dG(\theta_S \mid \alpha_S).$$

PROOF.

$$\begin{aligned} \bar{H}_{(n)}(\mathbf{x} \mid \alpha_S, S \in \mathcal{S}) &= \int \prod_{S \in \mathcal{S}} \bar{F}(\max_{j \in S} x_j \mid \theta_S) \prod_{S \in \mathcal{S}} dG(\theta_S \mid \alpha_S) \\ &= \prod_{S \in \mathcal{S}} \int \bar{F}(\max_{j \in S} x_j \mid \theta_S)\, dG(\theta_S \mid \alpha_S) = \prod_{S \in \mathcal{S}} \bar{H}(\max_{j \in S} x_j \mid \alpha_S). \quad \| \end{aligned}$$

If in the above proposition, $\mathcal{F}$ is a survival product family and $\mathcal{G}$ is a convolution family, then it follows from Theorem 3.7 and Theorem 3.13 that $\bar{\mathcal{H}}_{(n)}$ is a survival product family, and of course this is the most interesting case.

EXAMPLE 3.15. MULTIVARIATE LOGISTIC DISTRIBUTIONS. Let $\mathcal{F}$ consist of the iterated exponential extreme value distributions for minima as in Example 3.8. Then $\mathcal{F}$ is a survival product family and $\bar{\mathcal{F}}_{(n)}$ consists of distributions of the form

$$\bar{F}(\mathbf{x} \mid \theta_S, S \in \mathcal{S}) = \exp\left[-\sum_{S \in \mathcal{S}} \theta_S \exp(\max_{j \in S} x_j)\right].$$

If the θ_S are independent and have gamma distributions with respective shape and scale parameters r_S and λ_S, then the mixture H of (1) is given by

$$\bar{H}(\mathbf{x} \mid r_S, \lambda_S, S \in \mathcal{S}) = \prod_{S \in \mathcal{S}} \left[\frac{\lambda_S}{\lambda_S + \exp(\max_{i \in S} x_i)} \right]^{r_S}.$$

If the parameters λ_S, $S \in \mathcal{S}$ are either 0 or are equal to $\lambda > 0$, say, then the mixing gamma distributions form a convolution family, and in this case the distribution $\bar{H}(\cdot \mid r_S, \lambda, S \in \mathcal{S})$ form a survival product family as expected.

4. Appendix: Association and Total Positivity. The following result of Ahmed, León and Proschan (1978) shows that the positive dependency property of association is preserved under mixing.

THEOREM 4.1. *Let H be a mixture given by* (4). *If*

(27) *for each fixed θ, $F(x \mid \theta)$ is associated,*

(28) *G is associated,*

(29) *$\int \xi(\mathbf{x}) dF(\mathbf{x} \mid \theta)$ is increasing in θ for all increasing $\xi : \mathcal{R}^n \to \mathcal{R}$ such that the integral exists,*

then H is associated.

LEMMA 4.2. *Let H be a mixture given by* (2). *If*

$$\int \xi(x) dF_i(x \mid \theta) \text{ is increasing in } \theta \text{ for all increasing } \xi : \mathcal{R} \to \mathcal{R} \tag{30}$$

such that each integral exists, $i = 1, \ldots, n$, then (29) *holds where $F(\mathbf{x} \mid \theta) = \prod F_i(x_i \mid \theta)$.*

THEOREM 4.3. *Let H be a mixture given by* (2). *If G is associated and if condition (30) holds, then H is associated.*

COROLLARY 4.4. *Let H be a mixture given by* (2). *If*

$$F_i(x_i \mid \theta) \text{ is decreasing in } \theta \text{ for all } x_i, i = 1, \ldots, n, \tag{31}$$

then H is associated.

Total Positivity in Mixtures.

Total positivity is often encountered in mixtures (e.g., see Marshall and Olkin, 1979, Example 18.A.12). In the multivariate setting, multivariate total positivity in the following sense arises.

DEFINITION 4.5. (Karlin and Rinott, 1980). Let $\mathcal{X} = \mathcal{X}_1 \times \cdots \times \mathcal{X}_n$ where each $\mathcal{X}_i$ is totally ordered. For $\mathbf{x}, \mathbf{y} \in \mathcal{X}$, let

$$\mathbf{x} \vee \mathbf{y} = (\max(x_1, y_1), \ldots, \max(x_n, y_n)), \quad \mathbf{x} \wedge \mathbf{y} = (\min(x_1, y_1), \ldots, \min(x_n, y_n)).$$

A function $\psi : \mathcal{X} \to [0, \infty)$ is said to be *multivariate totally positive of order* 2 (MTP_2) if

$$\psi(\mathbf{x} \vee \mathbf{y})\, \psi(\mathbf{x} \wedge \mathbf{y}) \geq \psi(\mathbf{x})\, \psi(\mathbf{y}) \text{ for all } \mathbf{x}, \mathbf{y} \in \mathcal{X}.$$

PROPOSITION 4.6. (Kemperman, 1977; Karlin and Rinott, 1980, Proposition 2.1). *Suppose that $\psi : \mathcal{X} \to [0, \infty)$ is totally positive of order 2 (TP_2) in each pair of arguments, the remaining arguments being fixed. Suppose also that $\mathbf{x}, \mathbf{y} \in \mathcal{X}$ and $\psi(\mathbf{x})\psi(\mathbf{y}) > 0$ implies $\psi(\mathbf{u}) > 0$ for all $\mathbf{u}$ such that $\mathbf{x} \wedge \mathbf{y} \leq \mathbf{u} \leq \mathbf{x} \vee \mathbf{y}$. Then ψ is MTP_2 on $\mathcal{X}$.*

THEOREM 4.7. (Karlin and Rinott, 1980, p. 472). *If $\mathbf{X} = (X_1 \ldots, X_n)$ has a joint density that is MTP_2 and if A and B are upper Borel sets in $\mathcal{R}^n$ (i.e., $a \in A$ and $a' \geq a \Rightarrow a' \in A$, and similarly for B), then*

$$P\{\mathbf{X} \in A \vee B\}\; P\{\mathbf{X} \in A \wedge B\} \geq P\{\mathbf{X} \in A\}\; P\{\mathbf{X} \in B\}, \tag{32}$$

where $A \vee B = \{\mathbf{u} = \mathbf{a} \vee \mathbf{b} : \mathbf{a} \in A, \mathbf{b} \in B\}$ and $A \wedge B = \{\mathbf{u} = \mathbf{a} \wedge \mathbf{b} : \mathbf{a} \in A, \mathbf{b} \in B\}$.

According to Theorem 3.1 of Esary, Proschan and Walkup (1967), random variables $X_1, \ldots, X_n$ are associated if and only if for all upper Borel sets in $\mathcal{R}^n$,

$$P\{\mathbf{X} \in A \cap B\} \geq P\{\mathbf{X} \in A\} P\{\mathbf{X} \in B\}. \tag{33}$$

Note that $A \cap B = A \wedge B$. A comparison of (32) and (33) provides a proof of the following.

COROLLARY 4.8. (Fortuin, Kastelyn and Ginibre, 1971; Karlin and Rinott, 1980, Theorem 4.2). *If $\mathbf{X} = (X_1, \ldots, X_n)$ has an MTP_2 joint density, then $X_1, \ldots, X_n$ are associated.*

To a large extent, joint densities which are MTP_2 arise in mixtures as a consequence of the following theorem.

THEOREM 4.9. (Karlin and Rinott, 1980, Proposition 3.4). *Let $\mathcal{X} = \Pi_{i=1}^n \mathcal{X}_i$, $\mathcal{Y} = \Pi_{i=1}^n \mathcal{Y}_i$ and $\mathcal{Z} = \Pi_{i=1}^n \mathcal{Z}_i$, where each $\mathcal{X}_i, \mathcal{Y}_i$ and $\mathcal{Z}_i$ is totally ordered. If ψ_1 is MTP_2 on $\mathcal{X} \times \mathcal{Y}, \psi_2$ is MTP_2 on $\mathcal{Y} \times \mathcal{Z}$ and if*

$$\psi(\mathbf{x}, \mathbf{z}) = \int_{\mathcal{Y}} \psi_1(\mathbf{x}, \mathbf{y})\, \psi_2(\mathbf{y}, \mathbf{z})\, \Pi d\sigma_i(y_i),$$

where each σ_i is σ-finite, then ψ is MTP_2 on $\mathcal{X} \times \mathcal{Z}$.

PROPOSITION 4.10.

(i) If $f(\mathbf{x} \mid \theta)$ is MTP_2 in $(\mathbf{x}, \theta)$ and g is MTP_2, then

$$h(\mathbf{x}) = \int f(\mathbf{x} \mid \theta) g(\theta) d\theta$$

is MPT_2.

(ii) If $f(\mathbf{x} \mid \theta)$ is MTP_2 in $(\mathbf{x}, \theta)$ and if $g(\theta \mid \alpha)$ is MTP_2 in (θ, α), then

$$h(\mathbf{x} \mid \alpha) = \int f(\mathbf{x} \mid \theta) g(\theta \mid \alpha) d\theta$$

is MTP_2 in $(\mathbf{x}, \alpha)$.

PROOF. This is an immediate consequence of Theorem 4.9.

PROPOSITION 4.11.

(i) If $h(\mathbf{x}) = \int \Pi_{i=1}^n f_i(x_i \mid \theta_i) g(\boldsymbol{\theta}) d\boldsymbol{\theta}$, where each $f_i(x_i \mid \theta_i)$ is TP_2 in (x_i, θ_i) and if g is MTP_2, then h is MTP_2.

(ii) If $h(\mathbf{x} \mid \boldsymbol{\alpha}) = \int \Pi_{i=1}^n f_i(x_i \mid \theta_i) g(\boldsymbol{\theta} \mid \boldsymbol{\alpha}) d\boldsymbol{\theta}$, where each $f_i(x_i \mid \theta_i)$ is TP_2 in x_i, θ_i, and if $g(\boldsymbol{\theta} \mid \boldsymbol{\alpha})$ is MTP_2 in $(\boldsymbol{\theta}, \boldsymbol{\alpha})$, then h is MTP_2 in $(\mathbf{x}, \boldsymbol{\alpha})$.

PROOF. By Proposition 4.6, $\Pi_i f_i(x_i \mid \theta_i)$ is MTP$_2$ in $(\mathbf{x}, \boldsymbol{\theta})$ so this result follows from Proposition 4.10. ||

PROPOSITION 4.12. *Let $h(\mathbf{x}) = \int f(\mathbf{x} \mid \boldsymbol{\theta}) \Pi dG_i(\theta_i)$. If $f(\mathbf{x} \mid \boldsymbol{\theta})$ is MTP_2 in $(\mathbf{x}, \boldsymbol{\theta})$, then h is MTP_2.*

PROOF. This follows directly from Theorem 4.9 or from Proposition 4.11. ||

Just as MTP$_2$ densities arise so do distribution functions and survival functions.

PROPOSITION 4.13.

(i) Let $H(\mathbf{x}) = \int F(\mathbf{x} \mid \boldsymbol{\theta}) g(\boldsymbol{\theta}) d\boldsymbol{\theta}$ and suppose that g is MTP_2. If $F(\mathbf{x} \mid \boldsymbol{\theta})$ is MTP_2 in $(\mathbf{x}, \boldsymbol{\theta})$, then H is MTP_2; if $\bar{F}(\mathbf{x} \mid \boldsymbol{\theta})$ is MTP_2 in $(\mathbf{x}, \boldsymbol{\theta})$, then $\bar{H}$ is MTP_2.

(ii) Let $H(\mathbf{x} \mid \boldsymbol{\alpha}) = \int F(\mathbf{x} \mid \boldsymbol{\theta})\, g(\boldsymbol{\theta} \mid \boldsymbol{\alpha}) d\boldsymbol{\theta}$ and suppose that $g(\boldsymbol{\theta} \mid \boldsymbol{\alpha})$ is MTP_2 in $(\boldsymbol{\theta}, \boldsymbol{\alpha})$. If $F(\mathbf{x} \mid \boldsymbol{\theta})$ is MTP_2 in $(\mathbf{x}, \boldsymbol{\theta})$, then $H(\mathbf{x} \mid \boldsymbol{\alpha})$ is MTP_2 in $(\mathbf{x}, \boldsymbol{\alpha})$. A similar statement holds if F and H are replaced by $\bar{F}$ and $\bar{H}$.

PROOF. These results follow from Theorem 4.9. ||

PROPOSITION 4.14. *Let* $H(\mathbf{x}) = \int F(\mathbf{x} \mid \boldsymbol{\theta}) \Pi dG_i(\theta_i)$. *If* $F(\mathbf{x} \mid \boldsymbol{\theta})$ *is* MTP_2 *in* $(\mathbf{x}, \boldsymbol{\theta})$, *then* H *is* MTP_2. *A similar statement holds with* $\bar{F}$ *and* $\bar{H}$ *in place of* F *and* H.

PROOF. These results follow from Theorem 4.9. ||

Although higher order multivariate total positivity has to our knowledge not been defined, one counterpart to MTP_2 is the condition of higher order total positivity in pairs of arguments.

THEOREM 4.15. *If* $\bar{F}_i(x_i \mid \theta)$ *is totally positive of order* k (TP_k) *in* (x_i, θ), $i = 1, \ldots, n$, *then*

$$\bar{H}(\mathbf{x}) = \int \Pi_{i=1}^n \bar{F}_i(x_i \mid \theta)\, dG(\theta)$$

is TP_k *in each pair* x_j, x_ℓ, $1 \le j,\ \ell \le k$ $(j \ne \ell)$, *the other arguments being fixed. If* $F_i(x_i \mid \theta)$ *is* TP_k *in* (x_i, θ), $i = 1, \ldots, n$, *then*

$$H(\mathbf{x}) = \int \Pi_{i=1}^n F_i(x_i \mid \theta)\, dG(\theta)$$

is TP_k *in each pair* $x_j, x_\ell, j \ne \ell$, *the other arguments being fixed.*

If F_i *has a density* f_i *with respect to some measure that is* TP_k *in* (x_i, θ), $i = 1, \ldots, n$, *and if*

$$h(\mathbf{x}) = \int \prod_{i=1}^n f_i(x_i \mid \theta)\, dG(\theta),$$

then h *is* TP_k *in each pair* $x_j, x_\ell, j \ne \ell$, *the other arguments being fixed.*

PROOF. This is an immediate consequence of the basic composition formula (Karlin, 1968, p. 17). ||

It is not difficult to show that if h is TP_k in pairs of its arguments, then H and $\bar{H}$ both have this property.

When $n = 2$, even TP_2 is known to have useful implications.

A random vector (X_1, X_2) is said to be *right corner set increasing* (RCSI) if

$$P\{X_1 > x_1,\ X_2 > x_2 \mid X_1 > x_1',\ X_2 > x_2'\}$$

is increasing in x_1' and x_2' for all x_1, x_2. Shaked (1977) shows that the survival function of (X_1, X_2) is TP_2 if and only if (X_1, X_2) is RCSI. Barlow and Proschan (1975) show that if (X_1, X_2) is RCSI, then X_1 and X_2 are associated. By analogous arguments or by applying these results to $(-X_1, -X_2)$ it can be shown the distribution function of (X_1, X_2) is TP_2 if and only if (X_1, X_2) is *left corner set decreasing* (LCSD), i.e.,

$$P\{X_1 \le x_1,\ X_2 \le x_2 \mid X_1 \le x_1',\ X_2 \le x_2'\}$$

is decreasing in x_1' and x_2' for all x_1, x_2. Moreover, if (X_1, X_2) is LCSD then X_1 and X_2 are associated.

Thus we see that when $n = 2$, TP_2 of either the distribution function or the survival function implies association. For $n > 2$, corresponding results are false (C. Newman, 1986, private communication).

REFERENCES

ACZÉL, J. (1966). *Lectures on Functional Equations and Their Applications.* Academic Press, New York.

AHMED, L., LEÓN, R.V., and PROSCHAN, F. (1978). Generalized association with applications in multivariate statistics. Technical Report, Florida State University.

ARBOUS, A.G. and KERRICH, J.E. (1951). Accident statistics and the concept of accident proneness. *Biometrics* **7** 340–432.

ARNOLD, B.C. (1983). *Pareto Distributions.* International Co-operative Publishing House, Fairland, Maryland.

BARLOW, R.E. and PROSCHAN, F. (1975). *Statistical Theory of Reliability and Life Testing.* Holt, Rinehart and Winston, New York.

DWASS, M. and TEICHER, H. (1957). On infinitely divisible random vectors. *Ann. Math. Statist.* **28** 461–470.

ESARY, J.D., PROSCHAN, F. and WALKUP, D. (1967). Association of random variables, with applications. *Ann. Math. Statist.* **38** 1466–1474.

FELLER, W. (1968). *An Introduction to Probability Theory and Its Applications.* Vol. I (third edition), Wiley, New York.

FELLER, W. (1971). *An Introduction to Probability Theory and Its Applications.* Vol. II (second edition), Wiley, New York.

FORTUIN, C.M., KASTELYN, P.W. and GINIBRE, J. (1971). Correlation inequalities on some partially ordered sets. *Comm. Math. Phys.* **22** 89–103.

KARLIN, S. (1968). *Total Positivity,* Vol. 1. Stanford University Press, Stanford, CA.

KARLIN, S. and RINOTT, Y. (1980). Classes of orderings of measures and related correlation inequalities. I. Multivariate totally positive distributions. *J. Mult. Anal.* **10** 467–498.

KEILSON, J. and STEUTEL, F.W. (1974). Mixtures of distributions, moment inequalities and measures of exponentiality and normality. *Ann. Prob.* **2** 112–130.

KEMPERMAN, J.H.B. (1977). On the FKG-inequality for measures on a partially ordered space. *Indag. Math.* **39** 313–331.

KENT, J.T. (1981). Convolution mixtures of infinitely divisible distributions. *Math. Proc. Camb. Phil. Soc.* **90** 141–153.

MARSHALL, A.W. and OLKIN, I. (1967). A multivariate exponential distribution. *J. Amer. Statist. Assoc.* **62** 30–44.

MARSHALL, A.W. and OLKIN, I. (1979). *Inequalities: Theory of Majorization and Its Applications.* Academic Press, New York.

MARSHALL, A.W. and OLKIN, I. (1985). A family of bivariate distributions generated by the bivariate Bernoulli distribution. *J. Amer. Statist. Assoc.* **80** 332–338.

MARSHALL, A.W. and SHAKED, M. (1986). Multivariate new better than used distributions. *Math. Oper. Res.* **11** 110–116.

SHAKED, M. (1977). A family of concepts of dependence for bivariate distributions. *J. Amer. Statist. Assoc.* **72** 642–650.

SHAKED, M. (1977). A concept of positive for exchangeable random variables. *Ann. Statist.* **5** 505–515.

TAKAHASI, K. (1965). Note on the multivariate Burr's distribution. *Ann. Inst. Statist. Math.* **17** 257–260.

DEPARTMENT OF STATISTICS
UNIVERSITY OF BRITISH COLUMBIA
VANCOUVER, B.C.
CANADA V6T 1W5

STATISTICS AND EDUCATION DEPARTMENT
STANFORD UNIVERSITY
STANFORD, CA 94305

ISING MODELS AND DEPENDENT PERCOLATION

BY CHARLES M. NEWMAN[1]

University of Arizona

An intimate relation between Ising and certain dependent percolation models was discovered some twenty years ago by Kasteleyn and Fortuin and developed more recently by Swendsen and Wang. We review this relation and the role of stochastic domination within it. When the Ising model is not ferromagnetic (i.e., not positively dependent), the related percolation model is more complicated but still of interest.

1. Introduction. Although percolation and more recently Ising models have been of interest to probabilists and statisticians for some time, they have been largely unaware of the beautiful and useful relation which exists between the two types of systems. This relation, originally discovered by Kasteleyn and Fortuin (1969) was clarified by recent work of Swendsen and Wang (1987) on Ising simulation methods, and further explained by Edwards and Sokal (1988). Our purpose here is to describe the relation (Section 2), explain how it yields certain stochastic monotonicity properties (Section 3) of Fortuin (1972) and then mention some applications due to Aizenman, Chayes, Chayes and Newman (1987, 1988). We also discuss (Section 5) the situation when one goes beyond the case of ferromagnetic (i.e., positively dependent) Ising models. It should be noted that most (all?) of what is presented in this paper satisfies one or more of the following descriptions: old, already published, known to the experts. For more details related to Sections 3 and 4 (and much of Section 2), see Aizenman, Chayes, Chayes and Newman (1988).

2. Random Colored Graphs – Positively Dependent Case. Let Λ be a finite set of sites (or vertices) and $\mathcal{B}$ the corresponding set of all bonds (or edges); i.e., $\mathcal{B}$ is the set of pairs $b = \{x, y\}$ of sites. (For many applications Λ is a subset of some regular d-dimensional lattice, say $\mathbf{Z}^d$, and one takes $\Lambda \uparrow \mathbf{Z}^d$.) We will consider bond random variables n_b taking values 0 or 1 and site random variables T_x taking values in $\{1, \cdots, q\}$. The n_b's define a random graph with vertex set Λ in which

[1]Research supported in part by NSF Grant DMS 85-14834 and AFOSR Grant 88-0189 and by AFOSR Contract F49620-86-C0130 with the U.R.I. program at the Arizona Center for Mathematical Sciences.

AMS 1980 subject classifications. Primary 60K35, 82A25; secondary 05C15.

Key words and phrases. Percolation, Ising models, random graphs, stochastic domination, Potts models.

a bond b is occupied (i.e., the edge b occurs) when $n_b = 1$; connected components of the graph are called clusters. The T_x's define a coloring of the sites from a set of q allowed colors. Our focus throughout this paper will be on certain natural randomly colored random graphs and on the corresponding marginal distributions of the T_x's alone and the n_b's alone. The colorings we consider will always be symmetric (i.e., the joint distribution will be invariant under exchanges of colors) but clearly generalizations to nonsymmetric situations are possible. Another type of generalization we will avoid (but which can be made in order to deal with Ising models with other than "pair interactions") is to consider, in addition to edges, randomly occupied faces, etc.

We begin by simply taking all the T_x's and n_b's to be jointly independent with the T_x's symmetric and with

$$(1) \qquad Pr(n_b = 1) = p_b \quad \text{for each } b \in \mathcal{B};$$

where we assume $0 \le p_b < 1$. Such n_b's describe independent bond percolation; two important examples are nearest neighbor models (with $\Lambda \subset \mathbf{Z}^d$), where

$$(2) \qquad p_{\{x,y\}} = \begin{cases} p, & \text{if } x \text{ and } y \text{ are nearest neighbors} \\ 0, & \text{otherwise,} \end{cases}$$

and $1/r^2$ models (with $\Lambda \subset \mathbf{Z}^1$), where

$$(3) \qquad p_{\{x,y\}} = p_{|y-x|} \text{ with } \lim_{x\to\infty} x^2 p_x = \beta.$$

We next introduce some dependence into this simple model by conditioning on the event

$$(4) \qquad \{T_x = T_y \text{ for every } b = \{x,y\} \text{ with } n_b = 1\};$$

i.e., *the sites in each cluster must have the same color.* Let μ^q denote the resulting joint (conditional) distribution of the T_x's and n_b's for some given Λ and p_b's. We leave it as an exercise for the reader to verify that the marginal distributions μ_v^q for the T_x's and μ_e^q for the n_b's, have densities (relative to uniform distributions),

$$(5) \qquad \mu_v^q : \text{const.} \exp\left(\sum_{\{x,y\}} K_{xy} 1_{T_x = T_y}\right), \quad \text{where } 1 - e^{-K_{xy}} = p_{\{x,y\}},$$

$$(6) \qquad \mu_e^q : \text{const.}\, q^C \prod_{b:n_b=0} (1-p_b) \prod_{b':n_{b'}=1} p_{b'}, \quad \text{where } C = \text{no. of clusters.}$$

When $q = 2$, μ_v^q is the Gibbs distribution of an Ising model; this may look more familiar when expressed in ± 1 valued variables S_x ($S_x = +1$ for $T_x = 1$, $S_x = -1$ for $T_x = 2$):

(7) $$\mu_v^2 : \text{ const. } \exp\left(\frac{1}{2}\sum_{\{x,y\}} K_{xy}S_xS_y\right).$$

For $q = 3, 4, \cdots, \mu_v^q$ is the distribution of a Potts model. The random graph corresponding to (6) (Fortuin and Kasteleyn (1972)) is known as a Fortuin-Kasteleyn random cluster model (or simply FK model); it is a dependent percolation model which is perfectly well defined for non-integer values of q, even though μ_v^q and μ^q are not. Note that μ_e^q for $q = 1$ is just independent bond percolation.

Fortuin and Kasteleyn (1969, 1972) focused on the distribution μ_e^q (and not on the joint distribution μ^q as was implicitly done by Swendsen and Wang (1987)) and on how various Ising/Potts quantities can be expressed in terms of μ_e^q. For example, the reader may easily derive for $q = 2$:

(8) $$\text{Cov}(S_x, S_{x'}) = \mu_e^2(x \text{ and } x' \text{ are in the same cluster}).$$

Such identities allow one to study Ising/Potts behavior by percolation methods.

3. Stochastic Domination. Among the most useful of percolation theoretic techniques are inequalities. Recall that the earliest example of associated random variables (variables such that increasing functions of them are positively correlated) is independent bond percolation. This result of Harris (1960) extends to FK models with $q > 1$ as shown by Fortuin (1972):

(9) for $q \geq 1$, the n_b's with distribution μ_e^q are associated.

(9) can be easily proved by the "standard FKG method" of Fortuin, Kasteleyn and Ginibre (1971) and Sarkar (1969). We remark that for $q = 2$, the association of the n_b's is different from the association of the S_x's, which follows from the FKG method applied to μ_v^2.

To motivate the following stochastic monotonicity properties we note the elementary calculation,

(10) $$\mu_e^q(n_b = 1 \mid \{n_{b'} : b' \neq b\}) = p_b \text{ or } \hat{p}_b \equiv \frac{p_b/q}{1 - p_b + p_b/q}$$

where the first value is taken if x and y are in the same cluster even with $n_b = 0$ and the second value otherwise. It was shown by Fortuin (1972) that

(11) μ_e^q is stochastically decreasing in $q \geq 1$ for fixed p_b's

(i.e., the expectation of an increasing function is decreasing) and

(12) μ_e^q is stochastically increasing in $q \geq 1$ for fixed $\hat{p}_b$'s

These domination results follow from (9) and the fact that C, the number of clusters, is a decreasing function while C plus the number of occupied bonds is increasing; the reader can supply the details.

4. Applications of Domination. In Ising or Potts models (in which the limit $\Lambda \uparrow \mathbf{Z}^d$ has already been taken) one is typically interested in the absence or presence of long range order. For our purposes, we will take this as synonymous, in the case of Ising models, with the vanishing or nonvanishing of $\lim_{\|x\|\to\infty} \mathrm{Cov}(S_0, S_x)$. A phase transition is said to occur if the change of a parameter (e.g., the p in a nearest neighbor model) switches the model from absence to presence of long range order.

By using identities such as (8) (note that its right-hand side is the expectation of an increasing function) together with both monotonicity results ((11) and (12)), one can generally conclude that a phase transition occurs for every real $q \geq 1$, if and only if it occurs for *some* real $q \geq 1$. By taking the special q to be 1, one can thus reduce the occurrence of Ising/Potts phase transitions to an issue of *independent* percolation! This approach is presented in great detail by Aizenman, Chayes, Chayes and Newman (1988), where it is applied to $1/r^2$ models and to situations where Λ tends to a logarithmic wedge in $\mathbf{Z}^d$; applications to dilute Ising/Potts models are given in Aizenman, Chayes, Chayes and Newman (1987). We remark that because the special q is 1, this approach does not even utilize the association of the FK models except for Harris' original independent percolation result. We conclude this section with a little more detail about one application to $1/r^2$ Ising models.

Consider fan $1/r^2$ independent percolation model (see (3)). If $\beta \equiv \lim_{x\to\infty} x^2 p_x = 0$ (in fact if $\beta \leq 1$), it was shown by Aizenman and Newman (1986) that (after $\Lambda \uparrow \mathbf{Z}^1$) there are a.s. no infinite clusters regardless of the individual p_x values. Next consider the related Ising model with $K_{xy} = K_{|x-y|}$ (see (5)). As a consequence of the stochastic domination of μ_e^2 by μ_e^1 and the identity (8) relating Ising and FK models, it follows from the independent percolation result that

$$\text{(13)} \qquad \lim_{x\to\infty} x^2 K_x = 0 \text{ implies absence of (Ising) long range order.}$$

We note that this result, a long standing conjecture in Ising model theory, was also independently proved (by other methods) by Berbee (1989).

5. Random Colored Graphs – General Case. The K_{xy}'s appearing in (5) (or (7)) are automatically non-negative. Let us consider more generally distributions with density

$$\text{(14)} \qquad \text{const.} \exp\left(\sum_{\{x,y\}} J_{xy} 1_{T_x = T_y}\right), \quad J_{xy} \in \mathbf{R}.$$

When $J_{xy} \geq 0$ for all x, y the Ising or Potts model is called ferromagnetic and when $J_{xy} \leq 0$ it is antiferromagnetic. It is quite fashionable these days to consider models with both positive and negative J_{xy}'s; spin glasses are of this type with the J_{xy}'s themselves random. In this section, we describe the generalization of FK models to the non-ferromagnetic context. The first published discussion of non-ferromagnetic FK models we are aware of is by Kasai and Okiji (1986). (We thank A. van Enter for informing us of this reference.) The most general extension of FK models is that of Edwards and Sokal (1988). Here, we only consider Ising or Potts models as in (14).

First partition the set of all bonds $\mathcal{B}$ into $\mathcal{B}_F$, the ferromagnetic ones (or F-bonds) corresponding to $J_{xy} \geq 0$, and $\mathcal{B}_A$ the antiferromagnetic ones (or A-bonds) where $J_{xy} < 0$. Define $p_b = 1 - \exp(- \mid J_{xy} \mid)$ for $b = \{x, y\}$ and again begin with independent T_x's and n_b's as before, but this time condition on the event

$$\begin{aligned} \{T_x = T_y \quad &\text{for every} \quad b = \{x,y\} \in \mathcal{B}_F \text{ with } n_b = 1 \text{ and} \\ T_x \neq T_y \quad &\text{for every} \quad b \in \mathcal{B}_A \text{ with } n_b = 1\}. \end{aligned} \tag{15}$$

We denote the resulting joint distribution $\tilde{\mu}^q$ and the two marginals $\tilde{\mu}^q_v$ and $\tilde{\mu}^q_e$. $\tilde{\mu}^q_v$ can readily be seen to be precisely the Ising model (14). The density of $\tilde{\mu}^q_e$ can be expressed in the form,

$$\text{const.} \prod_C R_q(C) \prod_{b:n_b=0} (1 - p_b) \prod_{b':n_{b'}=1} p_{b'} \tag{16}$$

where the first product is over all clusters C of the n_b's and $R_q(C)$ is the number of allowed colorings of C according to the rules that endsites of F-bonds (respectively A-bonds) have the same (respectively different) colors. Here C is thought of as a connected subgraph in which each edge is labelled F or A.

In the purely ferromagnetic case, R_q is of course simply q and we are back to the usual FK model measure (6). In the purely antiferromagnetic case, $R_q(C)$ is the number of q-colorings of the graph C in the usual sense of graph colorings where adjacent vertices must have different colors. For a given cluster C, $R_q(C)$ may vanish if q is not sufficiently large; for example for C the complete graph on k vertices in the purely antiferromagnetic case,

$$R_q(C) = \frac{q(q-1)\cdots(q-k+1)}{k!}.$$

This formula shows something besides the vanishing of R_q for certain integer values of q; it shows that q cannot in general be taken nonintegral (as can be done in the standard FK models) since that can lead to negative values of R_q.

In addition to clusters which are ruled out for small q, there are ones ruled out for all q—namely if there is an occupied A-bond between two sites connected by a path of occupied F-bonds. If C is not one of these totally prohibited clusters, then $R_q(C)$ can be calculated as the ordinary number of q-colorings of a "reduced graph" whose vertices are the connected components of C obtained by only using F-bonds and in which an edge occurs whenever there is at least one A-bond in C between the two corresponding components.

For the remainder of our discussion, we restrict attention to Ising models ($q = 2$). Here it is clear that for any C, $R_2(C)$ either vanishes or else equals exactly 2. Borrowing terminology from spin glass theory, we will call any configuration of the n_b's with a cluster C having $R_2(C) = 0$ a *frustrated* configuration; i.e., a configuration of the n_b's is frustrated if there is *no* 2-coloring of the sites which satisfies the *occupied* F-bonds and A-bonds. Let us write U for the collection of unfrustrated configurations of the n_b's. Then the density of $\tilde{\mu}_e^2$ may be expressed as

$$\tilde{\mu}_e^2 : \text{const.}\, 1_U 2^C \prod_{b:n_b=0} (1-p_b) \prod_{b':n_{b'}=1} p_{b'}. \tag{17}$$

The Ising model covariance may be expressed in terms of the $\tilde{\mu}_e^2$ expectation (which we denote $\tilde{E}_e^2$) as (compare (8))

$$| \operatorname{Cov}(S_x, S_{x'}) | = | \tilde{E}_e^2(\eta(x,x')) | \leq \tilde{\mu}_e^2 \,(x \text{ and } x' \text{ are in the same cluster}) \tag{18}$$

where

$$\eta(x,x') = \begin{cases} 0, & \text{if } x \text{ and } x' \text{ are not in the same cluster} \\ -1, & \text{if there is a path of occupied bonds between} \\ & \quad x \text{ and } x' \text{ with an odd number of } A\text{-bonds} \\ +1, & \text{otherwise.} \end{cases} \tag{19}$$

Although this measure seems rather difficult to work with, it can at least be used to obtain some modest results. We present these now primarily as an exercise in using $\tilde{\mu}_e^2$ rather than because of their intrinsic merit. Better results will presumably require an analysis of $\tilde{E}_e^2(\eta(x,x'))$ deeper than the trivial inequality of (18). (We remark that a formula similar to (18) appearing as Equation (A·7) in Kasai and Okiji (1988) appears to be seriously incorrect in that η is not averaged.)

For given J_{xy}'s, we let μ_e^2 denote the ($q=2$) ferromagnetic FK measure, (6); its related Ising variables will be denoted S_x while those from $\tilde{\mu}_v^2$ will be denoted $\tilde{S}_x$. Any frustrated configuration of n_b's remains frustrated when an $n_{b'}$ is changed from 0 to 1 (whether b' be an F-bond or an A-bond); hence 1_U is a decreasing function on the n_b's. Since μ_e^2 is associated, it follows that

$$\tilde{\mu}_e^2 \text{ is stochastically dominated by } \mu_e^2. \tag{20}$$

We thus have from (18) (with tildes) and (8) that

$$| \operatorname{Cov}(\tilde{S}_x, \tilde{S}_{x'}) | \leq \operatorname{Cov}(S_x, S_{x'}). \tag{21}$$

We conclude that absence of long range order in the Ising model with J_{xy} replaced by $| J_{xy} |$ implies its absence in the original model. For example, (13) remains valid with K_x replaced by J_x regardless of the signs of the J_x's.

REFERENCES

AIZENMAN, M., CHAYES, J.T., CHAYES, L. and NEWMAN, C.M. (1987). The phase boundary in dilute and random Ising and Potts ferromagnets. *J. Phys. A: Math. Gen.* **20** L313–L318.

AIZENMAN, M., CHAYES, J.T., CHAYES, L. and NEWMAN, C.M. (1988). Discontinuity of the magnetization in one-dimensional $1/\mid x-y\mid^2$ Ising and Potts models. *J. Stat. Phys.* **50** 1–40.

AIZENMAN, M. and NEWMAN, C.M. (1986). Discontinuity of the percolation density in one-dimensional $1/\mid x-y\mid^2$ percolation models. *Commun. Math. Phys.* **107** 611–647.

BERBEE, H. (1989). Uniqueness of Gibbs measures and absorption probabilities. *Ann. Prob.* (to appear).

EDWARDS, R.G. and SOKAL, A.D. (1988). Generalization of the Fortuin-Kasteleyn-Swendsen-Wang representation and Monte Carlo algorithm. *Phys. Rev. D* **38** 2009–2012.

FORTUIN, C.M. (1972). On the random-cluster model. III. The simple random-cluster model. *Physica* **59** 545–570.

FORTUIN, C.M. and KASTELEYN, P.W. (1972). On the random-cluster model. I. Introduction and relation to other models. *Physica* **57** 536–564.

FORTUIN, C.M., KASTELEYN, P.W. and GINIBRE, J. (1971). Correlation inequalities on some partially ordered sets. *Commun. Math. Phys.* **22** 89–103.

HARRIS, T.E. (1960). A lower bound for the critical probability in a certain percolation process. *Proc. Camb. Phil. Soc.* **56** 13–20.

KASAI, Y. and OKIJI, A. (1986). Percolation problem of frozen-bonds in quenched $\pm J$ Ising spin system. *Prog. Theor. Phys.* **75** 1076–1086.

KASAI, Y. and OKIJI, A. (1988). Percolation problem describing $\pm J$ Ising spin glass system. *Prog. Theor. Phys.* **79** 1080–1094.

KASTELEYN, P.W. and FORTUIN, C.M. (1969). Phase transitions in lattice systems with random local properties. *J. Phys. Soc. Japan* **26** (Suppl.) 11–14.

SARKAR, T.K. (1969). Some lower bounds of reliability. Technical Report No. 124, Dept. of Operations Research and Statistics, Stanford University, Stanford, CA.

SWENDSEN, R.H. and WANG, J.-S. (1987). Nonuniversal critical dynamics in Monte Carlo simulations. *Phys. Rev. Lett.* **58** 86–88.

DEPARTMENT OF MATHEMATICS
BUILDING 89
UNIVERSITY OF ARIZONA
TUCSON, AZ 85721

AN ORDERING OF DEPENDENCE

By Marco Scarsini[1]

Università "La Sapienza"

An ordering of dependence is defined on the space of probability measures on a finite product space, with fixed marginals. The definition of this ordering involves the Lorenz curve of the likelihood ratio of a probability measure w.r.t. the product measure of the marginals. A minimal element w.r.t. this dependence ordering always exists and equals the product measure. Conditions for the existence of a maximal element are examined. The ordering is generalized to the case of infinite spaces. Comparison with some other orders of dependence is considered.

1. Introduction. The purpose of this paper is to study an ordering of dependence for pairs of random variables (r.v.'s) taking values in sets that are finite, but not necessarily numerical or (even partially) ordered. The basic idea is that, when two r.v.'s are stochastically independent, then there is no dependence. An ordering of dependence should express this fact by having its unique minimum in the case of independent r.v.'s. Independence is then a reference point and a pair of r.v.'s is more dependent than another if it is more distant than the other from the reference situation of independence, in a sense that will be specified later.

If we want to compare two r.v.'s (or, analogously, two probability measures on a finite product space) w.r.t. dependence, then it makes sense to start considering only probability measures having the same marginals. The study of dependence will be carried out by introducing a preorder on the space of probability measures on a finite product space having the same marginals, or (equivalently) on the space of all probability matrices of fixed dimensions having prescribed row and column sums. Since we are considering qualitative r.v.'s, any ordering of dependence should not depend on the order in which the different possible outcomes are considered, in other words, the ordering should be invariant w.r.t. permutations of lines (rows and/or columns) in the probability matrices.

---

[1]Partially supported by MPI.

AMS 1980 subject classification. Primary 62H20.

Key words and phrases. Probability measures with fixed marginals, generalized concentration curve.

I am greatly indebted to Eugenio Regazzini for giving me a copy of his unpublished papers on generalized concentration curve and dependence. I also wish to thank all the people who offered comments when this paper was presented at seminars in Stanford and Perugia and at the "Symposium on Dependence in Statistics and Probability" in Hidden Valley.

The tool that we will use is the concentration curve of a probability measure with respect to another, introduced by Cifarelli and Regazzini (1987). Our results are strictly related to some ideas of Ali and Silvey (1965a), (1965b). Their results are not expressed in terms of concentration curves, though.

2. Generalized Concentration Curve. Let P, Q be two probability measures on the power set of a finite space X. From now on, for $x \in X$, we will write $P(x)$ instead of $P(\{x\})$. Cifarelli and Regazzini (1987) have defined the concentration curve of a measure P w.r.t. Q, as follows. Let ℓ_P be the (generalized) likelihood ratio of P w.r.t. Q: for $x \in X$

$$\ell_P(x) = \begin{cases} \frac{P(x)}{Q(x)}, & \text{if } Q(x) \neq 0, \\ \infty, & \text{if } Q(x) = 0, \end{cases}$$

where it is assumed that P and Q never vanish simultaneoulsy.

It is clear that ℓ_P is a r.v. on $(X, 2^X, Q)$ with values in $[0, \infty]$, and, if $P \ll Q$, then

$$E(\ell_P) = \sum_{x \in X} \frac{P(x)}{Q(x)} Q(x) = 1.$$

If m is the distribution function of ℓ_P

$$m(t) = Q\{x \in X : \ell_P(x) \leq t\},$$

and m^{-1} is the right-continuous generalized inverse of m

$$m^{-1}(z) = \sup\{t : m(t) \leq z\},$$

then the concentration curve of P with respect to Q is the Lorenz curve of ℓ_P, that is

$$\phi_P(u) = \int_0^u m^{-1}(z)\, dz.$$

The rationale for calling ϕ_P the concentration function of P w.r.t. Q is the following. Let $A(u)$ be a set of the form

$$A(u) = \{x : \ell_P(x) \leq u\}.$$

If $Q(A(u)) = t$, then $P(A(u)) = \phi(t)$.

Therefore, for each set A containing the "poorest" points (in terms of ℓ_P), ϕ_P relates the probability mass concentrated on A by Q to the probability mass concentrated on A by P. For points outside the range of Q, a sort of randomization is performed. Actually, in the framework of hypothesis testing, ϕ is the so called α–β curve for testing the null hypothesis P versus the alternative Q, when randomization is allowed (see Lehmann (1986), pp. 76–77).

The following well known result will be used later.

PROPOSITION 2.1. *(Strassen (1965)) Let S, T be two r.v.'s such that $E(S) = E(T)$, and let $\psi_S(\psi_T)$ be the Lorenz curve of $S(T)$. Then $\psi_S(u) \leq \psi_T(u)$ $\forall u \in [0,1]$ iff $\exists S', Z'$ defined on a common probability space, such that $S \stackrel{\mathcal{L}}{=} S'$ and $T \stackrel{\mathcal{L}}{=} E(S'|Z')$.*

3. Definition of the Dependence Ordering. Let X, Y be finite sets and let Q_X, Q_Y be two probability measures on X and Y, respectively, such that $Q_X(x) > 0$ $\forall x \in X$ and $Q_Y(y) > 0$ $\forall y \in Y$. Let $\mathcal{P}(Q_X, Q_Y)$ be the set of all probability measures on $X \times Y$, whose marginals are Q_X, Q_Y. Let μ be the product measure of Q_X and Q_Y. For each $P \in \mathcal{P}(Q_X, Q_Y)$, we can define a likelihood ratio ℓ_P of P w.r.t. the product measure μ.

Since every probability measure on a finite product space can be represented by a matrix, we will use the same symbol P for the probability measure and the corresponding matrix, namely

$$P = \{p_{ij}\}, \qquad p_{ij} = P(x_i, y_j) \qquad x_i \in X, y_j \in Y.$$

Let $P_1, P_2 \in \mathcal{P}(Q_X, Q_Y)$. Using an idea introduced by Cifarelli and Regazzini (1986) to study measures of dependence, a dependence ordering $\stackrel{\mathrm{D}}{\succeq}$ is defined as follows

$$P_1 \stackrel{\mathrm{D}}{\succeq} P_2 \quad \text{iff} \quad \phi_{P_1}(u) \leq \phi_{P_2}(u) \quad \forall u \in [0,1].$$

The relation $\stackrel{\mathrm{D}}{\succeq}$ is a preorder (reflexive and transitive).

The rationale for the ordering $\stackrel{\mathrm{D}}{\succeq}$ is the following: If X, Y are independent, i.e. $P = \mu$, then $\ell_P \equiv 1$. The more X, Y are dependent, the more P differs from μ, the more ℓ_P is spread out, the lower ϕ_P is. In a situation of strong dependence between X and Y, many pairs (x, y) will have "small" probabilities, and some pairs will have "high" probabilities, where small and high is measured in terms of the corresponding mass concentrated by the product measure μ. Therefore ℓ_P assumes values far from one with high μ–probability, and ϕ_P tends to be low.

From now on, for the sake of brevity, we will use the symbol $\mathcal{P}(Q_X, Q_Y)$ to indicate the ordered space $(\mathcal{P}(Q_X, Q_Y), \stackrel{\mathrm{D}}{\succeq})$, unless otherwise stated.

PROPOSITION 3.1. *Let $P \in \mathcal{P}(Q_X, Q_Y)$ and let $R = \Pi_1 P \Pi_2$, where Π_1, Π_2 are permutation matrices. Then $\ell_P \stackrel{\mathcal{L}}{=} \ell_R$.*

PROOF. Let $L_P = \{\ell_P(x, y)\}$. Then $L_R = \Pi_1 L_P \Pi_2$. Hence the result. ||

PROPOSITION 3.2. *Let $P \in \mathcal{P}(Q_X, Q_Y)$ and let P^* be obtained by pooling two lines of P. Then*

$$\phi_{P^*}(u) \geq \phi_P(u) \qquad \forall u \in [0,1],$$

and

$$\phi_{P^*}(u) = \phi_P(u) \qquad \forall u \in [0,1]$$

iff the two lines are proportional.

PROOF. Without loss of generality, let x_1, x_2 be the two lines that are pooled, to form a new line x, say. Then

$$\begin{aligned}
\ell_{P^*}(x,y) &= \frac{P^*(x,y)}{Q_X(x)Q_Y(y)} \\
&= \frac{Q_X(x_1)}{Q_X(x_1)+Q_X(x_2)} \frac{P(x_1,y)}{Q_X(x_1)Q_Y(y)} \\
&\quad \frac{Q_X(x_2)}{Q_X(x_1)+Q_X(x_2)} \frac{P(x_2,y)}{Q_X(x_2)Q_Y(y)} \\
&= \alpha\ell_P(x_1,y)+(1-\alpha)\ell_P(x_2,y),
\end{aligned}$$

with

$$\alpha = \frac{Q_X(x_1)}{Q_X(x_1)+Q_X(x_2)}.$$

If we apply Proposition 1.1, we obtain $\phi_{P^*}(u) \geq \phi_P(u)$, $\forall u \in [0,1]$.

Of course, if the two lines are proportional, then

$$\ell_P(x_1,y) = \ell_P(x_2,y) = \ell_{P^*}(x,y);$$

therefore, $\phi_P = \phi_{P^*}$. ||

Related results can be found in Ali and Silvey (1965a), (1965b).

The space $\mathcal{P}(Q_X, Q_Y)$ has a minimum.

PROPOSITION 3.3. *Let $P^* \in \mathcal{P}(Q_X,Q_Y)$. Then $P \overset{D}{\succeq} P^* \quad \forall P \in \mathcal{P}(Q_X,Q_Y)$ iff $P^* = \mu$.*

PROOF. This is an immediate consequence of the fact that $\phi(u) = u\ \forall u \in [0,1]$ iff $P \equiv Q$ (Cifarelli and Regazzini (1987)). ||

PROPOSITION 3.4. *Let $P^* \in \mathcal{P}(Q_X,Q_Y)$ and $Q_X = Q_Y$. Then*

$$P^* \overset{D}{\succeq} P \quad \forall P \in \mathcal{P}(Q_X,Q_Y) \tag{1}$$

iff

(2) $$P^*(x,x) = Q_X(x) = Q_Y(x) \qquad \forall x \in X$$

(modulo marginal preserving permutations of lines in P^*).

PROOF. (2) $\Longrightarrow$ (1): If we order the x's according to the value of Q_X increasingly, we have that

$$\max_{\substack{(x,y)\in X^2 \\ P\in\mathcal{P}(Q_X,Q_Y)}} \ell_P(x,y) = \ell_{P^*}(x_1,x_1) = 1/Q_X(x_1)$$

where $P^*(x_1,x_1) = Q_X(x_1)$. Now, if $P^*(x_1,x_1) = Q_X(x_1)$, then

$$P^*(x_1,y) = P^*(y,x_1) = 0 \qquad \forall\, y \neq x_1.$$

Iterating the procedure, we have

$$\max_{\substack{(x,y)\in(X\setminus\{x_1\})^2 \\ P\in\mathcal{P}(Q_X,Q_Y)}} \ell_P(x,y) = \ell_{P^*}(x_2,x_2) = 1/Q_X(x_2)$$

where $P^*(x_2,x_2) = Q_X(x_2)$, and so on.

Therefore $P^* \overset{D}{\succeq} P \quad \forall P \in \mathcal{P}(Q_X,Q_Y)$, when P^* satisfies (2).

(1) $\Longrightarrow$ (2): Assume that $\tilde{P} \overset{D}{\succeq} P \quad \forall P \in \mathcal{P}(Q_X,Q_Y)$, and $\tilde{P} \neq P^*$ (modulo permutations of lines), where P^* satisfies (2). Then $\tilde{P} \overset{D}{\succeq} P^*$. Now, since $P^* \overset{D}{\succeq} P$ $\forall P \in \mathcal{P}(Q_X,Q_Y)$, then $\phi_{\tilde{P}} = \phi_{P^*}$. But, then

$$\phi_{\tilde{P}}(1-Q_X^2(x_1)) = \phi_{P^*}(1-Q_X^2(x_1)) = 1-Q_X(x_1),$$

which is possible only if

$$\tilde{P}(x_1,x_1) = P^*(x_1,x_1) = Q_X(x_1).$$

Furthermore,

$$\phi(1-Q_X^2(x_1)-Q_X^2(x_2)) - \phi_{P^*}(1-Q_X^2(x_1)-Q_X^2(x_2)) = 1-Q_X(x_1)-Q_X(x_2),$$

which is possible only if $\tilde{P}(x_2,x_2) = P^*(x_2,x_2) = Q_X(x_2)$, etc. Iteration of the argument gives $\tilde{P} = P^*$. $\quad\|$

Proposition 3.4 shows that, when $Q_X = Q_Y$, then there exists a maximum on $\mathcal{P}(Q_X,Q_Y)$. This maximum is unique modulo permutations of lines. This is not the only case in which a maximum exists, as the following proposition shows.

PROPOSITION 3.5. *Consider $\mathcal{P}(Q_X,Q_Y)$, with Q_X,Q_Y uniform on X,Y, respectively. Let card$(X) = M$, card$(Y) = N$, $N > M$. Let P^* be defined as follows.*

Partition P^* into n square matrices of the following form: P_1 is the largest NW square sub-matrix of $P_0 \stackrel{def}{=} P^*$. $R_n =$ is the matrix obtained by deleting the lines of P_n from P_{n-1}, where P_n is the largest NW square submatrix of R_{n-1}. Let the elements of each matrix P_n be zero outside the main diagonal. There exists only one possible configuration of this type such that $P^* \in \mathcal{P}(Q_X, Q_Y)$. Furthermore $P^* \stackrel{D}{\succeq} P \quad \forall P \in \mathcal{P}(Q_X, Q_Y)$. If $P \equiv P^*$, then P is obtained from P^* via permutation of lines.

PROOF.

$$\max_{\substack{(x,y)\in X\times Y \\ P\in\mathcal{P}(Q_X,Q_Y)}} \ell_P(x,y) = \ell_{P^*}(x_1,y_1) = Q_Y(y_1)$$

If $P^*(x_1, y_1) = Q_Y(y_1)$, then $P(x, y_1) = 0 \quad \forall x \neq x_1$. Under this constraint,

$$\max_{\substack{(x,y)\in X\times\{Y\setminus y_1\} \\ P\in\mathcal{P}(Q_X,Q_Y)}} \ell_P(x,y) = \ell_{P^*}(x_2,y_2) = Q_Y(y_2)$$

and so on, for $y_1, \ldots, y_M$.

Given these constraints, the argument can be repeated for the matrices R_1, ..., R_n, and the result follows. ||

4. Tetrachoric Tables. In the case of 2×2 tables, a (unique) maximum exists.

THEOREM 4.1. *Let $X = \{x_1, x_2\}$, $Y = \{y_1, y_2\}$.*

$$\begin{array}{lll} Q_X(x_1) = \alpha & Q_X(x_2) = \beta = 1 - \alpha & \alpha \leq \beta \\ Q_Y(y_1) = \gamma & Q_Y(y_2) = \delta = 1 - \gamma & \gamma \leq \delta \end{array}$$

If P^* has the following form

$$\begin{array}{ll} P^*(x_1, y_1) = \alpha \wedge \gamma & P^*(x_1, y_2) = (\alpha - \gamma)_+ \\ P^*(x_2, y_1) = (\gamma - \alpha)_+ & P^*(x_2, y_2) = \beta \wedge \delta, \end{array}$$

then P^* is the unique maximum w.r.t. $\stackrel{D}{\succeq}$.

PROOF. Without loss of generality, assume $\delta \leq \beta$ (whence $\alpha \leq \gamma$). Then

$$\begin{array}{ll} \phi_{P^*}(x) = 0 & 0 \leq x \leq \alpha\delta \\ = \frac{\gamma-\alpha}{\beta\gamma} x - \frac{\gamma-\alpha}{\beta\gamma}\alpha\delta & \alpha\delta \leq x \leq \alpha\delta + \beta\gamma \\ = \frac{x}{\gamma} - \frac{\delta}{\gamma}(\alpha + \gamma) & \alpha\delta + \beta\gamma \leq x \leq 1 - \beta\delta \\ = \frac{x}{\beta} - \frac{1}{\beta} + 1 & 1 - \beta\delta \leq x \leq 1. \end{array}$$

For any $P \in \mathcal{P}(Q_X, Q_Y)$, we have, for $i = 1, 2$,

$$\ell_P(x_i, y_1) \geq 1 \Longleftrightarrow \ell_P(x_i, y_2) \leq 1,$$

and, for $j = 1, 2$,

$$\ell_P(x_1, y_j) \geq 1 \Longleftrightarrow \ell_P(x_2, y_j) \leq 1.$$

Therefore, only two possible arrangements are possible:

$$\ell_P(x_2, y_2) \leq \ell_P(x_1, y_1) \leq \ell_P(x_2, y_1) \leq \ell_P(x_1, y_2) \tag{3}$$

or

$$\ell_P(x_1, y_2) \leq \ell_P(x_2, y_1) \leq \ell_P(x_1, y_1) \leq \ell_P(x_2, y_2). \tag{4}$$

Moreover, if we indicate $p_{ij} = P(x_i, y_j), \quad i, j = 1, 2,$

$$\begin{gathered} \alpha + \gamma - 1 \leq p_{11} \leq \alpha \\ 0 \leq p_{12} \leq 1 - \gamma \\ \gamma - \alpha \leq p_{21} \leq 1 - \alpha \\ 0 \leq p_{22} \leq 1 - \gamma. \end{gathered}$$

If P is such that (4) holds, then $\phi_P(x)$ is a broken line with corners in $x = \alpha\delta$, $\alpha\delta + \beta\gamma$, $1 - \beta\delta$.

$$\begin{gathered} \phi_P(\alpha\delta) = p_{12} \geq 0 = \phi_{P^*}(\alpha\delta) \\ \phi_P(\alpha\delta + \beta\gamma) = \gamma - \alpha + 2p_{12} \geq \gamma - \alpha = \phi_{P^*}(\alpha\delta + \beta\gamma) \\ \phi_P(1 - \beta\delta) = 1 - p_{22} \geq \gamma = \phi_{P^*}(1 - \beta\delta) \end{gathered}$$

If P is such that (3) holds, then $\phi_P(x)$ is a broken line with corners in $x = \beta\delta$, $\beta\delta + \alpha\gamma$, $1 - \alpha\delta$

$$\begin{gathered} \phi_P(\beta\delta) = p_{22} \geq 0 = \phi_{P^*}(\beta\delta) \\ \phi_P(\beta\delta + \alpha\gamma) = p_{11} + p_{22} \geq \alpha + \gamma - 1 \geq \phi_{P^*}(\beta\delta + \alpha\gamma) = (\gamma - \alpha)\tfrac{1-\beta\gamma}{\beta\gamma} \\ \phi_P(1 - \alpha\delta) = 1 - p_{12} \geq \gamma \geq \phi_{P^*}(1 - \alpha\delta) = \tfrac{1 - \alpha\delta - \delta(\alpha+\gamma)}{\gamma}. \end{gathered}$$

Since $\phi_P(x) \geq \phi_{P^*}(x)$ for every x, where ϕ_P has a corner, and since ϕ_P, ϕ_{P^*} are Lorenz curves (i.e. increasing, convex, with $\phi_P(0) = \phi_{P^*}(0) = 0$, $\phi_P(1) = \phi_{P^*}(1) = 1$), then $\phi_P(x) \geq \phi_{P^*}(x)\ \forall\ x \in [0, 1]$. Therefore, P^* is the maximum w.r.t. $\overset{\mathrm{D}}{\succeq}$. ||

It could be proven that, in the case of $2 \times n$ matrices, a maximum exists, but, in general, it is not unique (not even modulo permutations). No such maximum exists in general.

5. Comparison with Different Orderings. Joe (1985) proposed an ordering of dependence for contingency tables with fixed row and column sums. We indicate this ordering by $\overset{\mathrm{J}}{\succeq}$. Let $P_1,\ P_2 \in \mathcal{P}(Q_X, Q_Y)$. $P_1 \overset{\mathrm{J}}{\succeq} P_2$ iff $\mathrm{vec}(P_1) \overset{\mathrm{m}}{\succeq} \mathrm{vec}(P_2)$, where vec$(P)$ is the vector obtained by piling the columns of P, and $\overset{\mathrm{m}}{\succeq}$ is the usual majorization ordering (see Marshall and Olkin (1979)).

Joe's ordering has a major drawback as an ordering of dependence: it does not have a unique minimum corresponding to the product measure. This is due to the fact that the values in the probability matrices are not weighed according to their marginals. The ordering $\overset{\mathrm{J}}{\succeq}$ performs well in this respect only when the marginals are uniform. This suggests the connection between the orderings $\overset{\mathrm{D}}{\succeq}$ and $\overset{\mathrm{J}}{\succeq}$, at least when Q_X and Q_Y assume only rational values, which is not a restrictive assumption for contingency tables.

THEOREM 5.1. *Let Q_X, Q_Y assume only rational values. Let $P_1, P_2 \in \mathcal{P}(Q_X, Q_Y)$. Split rows and columns of P_1, P_2 in such a way that their marginals become uniform. (This is always possible, given rationality of the values of Q_X, Q_Y). Call $\tilde{P}_1, \tilde{P}_2 \in \mathcal{P}(\tilde{U}_n, \tilde{U}_m)$ these new matrices. Then $P_1 \overset{D}{\succeq} P_2$ iff $\tilde{P}_1 \overset{J}{\succeq} \tilde{P}_2$.*

PROOF. If the marginals are uniform, then $\ell_{\tilde{P}_1}$ assumes each value $\ell_{\tilde{P}_1}(x_i, y_j)$ with probability $(mn)^{-1}$. Analogously for $\ell_{\tilde{P}_2}$. Furthermore

$$\ell_{\tilde{P}_h}(i,j) = mnP_h(i,j) \quad h = 1,2.$$

Therefore, if $\overset{\mathrm{L}}{\succeq}$ is the Lorenz ordering, we have

$$\ell_{\tilde{P}_1} \overset{\mathrm{L}}{\succeq} \ell_{\tilde{P}_2} \text{ iff } \ell_{\tilde{P}_1} \overset{\mathrm{m}}{\succeq} \ell_{\tilde{P}_2} \text{ iff } \mathrm{vec}(\tilde{P}_1) \overset{\mathrm{m}}{\succeq} \mathrm{vec}(\tilde{P}_2) \text{ iff } \tilde{P}_1 \overset{\mathrm{J}}{\succeq} \tilde{P}_2.$$

Since $\phi_{P_h} = \phi_{\tilde{P}_h}$, for $h = 1, 2$, then the result follows. ||

Theorem 5.1 shows that considering the likelihood ratio as the keypoint to study dependence is the same as transforming the matrix so that the marginals are uniform, and then vectoralize it. The idea of rendering the marginals uniform, when studying dependence, has been applied to the study of concordance (positive quadrant dependence).

6. Concordance and Dependence. Consider two linearly ordered measurable spaces $(X, \mathcal{X})$, $(Y, \mathcal{Y})$, and the class $\mathcal{P}(Q_X, Y_Y)$ of probability measures on $(X \times Y, \mathcal{X} \otimes \mathcal{Y})$ with marginals Q_X, Q_Y. For $P_1, P_2 \in \mathcal{P}(Q_X, Q_Y)$, P_1 is said more concordant (or more positive quadrant dependent) than P_2 ($P_1 \overset{\mathrm{c}}{\succeq} P_2$) iff

$$P_1\{(\xi, v) : \xi > x; v > y\} \geq P_2\{(\xi, v) : \xi > x; v > y\} \qquad \forall (x, y) \in X \times Y$$

(see for instance Yanagimoto and Okamoto (1969), Tchen (1980), Scarsini (1984), Kimeldorf and Sampson (1987)).

The class $\mathcal{P}(Q_X, Q_Y)$ has a minimum P_c^- and a maximum P_c^+ w.r.t. $\overset{\mathrm{c}}{\succeq}$, which are referred to as Fréchet bounds.

$$P_c^+\{(\xi,v):\xi>x;v>y\}=\min(Q_X\{\xi:\xi>x\},Q_Y\{v:v>y\}),$$
$$P_c^-\{(\xi,v):\xi>x;v>y\}=\max(Q_X\{\xi:\xi>x\}+Q_Y\{v:v>y\}-1,0).$$

In general P^+ and P_c^+ differ and there does not exist any permutation of rows and columns that make the two coincide.

EXAMPLE.

P^+

	0	1	2	
0	1/4			1/4
1		1/4		1/4
2			1/4	1/4
3	1/12	1/12	1/12	1/4
	1/3	1/3	1/3	

P_c^+

	0	1	2	
0	1/4			1/4
1	1/12	1/6		1/4
2		1/6	1/12	1/4
3			1/4	1/4
	1/3	1/3	1/3	

No permutation of lines leads from P^+ to P_c^+ or vice versa.

7. Infinite Spaces. Generalization to the case of infinite spaces requires some care in the definition of the generalized concentration curve and poses some problems in the interpretation of the ordering.

For the definition of the generalized concentration curve, again we use the results provided by Cifarelli and Regazzini (1987). Let $(X,\mathcal{X},Q_X)$, $(Y,\mathcal{Y},Q_Y)$ be two probability spaces and let $\mathcal{X},\mathcal{Y}$ contain the singletons. Let μ be the product measure of Q_X,Q_Y on $(X\times Y,\mathcal{X}\otimes\mathcal{Y})$. If $P\in\mathcal{P}(Q_X,Q_Y)$, there exists a partition $\{N,N^c\}\subset\mathcal{X}\otimes\mathcal{Y}$ and a nonnegative real function h on $X\times Y$, such that

$$P(A)=\int_{A\cap N^c}h(\mathbf{z})\;\mu(d\mathbf{z})+P_S(A\cap N)\;\;\forall\;A\in\mathcal{X}\otimes\mathcal{Y}$$

$$\mu(N)=0\;\;P_S(N)=P_S(X\times Y);$$

N is unique, modulo μ–null sets.

$$\ell_P(x,y)=\begin{cases}h(x,y), & \text{if } (x,y)\in N^c;\\ \infty, & \text{if } (x,y)\in N.\end{cases}$$

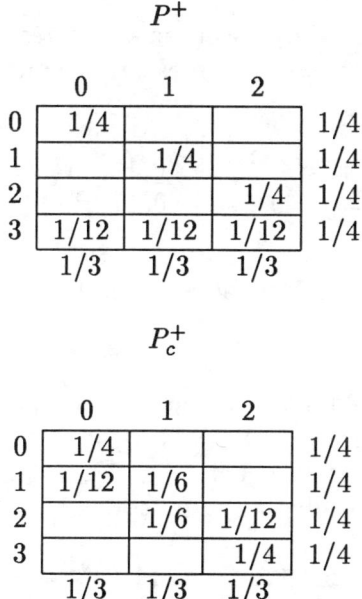

As before, let m be the d.f. of ℓ_P

$$m(t) = \mu\{(x, y) \in X \times Y : \ell_P(x, y) \leq t\}.$$

Then the generalized concentration curve of P w.r.t. μ is

$$\phi_P(u) = \int_0^u m^{-1}(t)\, dt \qquad u \in [0, 1)$$

and we define $\phi_P(1) = 1$. Then ϕ_P is convex and increasing on $[0, 1]$ and continuous on $[0, 1)$. The jump $\phi_P(1) - \phi_P(1^-)$ represents the mass of P_S, which can assume any value in $[0, 1]$. We have

$$\begin{array}{llll} \phi_P(u) = u & \forall u \in [0,1] & \text{iff} & P = \mu \\ \phi_P(u) = 0 & \forall u \in [0,1) & \text{iff} & P \perp \mu. \end{array}$$

The dependence ordering $\overset{D}{\succeq}$ is defined as in Section 2. For $P_1, P_2 \in \mathcal{P}(Q_X, Q_Y)$, $P_1 \overset{D}{\succeq} P_2$ iff $\phi_{P_1}(u) \leq \phi_{P_2}(u)$, $\forall u \in [0, 1]$.

Some phenomena should warn against an acritical use of the ordering $\overset{D}{\succeq}$ in this general situation. Let $X = Y = [0, 1]$, $\mathcal{X} = \mathcal{Y} = \text{Bor}([0, 1])$. Let $Q_X = Q_Y =$ Lebesgue measure on $[0, 1]$. Let $P_k \in \mathcal{P}(Q_X, Q_Y)$ be defined as follows: P_k concentrates its mass uniformly on the functions

$$y = x + c, \quad c = \pm(1/2)^k \quad k \in I\!N,$$

with $(x, y) \in [0, 1]^2$. Since $P_k \perp \mu$, $\forall\ k \in I\!N$, it holds that $\phi_{P_k}(u) = 0\ \forall\ u \in [0, 1)$. But $P_k \xrightarrow{\mathcal{L}} \mu$. This fact shows that $\overset{D}{\succeq}$ indicates how concentrated P is on sets of small μ–probability, rather than how dependent P is. These two concepts basically coincide when X and Y are finite, but they differ in general.

Joe (1987) considered an ordering $\overset{J}{\succeq}$ of dependence for general probability measures on product spaces. This ordering generalizes the ordering $\overset{J}{\succeq}$ for contingency tables. The main difference between $\overset{D}{\succeq}$ and $\overset{J}{\succeq}$ is that, in defining $\overset{J}{\succeq}$, Joe does not consider the likelihood ratio of P w.r.t. μ, but a density w.r.t. a generic product measure (usually Lebesgue or counting measure). This implies the drawbacks that we noticed for the discrete case in Section 5.

8. Concluding Remarks. We have introduced an ordering of dependence on the set of probability measures on a product space, with fixed marginals. We have considered only products of two spaces. The generalization to n–fold product spaces is straightforward.

The ordering $\overset{D}{\succeq}$ is based on the Lorenz curve of the likelihood ratio of a probability measure w.r.t the product measure of its marginals. Consider two probability measures $P_1 \in \mathcal{P}(Q_X, Q_Y)$, $P_2 \in \mathcal{P}(Q_W, Q_Z)$, where X, Y, W, Z are finite and Q_X, Q_Y, Q_W, Q_Z assume only rational values. The argument used in Section 5 shows that, by splitting lines of P_1, P_2, it is possible to obtain $\tilde{P}_1, \tilde{P}_2 \in \mathcal{P}(U_n, U_m)$, with U_n, U_m uniform, and such that $\ell_{P_1} \stackrel{\mathcal{L}}{=} \ell_{\tilde{P}_1}$, $\ell_{P_2} \stackrel{\mathcal{L}}{=} \ell_{\tilde{P}_2}$. Comparison of $\tilde{P}_1$, $\tilde{P}_2$ w.r.t. $\stackrel{\mathrm{D}}{\succeq}$ induces an analogous comparison for P_1, P_2. This suggests to remove the constraint of considering only probability measures with fixed marginals.

It is actually possible to show that, for any probability space $(X \times Y, \mathcal{X} \otimes \mathcal{Y}, P)$, with $P \in \mathcal{P}(Q_X, Q_Y)$, there exists $([0,1]^2, \mathrm{Bor}([0,1]^2), \pi)$ such that π has uniform marginals and $\ell_P \stackrel{\mathcal{L}}{=} \ell_\pi$.

Transforming a probability measure on a product space into a probability measure on the unit square with uniform marginals (in a suitable dependence preserving way) is a common idea in the study of concordance of random variables. It is interesting to see that it appears here, given that concordance and dependence are different concepts. The linear order structure on the spaces X, Y is a basic ingredient for the definition of concordance, whereas any such structure is neglected when dealing with dependence.

Once more we want to emphasize that, while in the case of finite spaces the preorder $\stackrel{\mathrm{D}}{\succeq}$ is a bona fide dependence ordering, in the general case the intuitive rationale for the ordering fails, especially when a probability measure is not dominated by the product measure of its marginals.

REFERENCES

ALI, S.M. and SILVEY, S.D. (1965a). Association between random variables and the dispersion of a Radon-Nikodym derivative. *J. Roy. Statist. Soc.* B **27** 100–107.

ALI, S.M., and SILVEY, S.D. (1965b). A further result on the relevance of the dispersion of a Radon-Nikodym derivative to the problem of measuring association. *J. Roy. Statist. Soc.* B **27** 108–110.

CIFARELLI, D.M. and REGAZZINI, E. (1986). Sulla funzione di concentrazione e sul suo ruolo nella statistica descrittiva. In *Atti della XXXIII Riunione Scientifica della Società Italiana di Statistica*, Bari, 347–352.

CIFARELLI, D.M. and REGAZZINI, E. (1987). On a general definition of concentration function. *Sankhyā B* **49** 307–319.

JOE, H. (1985). An ordering of dependence for contingency tables. *Linear Algebra and Its Applications* **70** 89–103.

JOE, H. (1987). Majorization, randomness and dependence for multivariate distributions. *Ann. Prob.* **15** 1217–1225.

KIMELDORF, G. and SAMPSON, A.R. (1987). Positive dependence orderings. *Ann. Inst. Stat. Math.* **39** 113–128.

LEHMANN, E.L. (1986). *Testing Statistical Hypotheses.* 2nd Edition. Wiley, New York.

MARSHALL, A.W. and OLKIN, I. (1979). *Inequalities: Theory of Majorization and Its Applications.* Academic Press, New York.

REGAZZINI, E. (1988). A few remarks on concentration comparisons. General Theory. Dipartimento di Matematica, Università di Milano.

SCARSINI, M. (1984). On measures of concordance. *Stochastica* **8** 201–218.

STRASSEN, V. (1965). The existence of probability measures with given marginals. *Ann. Math. Statist.* **36** 423–439.

TCHEN, A.H. (1980) Inequalities for distributions with given marginals. *Ann. Prob.* **8** 814–827.

YANAGIMOTO, T. and OKAMOTO, M. (1969). Partial orderings on permutations and monotonicity of a rank correlation statistic. *Ann. Inst. Statist. Math.* **21**, 489–506.

DIPARTIMENTO DI SCIENZE ATTUARIALI
UNIVERSITA LA SAPIENZA
VIA DEL CASTRO LAURENZIANO 9
I–00161 ROMA, ITALY

DYNAMIC CONSTRUCTION AND SIMULATION OF RANDOM VECTORS

BY MOSHE SHAKED[1] AND J. GEORGE SHANTHIKUMAR[1]

University of Arizona and University of California, Berkeley

In this paper described is a novel method of generation of nonnegative random variables $T_1, \ldots, T_n$ which may be dependent and which have an absolutely continuous joint distribution. In this method first $\min(T_1, \ldots, T_n)$ is generated and then one of the indices $1, \ldots, n$ (j_1, say) is chosen and T_{j_1}, is determined. Once $T_{j_1}, \ldots, T_{j_k}$ have been determined, then $\min_{j \epsilon \{1,\ldots,n\}-\{j_1,\ldots,j_k\}}(T_j)$ is generated and one of the remaining indices (j_{k+1}, say) is chosen and $T_{j_{k+1}}$ is determined. The novel method has a clear intuitive meaning, mainly for applications in reliability theory. The new method is applied to obtain stochastic comparisons of two absolutely continuous random vectors consisting of nonnegative random variables. Also, the use of the new method is illustrated in obtaining some multivariate aging properties and positive dependence properties of vectors of random lifetimes.

1. Introduction. Consider an absolutely continuous nonnegative random variable T with distribution function F, survival function $\bar{F} = 1 - F$ and hazard function $\Lambda = -\log \bar{F}$. The random variable T can be thought of as a lifetime of a device. The hazard rate (or the instantaneous failure rate of the device) at time t is defined as

$$\lambda(t) = \frac{f(t)}{P(T \geq t)} = \frac{f(t)}{\bar{F}(t)} = \frac{d}{dt}\Lambda(t), \; t \geq 0,$$

where $f = \frac{d}{dt}F$ is the density function of T. It is well known (and easy to verify) that T is stochastically equal to (that is, has the same distribution as) the time of the first epoch of a nonhomogeneous Poisson process on $[0, \infty)$ with intensity function λ. Thus, in order to generate a random variable $\hat{T}$ which has the same

[1]Supported by the Air Force Office of Scientific Research, U.S.A.F., under Grant AFOSR-84-0205. Reproduction in whole or in part is permitted for any purpose of the United States Government.

AMS subject classification. 60K10.

Key words and phrases. Reliability theory, nonhomogeneous Poisson process, instantaneous failure rate, stochastic ordering, multivariate IFR.

We thank the referee and the editors for useful comments.

distribution as T, one can generate a nonhomogeneous Poisson process with intensity function λ and let $\hat{T}$ be the time of the first epoch in that process. This can be done by generating a standard exponential random variable S and define $\hat{T}$ by

$$\hat{T} = \inf\{t : \Lambda(t) > S\} = \Lambda^{-1}(S);$$

see, e.g., Lewis and Shedler (1979), Ross (1985), and Shanthikumar (1986).

The purpose of this paper is to give a multivariate analog of this univariate result. This is done in Section 2 where we introduce a method (called the *dynamic construction*) which, for every nonnegative absolutely continuous random vector $\mathbf{T} = (T_1, \ldots, T_n)$, constructs a random vector $\hat{\mathbf{T}} = (\hat{T}_1, \ldots, \hat{T}_n)$ of times of first epoch of some nonhomogeneous Poisson processes such that $\hat{\mathbf{T}} \overset{\text{st}}{=} \mathbf{T}$ (in this paper $\overset{\text{st}}{=}$ denotes stochastic equality).

A random variable X is said to be *stochastically smaller* than a random variable Y (denoted $X \overset{\text{st}}{\leq} Y$) if, for every t, $P\{X > t\} \leq P\{Y > t\}$. A random vector $\mathbf{X} = (X_1, \ldots, X_n)$ is said to be stochastically smaller than a random vector $\mathbf{Y} = (Y_1, \ldots, Y_n)$ [denoted $\mathbf{X} \overset{\text{st}}{\leq} \mathbf{Y}$] if $g(\mathbf{X}) \overset{\text{st}}{\leq} g(\mathbf{Y})$ for every increasing Borel measurable real function g. A function g is called increasing if $g(x_1, \ldots, x_n) \leq g(y_1, \ldots, y_n)$ whenever $(x_1, \ldots, x_n) \leq (y_1, \ldots, y_n)$. [In this paper 'increasing' and 'decreasing' are not used in the strict sense. For vectors $\mathbf{x} = (x_1, \ldots, x_n)$ and $\mathbf{y} = (y_1, \ldots, y_n)$ we denote $\mathbf{x} \leq \mathbf{y}$ to mean $x_i \leq y_i,\ i, \ldots, n$.] It is well known that $\mathbf{X} \overset{\text{st}}{\leq} \mathbf{Y}$ if and only if

$$Eg(\mathbf{X}) \leq Eg(\mathbf{Y}) \tag{1}$$

for every increasing Borel-measurable real function g for which the expectations exist. Another condition which is equivalent to $\mathbf{X} \overset{\text{st}}{\leq} \mathbf{Y}$, is

$$P\{\mathbf{X} \epsilon U\} \leq P\{\mathbf{Y} \epsilon U\}$$

for every Borel set U which has an increasing indicator function [such sets are called increasing (or upper) sets].

If X and Y are nonnegative absolute continuous random variables with hazard rate functions μ and η, respectively, then it is well known (and easy to verify) that

$$[\mu(t) \geq \eta(t),\ t \geq 0] \Longrightarrow X \overset{\text{st}}{\leq} Y. \tag{2}$$

It follows that if $\mathbf{X} = (X_1, \ldots, X_n)$ and $\mathbf{Y} = (Y_1, \ldots, Y_n)$ are vectors of independent nonnegative absolutely continuous random variables such that the hazard rate functions of X_i and Y_i are μ_i and η_i, respectively, $i = 1, \ldots, n$, then

$$[\mu_i(t) \geq \eta_i(t),\ t \geq 0,\ i = 1, \ldots, n] \Longrightarrow \mathbf{X} \overset{\text{st}}{\leq} \mathbf{Y}. \tag{3}$$

It is not hard to show that the converse of (2), and hence also of (3), is false.

In this paper we extend (3) to random vectors $\mathbf{X}$ and $\mathbf{Y}$ which may have dependent components. This is done in Section 3 where random vectors $\hat{\mathbf{X}}$ and $\hat{\mathbf{Y}}$ are constructed simultaneously (on a common probability space), using the dynamic construction, such that $\hat{\mathbf{X}}$ and $\hat{\mathbf{Y}}$ satisfy: $\hat{\mathbf{X}} \stackrel{\mathrm{st}}{=} \mathbf{X}$, $\hat{\mathbf{Y}} \stackrel{\mathrm{st}}{=} \mathbf{Y}$ and $P\{\hat{\mathbf{X}} \leq \hat{\mathbf{Y}}\} = 1$. The results are similar to those of Shaked and Shanthikumar (1987a), but the proofs are different.

The rest of the paper (Sections 4 and 5) consists of applications of the results of Section 3. In Section 4 we identify some conditions on the hazard rates of a random vector $\mathbf{T}$ (the hazard rates are defined in Section 2) which imply that $\mathbf{T}$ satisfies the MIHR$|\ \mathcal{F}_t$ property of Arjas (1981a). In Section 5 we show that another set of conditions on the hazard rates imply that $\mathbf{T}$ has the WBF (weakened by failures) property of Arjas and Norros (1984). In particular, such random variables are associated in the sense of Esary, Proschan, and Walkup (1967).

In this paper, for $t_* \geq 0$, we will consider nonhomogeneous Poisson processes on $[t_*, \infty)$ with intensity function λ defined on $[t_*, \infty)$. By this phrase we mean nonhomogeneous Poisson processes which start counting at time t_*, or, equivalently, nonhomogeneous Poisson processes with intensity 0 on $[0, t_*)$ and intensity λ on $[t_*, \infty)$.

2. The Dynamic Construction. For the purpose of generating a random vector $\hat{\mathbf{T}}$, such that $\hat{\mathbf{T}} \stackrel{\mathrm{st}}{=} \mathbf{T}$ for some given $\mathbf{T}$, two alternative constructions have been used. They are the *standard construction* (see, e.g., Arjas and Lehtonen (1978)) and the *total hazard construction* (see, e.g., Norros (1986) and Shaked and Shanthikumar (1986a, 1987b)). If $\mathbf{T}$ is n-dimensional, then each of these constructions requires n uniform random variables in order to generate $\hat{\mathbf{T}}$. The dynamic construction, described below, requires more than n uniform random variables but it has an intuitive meaning which has theoretical and practical advantages, especially in reliability theory.

In the dynamic construction, the random variables $T_1, \ldots, T_n$, are thought of as the lifetimes of n components numbered $1, 2, \ldots, n$. The dynamic point of view can be described as follows: Let $t \geq 0$ be zero or an observed time of failure of one of the components. Assume that at that time t, it is known which components are still alive and the failure times of the components which fail before or at time t. Given this information, the dynamic construction considers then the conditional distribution of the time to next failure, t' say ($t' > t$), and the conditional probability that this next failure is of a particular component of those still alive at time t. The time t' is a new starting point at which the conditional distribution of the following failure and the identity of the next failed component are considered. This is done inductively until all n components have failed.

Given that $t \geq 0$ is a failure time of a component and that components $i_1, \ldots, i_k$ are still alive then [if $t = 0$ then $k = n$ and $\{i_1, \ldots, i_n\} = \{1, \ldots, n\}$], the next failure time is $\min(T_{i_1}, \ldots, T_{i_k})$ and, in the dynamic construction below, this time is described (as in the univariate case, see Section 1) as the time of the first epoch of a nonhomogeneous Poisson process, starting at time t, with intensity function depending on the observed 'history' up to time t. Once $\min(T_{i_1},\ldots,T_{i_k})$ is observed, at time t', say, the identity of the failed component (which must be one of $i_1,\ldots,i_k$) is chosen according to the conditional distribution [on $\{i_1,\ldots,i_k\}$] of the component identities given in the history up to time $t'-$ and that a failure has occurred at time t'.

For example, in the bivariate case there are two components, 1 and 2, which start to live at time 0. First consider the distribution of $\min(T_1,T_2)$ with hazard rate function λ, say, on $[0,\infty)$. It is the distribution of the time of the first epoch in a nonhomogeneous Poisson process with intensity function λ. Next, given that $\min(T_1,T_2)=t$, say, choose an i from $\{1,2\}$ according to the probability $p_t(i)$, $i=1,2$, where $p_t(i)$ is the conditional probability [given $\min(T_1,T_2)=t$] that component i fails at time t [that is, that in fact $\min(T_1,T_2)=T_i$]. Note that $p_t(1)+p_t(2)=1$ for all t. This way, one of the T_i's (the smallest of the two) is stochastically represented as the time of the first epoch of a nonhomogeneous Poisson process. Finally, given that $\min(T_1,T_2)=t$ and that $\min(T_1,T_2)=T_i$ for some $i\epsilon\{1,2\}$, let λ_{3-i} (which may depend on t) be the (conditional) hazard rate function of the surviving component on $[t,\infty)$. The random variable T_{3-i} can also be stochastically represented as the time of the first epoch of a nonhomogeneous Poisson process (with intensity function λ_{3-i}). It is not hard to see that the functions $\lambda(\cdot)$, $p_\cdot(1)$, $p_\cdot(2)$, $\lambda_1(\cdot)$ and $\lambda_2(\cdot)$ determine the distribution F of (T_1,T_2) [see, e.g., Cox (1972) or Shaked and Shanthikumar (1986b)]. In the bivariate case these functions (and similar ones in the case $n>2$) will be the building blocks of the dynamic construction described below.

In general, let $\mathbf{T}=(T_1,\ldots,T_n)$ be a nonnegative absolutely continuous random vector to be thought of as a vector of lifetimes of n components. For $I=\{i_1,\ldots,i_k\}\subset\{1,\ldots,n\}$, let $\mathbf{t}_I$ denote $(t_{i_1},\ldots,t_{i_k})$. The complement of I will be denoted by $\bar{I}=\{1,\ldots,n\}-I$ and if $\bar{I}=\{j_1,\ldots,j_{n-k}\}$ then $\mathbf{t}_{\bar{I}}=(t_{j_1}\ldots,t_{j_{n-k}})$. Let $\mathbf{e}=(1,\ldots,1)$. The length of $\mathbf{e}$ will vary from one formula to another, but it will always be possible to determine it from the expression in which $\mathbf{e}$ appears.

We will often consider the conditional distribution of $\mathbf{T}_I$ given that $\mathbf{T}_{\bar{I}}=\mathbf{t}_{\bar{I}}$ and that $\mathbf{T}_I\geq t\mathbf{e}$ for some $\mathbf{t}_{\bar{I}}\geq 0\mathbf{e}$ and $t\geq\bigvee_{i\epsilon\bar{I}}t_i\equiv\max\{t_i: i\epsilon\bar{I}\}$. Then, for $i\epsilon I$, the conditional density of T_i, at time t, given the above information, will be called the *conditional hazard rate* of T_i (or the *conditional instantaneous failure rate* of component i) at time t. It will be denoted by $\lambda_i(t\mid\mathbf{T}_{\bar{I}}=\mathbf{t}_{\bar{I}},\ \mathbf{T}_I\geq t\mathbf{e})$. Formally, for $i\epsilon I$,

$$\begin{aligned}&\lambda_i(t\mid\mathbf{T}_{\bar{I}}=\mathbf{t}_{\bar{I}},\ \mathbf{T}_I\geq t\mathbf{e})\\ &=\lim_{\Delta t\to 0}\frac{1}{\Delta t}P\{t\leq T_i\leq t+\Delta t\mid\mathbf{T}_{\bar{I}}=\mathbf{t}_{\bar{I}},\ \mathbf{T}_I\geq t\mathbf{e}\}.\end{aligned}\tag{4}$$

The absolute continuity of $\mathbf{T}$ ensures that this limit exists. To save space we sometimes suppress the condition $\mathbf{T}_I\geq t\mathbf{e}$ and just write $\lambda_i(t\mid\mathbf{T}_{\bar{I}}=\mathbf{t}_{\bar{I}},\ \cdot)$ but the reader should keep in mind that '$\cdot$' means $\mathbf{T}_I \geq t\mathbf{e}$ with t being the same as the first argument of λ_i. The function $\lambda_i(\cdot \mid \mathbf{T}_{\bar{I}} = \mathbf{t}_{\bar{I}},\ \cdot)$ will be of interest to us only on the (random) interval $\left(\max_{j\epsilon\bar{I}} T_j,\ \min_{i\epsilon I} T_i\right)$, however, to avoid a discussion of such random hazard rate functions [such a discussion can be found in Arjas (1981b)] we do not emphasize this point here. Note however, that $\lambda(\cdot \mid \mathbf{T}_{\bar{I}} = \mathbf{t}_{\bar{I}}, \cdot)$ of (4) is well defined for almost every $t \geq \bigvee_{j\epsilon\bar{I}} t_j$.

The absolute continuity implies that, with probability one, no two failures can occur at the same time. Thus, for $t \geq \bigvee_{i\epsilon\bar{I}} t_i$,

$$\lambda_I(t \mid \mathbf{T}_{\bar{I}} = \mathbf{t}_{\bar{I}},\ \mathbf{T}_I \geq t\mathbf{e}) \equiv \sum_{i\epsilon I} \lambda_i(t \mid \mathbf{T}_{\bar{I}} \geq \mathbf{t}_{\bar{I}},\ \mathbf{T}_I \geq t\mathbf{e}) \tag{5}$$

is the conditional hazard rate of $\min_{i\epsilon I}(T_i)$ at time t. For $i\epsilon I$, denote

$$\begin{aligned} & p_t(i \mid \mathbf{T}_{\bar{I}} = \mathbf{t}_{\bar{I}},\ \mathbf{T}_I \geq t\mathbf{e}) \\ & \equiv \frac{\lambda_i(t \mid \mathbf{T}_{\bar{I}} = \mathbf{t}_{\bar{I}},\ \mathbf{T}_I \geq t\mathbf{e})}{\lambda_I(t \mid \mathbf{T}_{\bar{I}} = \mathbf{t}_{\bar{I}},\ \mathbf{T}_I \geq t\mathbf{e})} \\ & = P\left\{\min_{j\epsilon I} T_j = T_i \mid \mathbf{T}_{\bar{I}} = \mathbf{t}_{\bar{I}},\ \mathbf{T}_I \geq t\mathbf{e},\ \min_{j\epsilon I}(T_j) = t\right\}. \end{aligned} \tag{6}$$

In the sequel we will also suppress sometimes the condition $\mathbf{T}_I \geq t\mathbf{e}$ in (5) and (6) and just write $\lambda_I(t \mid \mathbf{T}_{\bar{I}} = \mathbf{t}_{\bar{I}}, \cdot)$ and $p_t(i \mid \mathbf{T}_{\bar{I}} = \mathbf{t}_{\bar{I}}, \cdot)$. Note that for $t \geq \bigvee_{j\epsilon\bar{I}} t_j$, $\sum_{i\epsilon I} p_t(i \mid \mathbf{T}_{\bar{I}} = \mathbf{t}_{\bar{I}}, \cdot) = 1$ and that

$$\lambda_i(t \mid \mathbf{T}_{\bar{I}} = \mathbf{t}_{\bar{I}}, \cdot) = p_t(i \mid \mathbf{T}_{\bar{I}} = \mathbf{t}_{\bar{I}}, \cdot)\lambda_I(t \mid \mathbf{T}_{\bar{I}} = \mathbf{t}_{\bar{I}}, \cdot). \tag{7}$$

For a nonnegative random vector $(T_1, \ldots, T_n)$ the *dynamic construction* of $(\hat{T}_1, \ldots, \hat{T}_n)$, such that $(\hat{T}_1, \ldots, \hat{T}_n) \stackrel{\text{st}}{=} (T_1, \ldots, T_n)$, consists of the following n steps:

Step 1: Consider n independent nonhomogeneous Poisson processes on $[0, \infty)$ indexed by $i\epsilon\{1, \ldots, n\}$ with intensity function $\lambda_i(t \mid \mathbf{T} \geq t\mathbf{e})$, $t \geq 0$, $i = 1, \ldots, n$. If Process j_1 yields the first epoch (out of all the n processes) then let the time of this epoch be $\hat{T}_{j_1}$.

Step 2: Given that Step 1 resulted in $\hat{T}_{j_1} = t_{j_1}$, consider $n-1$ independent nonhomogeneous Poisson processes on $[t_{j_1}, \infty)$ indexed by $i\epsilon I \equiv \{1, \ldots, n\} - \{j_1\}$. For $i\epsilon I$ let the intensity function of Process i be $\lambda_i(t \mid T_{j_1} = t_{j_1}, \mathbf{T}_I \geq t\mathbf{e})$, $t \geq t_{j_1}$. If Process $j_2(\epsilon I)$ yields the first epoch (out of all the $n-1$ processes) then let the time of this epoch be $\hat{T}_{j_2}$.

...

...

...

Step k + 1: Given that Steps $1, \ldots k$ resulted in $\hat{T}_{j_1} = t_{j_1}, \ldots, \hat{T}_{j_k} = t_{j_k}$, let $I = \{1, \ldots, n\} - \{j_1, \ldots, j_k\}$ and consider $n - k$ independent nonhomogeneous

Poisson processes on $[\bigvee_{j\epsilon\bar{I}} t_j, \infty)$ indexed by $i\epsilon I$. For $i\epsilon I$ let the intensity function of Process i be $\lambda_i(\cdot \mid \mathbf{T}_{\bar{I}} = \mathbf{t}_{\bar{I}}, \cdot)$ on $[\bigvee_{j\epsilon\bar{I}} t_j, \infty)$. If Process $j_{k+1}(\epsilon I)$ yields the first epoch (out of the $n-k$ processes) then let $\hat{T}_{j_{k+1}}$ be the time of this epoch.

...

...

...

Step n: Given that Steps $1, \ldots, n-1$ resulted in $\hat{T}_{j_1} = t_{j_1}, \ldots, \hat{T}_{j_{n-1}} = t_{j_{n-1}}$, let $I = \{j_n\} \equiv \{1, \ldots, n\} - \{j_1, \ldots, j_{n-1}\}$ and consider a nonhomogeneous Poisson process on $[\bigvee_{i\epsilon\bar{I}} t_i, \infty)$ with intensity functions $\lambda_{j_n}(t \mid \mathbf{T}_{\bar{I}} = \mathbf{t}_{\bar{I}},\ T_{j_n} \ge t)$. Let the time of the first epoch of this process be $\hat{T}_{j_n}$.

The verification that $(\hat{T}_1, \ldots, \hat{T}_n) \stackrel{\text{st}}{=} (T_1, \ldots, T_n)$ is straightforward. It follows from the following representation of the joint density $f(t_1, \ldots, t_n)$ of $(T_1, \ldots, T_n)$: For $0 \le t_1 \le \cdots \le t_n$,

$$
\begin{aligned}
f(t_1, \ldots, t_n) = {} & \lambda_1(t_1 \mid \mathbf{T} \ge t_1\mathbf{e}) \exp\{-\Lambda_{\{1,\ldots,n\}}(t_1)\} \\
& \times\{\prod_{i=2}^{n}[\lambda_i(t_i \mid \mathbf{T}_{\{1,\ldots,i-1\}} = \mathbf{t}_{\{1,\ldots,i-1\}},\ \cdot) \\
& \times \exp\{-(\Lambda_{\{i,\ldots,n\}}(t_i \mid \mathbf{T}_{\{1,\ldots,i-1\}} = \mathbf{t}_{\{1,\ldots,i-1\}},\ \mathbf{T}_{\{i,\ldots,n\}} \ge t_{i-1}\mathbf{e}))\}]\},
\end{aligned}
$$

and similar expressions are valid when $0 \le t_{\pi(1)} \le \cdots \le t_{\pi(n)}$ for any permutation π of $(1, \ldots, n)$ [see, e.g., Cox (1972) or Lemma 1.1 of Shaked and Shanthikumar (1986b)]. Here $\Lambda_{\{1,\ldots,n\}}(t) \equiv \int_0^t \lambda_{\{1,\ldots,n\}}(u \mid \mathbf{T} \ge u\mathbf{e})du$, for $t \ge 0$, and $\Lambda_I(t \mid \mathbf{T}_{\bar{I}} = \mathbf{t}_{\bar{I}},\ \mathbf{T}_I \ge (\bigvee_{i\epsilon\bar{I}} t_i)\mathbf{e}) \equiv \int_{\bigvee_{i\epsilon\bar{I}} t_i}^t \lambda_I(u \mid \mathbf{T}_{\bar{I}} = \mathbf{t}_{\bar{I}},\ \mathbf{T}_I \ge u\mathbf{e})du$ for $I \subset \{1, \ldots, n\}$ and $t \ge \bigvee_{i\epsilon\bar{I}} t_i$.

For some purposes the following *modified dynamic construction* is more useful than the dynamic construction described above:

Step 1: Consider a nonhomogeneous Poisson process on $[0, \infty)$ with intensity function $\lambda_{\{1,\ldots,n\}}(t \mid \mathbf{T} \ge t\mathbf{e})$, $t \ge 0$. At the time s, say, of the first epoch, choose an index i with probability $P\{\text{the index } i \text{ is chosen}\} = p_s(i \mid \mathbf{T} \ge s\mathbf{e})$, $i = 1, \ldots, n$. If the chosen i is j_1 then let $\hat{T}_{j_1} = s$.

...

...

...

Step k + 1: Given that Steps $1, \ldots, k$ resulted in $\hat{T}_{j_1} = t_{j_1}, \ldots, \hat{T}_{j_k} = t_{j_k}$ let $I = \{1, \ldots, n\} - \{j_1, \ldots, j_k\}$. Consider a nonhomogeneous Poisson process on $(\bigvee_{i\epsilon\bar{I}} t_i, \infty)$ with intensity function $\lambda_I(\cdot \mid \mathbf{T}_{\bar{I}} = \mathbf{t}_{\bar{I}}, \cdot)$. At the time s, say, of the first epoch, choose an index from I with probability $p_s(i \mid \mathbf{T}_{\bar{I}} = \mathbf{t}_{\bar{I}}, \cdot)$, $i\epsilon I$. If the chosen index i is j_{k+1} then let $\hat{T}_{j_{k+1}} = s$.

...

...

...

Step n: The last step is the same as in the dynamic construction.

3. Stochastic Ordering Via Conditions on the Hazard Rates. Let $\mathbf{X} = (X_1, \ldots, X_n)$ and $\mathbf{Y} = (Y_1, \ldots, Y_n)$ be two nonnegative absolutely continuous random vectors. For any set $I \subset \{1, \ldots, n\}$ and fixed $\mathbf{t}_{\bar{I}} \geq 0\mathbf{e}$, $t \geq \bigvee_{j \in \bar{I}} t_j$ and $i \in I$, let the conditional hazard rates of X_i and Y_i be defined (as in (4)) by

$$(8) \qquad \begin{aligned} &\mu_i(t \mid \mathbf{X}_{\bar{I}} = \mathbf{t}_{\bar{I}},\ \mathbf{X}_I \geq t\mathbf{e}) = \mu_i(t \mid \mathbf{X}_{\bar{I}} = \mathbf{t}_{\bar{I}}, \cdot) \\ &= \lim_{\Delta t \to 0} \frac{1}{\Delta t} P\{t \leq X_i \leq t + \Delta t \mid \mathbf{X}_{\bar{I}} = \mathbf{t}_{\bar{I}},\ \mathbf{X}_I \geq t\mathbf{e}\} \end{aligned}$$

and

$$(9) \qquad \begin{aligned} &\eta_i(t \mid \mathbf{Y}_{\bar{I}} = \mathbf{t}_{\bar{I}},\ \mathbf{Y}_I \geq t\mathbf{e}) = \eta_i(t \mid \mathbf{Y}_{\bar{I}} = \mathbf{t}_{\bar{I}}, \cdot) \\ &= \lim_{\Delta t \to 0} \frac{1}{\Delta t} P\{t \leq Y_i \leq t + \Delta t \mid \mathbf{Y}_{\bar{I}} = \mathbf{t}_{\bar{I}},\ \mathbf{Y}_I \geq t\mathbf{e}\}. \end{aligned}$$

In this section we find sufficient conditions on the μ_i's and η_i's which imply that $\mathbf{X} \overset{\text{st}}{\leq} \mathbf{Y}$. The following main result of this section is a multivariate analog of (2). Here, and in the remainder of the paper, for $I \subset \{1, \ldots, n\}$ and $\mathbf{t}_{\bar{I}} \geq 0\mathbf{e}$, we denote $M(\mathbf{t}_I) = \bigvee_{i \in I} t_i$ where $\bigvee_{i \in I} t_i = 0$ if $I = \emptyset$. Similarly, for (not necessarily disjoint) sets $I, J, K, \cdots \subset \{1, \ldots, n\}$ and $\mathbf{t}_I \geq 0\mathbf{e}$, $\mathbf{t}_J \geq 0\mathbf{e}$, $\mathbf{t}_K \geq 0\mathbf{e}, \ldots$, we denote $M(\mathbf{t}_I,\ \mathbf{t}_J,\ \mathbf{t}_K, \ldots) = (\bigvee_{i \in I} t_i) \vee \left(\bigvee_{j \in J} t_j\right) \vee (\bigvee_{k \in K} t_k) \vee \ldots$.

THEOREM 3.1. *Let* $\mathbf{X}$ *and* $\mathbf{Y}$ *have conditional hazard rates as in (8) and (9). If for all disjoint sets* $I, J \subset \{1, \ldots, n\}$ *such that* $\overline{I \cup J} \neq \emptyset$ *and for all fixed* $\mathbf{t}_J \geq 0\mathbf{e}$, *the following holds:*

$$(10) \qquad \begin{aligned} &\mu_k(M(\mathbf{t}_J, \tilde{\mathbf{t}}_I) + u \mid \mathbf{X}_I = \mathbf{t}_I,\ \mathbf{X}_J = \mathbf{t}_J,\ X_{\overline{I \cup J}} \geq (M(\mathbf{t}_J,\ \tilde{\mathbf{t}}_I) + u)\mathbf{e}) \\ &\geq \eta_k(M(\mathbf{t}_J, \tilde{\mathbf{t}}_I) + u \mid \mathbf{Y}_I = \tilde{\mathbf{t}}_I,\ \mathbf{Y}_{\bar{I}} \geq (M(\mathbf{t}_J, \tilde{\mathbf{t}}_I) + u)\mathbf{e}) \end{aligned}$$

whenever $\mathbf{t}_I \leq \tilde{\mathbf{t}}_I$, $u \geq 0$ *and* $k \in \overline{I \cup J}$ *(J may be the empty set), then* $\mathbf{X} \overset{st}{\leq} \mathbf{Y}$.

REMARK 3.2. Roughly speaking, the events on which the failure rates are conditioned in (10) are two histories of the same length $M(\mathbf{t}_J, \tilde{\mathbf{t}}_I)$. The history on the left hand side of (10) has more failures than the history on the right hand side, and, for components which failed in both histories, the failure times in the former are earlier than in the latter. Condition (10) says that whenever the histories of $\mathbf{X}$ and $\mathbf{Y}$ can thus be compared, then the failure rate, under the law of $\mathbf{X}$, of each surviving component in the history of $\mathbf{X}$, is larger than the failure rate of the same component under the law of $\mathbf{Y}$.

REMARK 3.3. If $\mathbf{X}$ and $\mathbf{Y}$ are vectors of independent random variables with hazard rate functions satisfying the condition in (3) then (10) holds. Thus Theorem 3.1 contains Result (3) as a special case.

The proof of Theorem 3.1 is given in the Appendix.

In Sections 4 and 5 we need a slight generalization of Theorem 3.1. We will need to compare random vectors $\mathbf{X} = (X_1, \ldots, X_n)$ and $\mathbf{Y} = (Y_1, \ldots, Y_n)$ where $\mathbf{Y}$ is nonnegative and absolutely continuous on $[0, \infty)^n$, but $\mathbf{X}$ may have some components which are identically zero, that is, $\mathbf{X}$ is of the form [relabelling components if necessary] $\mathbf{X} = (0, \ldots, 0,\ X_{\ell+1}, \ldots, X_n)$ for some $\ell \epsilon \{0, 1, \ldots, n-1\}$, where $(X_{\ell+1}, \ldots, X_n)$ is nonnegative and absolutely continuous on $[0, \infty)^{n-\ell}$. We will condition on events of the form $\{\mathbf{X}_I = \mathbf{t}_I,\ \mathbf{X}_{\bar{I}} \geq t\mathbf{e}\}$ $t \geq M(\mathbf{t}_I)$ where $I \supset \{1, \ldots, \ell\}$ and $\mathbf{t}_{\{1,\ldots,\ell\}} = 0\mathbf{e}$. For $i \epsilon \bar{I}$, the conditional hazard rate at time t of X_i, given $\mathbf{X}_I = \mathbf{t}_I$ and $\mathbf{X}_{\bar{I}} \geq t\mathbf{e}$, denoted by $\mu_i(t \mid \mathbf{X}_I = \mathbf{t}_I,\ \mathbf{X}_{\bar{I}} \geq t\mathbf{e})$, $t \geq M(\mathbf{t}_I)$, is well defined as in (8). Condition (10) then is well defined for every $I \supset \{1, \ldots, \ell\}$.

THEOREM 3.4. *Let* $\mathbf{X}$ *and* $\mathbf{Y}$ *be as described above. If for all disjoint sets* I, $J \subset \{1, \ldots, n\}$ *such that* $(I \cup J) \supset \{1, \ldots, \ell\}$ *and* $\overline{I \cup J} \neq \emptyset$ *and for all fixed* $\mathbf{t}_J$, *condition (10) holds whenever* $\mathbf{t}_I \leq \tilde{\mathbf{t}}_I$ $[\mathbf{t}_{\{1,\ldots,\ell\}} = 0\mathbf{e}]$, *and* $k \epsilon \overline{I \cup J}$, *then* $\mathbf{X} \overset{st}{\leq} \mathbf{Y}$.

The proof of Theorem 3.4 is similar to the proof of Theorem 3.1. Instead of starting with Step 1 one defines $\hat{X}_1 = \cdots = \hat{X}_\ell \equiv 0$ and then starts the construction (described in the proof of Theorem 3.1) in step $(\ell + 1).1$. We omit the details.

4. Hazard Rates and the MIHR| $\mathcal{F}_t$ Property. Let $T_1, \ldots, T_n$ be nonnegative random variables to be thought of as lifetimes of components numbered $1, \ldots, n$. Let Z_i be the life indicator of component i, that is,

$$\begin{aligned} Z_i(t) &= 1 \text{ if } t < T_i, \\ &= 0 \text{ if } t \geq T_i. \end{aligned}$$

For $t \geq 0$, let $\mathcal{F}_t$ be the σ-field generated by $\{Z_i(s) : 0 \leq s \leq t,\ i = 1, \ldots, n\}$, that is,

$$\mathcal{F}_t = \sigma(Z_i(s) : 0 \leq s \leq t,\ 1 \leq i \leq n). \tag{11}$$

We will condition on sets in $\mathcal{F}_t$ of the form

$$A_t = \{\mathbf{T}_I = \mathbf{t}_I,\ \mathbf{T}_{\bar{I}} \geq t\mathbf{e}\},$$

where $I \subset \{1, \ldots, n\}$ and $t \geq M(\mathbf{t}_I)$. Thus A_t is an observed history of components $1, \ldots, n$ until time t. It describes which components are still alive at time t and the failure times of the components that are already dead at time t.

Let θ_t denote a shift by t in time and define

$$\theta_t T_i = (T_i - t)^+ = \max(T_i - t,\, 0), \;\; i = 1, \ldots, n, \;\; t \geq 0.$$

Denote $\mathbf{T} = (T_1, \ldots, T_n)$ and $\theta_t \mathbf{T} = (\theta_t T_1, \ldots, \theta_t T_n)$.

Arjas (1981a) considered the class of multivariate increasing hazard rate (MIHR) random vectors described in the following definition.

DEFINITION 4.1. The random vector $(T_1, \ldots, T_n)$ is called MIHR relative to $(\mathcal{F}_t)_{t \geq 0}$ [denoted by MIHR| $(\mathcal{F}_t)_{t \geq 0}$ or just by MIHR| $\mathcal{F}_t$] if for all $t \leq t'$ and all Borel upper sets U in R^n,

$$P\{\theta_t \mathbf{T} \epsilon U \mid \mathcal{F}_t\} \geq P\{\theta_{t'} \mathbf{T} \epsilon U \mid \mathcal{F}_{t'}\} \quad \text{a.s.} \tag{12}$$

For the random vector $\mathbf{T}$ let the conditional hazard rates be defined as in (4).

THEOREM 4.2. *Suppose that the conditional hazard rates of* $\mathbf{T}$ *satisfy:*

(i) For disjoint sets I, $J \subset \{1, \ldots, n\}$, $J \neq \emptyset$ *and fixed* $\mathbf{t}_{I \cup J} \geq 0\mathbf{e}$, $\tilde{\mathbf{t}}_I \geq 0\mathbf{e}$, $k \epsilon \overline{I \cup J}$,

$$\begin{aligned} &\lambda_k(M(\mathbf{t}_I, \tilde{\mathbf{t}}_I, \mathbf{t}_J) + u \mid \mathbf{T}_I = \mathbf{t}_I,\ \mathbf{T}_J = \mathbf{t}_J, \cdot) \\ &\geq \lambda_k(M(\mathbf{t}_I, \tilde{\mathbf{t}}_I, \mathbf{t}_J) + u \mid \mathbf{T}_I = \tilde{\mathbf{t}}_I, \cdot),\ u \geq 0. \end{aligned} \tag{13}$$

(ii) For disjoint sets I, $J \subset \{1, \ldots, n\}$ *and fixed* $k \epsilon \overline{I \cup J}$, $t \leq t'$, $\mathbf{t}_I \leq t\mathbf{e}$, $\tilde{\mathbf{t}}_J \geq t\mathbf{e}$ *and* $\mathbf{t}_J \geq t\mathbf{e}$ *such that* $\tilde{\mathbf{t}}_J - t\mathbf{e} \geq \mathbf{t}_J - t'\mathbf{e}$,

$$\begin{aligned} &\lambda_k(t' + M(\tilde{\mathbf{t}}_J) - t + u \mid \mathbf{T}_I = \mathbf{t}_I,\ \mathbf{T}_J = \mathbf{t}_J, \cdot) \\ &\geq \lambda_k(M(\tilde{\mathbf{t}}_J) + u \mid \mathbf{T}_I = \mathbf{t}_I,\ \mathbf{T}_J = \tilde{\mathbf{t}}_J, \cdot),\ \ u \geq 0. \end{aligned} \tag{14}$$

Then $\mathbf{T}$ *is MIHR|* $\mathcal{F}_t$.

REMARKS 4.3.

(a) Condition (i) means that the smaller is the working set, the larger are the instantaneous failure rates of the surviving components.

(b) In Condition (ii) two histories are compared. The 'future' of one starts at $M(\tilde{\mathbf{t}}_J) = t + M(\tilde{\mathbf{t}}_J) - t$ and the future of the other starts at $t' + M(\tilde{\mathbf{t}}_J) - t$. In both histories $\mathbf{t}_I$ are identical. The (known) $\mathbf{T}_J$ may be different, however the (known) $\mathbf{T}_J$ in the respective histories satisfy $\theta_t \tilde{\mathbf{t}}_J \geq \theta_{t'} \mathbf{t}_J$. Condition (ii) states that then $\theta_{M(\tilde{\mathbf{t}}_J)} \mathbf{T}_{\overline{I \cup J}}$ (which are the 'future' of one history) have smaller instantaneous failure rates than $\theta_{t' + M(\tilde{\mathbf{t}}_J) - t} \mathbf{T}_{\overline{I \cup J}}$ [which are the 'future' of the other history].

(c) Substituting $t = 0$, $\tilde{\mathbf{t}}_J = \mathbf{t}_J$, $I = \emptyset$ in (14) we see that for every set $J \subset \{1, \ldots, n\}$ and every $t' \geq 0$ and $k \epsilon J$,

$$\begin{aligned} &\lambda_k(M(\mathbf{t}_J) + t' + u \mid \mathbf{T}_J = \mathbf{t}_J, \cdot) \\ (15) \qquad &\geq \lambda_k(M(\mathbf{t}_J) + u \mid \mathbf{T}_J = \mathbf{t}_J, \cdot), \quad u \geq 0. \end{aligned}$$

That is, all the conditional hazard rate functions are increasing between failures. Our condition (ii) requires more than this local IHR property.

(d) In the proof of Theorem 4.2, Condition (i) is used to compare two conditional hazard rates given (two) histories in which different numbers of components are known to have failed already. In contrast, Condition (ii) is used to compare two conditional failure rates given (two) histories in both of which the same components (indexed by $i \epsilon I \cup J$) have already failed and the components indexed by $i \epsilon \overline{I \cup J}$ are still alive.

(e) If $T_1, \ldots, T_n$ are independent, absolutely continuous IHR random variables then it is easily seen that (i) and (ii) hold. Thus Theorem 4.2 agrees with a result of Arjas (1981a) which states that independent IHR random variables are MIHR$|\ \mathcal{F}_t$.

Proof of Theorem 4.2.: Fix t and t' ($t' \geq t \geq 0$). Let

$$\begin{aligned} (16) \qquad D_t &= \{\mathbf{T}_B = \mathbf{t}_B,\ \mathbf{T}_{\bar{B}} \geq t\mathbf{e}\} \\ (17) \qquad E_{t'} &= \{\mathbf{T}_B = \mathbf{t}_B,\ \mathbf{T}_C = \mathbf{t}_C,\ T_{\overline{B \cup C}} \geq t'\mathbf{e}\} \end{aligned}$$

where B, $C \subset \{1, \ldots, n\}$ are disjoint and $\mathbf{t}_B \leq t\mathbf{e}$, $t\mathbf{e} < \mathbf{t}_C \leq t'\mathbf{e}$. Denote the cardinality of $\bar{B}$ by $\tilde{n}(\leq n)$. Notice that given D_t [respectively, $E_{t'}$], $\theta_t \mathbf{T}_B = 0\mathbf{e}$ [respectively, $\theta_{t'} \mathbf{T}_B = 0\mathbf{e}$]. Let $\mathbf{V}$ [respectively, $\mathbf{W}$] be an $\tilde{n}$-dimensional random vector distributed according to the conditional distribution of $\theta_t \mathbf{T}_{\bar{B}}$ given D_t [respectively, $\theta_{t'} \mathbf{T}_{\bar{B}}$ given $E_{t'}$]. We will show that

$$(18) \qquad \mathbf{V} \overset{\text{st}}{\geq} \mathbf{W}$$

and then (12) follows.

Notice that when $C \neq \emptyset$ then, with probability one, $\mathbf{W}_C = 0\mathbf{e}$ whereas $\mathbf{V}_C \geq 0\mathbf{e}$. Hence in order to prove (18) we will use Theorem 3.1 when $C = \emptyset$ and Theorem 3.4 when $C \neq \emptyset$.

Given, for some $I \subset \bar{B}$, that $\mathbf{V}_I = \mathbf{s}_I$ and $\mathbf{V}_{\overline{I \cup B}} \geq s\mathbf{e}$ (where $s\mathbf{e} \geq \mathbf{s}_I$), the conditional hazard rate of V_k at s (where $k \epsilon \overline{I \cup B}$) is

$$\begin{aligned} &\eta_k(s \mid \mathbf{V}_I = \mathbf{s}_I,\ \mathbf{V}_{\overline{B \cup I}} \geq s\mathbf{e}) \\ (19) \qquad &= \lambda_k(t + s \mid \mathbf{T}_B = \mathbf{t}_B,\ \mathbf{T}_I = \mathbf{s}_I + t\mathbf{e}, \cdot),\ s \geq M(\mathbf{s}_I), \end{aligned}$$

where λ_k is the conditional hazard rate function of T_k as defined in (4). Similarly, given, for some $i \subset \overline{B \cup C}$, that $\mathbf{W}_I = \mathbf{s}_I$ and $\mathbf{W}_{\overline{I \cup B \cup C}} \geq s\mathbf{e}$ (where $s\mathbf{e} \geq \mathbf{s}_I$), the conditional hazard rate of W_k at s (where $k \epsilon \overline{I \cup B \cup C}$) is

$$
\begin{aligned}
&\mu_k(s \mid \mathbf{W}_I = \mathbf{s}_I,\ \mathbf{W}_{\overline{B \cup C \cup I}} \geq s\mathbf{e}) \\
&= \lambda_k(t' + s \mid \mathbf{T}_B = \mathbf{t}_B,\ \mathbf{T}_C = \mathbf{t}_C,\ \mathbf{T}_I = \mathbf{s}_I + t'\mathbf{e}, \\
&\mathbf{T}_{\overline{B \cup C \cup I}} \geq (s + t')\mathbf{e}),\ s \geq M(\mathbf{s}_I).
\end{aligned} \tag{20}
$$

In order to prove (18) we will show that the η_k's and μ_k's defined in (19) and (20) satisfy the $\tilde{n}$-dimensional version of (10), that is, for all disjoint sets I, $J \subset \bar{B}$, such that $\bar{B} \cap (\overline{I \cup J}) \neq \emptyset$ and $(I \cup J) \supset C$,

$$
\begin{aligned}
&\mu_k(M(\mathbf{t}_J,\ \tilde{\mathbf{t}}_I) + u \mid \mathbf{W}_I = \mathbf{t}_I,\ \mathbf{W}_J = \mathbf{t}_J,\ W_{\overline{B \cup I \cup J}} \geq (M(\mathbf{t}_J, \tilde{\mathbf{t}}_I) + u)\mathbf{e}) \\
&\geq \eta_k(M(\mathbf{t}_J, \tilde{\mathbf{t}}_I) + u \mid \mathbf{V}_I = \tilde{\mathbf{t}}_I,\ V_{\overline{B \cup I}} \geq (M(\mathbf{t}_J, \tilde{\mathbf{t}}_I) + u)\mathbf{e}),
\end{aligned} \tag{21}
$$

whenever $\mathbf{t}_I \leq \tilde{\mathbf{t}}_I$, $u \geq 0$ and $k \epsilon \overline{I \cup J \cup B}$.

Two cases will be considered.

Case 1: $C = \emptyset$. First we show that (21) holds when $J \neq \emptyset$. Let I and J be as in (21) and let $\mathbf{t}_I \leq \tilde{\mathbf{t}}_I$, $u \geq 0$ and $k \epsilon \overline{I \cup J \cup B}$. Then, from (19) and (20) [here LHS and RHS stand for 'left hand side' and 'right hand side'],

$$
\begin{aligned}
\text{LHS}(21) &= \lambda_k(M(t'\mathbf{e} + \mathbf{t}_J,\ t'\mathbf{e} + \tilde{\mathbf{t}}_I) + u \mid \mathbf{T}_B = \mathbf{t}_B,\ \mathbf{T}_I = \mathbf{t}_I + t'\mathbf{e}, \\
&\qquad \mathbf{T}_J = \mathbf{t}_J + t'\mathbf{e},\ \cdot), \\
\text{RHS}(21) &= \lambda_k(M(t\mathbf{e} + \mathbf{t}_J,\ t\mathbf{e} + \tilde{\mathbf{t}}_I) + u \mid \mathbf{T}_B = \mathbf{t}_B,\ \mathbf{T}_I = \tilde{\mathbf{t}}_I + t\mathbf{e},\ \cdot).
\end{aligned}
$$

Hence

$$
\begin{aligned}
&\text{RHS}(21) \leq \lambda_k(M(t'\mathbf{e} + \mathbf{t}_J, t'\mathbf{e} + \tilde{\mathbf{t}}_I) + u \mid \mathbf{T}_B = \mathbf{t}_B,\ \mathbf{T}_I = \tilde{\mathbf{t}}_I + t\mathbf{e},\ \cdot) \\
&\leq \text{LHS}(21),
\end{aligned} \tag{22}
$$

where the first inequality follows from (15) and the second from (13).

Now we show that (21) holds also when $J = \emptyset$ (recall that we still assume $C = \emptyset$). Let $I \subset \bar{B}$,

$$
\mathbf{t}_I \leq \tilde{\mathbf{t}}_I, \tag{23}
$$

and $k \epsilon \overline{I \cup B}$. Then

$$
\begin{aligned}
&\text{LHS}(21) = \mu_k(M(\tilde{\mathbf{t}}_I) + u \mid \mathbf{W}_I = \mathbf{t}_I,\ \mathbf{W}_{\overline{I \cup B}} \geq (M(\tilde{\mathbf{t}}_I) + u)\mathbf{e}) \\
&= \lambda_k(M(t'\mathbf{e} + \tilde{\mathbf{t}}_I) + u \mid \mathbf{T}_B = \mathbf{t}_B,\ \mathbf{T}_I = t'\mathbf{e} + \mathbf{t}_I,\ \cdot),
\end{aligned} \tag{24}
$$

$$\text{RHS}(21) = \eta_k(M(\tilde{\mathbf{t}}_I) + u \mid \mathbf{V}_I = \tilde{\mathbf{t}}_I,\ \mathbf{V}_{\overline{I \cup B}} \geq (M(\tilde{\mathbf{t}}_I) + u)\mathbf{e})$$
$$= \lambda_k(M(t\mathbf{e} + \tilde{\mathbf{t}}_I) + u \mid \mathbf{T}_B = \mathbf{t}_B,\ \mathbf{T}_I = t\mathbf{e} + \tilde{\mathbf{t}}_I,\ \cdot), \tag{25}$$

Now, in (14) plug:

$$\begin{aligned} B \quad & \text{in place of } I, \\ I \quad & \text{in place of } J, \\ \mathbf{t}_I + t'\mathbf{e} \quad & \text{in place of } \mathbf{t}_J, \\ \tilde{\mathbf{t}}_I + t\mathbf{e} \quad & \text{in place of } \tilde{\mathbf{t}}_J. \end{aligned} \tag{26}$$

The resulting LHS (14) is equal to (24) and the resulting RHS (14) is equal to (25). By assumption, (14) holds if t, t', $\mathbf{t}_J$ and $\tilde{\mathbf{t}}_J$ *there* (in (14)) satisfy $\tilde{\mathbf{t}}_J - t\mathbf{e} \geq \mathbf{t}_J - t'\mathbf{e}$. This inequality translates (through substitution (26)) to $\mathbf{t}_I \leq \tilde{\mathbf{t}}_I$ *here* (in (24) and (25)); the latter is true by (23). Thus, from (ii) we obtain that LHS (21) [i.e., (24)] $\geq$ RHS (21) [i.e., (25)].

Case 2: $C \neq \emptyset$. Let I, $J \subset \bar{B}$ be disjoint sets such that $\bar{B} \cap (\overline{I \cup J}) \neq \emptyset$ and $(I \cup J) \supset C$. In LHS (21) we only have to condition on $\mathbf{W}_{I \cap \bar{C}}$ and $\mathbf{W}_{J \cap \bar{C}}$ because $\mathbf{W}_C = 0$. Let $\tilde{\mathbf{t}}_{I \cap C} \geq 0\mathbf{e}$, $\mathbf{t}_{J \cap \bar{C}} \geq 0\mathbf{e}$, $\tilde{\mathbf{t}}_{I \cap \bar{C}} \geq \mathbf{t}_{I \cap \bar{C}} \geq 0\mathbf{e}$, $u \geq 0$ and $k \epsilon \overline{I \cup J \cup B}$. Then

$$\begin{aligned} \text{LHS}(21) = \lambda_k(M(&\mathbf{t}_{J \cap \bar{C}} + t'\mathbf{e},\ \tilde{\mathbf{t}}_{I \cap \bar{C}} + t'\mathbf{e}) + u \mid \\ &\mathbf{T}_B = \mathbf{t}_B,\ \mathbf{T}_C = \mathbf{t}_C,\ \mathbf{T}_{I \cap \bar{C}} = \mathbf{t}_{I \cap \bar{C}} + t'\mathbf{e}, \\ &\mathbf{T}_{J \cap \bar{C}} = \mathbf{t}_{J \cap \bar{C}} + t'\mathbf{e},\ \cdot), \end{aligned} \tag{27}$$

$$\begin{aligned} \text{RHS}(21) = \lambda_k(M(&\mathbf{t}_{J \cap \bar{C}} + t\mathbf{e},\ \tilde{\mathbf{t}}_{I \cap \bar{C}} \\ &+ t\mathbf{e}) + u \mid \mathbf{T}_B = \mathbf{T}_B,\ \mathbf{T}_I = \tilde{\mathbf{t}}_I + t\mathbf{e},\ \cdot). \end{aligned} \tag{28}$$

If $J \neq \emptyset$ then in LHS (21) [of (27)] it is given that more components have already failed than in the condition given in RHS (21) [of (28)]. So the fact that (27) $\geq$ (28) follows from (15) and (i) as in (22).

If $J = \emptyset$ then in both (27) and (28) it is given that the same number of components have failed (though possibly at different times). For this case the proof of (27) $\geq$ (28) uses (ii) and is similar to (though notationally somewhat more involved than) the proof of (24) $\geq$ (25). We omit the details. ||

REMARK 4.4. Condition (i) of Theorem 4.2 looks simple (see Remark 4.3 (a)) but it is stronger than what is really required. A careful study of the proof of Theorem 4.2 shows that the following condition (which is weaker than (i) and (ii) combined) implies that $\mathbf{T}$ is MIHR$|\ \mathcal{F}_t$:

(iii) For disjoint sets I, J, $L \subset \{1,\dots,n\}$ and fixed $k \epsilon \overline{I \cup J \cup L}$, $t \le t'$, $\mathbf{t}_I \le t\mathbf{e}$, $\tilde{\mathbf{t}}_J \ge t\mathbf{e}$, $\mathbf{t}_J \ge t\mathbf{e}$ and $\mathbf{t}_L \ge t\mathbf{e}$ such that $\tilde{\mathbf{t}}_J - t\mathbf{e} \ge \mathbf{t}_J - t'\mathbf{e}$ (L may be empty),

$$\begin{aligned} &\lambda_k(t' - t + M(\tilde{\mathbf{t}}_J,\ \mathbf{t}_L) + u \mid \mathbf{T}_I = \mathbf{t}_I,\ \mathbf{T}_J = \mathbf{t}_J,\ \mathbf{T}_L = \mathbf{t}_L,\ \cdot) \\ &\ge \lambda_k(M(\tilde{\mathbf{t}}_J,\ \mathbf{t}_L) + u \mid \mathbf{T}_I = \mathbf{t}_I,\ \mathbf{T}_J = \tilde{\mathbf{t}}_J,\ \cdot),\quad u \ge 0. \end{aligned} \tag{29}$$

To see that indeed (iii) is weaker than (i) and (ii) combined, note that if $L \neq \emptyset$ then (29) follows from (13) and (15). If $L = \emptyset$ then (29) is the same as (14).

5. Hazard Rates and the WBF Property. Let $T_1,\dots,T_n$ be nonnegative random lifetimes as in Section 4. Fix $t > 0$, $B \subset \{1,\dots,n\}$ [such that $\bar{B} \neq \emptyset$], $\ell \epsilon \bar{B}$ and $\mathbf{t}_B \le t\mathbf{e}$. Consider the two histories

$$D = \{\mathbf{T}_B = \mathbf{t}_B,\ \mathbf{T}_{\bar{B}} \ge t\mathbf{e}\}, \tag{30}$$

$$E = \{\mathbf{T}_B = \mathbf{t}_B,\ T_\ell = t,\ T_{\bar{B}-\{\ell\}} \ge t\mathbf{e}\}. \tag{31}$$

Let $\mathbf{V}$ [respectively, $\mathbf{W}$] be distributed according to the conditional distribution of $\theta_t \mathbf{T}_{\bar{B}}$ given D [respectively, E]. Arjas and Norros (1984) studied random vectors $\mathbf{T} = (T_1,\dots,T_n)$ which have the property given in the following definition.

Definition 5.1. The random vector $\mathbf{T} = (T_1,\dots,T_n)$ is said to be *weakened by failures* (WBF) if for all $t \ge 0$, $B \subset \{1,\dots,n\}$ [such that $\bar{B} \neq \emptyset$], $\ell \epsilon \bar{B}$ and $\mathbf{t}_B \le t\mathbf{e}$, the random vectors $\mathbf{V}$ and $\mathbf{W}$ satisfy

$$\mathbf{V} \overset{\text{st}}{\ge} \mathbf{W}. \tag{32}$$

Theorem 5.2. *Suppose the conditional hazard rates of* $\mathbf{T}$ *satisfy (i) of Theorem 4.2 and*

(iv) For every set $I \subset \{1,\dots,n\}$ *and fixed* $\mathbf{t}_I$, $\tilde{\mathbf{t}}_I$ *[such that* $\mathbf{t}_I \le \tilde{\mathbf{t}}_I$*] and* $k \epsilon \bar{I}$,

$$\lambda_k(M(\tilde{\mathbf{t}}_I) + u \mid \mathbf{T}_I = \mathbf{t}_I,\ \cdot) \ge \lambda_k(M(\tilde{\mathbf{t}}_I) + u \mid \mathbf{T}_I = \tilde{\mathbf{t}}_I,\ \cdot),\quad u \ge 0.$$

Then $\mathbf{T}$ *has the WBF-property.*

Remark 5.3. In Condition (iv) two histories of the same length and with the same number of failures are compared. The history with the earlier failure times yields higher failure rates for the surviving components.

Proof of Theorem 5.2: Suppose the cardinality of $\bar{B}$ is $\tilde{n}(< n)$. Since $W_\ell = 0$ with probability one, whereas $V_\ell \ge 0$, use will be made of Theorem 3.4.

For $I \subset \bar{B}$, given $\mathbf{V}_I = \mathbf{s}_I$ and $V_{\overline{I \cup B}} \ge s\mathbf{e}$ (where $s \ge M(\mathbf{s}_I)$), the conditional hazard rate of V_k at time s, where $k \epsilon \overline{B \cup I}$, is

$$(33)\quad \eta_k(s \mid \mathbf{V}_I = \mathbf{s}_I,\ \mathbf{V}_{\overline{B \cup I}} \geq s\mathbf{e}) = \lambda_k(t+s \mid \mathbf{T}_B = \mathbf{t}_B,\ \mathbf{T}_I = \mathbf{s}_I + t\mathbf{e},\ \cdot),$$

where λ_k is defined as in (4). Similarly, for $I \subset \overline{B \cup \{\ell\}}$, given $\mathbf{W}_I = \mathbf{s}_I$, $W_{\overline{B \cup I \cup \{\ell\}}} \geq s\mathbf{e}$ (where $s \geq M(\mathbf{s}_I)$), the conditional hazard rate of W_k at s, where $k \epsilon \overline{B \cup I \cup \{\ell\}}$ is

$$\begin{aligned} &\mu_k(s \mid \mathbf{W}_I = \mathbf{s}_I,\ \mathbf{W}_{\overline{B \cup I \cup \{\ell\}}} \geq s\mathbf{e}) \\ (34)\qquad &= \lambda_k(t+s \mid \mathbf{T}_B = \mathbf{t}_B,\ T_\ell = t,\ \mathbf{T}_I = \mathbf{s}_I + t\mathbf{e},\ \cdot). \end{aligned}$$

In order to prove (32) we will show that the η_k's and μ_k's defined in (33) and (34) satisfy the $\tilde{n}$-dimensional version of (10) required in Theorem 3.4. That is, for disjoint sets $I, J \subset \bar{B}$, such that $\bar{B} \cap (\overline{I \cup J}) \neq \emptyset$ and $\ell \epsilon I \cup J$,

$$\begin{aligned} &\mu_k(M(\mathbf{t}_J, \tilde{\mathbf{t}}_I) + u \mid \mathbf{W}_I = \mathbf{t}_I,\ \mathbf{W}_J = \mathbf{t}_J,\ \mathbf{W}_{\overline{B \cup I \cup J}} \geq (M(\mathbf{t}_J, \tilde{\mathbf{t}}_I) + u)\mathbf{e}) \\ (35)\qquad &\geq \eta_k(M(\mathbf{t}_J, \tilde{\mathbf{t}}_I) + u \mid \mathbf{V}_I = \tilde{\mathbf{t}}_I,\ \mathbf{V}_{\overline{B \cup I}} \geq (M(\mathbf{t}_J, \tilde{\mathbf{t}}_I) + u)\mathbf{e}), \end{aligned}$$

whenever $\mathbf{t}_I \leq \tilde{\mathbf{t}}_I$, $u \geq 0$, $k \epsilon \overline{I \cup J \cup B}$. Since $\ell \epsilon I \cup J$, the LHS (35) is well defined only if $t_\ell = 0$, $\tilde{t}_\ell \geq 0$.

If $\ell \notin I$ then $J \neq \emptyset$ because $\ell \epsilon J$. Then from (i) it follows that

$$\begin{aligned} &\lambda_k(M(\mathbf{t}_J, \tilde{\mathbf{t}}_I) + t + u \mid \mathbf{T}_B, = \mathbf{t}_B,\ \mathbf{T}_I = \mathbf{t}_I + t\mathbf{e},\ \mathbf{T}_J = \mathbf{t}_J + t\mathbf{e}, \cdot) \\ (36)\qquad &\geq \lambda_k(M(\mathbf{t}_J, \tilde{\mathbf{t}}_I) + t + u \mid \mathbf{T}_B, = \mathbf{t}_B,\ \mathbf{T}_I = \mathbf{t}_I + t\mathbf{e},\ \cdot). \end{aligned}$$

But in this case $[J \neq \emptyset]$ (36) is equivalent to (35).

If $\ell \epsilon I$ and $J \neq \emptyset$ then, in a similar manner, one can again obtain (35) using (i).

If $J = \emptyset$ (then of course $\ell \epsilon I$) then (35) is equivalent to

$$\begin{aligned} &\lambda_k(M(\tilde{\mathbf{t}}_I) + t + u \mid \mathbf{T}_B = \mathbf{t}_B,\ \mathbf{T}_I = \mathbf{t}_I + t\mathbf{e},\ \cdot) \\ (37)\qquad &\geq \lambda_k(M(\tilde{\mathbf{t}}_I) + t + u \mid \mathbf{T}_B = \mathbf{t}_B,\ \mathbf{T}_I = \tilde{\mathbf{t}}_I + t\mathbf{e},\ \cdot). \end{aligned}$$

But (37) follows from (iv).

Thus (35) holds for every choice of $I, J \subset \bar{B}$ [such that $\bar{B} \cap (\overline{I \cup J}) \neq \emptyset$ and $\ell \epsilon (I \cup J)$], $\mathbf{t}_I \leq \tilde{\mathbf{t}}_I$, $u \geq 0$ and $k \epsilon \overline{I \cup J \cup B}$ and the stochastic comparison (32) now follows from Theorem 3.4. ||

From Theorem 1 of Arjas and Norros (1984) it follows that if an absolutely continuous $\mathbf{T}$ has the WBF property then $\mathbf{T}$ is a vector of associated random variables (in the sense of Esary, Proschan, and Walkup (1967)). Thus from Theorem 5.2 we obtain:

THEOREM 5.4. *Suppose the conditional hazard rate functions of* $\mathbf{T}$ *satisfy (i) of Theorem 4.2 and (iv) of Theorem 5.2. Then* $\mathbf{T}$ *is a vector of associated random variables.*

REMARK 5.5. In the proof of Theorem 5.2, Condition (i) is applied only with $\mathbf{t}_I \le \tilde{\mathbf{t}}_I$. Thus the conclusions of Theorems 5.2 and 5.4 are valid under the following condition (which is weaker than (i) and (iv) combined):

(v) For disjoint sets $I, J \subset \{1, \ldots, n\}$ and fixed $\mathbf{t}_I$, $\tilde{\mathbf{t}}_I$, $\mathbf{t}_J$ [such that $\mathbf{t}_I \le \tilde{\mathbf{t}}_I$] and $k \in \overline{I \cup J}$ (J may be empty),

$$\begin{aligned}&\lambda_k(M(\tilde{\mathbf{t}}_I, \mathbf{t}_J) + u \mid \mathbf{T}_I = \mathbf{t}_I,\ \mathbf{T}_J = \mathbf{t}_J\ \cdot)\\ &\ge \lambda_k(M(\tilde{\mathbf{t}}_I, \mathbf{t}_J) + u \mid \mathbf{T}_I = \tilde{\mathbf{t}}_I,\ \cdot), u \ge 0.\end{aligned}$$

The results of Sections 4 and 5 can be applied to various stochastic models. We will not give details of the applications here because they are similar to the applications given in Section 6 in Shaked and Shanthikumar (1987a).

Appendix: Proof of Theorem 3.1. Using the dynamic construction described in Section 2 we will construct simultaneously two random vectors $\hat{\mathbf{X}}$ and $\hat{\mathbf{Y}}$ on a common probability space such that

$$\hat{\mathbf{X}} \stackrel{\text{st}}{=} \mathbf{X}, \tag{38}$$

$$\hat{\mathbf{Y}} \stackrel{\text{st}}{=} \mathbf{Y}, \tag{39}$$

and

$$\hat{\mathbf{X}} \le \mathbf{Y} \text{ with probability one.} \tag{40}$$

Then, for every increasing Borel measurable function g,

$$Eg(\mathbf{X}) = Eg(\hat{\mathbf{X}}) \le Eg(\hat{\mathbf{Y}}) = Eg(\mathbf{Y}),$$

provided the expectations exist, and the desired result then follows from (1).

We describe the construction of $\hat{\mathbf{X}}$ and $\hat{\mathbf{Y}}$ according to the steps (Step 1 through Step n) of the construction of $\hat{\mathbf{X}}$.

Step 1: Consider n independent nonhomogeneous Poisson processes on $[0, \infty)$ indexed by $i \in \{1, \ldots, n\}$ with intensity functions $\mu_i(t \mid \mathbf{X} \ge t\mathbf{e})$, $t \ge 0$, $i = 1, \ldots, n$. If Process j_1 yields the first epoch (out of all n processes) at time t_{j_1}, say, then let this time be $\hat{X}_{j_1}$. Also, with probability $\eta_{j_1}(t_{j_1} \mid \mathbf{Y} \ge t_{j_1}\mathbf{e})/\mu_{j_1}(t_{j_1} \mid \mathbf{X} \ge t_{j_1}\mathbf{e})$ let this time, t_{j_1}, be also $\hat{Y}_{j_1}$ and with probability $1 - \eta_{j_1}(t_{j_1} \mid \mathbf{Y} \ge t_{j_1}\mathbf{e})/\mu_{j_1}(t_{j_1} \mid \mathbf{X} \ge t_{j_1}\mathbf{e})$ delay the determination of $\hat{Y}_{j_1}$ for a later step. If $\hat{X}_{j_1} = \hat{Y}_{j_1}$ then go to Step 2.2. If $\hat{Y}_{j_1}$ has not yet been determined then go to Step 2.1.

The proof of Theorem 3.1 continues after Remark A.1.

REMARK A.1.

(a) For Step 1 to be sensible it is required that

$$\eta_{j_1}(t_{j_1} \mid \mathbf{Y} \geq t_{j_1}\mathbf{e})/\mu_{j_1}(t_{j_1} \mid \mathbf{X} \geq t_{j_1}\mathbf{e}) \leq 1, \text{ but this follows from (10).}$$

(b) At the conclusion of Step 1, $\hat{X}_{j_1}$ has been determined (and perhaps also $\hat{Y}_{j_1}$) and

(*.1) $$\hat{X}_{j_1} \leq \hat{Y}_{j_1} \text{ with probability one,}$$

because either $\hat{Y}_{j_1} = \hat{X}_{j_1}$ or $\hat{Y}_{j_1}$ is going to be determined at (and then be equal to) some time after t_{j_1}.

(c) The next steps will be indexed by a pair $m.\ell$ ($\ell \leq m$). For $1 \leq \ell < m \leq n$, Step $m.\ell$ produces one of the following:

(α) Generates just the m-th $\hat{X}$ (and the procedure then proceeds to Step $(m+1).\ell$).

(β) Generates just the ℓ-th $\hat{Y}$ (and the procedure then proceeds to Step $m.(\ell+1)$).

(γ) Generates both the m-th $\hat{X}$ and the ℓ-th $\hat{Y}$ (and the procedure then proceeds to Step $(m+1).(\ell+1)$).

For $\ell = m \epsilon \{1, \ldots, n\}$, Step $m.m$ produces either (α) or (γ). For $\ell < m = n+1$, Step $(n+1).\ell$ produces only (β). The procedure ends upon entrance to Step $(n+1).(n+1)$ which is vacuous.

PROOF OF THEOREM 3.1 (continued):

Step 2.1: It is given, for some $j_1 \epsilon \{1, \ldots, n\}$ and $t_{j_1} \geq 0$, that $\hat{X}_{j_1} = t_{j_1}$, $\hat{\mathbf{X}}_{\{1,\ldots,n\}-\{j_1\}} \geq t_{j_1}\mathbf{e}$ and $\hat{\mathbf{Y}} \geq t_{j_1}\mathbf{e}$. Consider $(n-1)+1$ independent nonhomogeneous Poisson processes on $[t_{j_1}, \infty)$. Let the first $n-1$ processes be called *processes of type 1* and call the last one a *process of type 2*. The $n-1$ type 1 processes are indexed by $i \epsilon I \equiv \{1, \ldots, n\} - \{j_1\}$. For $i \epsilon I$ let the intensity function of the type 1 process i be $\mu_i(t \mid X_{j_1} = t_{j_1}, \mathbf{X}_I \geq t\mathbf{e})$, $t \geq t_{j_1}$. Let the intensity of the type 2 process be $\eta_{j_1}(t \mid \mathbf{Y} \geq t\mathbf{e})$, $t \geq t_{j_1}$. If the type 2 process yields the first epoch (out of all n processes) at time $\tilde{t}_{j_1}$, say, then let the time of this epoch be $\hat{Y}_{j_1}$ and go to Step 2.2. If the type 1 process $j_2(\epsilon I)$ yields the first epoch (out of all n processes) at time t_{j_2}, say, then let the time of this epoch be $\hat{X}_{j_2}$. Also, in this case, with probability $\eta_{j_2}(t_{j_2} \mid \mathbf{Y} \geq t_{j_2}\mathbf{e})/\mu_{j_2}(t_{j_2} \mid X_{j_1} = t_{j_1}, \mathbf{X}_I \geq t_{j_2}\mathbf{e})$ [which is ≤ 1 by (10)] let the time of this epoch, t_{j_2}, be also $\hat{Y}_{j_2}$ and go to Step 3.2. With probability $1 - \eta_{j_2}(t_{j_2} \mid \mathbf{Y} \geq t_{j_2}\mathbf{e})/\mu_{j_2}(t_{j_2} \mid X_{j_1} = t_{j_1}, \mathbf{X}_I \geq t_{j_2}\mathbf{e})$ delay the determination of $\hat{Y}_{j_2}$ for a later step and go to Step 3.1.

Step 2.2: It is given, for some $j_1 \epsilon \{1, \ldots, n\}$ and fixed $t_{j_1} \leq \tilde{t}_{j_1}$, that $\hat{X}_{j_1} = t_{j_1}$, $\hat{Y}_{j_1} = \tilde{t}_{j_1}$, $\mathbf{X}_{\{1,\ldots,n\}-\{j_1\}} \geq \tilde{t}_{j_1}\mathbf{e}$ and $\mathbf{Y}_{\{1,\ldots,n\}-\{j_1\}} \geq \tilde{t}_{j_1}\mathbf{e}$. Consider $n-1$ independent Poisson processes on $[\tilde{t}_{j_1}, \infty)$ indexed by $i \epsilon I = \{1, \ldots, n\} - \{j_1\}$. For $i \epsilon I$ let the intensity function of Process i be $\mu_i(t \mid \mathbf{X}_{j_1} = t_{j_1}, \mathbf{X}_I \geq t\mathbf{e})$, $t \geq \tilde{t}_{j_1}$. If Process j_2 yields the first epoch (out of all the $n-1$ processes) at time t_{j_2}, say, then let the time of this epoch be $\hat{X}_{j_2}$. Also with probability $\eta_{j_2}(t_{j_2} \mid Y_{j_1} = \tilde{t}_{j_1}, \mathbf{Y}_I \geq t_{j_2}\mathbf{e})/\mu_{j_2}(t_{j_2} \mid X_{j_1} = t_{j_1}, \mathbf{X}_I \geq t_{j_2}\mathbf{e})$ let the time of this epoch also be $\hat{Y}_{j_2}$ and with probability $1 - \eta_{j_2}(t_{j_2} \mid Y_{j_1} = \tilde{t}_{j_1}, \mathbf{Y}_I \geq t_{j_2}\mathbf{e})/\mu_{j_2}(t_{j_2} \mid X_{j_1} = t_{j_1}, \mathbf{X}_I \geq t_{j_2}\mathbf{e})$ delay the determination of $\hat{Y}_{j_2}$ for a later step. Again, the last sentence is sensible because, by (10), $\eta_{j_2}(t_{j_2} \mid Y_{j_1} = \tilde{t}_{j_1}, \mathbf{Y}_I \geq t_{j_2}\mathbf{e})/\mu_{j_2}(t_{j_2} \mid X_{j_1} = t_{j_1}, \mathbf{X}_I \geq t_{j_2}\mathbf{e}) \leq 1$. If $\hat{X}_{j_2} = \hat{Y}_{j_2}$ then go to Step 3.3. If $\hat{Y}_{j_2}$ has not yet been determined then go to Step 3.2.

The proof of Theorem 3.1 continues after Remark A.2.

REMARK A.2. After the conclusion of Step 2 (that is, just after the last step of the form 2.i, for some $i \epsilon \{1, 2\}$, has been executed) $\hat{X}_{j_1}$ and $\hat{X}_{j_2}$ (and perhaps also $\hat{Y}_{j_1}$ and $\hat{Y}_{j_2}$) have been determined. In addition to (*.1) we also have

$$(*.2) \qquad \hat{X}_{j_2} \leq \hat{Y}_{j_2} \text{ with probability one.}$$

PROOF OF THEOREM 3.1 (continued):

STEP $(m+1).(\ell+1)$ [for $\ell \leq m \leq n$, $\ell < n$]: It is given, for some $I = \{j_1, \ldots, j_\ell\} \subset \{1, \ldots, n\}$, $J = \{j_{\ell+1}, \ldots, j_m\} \subset \{1, \ldots, n\}$ [such that $I \cap J = \emptyset$] and for some fixed $0\mathbf{e} \leq \mathbf{t}_I \leq \tilde{\mathbf{t}}_I$ and $\mathbf{t}_J \geq 0\mathbf{e}$, that $\mathbf{X}_I = \mathbf{t}_I$, $\mathbf{Y}_I = \tilde{\mathbf{t}}_I$, $\mathbf{X}_J = \mathbf{t}_J$, $\mathbf{X}_{\overline{I \cup J}} \geq (M(\mathbf{t}_J, \tilde{\mathbf{t}}_I))\mathbf{e}$ and $\mathbf{Y}_I \geq (M(\mathbf{t}_J, \tilde{\mathbf{t}}_I))\mathbf{e}$ [note that if $\ell = m$ then $J = \emptyset$ and if $m+1 = n+1$ then $\overline{I \cup J} = \emptyset$]. Consider $(n-m)$ nonhomogeneous Poisson processes of type 1 indexed by $k \epsilon \overline{I \cup J}$ and $(m-\ell)$ nonhomogeneous Poisson processes of type 2 indexed by $j \epsilon J$. All processes are independent and on $[M(\mathbf{t}_J, \tilde{\mathbf{t}}_I), \infty)$. For $k \epsilon \overline{I \cup J}$ let the intensity function of type 1 process k be $\mu_k(t \mid \mathbf{X}_I = \mathbf{t}_I, \mathbf{X}_J = \mathbf{t}_J, \cdot)$. For $j \epsilon J$ let the intensity function of type 2 process j be $\eta_j(t \mid \mathbf{Y}_I = \tilde{\mathbf{t}}_I, \cdot)$. If type 2 process $j_{\ell+1} \epsilon J$ yields the first epoch (out of all $n-\ell = (n-m)+(m-\ell)$ processes) at time $\tilde{t}_{j_{\ell+1}}$, say, then let the time of this epoch be $\hat{Y}_{j_{\ell+1}}$ and go to set $(m+1).(\ell+2)$. If type 1 process $j_{m+1} \epsilon \overline{I \cup J}$ yields the first epoch (out of all $n-\ell$ processes) at time $t_{j_{m+1}}$, say, then let the time of this epoch be $\hat{X}_{j_{m+1}}$. Also, in this case, with probability $\eta_{j_{m+1}}(t_{j_{m+1}} \mid \mathbf{Y}_I = \tilde{\mathbf{t}}_I, \cdot)/\mu_{j_{m+1}}(t_{j_{m+1}} \mid \mathbf{X}_I = \mathbf{t}_I, \mathbf{X}_J = \mathbf{t}_J, \cdot)$ [which is ≤ 1 by (10)] let the time of this epoch, $t_{j_{m+1}}$, be also $Y_{j_{m+1}}$ and go to Step $(m+2).(\ell+2)$. With probability $1 - \eta_{j_{m+1}}(t_{j_{m+1}} \mid \mathbf{Y}_I = \tilde{\mathbf{t}}_I, \cdot)/\mu_{j_{m+1}}(t_{j_{m+1}} \mid \mathbf{X}_I = \mathbf{t}_I, \mathbf{X}_J = \mathbf{t}_J, \cdot)$ delay the determination of $\hat{Y}_{j_{m+1}}$ for a later step and go to Step $(m+2).(\ell+1)$.

If $m \leq n-1$, then at the conclusion of Step $m+1$ (that is, just after the last step of the form $(m+1).i$ for some $i \epsilon \{1, \ldots, m+1\}$ has been executed) $\hat{X}_{j_1}, \ldots, \hat{X}_{j_{m+1}}$ (and some of $\hat{Y}_{j_1}, \ldots, \hat{Y}_{j_{m+1}}$) have been determined. In addition to $(*.1), (*.2), \ldots, (*.m)$ we also have

(*.m+1) $\hat{X}_{j_{m+1}} \le \hat{Y}_{j_{m+1}}$ with probability one.

Executing all the steps in sequence (the last step must be the one before entrance to Step $(n+1).(n+1)$) we obtain $(*.1), \ldots, (*.n)$. From this it follows that (40) holds. Notice that at the conclusion of Step m, the m-th $\hat{X}_j$ has been determined as described in the dynamic construction in Section 2, $m = 1, 2, \ldots, n$. That is, $\mathbf{X}$ and $\hat{\mathbf{X}}$ have the same instantaneous failure rates. Hence, by Lemma 1.1 of Shaked and Shanthikumar (1986b) we have (38). Using well known results about thinning of nonhomogeneous Poisson processes (see e.g. Savits (1988)) it is seen that for each ℓ, just after the last step of the form $i.\ell$, the ℓ-th $\hat{Y}_j$ have been determined as described in the dynamic construction in Section 2. Hence, again by Lemma 1.1 of Shaked and Shanthikumar (1986b), we have (39).

REFERENCES

ARJAS, E. (1981a). A stochastic process approach to multivariate reliability: Notions based on conditional stochastic order. *Math. Oper. Res.* **6** 263–276.

ARJAS, E. (1981b). The failure and hazard processes in multivariate reliability systems. *Math. Oper. Res.* **6** 551–562.

ARJAS, E. and LEHTONEN, T. (1978). Approximating many server queues by means of single server queues. *Math. Oper. Res.* **3** 205–223.

ARJAS, E. and NORROS, I. (1984). Life lengths and association: A dynamic approach. *Math. Oper. Res.* **9** 151–158.

COX, D.R. (1972). Regression models and life tables (with discussion). *J. Roy. Statist. Soc.* **B** 187–220.

ESARY, J.D., PROSCHAN, F., and WALKUP, D.W. (1967). Association of random variables, with applications. *Ann. Math. Statist.* **38** 1466–1474.

LEWIS, P.A.W. and SHEDLER, G.S. (1979). Simulation of nonhomogeneous Poisson processes by thinning. *Nav. Res. Logist. Quart.* **26** 403–413.

NORROS, I. (1986). A compensator representation of multivariate life length distributions, with applications. *Scand. J. Statist.* **13** 99–112.

ROSS, S.M. (1985). *Introduction to Probability Models.* Academic Press, New York.

SAVITS, T.H. (1988). Some multivariate distributions derived from a non-fatal shock model. *J. Appl. Prob.* **25** 383-390.

SHAKED, M. and SHANTHIKUMAR, J.G. (1986a). The total hazard construction, antithetic variates and simulation of stochastic systems. *Commun. Statist.-Stochastic Models* **2** 237–249.

SHAKED, M. and SHANTHIKUMAR, J.G. (1986b). Multivariate imperfect repair. *Oper. Res.* **34** 437–448.

SHAKED, M. and SHANTHIKUMAR, J.G. (1987a). Multivariate hazard rates and stochastic ordering. *Adv. Appl. Prob.* **19** 123–137.

SHAKED, M. and SHANTHIKUMAR, J.G. (1987b). The multivariate hazard construction. *Stoch. Proc. Appl.* **24** 241–258.

SHANTHIKUMAR, J.G. (1986). Uniformization and hybrid simulation/analytic models of renewal processes. *Oper. Res.* **34** 573–580.

Department of Mathematics
Building 89
University of Arizona
Tucson, AZ 85721

School of Business Administration
University of California, Berkeley
Berkeley, CA 94720

MODELS FOR DEPENDENT LIFELENGTHS INDUCED BY COMMON ENVIRONMENTS

By Nozer D. Singpurwalla[1] and Mark A. Youngren

George Washington University and US Army Concepts Analysis Agency

Multivariate distributions for the lifelengths of the components of a system operating under a common environment, when the environment has a different effect on each component, and when the environment is dynamic, are derived. Modelling of the dynamic environment is by a gamma process.

1. Introduction. Multivariate distributions for the lifelengths of biological and engineering systems have been proposed by Freund (1961), Downton (1970), Marshall and Olkin (1967), Lindley and Singpurwalla (1986), and Lee and Gross (1990). In this paper we build upon the theme proposed by Lindley and Singpurwalla, and generate classes of multivariate distributions which may lead to improved assessments of system reliability.

As a motivating scenario, suppose that we have an m-component, parallel redundant system, and suppose that the lifelengths of these components are judged exponential with known scale parameters $\lambda_{10}, \lambda_{20}, \ldots, \lambda_{m0}$ when they are tested in a laboratory individually. The λ_{io}'s [or more generally, the $\lambda_{io}(t)$'s, if the lifelengths are judged to be other than exponential] will be referred to as the *baseline failure rates* of the m components. Suppose that the effect of the common operating environment—when assumed to be static over time—is to modulate each λ_{io} by a common factor η, where η is unknown and has distribution G, so that the reliabilities become $\exp\{-\int_0^t \eta\lambda_{io}(u)du\}$. Uncertainty about η induces dependence among the component lifelengths $T_1, \ldots, T_m$. The T_i's, $i = 1, \ldots, m$, have a multivariate distribution whose nature is prescribed by the form of G. When the operating environment is dynamic, η becomes a function of time t, say $\eta(t)$; we will refer to η, or more generally $\eta(t)$, $t \geq 0$, as the *environmental factor function*—henceforth EFF. It is important to bear in mind that the EFF is merely a parameter that has little, if any, physical meaning. It is introduced for convenience with the aim of capturing our opinion about the effects of the environment on the failure rate of each component.

[1]Research supported by Grant DAAL03-87-K-0056, the U.S. Army Research Office and Contract N00014-85-K-0202, Project NR 042-372, Office of Naval Research.

AMS 1980 subject classifications. Primary 60K10; secondary 62N05.

Key words and phrases. Multivariate life distributions, gamma process, failure rate, environmental effects.

When $m = 2$, $\eta(t) = \eta$ and G is a gamma distribution with scale β and shape α, T_1 and T_2 will have a bivariate Pareto distribution (Johnson and Kotz [1972] p. 285) with a joint survival function

$$(1) \quad \bar{F}(t_1,t_2) \stackrel{\text{def}}{=} P(T_1 > t_1, T_2 > t_2) = \left(\frac{\beta}{\lambda_{10}t_1 + \lambda_{20}t_2 + \beta}\right)^{\alpha}, \quad t_1, t_2 > 0.$$

The above distribution, which can be transformed to a bivariate logistic distribution, was motivated by Lindley and Singpurwalla; it can be shown to be a special case of the Dirichlet distribution. Currit and Singpurwalla (1988) compared the behavior of $\bar{F}(t_1,t_2)$ with $\exp(-(\lambda_{10}t_1 + \lambda_{20}t_2))$, the survival function obtained under the assumption that T_1 and T_2 are independent and exponentially distributed with parameters λ_{10} and λ_{20}, respectively, and showed that the two could lead to drastically different results. The aim of this paper is to consider extensions of (1) along the several lines described below.

2. Multiple Environmental Factor Functions with Dependence. A natural way to expand upon the previous theme is to assume that each λ_{io} is modulated by η_i, $i = 1, \ldots, m$, and that the uncertainty about the η_i's is described by a meaningful multivariate distribution. Dependencies between the η_i's can be motivated when some factors which constitute the environment—such as temperature—may have an identical effect on all the components, whereas the other factors—such as humidity—may have different effects on the different components. A plausible model for describing dependencies among the η_i's is due to Cherian (1941) and David and Fix (1961)—henceforth C–D–F.

Let $m = 2$ and assume that $\eta_i = X_0 + X_i$, $i = 1, 2$, where the random quantity X_0 captures the contribution of the common factors on both the components, and X_i captures the contribution of the other factors on component i. In the C–D–F model, X_0, X_1 and X_2 are assumed to be independent, each having a gamma distribution with scale (shape) parameter $\beta_i(\alpha_i)$, $i = 0, 1, 2$, respectively. Clearly, η_1 and η_2 are dependent and have a joint density which may be easily derived (see Johnson and Kotz (1972), pp. 216–220).

It is easy to verify that under the above scenario,

$$(2) \quad \bar{F}_{\text{CDF}}(t_1,t_2) = \left(\frac{\beta_0}{\lambda_{10}t_1 + \lambda_{20}t_2 + \beta_0}\right)^{\alpha_0} \Pi_{i=1}^{2} \left(\frac{\beta_i}{\lambda_{10}t_i + \beta_i}\right)^{\alpha_i}, \quad t_1, t_2 > 0.$$

2.1. Inequalities for Survival Functions with Increasing Degrees of Dependence. The nature of the dependence between T_1 and T_2 depends on the dependence between η_1 and η_2. In the case of (1), $\eta_1 = \eta_2 = \eta$ and so the dependence between η_1 and η_2 is the strongest possible. The C–D–F case and the independent case are increasingly less dependent. To facilitate the construction of three pairs of random quantities (η_1, η_2), (η_1, η_2') and (η_1, η_2'') with decreasing degrees of dependence, four mutually independent random quantities X_0, X_0', X_1 and X_1' are introduced with

$X_0\ (X_1) \stackrel{d}{=} X_0'(X_1')$, where the notation "$X \stackrel{d}{=} Y$" indicates that X has the same distribution as Y. Let $\eta_1 = \eta_2 = X_0 + X_1$, $\eta_2' = X_0 + X_1'$ and $\eta_2'' = X_0' + X_1'$ and suppose that X_i and X_i' have a gamma distribution with shape α_i and scale β_i, $i = 0, 1$. Clearly, $\eta_2 \stackrel{d}{=} \eta_2' \stackrel{d}{=} \eta_2''$ but the pairs (η_1, η_2), (η_1, η_2'), (η_1, η_2'') are increasingly less dependent.

It is now easy to verify that the pair (η_1, η_2) [where η_1 and η_2 are identical] will result in the bivariate survival function of the form given by (1); specifically

$$\bar{F}_{\mathrm{LS}}(t_1, t_2) = \Pi_{i=0}^{1} \left(\frac{\beta_i}{\beta_i + \lambda_{10} t_1 + \lambda_{20} t_2} \right)^{\alpha_i}, \quad t_1, t_2 \geq 0.$$

When $\alpha_1 = \alpha_2$, the pair (η_1, η_2') will lead to the bivariate survival function (2), which because of its derivation via the C–D–F distribution will be denoted $\bar{F}_{\mathrm{CDF}}(t_1, t_2)$. Finally, since η_1 and η_2'' are independent, the resulting survival function is

$$\bar{F}_I(t_1, t_2) = \Pi_{\ell=1}^{2} \left(\frac{\beta_0}{\beta_0 + \lambda_{\ell 0} t_\ell} \right)^{\alpha_0} \Pi_{m=1}^{2} \left(\frac{\beta_1}{\beta_1 + \lambda_{m0} t_m} \right)^{\alpha_1}, \quad t_1, t_2 \geq 0.$$

Let "$(X_1, Y_1) \stackrel{D}{>} (X_2, Y_2)$" denote the fact that the pair (X_1, Y_1) is more dependent than the pair (X_2, Y_2). Then, by construction, $(\eta_1, \eta_2) \stackrel{D}{>} (\eta_1, \eta_2') \stackrel{D}{>} (\eta_1, \eta_2'')$, and now it is easy to verify

THEOREM 2.1.

$$\bar{F}_I(t_1, t_2) < \bar{F}_{\mathrm{CDF}}(t_1, t_2) < \bar{F}_{\mathrm{LS}}(t_1, t_2), \text{ for } t_1, t_2 > 0.$$

Thus, for any fixed t_1, $t_2 > 0$, the bivariate survival function of 2 component parallel redundant systems increases as the degree of dependence between their EFF's increases. The inequality generalizes for the case of m components. When we set $t_1 = t_2 = \ldots = t_m$, we obtain inequalities for the system reliability function of series systems.

3. Dependencies Induced By Dynamic Environments. The material in the previous two sections assumed that the EFF is constant over time, so that $\eta_i(t) = \eta_i$, $i = 1, \ldots, m$. This assumption is not meaningful when the environment is dynamic as is often the case. As a starting scenario, suppose that $\eta_i(t) = \eta(t)$, $t \geq 0$ and $i = 1, \ldots, m$, and suppose that our uncertainty about $\eta(t)$ is described by a continuous time stochastic process, called the *gamma process*. The gamma process for the EFF produces useful results, and can be motivated as the limit of a piecewise constant EFF with independent gamma distributed innovations.

3.1. Motivating the Gamma Process. Suppose that $\eta(t)$ is a piecewise constant right continuous function over specified time intervals $[t_j, t_{j+1})$, $j = 0, 1, \ldots$, where $t_0 \equiv 0$. Specifically, let $\eta(t) = \eta_j$, $t \in [t_j, t_{j+1})$, with the η_j's unknown. Suppose that the environment is composed of a known number of at most $s + 1$ distinct stresses, each having the same effect on all the m components, with the k–th stress contributing an innovation C_k to η_j. The parameter η_j changes to η_{j+1} when one or more of the innovations C_k appears or disappears. Suppose further that the effects of the innovations C_k are additive, so that $\eta_j = \sum_{k=0}^{s} I_k(t_j)C_k$, $j = 0, 1, \ldots$, where $I_k(t_j) = 1$, if the k–th stress is present in $[t_j, t_{j+1})$, and is 0 otherwise. The innovations C_k and the variables $I_k(t_j)$ are assumed to be mutually independent for all $j = 0, 1, \ldots$ and all $k = 0, 1, \ldots, s$. If N_j, the number of stresses during $[t_j, t_{j+1})$ is known, but their identities are unknown, and if each C_k is assumed to have a gamma distribution with parameters α and β, then the η_j's are independent gamma distributed variables with parameter $N_j\alpha$ and β. It can now be shown [cf. Youngren (1988), p. 54], that in the limit, as $\Delta t_j \stackrel{\text{def}}{=} (t_{j+1} - t_j) \to 0$, the cumulative failure rate of the i-th component at time t, $0 \le t_n < t \le t_{n+1}$, is a gamma process. We denote the cumulative failure rate of the i-th component as

$$\Lambda_i(t) = \lambda_{io}\left[\sum_{j=0}^{n-1} \eta_j(t_{j+1} - t_j) + \eta_n(t - t_n)\right];$$

recall that λ_{io} is the baseline failure rate of the i-th component.

Instead of assuming that the η_j's are independent as is done above, suppose that the η_j's have a time dependent structure as follows. Let

$$\begin{aligned} I_k(t_j) &= 0,\ 0 \le j < k, \quad j = 0, 1, \ldots, j; \\ &= 1,\ j \ge k, \qquad\quad k = 0, 1, \ldots, s; \text{ then} \end{aligned}$$

$\eta_j = \sum_{k=0}^{j} C_k$, and if one's uncertainty about the C_k's is described via a gamma distribution with parameters α_j and β, then here again it can be shown [cf. Youngren (1988), p. 60], that when $\Delta t_j \to 0$, $\eta(t)$ is a gamma process for any $t \ge 0$.

3.2. Preliminaries on Gamma Processes. The gamma process is nonnegative, nondecreasing in time and possesses independent increments. It has been studied by Ferguson and Klass (1972), Çinlar (1980), and Dykstra and Laud (1981); the use of gamma processes in survival analysis is primarily due to Ferguson (1973), Ferguson and Phadia (1979), and Kalbfleisch (1978).

DEFINITION 3.1. Let $\alpha(t)$ be a nondecreasing left-continuous real valued function on $[0, \infty)$ with $\alpha(0) = 0$, and let $\beta \in (0, \infty)$. A stochastic process $(Y(t),\ t \ge 0)$ is said to be a *gamma process* with parameters $\alpha(t)$ and β, denoted "$Y(t) \in G_{pr}(\alpha(t), \beta)$", if:

1. $Y(0) = 0$
2. $Y(t)$ has independent increments, and
3. $Y(t) - Y(s) \sim \gamma(\alpha(t) - \alpha(s), \frac{1}{\beta})$ for any $0 \le s \le t$.

Dykstra and Laud (1981) extend the gamma process to include a time-varying scale parameter $\beta(t)$.

DEFINITION 3.2. Let $\beta(t)$, $t \geq 0$ be a positive right-continuous real valued function, and let $Y(t) \in G_{pr}(\alpha(t), 1)$. The process $Z(t) \stackrel{\text{def}}{=} \int_0^t \beta(s)\, dY(s)$ is an *extended gamma process* denoted "$Z(t) \in G_{pr}(\alpha(t), \beta(t))$."

Note that the gamma process is a special case of the extended gamma process, where $\beta(t) = \beta \ \forall t$.

Dykstra and Laud (1981) give the following properties of the extended gamma process. Let $Z(t) \in G_{pr}(\alpha(t), \beta(t))$. Then

$$\begin{aligned} E[Z(t)] &= \int_0^t \beta(u)\, d\alpha(u), \quad \text{Var}[Z(t)] = \int_0^t \beta^2(u)\, d\alpha(u), \text{ and} \\ G^*_{Z(t)}(s) &= \exp\left[-\int_0^t \log(1 + s\beta(u))\, d\alpha(u)\right], \ s > 0, \end{aligned}$$

where $G^*_{Z(t)}$ is the Laplace Stieltjes transform (LST) of the distribution of $Z(t)$.

3.3. Modelling the EFF as a Gamma Process. Suppose that $\eta(t)$ is described by a gamma process with parameters $(\alpha(t), \frac{1}{\beta})$. If the baseline failure rate is a continuous, positive, real valued function of time, then the following theorem is used to derive the bivariate and marginal survival functions.

THEOREM. *Let $\eta(t) \in G_{pr}(\alpha(t), \frac{1}{\beta})$, let $\lambda_0(t)$ be a known continuous positive real valued function and let $\Lambda(t) = \int_0^t \lambda_0(u)\, \eta(u)\, du$. Then the univariate survival function is*

$$\bar{F}(t) = \exp\left\{-\int_0^t \log[1 + \frac{1}{\beta}\int_u^t \lambda_0(s)\, ds] d\alpha(u)\right\}.$$

The proof of this theorem is based on Dykstra and Laud (1981), and is given by Youngren (1988).

The bivariate survival function for $0 \leq t_1 \leq t_2$ follows directly from the above theorem. Specifically, $\bar{F}(t_1, t_2) =$

$$\begin{aligned} &\exp\left\{-\int_0^{t_1} \log[1 + \frac{1}{\beta}\int_u^{t_1} (\lambda_{10}(s) + \lambda_{20}(s))\, ds] d\alpha(u)\right\} \\ \times\ &\exp\left\{-\int_{t_1}^{t_2} \log[1 + \frac{1}{\beta}\int_u^{t_2} \lambda_{20}(s)\, ds] d\alpha(u)\right\}. \end{aligned}$$

We can choose plausible functional forms for $\alpha(t)$ and $\lambda_{io}(t)$, suggested by the physical model of the environment, that enable us to obtain closed form solutions for the survival functions. If we use our time-dependent model of Section 3.1, wherein $\eta(t) = \sum_{k=0}^{j} C_k$, for $t \in [t_j, t_{j+1})$, then for $s \in [t_\ell, t_{\ell+1})$, $\eta(t) - \eta(s) = \sum_{k=\ell+1}^{j} C_k$ is distributed as gamma with a shape parameter that depends on the length of the interval $(t - s)$. This leads, in the limit as $\Delta t \to 0$, to a gamma process with a linear shape function $\alpha(t)$, say $\alpha(t) = \alpha_1 t$, for some $\alpha_1 > 0$, $t \geq 0$.

Assume that $\eta(t) \in G_{pr}(\alpha_1 t, \frac{1}{\beta})$, which implies that the component failure rate is $\lambda_i(t) = \lambda_{io}\eta(t) \in G_{pr}(\alpha_1 t, \frac{\lambda_{io}}{\beta})$. For convenience let $\beta_{11} \stackrel{\text{def}}{=} \frac{\lambda_{1o}}{\beta}$ and $\beta_{21} \stackrel{\text{def}}{=} \frac{\lambda_{2o}}{\beta}$; then the bivariate survival function for $0 \leq t_1 \leq t_2$ is

$$\begin{aligned} \bar{F}(t_1, t_2) &= (1 + (\beta_{11} + \beta_{21})t_1)^{\frac{\alpha_1}{(\beta_{11}+\beta_{21})}} (1 + \beta_{21}(t_2 - t_1))^{\frac{\alpha_1}{\beta_{21}}} \\ &\times \left[\frac{1 + \beta_{21}(t_2 - t_1)}{1 + (\beta_{11} + \beta_{21})t_1}\right]^{\alpha_1 t_1} \cdot \left[\frac{e}{1 + \beta_{21}(t_2 - t_1)}\right]^{\alpha_1 t_2}, \text{ with} \end{aligned}$$

marginal failure rate functions $r_i(t_i) = \alpha_1 \log[1 + \beta_{i1} t_i]$, $i = 1, 2$.

It is interesting to note that the marginal distributions can also be obtained using an extended gamma process for $\lambda_i(t)$ with shape parameter $\alpha(t) = \alpha_1 t$, and scale parameter $\frac{\lambda_{io}(t)}{\beta} = \beta_{i1} t$.

REFERENCES

CHERIAN, K.C. (1941). A bivariate correlated gamma type distribution function. *J. Ind. Math. Soc.* **5** 133–144.

ÇINLAR, E. (1980). On a generalization of gamma processes. *J. Appl. Prob.* **17** 467–480.

CURRIT, A. and SINGPURWALLA, N.D. (1988). On the reliability function of a system of components sharing a common environment. *J. Appl. Prob.* **25**, No. 4, 763–771.

DAVID, F.N. and FIX, E. (1961). Rank correlation and regressions in a non-normal surface. In *Fourth Berkeley Symposium on Mathematical Statistics and Probability* (J. Neyman, ed.), University of California Press, **1** 177–197.

DOWNTON, F. (1970). Bivariate exponential distributions in reliability theory. *J. Roy. Statist. Soc.* B **32** 408–417.

DYKSTRA, R.L. and LAUD, P. (1981). A Bayesian nonparametric approach to reliability. *Ann. Statist.* **9**, No. 2, 356–367.

FERGUSON, T.S. (1973). A Bayesian analysis of some nonparametric problems. *Ann. Statist.* **1**, No. 2, 209–230.

FERGUSON, T.S. and KLASS, M.J. (1972). A representation of independent increment processes without Gaussian components. *Ann. Math. Statist.* **43**, No. 5, 1634–1643.

FERGUSON, T.S. and PHADIA, E.G. (1979). Bayesian nonparametric estimation based on censored data. *Ann. Statist.* **7**, No. 1, 163–186.

FREUND, J.E. (1961). A bivariate extension of the exponential distribution. *J. Amer. Statist. Assoc.* **56** 971–977.

JOHNSON, N.L. and KOTZ, S. (1972). *Distributions in Statistics: Continuous Multivariate Distributions.* John Wiley and Sons, Inc., New York.

KALBFLEISCH, J.D. (1978). Nonparametric Bayesian analysis of survival time data. *J. Roy. Statist. Soc.* B **40**, No. 2, 214–221.

LEE, M.-L.T. and GROSS, A.J. (1990). Lifetime distributions under unknown environment. *J. Statist. Plan. Infer.*, to appear.

LINDLEY, D.V. and SINGPURWALLA, N.D. (1986). Multivariate distributions for the lifelengths of components of a system sharing a common environment. *J. Appl. Prob.* **23**, No. 2, 418–431.

MARSHALL, A.W. and OLKIN, I. (1967). A multivariate exponential distribution. *J. Amer. Statist. Assoc.* **62** 30–44.

YOUNGREN, M.A. (1988). Dependent lifelengths induced by dynamic environments. D.Sc. dissertation, George Washington University.

DEPARTMENT OF OPERATIONS RESEARCH
GEORGE WASHINGTON UNIVERSITY
WASHINGTON, DC 20052

U.S. ARMY CONCEPTS ANALYSIS AGENCY
8120 WOODMONT AVE.
BETHESDA, MD 20814–2797

SOME COMMENTS ON POSITIVE QUADRANT DEPENDENCE IN THREE DIMENSIONS

By K. Subramanyam[1]

University of North Carolina at Wilmington

An extreme point analysis has been performed on two natural definitions of positive quadrant dependence of three random variables. This analysis helps us to understand how much these two notions of dependence are different. In the case of two random variables these two notions of dependence are equivalent.

1. Introduction. Let X and Y be two random variables with some joint probability distribution function F. X and Y (or F) are said to be positively quadrant dependent (PQD) if

$$\Pr(X \le x, Y \le y) \ge \Pr(X \le x)\Pr(Y \le y) \tag{1}$$

for all real numbers x and y. The condition (1) is equivalent to

$$\Pr(X \ge x, Y \ge y) \ge \Pr(X \ge x)\Pr(Y \ge y) \tag{2}$$

for all x and y. See Lehmann (1966, p. 1138).

One faces problems if one wishes to extend the notion of positive quadrant dependence to more than two random variables. If X, Y, and Z are three random variables, one could say that X, Y, and Z are PQD by adapting either of the conditions (1) or (2) in a natural way. To be more precise, we say that X, Y, and Z are positively lower orthant dependent (PLOD) if

$$\Pr(X \le x, Y \le y, Z \le z) \ge \Pr(X \le x)\Pr(Y \le y)\Pr(Z \le z) \tag{3}$$

for all x, y, and z; and we say that X, Y, and Z are positively upper orthant dependent (PUOD) if

$$\Pr(X \ge x, Y \ge y, Z \ge z) \ge \Pr(X \ge x)\Pr(Y \ge y)\Pr(Z \ge z) \tag{4}$$

---

[1]Supported by AFOSR Contract AFSO-88-0030.

AMS 1980 subject classifications. Primary 60E05, secondary 62H05.

Key words and phrases. Positive upper orthant dependence, positive lower orthant dependence, convex set, extreme points.

The author is thankful to the referee and the editors for their comments which led to an improvement in the presentation of the paper.

for all x, y, and z.

These two concepts have been examined by Ahmed, Langberg, Léon and Proschan (1978) and by several authors cited in that paper. See also Block and Ting (1981), and Chhetry, Kimeldorf and Sampson (1989).

In this paper, we discuss the ramifications of the definitions of PLOD and PUOD. These two notions of PLOD and PUOD are not equivalent. Ahmed, Langberg, Léon and Proschan (1978) gave an example of a trivariate distribution which is PUOD, but not PLOD.

The main goal of this paper is to examine how different are these two notions of dependence. More precisely, we want to perform extreme point analysis on these two notions of dependence. In some special cases, extreme point analysis helps us to characterize all trivariate distributions which are both PLOD and PUOD.

2. Extreme Point Analysis. We consider the case where each of X, Y, and Z assumes only two values 1 and 2, say. Let $P_{ijk} = \Pr(X = i, Y = j, Z = k)$, $i = 1,2; j = 1,2,; k = 1,2$. The joint probability law of X, Y, and Z is written, for convenience,

$$P = \begin{bmatrix} P_{111} & P_{112} & P_{121} & P_{122} \\ P_{211} & P_{212} & P_{221} & P_{222} \end{bmatrix}$$

In terms of this new notation, P is PLOD if

$$P_{111} \geq p_1 q_1 r_1 \tag{5}$$
$$P_{111} + P_{112} \geq p_1 q_1 \tag{6}$$
$$P_{111} + P_{121} \geq p_1 r_1 \tag{7}$$
$$P_{111} + P_{211} \geq q_1 r_1 \tag{8}$$

and P is PUOD IS

$$P_{222} \geq p_2 q_2 r_2 \tag{9}$$
$$P_{222} + P_{221} \geq p_2 q_2 \tag{10}$$
$$P_{222} + P_{212} \geq p_2 r_2 \tag{11}$$
$$P_{222} + P_{122} \geq q_2 r_2 \tag{12}$$

where $p_1 = \Pr(X = 1)$; $q_1 = \Pr(Y = 1)$; $r_1 = \Pr(Z = 1)$; $p_2 = 1 - p_1$; $q_2 = 1 - q_1$; and $r_2 = 1 - r_1$.

Let $0 < p_1 < 1$, $0 < q_1 < 1$, and $0 < r_1 < 1$ be three fixed numbers. Let $M_{\text{PLOD}}(p_1, q_1, r_1)$ be the collection of all trivariate distributions $P = (P_{ijk})$ with support contained in $\{(i,j,k); i = 1,2, j = 1,2, \text{and } k = 1,2\}$ such that P is PLOD, and the marginal distributions of X, Y, and Z under P are $p_1, 1 - p_1$; $q_1, 1 - q_1$; and $r_1, 1 - r_1$, respectively. The set $M_{\text{PUOD}}(p_1, q_1, r_1)$ is defined analogously. The following result is obvious.

THEOREM 1. *The sets* $M_{PLOD}(p_1, q_1, r_1)$ *and* $M_{PUOD}(p_1, q_1, r_1)$ *are compact and convex. More strongly, they are simplexes, i.e., each of these sets is bounded and a finite intersection of hyperplanes.*

Nguyen and Sampson (1985) have looked into properties of sets of the above type for bivariate distributions with fixed marginals. Subramanyam and Bhaskara Rao (1986) have developed an algebraic method for identifying the extreme points of sets of the above type in the context of bivariate distributions.

Being simplexes, the sets $M_{\text{PLOD}}(p_1, q_1, r_1)$ and $M_{\text{PUOD}}(p_1, q_1, r_1)$ have each a finite number of extreme points. Once we identify the extreme points of the set $M_{\text{PLOD}}(p_1, q_1, r_1)$ say, we can express every member of $M_{\text{PLOD}}(p_1, q_1, r_1)$ as a convex combination of its extreme points. We describe now a method of identifying the extreme points of $M_{\text{PLOD}}(p_1, q_1, r_1)$ as well as $M_{\text{PUOD}}(p_1, q_1, r_1)$. First, we take up the case of $M_{\text{PLOD}}(p_1, q_1, r_1)$. Any $P = (P_{ijk}) \varepsilon M_{\text{PLOD}}(p_1, q_1, r_1)$ must satisfy the inequalities (5), (6), (7), and (8). Also, due to marginality restrictions, we should have

(13) $$P_{111} + P_{112} + P_{121} \leq p_1$$
(14) $$P_{111} + P_{112} + P_{211} \leq q_1$$
(15) $$P_{111} + P_{121} + P_{211} \leq r_1.$$

The following are the natural nonnegativity conditions.

(16) $$P_{112} \geq 0$$
(17) $$P_{121} \geq 0$$
(18) $$P_{211} \geq 0$$

All these inequalities (5) to (8) and (13) to (18) involve $P_{111}, P_{112}, P_{121}, P_{211}$ only. If some four numbers $P_{111}, P_{112}, P_{121}, P_{211}$ satisfy the inequalities (5) to (8) and (13) to (18), then one could define

(19) $$P_{122} = p_1 - (P_{111} + P_{112} + P_{121}),$$
(20) $$P_{212} = q_1 - (P_{111} + P_{112} + P_{211}),$$
(21) $$P_{221} = r_1 - (P_{111} + P_{121} + P_{211}),$$

and

(22) $$P_{222} = 1 - p_1 - q_1 - r_1 + P_{111} + (P_{111} + P_{112} + P_{121} + P_{211}).$$

The numbers P_{122}, P_{212}, and P_{211} will be nonnegative. If $P_{222} \geq 0$, then

$$P = (P_{ijk}) \varepsilon M_{\text{PLOD}}(p_1, q_1, r_1).$$

A standard method of identifying the extreme points of $M_{\text{PLOD}}(p_1, q_1, r_1)$ is as follows. Select 4 inequalities from (5) to (8) and (13) to (18). Replace the inequality signs by equality signs. Solve the resultant system of 4 linear equations in 4 unknowns P_{111}, P_{112}, P_{121}, and P_{211}. If there is a solution, and this solution satisfies the remaining inequalities, determine P_{122}, P_{212}, P_{221}, and P_{222} as per the equations (19), (20), (21), and (22). If $P_{222} \geq 0$, then

$$P = (P_{ijk}) \, \varepsilon \, M_{\text{PLOD}}(p_1, q_1, r_1)$$

It is easy to check that this P is an extreme point of $M_{\text{PLOD}}(p_1, q_1, r_1)$, and every extreme point of $M_{\text{PLOD}}(p_1, q_1, r_1)$ arises this way. For ideas concerning this approach, one may refer to Subramanyam and Bhaskara Rao (1986). A computer program is easy to write which will identify the extreme points of $M_{\text{PLOD}}(p_1, q_1, r_1)$.

In this context, define the joint distribution function

$$F_U(x, y, z) = F_1(x) \wedge F_2(y) \wedge F_3(z)$$

for all x, y, and z, where $F_1(x) = 0$ if $x < 1$, $= p_1$ if $1 \leq x < 2$, and $= 1$ if $x \geq 2$; $F_2(y) = 0$ if $y < 1$, $= q_1$ if $1 \leq y < 2$, and $= 1$ if $y \geq 2$; and $F_3(z) = 0$ if $z < 1$, $= r_1$ if $1 \leq z < 2$, and $= 1$ if $z \geq 2$; and for any two numbers u, v, $u \wedge v$ stands for the minimum of the numbers u and v. $F_U(x, y, z)$ is the upper Fréchet bound with marginals F_1, F_2, and F_3. An explicit computation shows that the corresponding distribution P_U has the following entries

$$\begin{aligned}
P_{111} &= p_1 \wedge q_1 \wedge r_1; P_{112} = p_1 \wedge q_1 - P_{111}; P_{121} = p_1 \wedge r_1 - P_{111}; \\
P_{211} &= q_1 \wedge r_1 - P_{111}; P_{221} = r_1 - P_{211} - P_{121} - P_{111}; \\
P_{212} &= q_1 - P_{112} - P_{211} - P_{111}; P_{122} = p_1 - P_{121} - P_{112} - P_{111}; \\
P_{222} &= 1 - P_{111} - P_{112} - P_{121} - P_{211} - P_{122} - P_{212} - P_{221}
\end{aligned}$$

It can be verified that the bound is PLOD, as well as PUOD. Furthermore, it is an extreme point.

Pursuing the above approach, we have isolated the extreme points of $M_{\text{PLOD}}(p_1, q_1, r_1)$ and $M_{\text{PUOD}}(p_1, q_1, r_1)$ when $p_1 = q_1 = r_1 = 1/2$, given in Table 1. The above extreme point analyses of the sets $M_{\text{PLOD}}(\frac{1}{2}, \frac{1}{2}, \frac{1}{2})$ and $M_{\text{PUOD}}(\frac{1}{2}, \frac{1}{2}, \frac{1}{2})$ reveal the following insights.

1. The extreme points of $M_{\text{PLOD}}(\frac{1}{2}, \frac{1}{2}, \frac{1}{2})$ and $M_{\text{PUOD}}(\frac{1}{2}, \frac{1}{2}, \frac{1}{2})$ fall into three distinct categories. The first five extreme points are common to both the sets. Observe that

$$\begin{aligned}
P_6 &= \frac{1}{2} P_4 + \frac{1}{2} P_{15} \\
P_8 &= \frac{1}{2} P_2 + \frac{1}{2} P_{15}
\end{aligned}$$

Table 1. Extreme Points of $M_{\text{PLOD}}(\frac{1}{2},\frac{1}{2},\frac{1}{2})$ and $M_{\text{PUOD}}(\frac{1}{2},\frac{1}{2},\frac{1}{2})$

Serial No.	$M_{\text{PLOD}}(\frac{1}{2},\frac{1}{2},\frac{1}{2})$	$M_{\text{PUOD}}(\frac{1}{2},\frac{1}{2},\frac{1}{2})$
1.	$P_1 = \frac{1}{8}\begin{bmatrix} 1 & 1 & 1 & 1 \\ 1 & 1 & 1 & 1 \end{bmatrix}$	$P_1 = \frac{1}{8}\begin{bmatrix} 1 & 1 & 1 & 1 \\ 1 & 1 & 1 & 1 \end{bmatrix}$
2.	$P_2 = \frac{1}{8}\begin{bmatrix} 2 & 0 & 2 & 0 \\ 0 & 2 & 0 & 2 \end{bmatrix}$	$P_2 = \frac{1}{8}\begin{bmatrix} 2 & 0 & 2 & 0 \\ 0 & 2 & 0 & 2 \end{bmatrix}$
3.	$P_3 = \frac{1}{8}\begin{bmatrix} 2 & 2 & 0 & 0 \\ 0 & 0 & 2 & 2 \end{bmatrix}$	$P_3 = \frac{1}{8}\begin{bmatrix} 2 & 2 & 0 & 0 \\ 0 & 0 & 2 & 2 \end{bmatrix}$
4.	$P_4 = \frac{1}{8}\begin{bmatrix} 2 & 0 & 0 & 2 \\ 2 & 0 & 0 & 2 \end{bmatrix}$	$P_4 = \frac{1}{8}\begin{bmatrix} 2 & 0 & 0 & 2 \\ 2 & 0 & 0 & 2 \end{bmatrix}$
5.	$P_5 = \frac{1}{8}\begin{bmatrix} 4 & 0 & 0 & 0 \\ 0 & 0 & 0 & 4 \end{bmatrix}$	$P_5 = \frac{1}{8}\begin{bmatrix} 4 & 0 & 0 & 0 \\ 0 & 0 & 0 & 4 \end{bmatrix}$
6.	$P_6 = \frac{1}{8}\begin{bmatrix} 1 & 1 & 1 & 1 \\ 2 & 0 & 0 & 2 \end{bmatrix}$	$P_7 = \frac{1}{8}\begin{bmatrix} 2 & 0 & 0 & 2 \\ 1 & 1 & 1 & 1 \end{bmatrix}$
7.	$P_8 = \frac{1}{8}\begin{bmatrix} 1 & 1 & 2 & 0 \\ 1 & 1 & 0 & 2 \end{bmatrix}$	$P_9 = \frac{1}{8}\begin{bmatrix} 2 & 0 & 1 & 1 \\ 0 & 2 & 1 & 1 \end{bmatrix}$
8.	$P_{10} = \frac{1}{8}\begin{bmatrix} 1 & 2 & 1 & 0 \\ 1 & 0 & 1 & 2 \end{bmatrix}$	$P_{11} = \frac{1}{8}\begin{bmatrix} 2 & 1 & 0 & 1 \\ 0 & 1 & 2 & 1 \end{bmatrix}$
9.	$P_{12} = \frac{1}{8}\begin{bmatrix} 1 & \frac{3}{2} & \frac{3}{2} & 0 \\ \frac{3}{2} & 0 & 0 & \frac{5}{2} \end{bmatrix}$	$P_{13} = \frac{1}{8}\begin{bmatrix} \frac{5}{2} & 0 & 0 & \frac{3}{2} \\ 0 & \frac{3}{2} & \frac{3}{2} & 1 \end{bmatrix}$
10.	$P_{14} = \frac{1}{8}\begin{bmatrix} 2 & 0 & 0 & 2 \\ 0 & 2 & 2 & 0 \end{bmatrix}$	$P_{15} = \frac{1}{8}\begin{bmatrix} 0 & 2 & 2 & 0 \\ 2 & 0 & 0 & 2 \end{bmatrix}$

$$P_{10} = \frac{1}{2}P_3 + \frac{1}{2}P_{15}$$
$$P_{12} = \frac{1}{4}P_5 + \frac{3}{4}P_{15}$$

Consequently, $P_6, P_8, P_{10}, P_{12} \,\varepsilon\, M_{\text{PUOD}}(\frac{1}{2},\frac{1}{2},\frac{1}{2})$. Also observe that

$$P_7 = \frac{1}{2}P_4 + \frac{1}{2}P_{14}$$
$$P_9 = \frac{1}{2}P_2 + \frac{1}{2}P_{14}$$
$$P_{11} = \frac{1}{2}P_3 + \frac{1}{2}P_{14}$$
$$P_{13} = \frac{1}{4}P_5 + \frac{3}{4}P_{14}$$

Consequently, $P_7, P_9, P_{11}, P_{13} \,\varepsilon\, M_{\text{PLOD}}(\frac{1}{2},\frac{1}{2},\frac{1}{2})$, and $P_i \,\varepsilon\, M_{\text{PLOD}}(\frac{1}{2},\frac{1}{2},\frac{1}{2}) \cap M_{\text{PUOD}}(\frac{1}{2},\frac{1}{2},\frac{1}{2}) = M_{\text{POD}}(\frac{1}{2},\frac{1}{2},\frac{1}{2})$ for $i = 1,2,\ldots,12,13$. The extreme point trivariate distribution P_{14} of $M_{\text{PLOD}}(\frac{1}{2},\frac{1}{2},\frac{1}{2})$ is not PUOD, because of (9). The extreme point trivariate distribution P_{15} of $M_{\text{PUOD}}(\frac{1}{2},\frac{1}{2},\frac{1}{2})$ is not PLOD, because of (5).

2. Because of the symmetry present in the probabilities $p_1 = \frac{1}{2} = p_2$, $q_1 = \frac{1}{2} = q_2$, and $r_1 = \frac{1}{2} = r_2$, the extreme points of $M_{\text{PUOD}}(\frac{1}{2},\frac{1}{2},\frac{1}{2})$ can be obtained from those of $M_{\text{PLOD}}(\frac{1}{2},\frac{1}{2},\frac{1}{2})$ by flipping 1 and 2 among the indices of P_{ijk}'s of P_i's, $i = 1,2,3,4,5,6,8,10,12,14$.

3. The distributions P_i's, $i = 1,2,\ldots,12,13$ are extreme points of $M_{\text{PLOD}}(\frac{1}{2},\frac{1}{2},\frac{1}{2}) \cap M_{\text{PUOD}}(\frac{1}{2},\frac{1}{2},\frac{1}{2})$.

4. If one wishes to construct a trivariate distribution P which is PLOD but not PUOD, one could use P_{14} as a building block. Look for convex combinations of P_{14} and some or all of $P_1,P_2,P_3,P_4,P_5,P_6,P_8,P_{10},P_{12}$. For instance, any convex combination $\lambda P_1 + (1-\lambda)P_{14}$ with $0 \leq \lambda < 1$ is PLOD but not PUOD, because of (9).

5. Note that the joint distribution P_5 is the upper Fréchet bound.

3. Concluding Remarks. The extreme point analysis of two natural definitions of positive quadrant dependence in three dimensions reveals that these two notions of dependence are not violently different in this $2 \times 2 \times 2$ case. Extreme point analysis is useful in evaluating the power function of any test proposed for testing independence of X, Y, and Z against strict positive quadrant dependence of X, Y, and Z. For details, in the case of 2 dimensions, see Subramanyam and Bhaskara Rao (1986). Also, certain measures of dependence can be shown to be affine functions over the sets M_{PLOD} and M_{PUOD}. This affine function property is useful to evaluate asymptotic power of tests based on these measures of dependence. All these ideas and an algebraic method for isolating extreme points of the sets M_{PLOD} and M_{PUOD} will be the subject matter of a forthcoming report.

REFERENCES

AHMED, A.N., LANGBERG, N.A., LÉON, R.V. and PROSCHAN, F. (1978). Two concepts of positive dependence with applications in multivariate analysis. Department of Statistics, Florida State University, AFOSR Technical Report No. 78–6.

BLOCK, H.W. and TING, M.L. (1981). Some concepts of multivariate dependence. *Commun. Statist.-Theor. Meth.* **A10** 749–762.

CHHETRY, D., KIMELDORF, G. and SAMPSON, A.R. (1989). Concepts of setwise dependence. *Probability in the Engineering and Information Sciences* **3** 367–380.

LEHMANN, E.L. (1966). Some concepts of dependence. *Ann. Math. Stat.* **37** 1137–1153.

NGUYEN, T.T. and SAMPSON, A.R. (1985). The geometry of certain fixed marginal probability distributions. *Linear Algebra and Its Applications* **70** 73–87.

SUBRAMANYAM, K. and BHASKARA RAO, M. (1986). Extreme point methods in the study of classes of bivariate distributions and some applications to contingency tables. Technical Report No. 86–12, Center for Multivariate Analysis, University of Pittsburgh.

DEPARTMENT OF MATHEMATICAL SCIENCES
UNIVERSITY OF NORTH CAROLINA
WILMINGTON, NC 28403

ON THE STRUCTURE OF $2 \times \infty$ BIVARIATE DISTRIBUTIONS WHICH ARE TOTALLY POSITIVE OF ORDER TWO

BY K. SUBRAMANYAM[1] AND M. BHASKARA RAO[1]

University of North Carolina at Wilmington and North Dakota State University

Let X and Y be two random variables such that X takes only two values 1 and 2. The notion of total positivity of order two for the joint probability distribution of X and Y is discussed in this paper from the viewpoint of convex analysis. The set of all $2 \times \infty$ probability measures which are totally positive of order two and with fixed second marginal probability measure is shown to be convex. Some of the extreme points of this set are explicitly spelled out, and an integral representation theorem in terms of extreme points is presented in a special case.

1. Introduction. Let X and Y be two random variables having a joint probability density function $f(\cdot,\cdot)$ with respect to some product probability measure λ on the Borel σ-field of R^2. The random variables X and Y are said to be totally positive of order two if the determinants

$$\begin{vmatrix} f(x,y) & f(x,y') \\ f(x',y) & f(x',y') \end{vmatrix}$$

are nonnegative for $-\infty < x \le x' < \infty$ and $-\infty < y \le y' < \infty$ a.e. $[\lambda]$. See Karlin (1968, p. 12). For its relation with other notions of dependence and further ramifications, see Barlow and Proschan (1981). See also Lehmann (1966).

The main purpose of this article is to perform extreme point analysis on the notion of total positivity of order two. What this means is that we look at the set of all bivariate probability density functions, examine convexity of this set, and if convex, enumerate all its extreme points. This kind of analysis was carried out on a limited scale in Subramanyam and Bhaskara Rao (1988). It was shown that the set of all bivariate probability density functions which are totally positive of order

[1]Supported by AFOSR contract AFSO–88–0030.

AMS 1980 subject classifications. Primary 60E05; secondary 62H05.

Key words and phrases. Bivariate distribution, total positivity of order two, convex set, compact set, extreme points.

The authors are very much grateful to the referee for many of his perceptive comments which led to a substantial improvement in the presentation of the paper.

two is not convex. The attention then was focused on the set of all $2 \times n$ bivariate distributions

$$\begin{pmatrix} p_{11} & p_{12} & \cdots & p_{1n} \\ p_{21} & p_{22} & \cdots & p_{2n} \end{pmatrix}$$

which are totally positive of order two and with fixed column marginals $p_{11}+p_{21} = q_1$, $p_{12}+p_{22} = q_2, \ldots, p_{1n}+p_{2n} = q_n$. The extreme points of this convex set were explicitly enumerated. The analysis of this convex set was found to be useful in testing certain hypotheses of independence and total positivity of order two.

The main thrust of this paper is in analyzing total positivity of order two in the realm of $2 \times \infty$ bivariate distributions. In Section 2, the set $M_\mu(\mathrm{TP}_2)$ of all $2 \times \infty$ probability measures which are totally positive of order two and with fixed second marginal probability measure μ is shown to be convex. Some of the extreme points of the set $M_\mu(\mathrm{TP}_2)$ are explicitly spelled out, and an integral representation of any given λ in $M_\mu(\mathrm{TP}_2)$ in terms of extreme points is presented in a special case. Some open questions are raised on the extreme points of the set $M_\mu(\mathrm{TP}_2)$.

2. Main Results. To begin with, we frame the definition of TP_2 in the language of probability measures. The basic notation is as follows. Let $\Omega = \{1,2\} \times R$. The space Ω consists of two lines $x = 1$ and $x = 2$ in R^2.

Let $\mathcal{C}$ be the Borel σ-field on R. We equip Ω with the following σ-field.

$$\mathcal{B} = \{A \subset \Omega;\ A = \{1\} \times B_1 \cup \{2\} \times B_2 \text{ for some } B_1 \text{ and } B_2 \text{ in } \mathcal{C}\}.$$

The above representation of A is unique.

For any probability measure μ on $\mathcal{B}$, let μ_1 and μ_2 denote the first and second marginal probability measures of μ on $\{1,2\}$ and $\mathcal{C}$, respectively, i.e.,

$$\begin{aligned} \mu_1(\{1\}) &= \mu(\{1\} \times R), \\ \mu_1(\{2\}) &= \mu(\{2\} \times R), \end{aligned}$$

and

$$\mu_2(B) = \mu(\{1,2\} \times B), \quad B \in \mathcal{C}.$$

For any two probability measures τ and ν on $\{1,2\}$ and $\mathcal{C}$, respectively, let $\tau \otimes \nu$ denote the product probability measure on $\mathcal{B}$. The probability measure $\tau \otimes \nu$ has the following explicit formula. For any $A = \{1\} \times B_1 \cup \{2\} \times B_2$ in $\mathcal{B}$ with B_1, $B_2 \in \mathcal{C}$,

$$(\tau \otimes \nu)(A) = \tau(\{1\})\nu(B_1) + \tau(\{2\})\nu(B_2).$$

For any two probability measures μ and λ, we use the notation $\mu << \lambda$ if μ is absolutely continuous with respect to λ.

For basic ideas on absolute continuity and product measures, see Halmos (1950).

DEFINITION. A probability measure μ on $\mathcal{B}$ is said to be totally positive of order two (TP_2) if the following determinants

$$D(y,y') = \begin{vmatrix} f(1,y) & f(1,y') \\ f(2,y) & f(2,y') \end{vmatrix}$$

are nonnegative almost surely $-\infty < y < y' < \infty$, where f is a version of the Radon-Nikodym derivative of μ with respect to some product probability measure $\tau \otimes \nu$ on $\mathcal{B}$ for which $\mu << \tau \otimes \nu$.

Some comments are in order on the above definition.

1. In the parlance of statistical theory, f is called a probability density function. One may wonder why one needs the dominating measure $\tau \otimes \nu$ to be a product measure in the above definition. If we were to allow any measure to dominate μ so as to get a density function, we could as well take μ itself as the dominating measure which gives the density function $f \equiv 1$. Then μ is TP_2 always! For the above definition to be nontrivial, we need to take the dominating measure to be a product measure. Moreover, the idea that μ is TP_2 is a deviation from independence, and to facilitate to measure the extent of deviation from independence one has to incorporate a product probability measure in the definition of TP_2. Thus a product probability measure enters the definition of TP_2 in the form of a dominating measure.

2. The statement that the determinants $D(y,y')$ are nonnegative almost surely $-\infty < y < y' < \infty$ requires some explanation. Let

$$U = \{(y,y') : -\infty < y < y' < \infty\}.$$

Let ν_U be the probability measure on the Borel σ-field of U defined by

$$\nu_U(A) = [\nu \otimes \nu(A)]/[\nu \otimes \nu(U)]$$

for every Borel subset A of U. If we let $A = \{(y,y') \varepsilon U : D(y,y') \geq 0\}$, then the TP_2 condition is equivalent to $\nu_U(A) = 1$.

3. It is assumed that $0 < \mu(\{1\}) < 1$ since, otherwise, μ is trivially TP_2.

4. The definition that μ being TP_2 can be rephrased purely in terms of μ dispensing totally with the necessity of working with probability density functions. The following is a result in that direction.

THEOREM 1. *A probability measure μ on $\mathcal{B}$ is TP_2 if and only if*

$$\mu(\{1\} \times [a,b])\mu(\{2\} \times [c,d]) \geq \mu(\{1\} \times [c,d])\mu(\{2\} \times [a,b])$$
$$\text{(1)} \qquad \textit{for all } -\infty < a \leq b < c \leq d < \infty.$$

Block, Savits, and Shaked (1982) frame the definition of TP_2 for probability measures. Their remarks (iii) and (iv) on page 767 are more or less tantamount to the statement of the above theorem. We will not give a proof of this result here.

REMARK. In the above theorem, one can have either open intervals or semi-open intervals in (1). In the terminology of random variables, the notion of TP_2 has the following description. Let X and Y be two random variables such that X takes values 1 and 2. Then X and Y are TP_2 if

$$\begin{aligned}&P(X=1,\ a<Y\leq b)P(X=2,\ c<Y\leq d)\\&\quad\geq P(X=1,\ c<Y\leq d)P(X=2,\ a<Y\leq b)\\&\quad\text{for all } -\infty<a\leq b<c\leq d<\infty.\end{aligned}$$

This implies that

$$P(X=1,\ Y\leq y)/P(X=2,\ Y\leq y)$$

is a decreasing function of y.

5. A natural product probability measure dominating μ is $\tau\otimes\mu_2$, where τ is a nontrivial measure on $\{1,2\}$. Since all such product measures are mutually absolutely continuous, it will be convenient for us to let $\tau(\{1\})=\frac{1}{2}=\tau(\{2\})$.

Convexity Property. The set of all probability measures on $\mathcal{B}$ each of which is TP_2 is not convex. Examples are easy to construct. See Subramanyam and Bhaskara Rao (1988). We look at the following subset. Let ν be a fixed probability measure on $\mathcal{C}$. Let $M_\nu(TP_2)$ be the set of all probability measures μ on $\mathcal{C}$ such that μ is TP_2 and $\mu_2=\nu$. We confine our attention to TP_2 measures whose second marginal is a fixed probability measure ν. For all μ in $M_\nu(TP_2)$, we take Radon-Nikodym derivatives with respect to the fixed product probability measure $\tau\otimes\nu$, where $\tau(\{1\})=\frac{1}{2}=\tau(\{2\})$.

We now study some of the properties of $M_\nu(TP_2)$.

PROPOSITION 1. *Let $\mu\in M_\nu(TP_2)$. Let f be a version of the Radon-Nikodym derivative of μ with respect to $\tau\otimes\nu$. Then*

$$\frac{1}{2}f(1,y)+\frac{1}{2}f(2,y)=1 \text{ for almost all } y\ [\nu].$$

PROOF. Observe that for every B in $\mathcal{C}$,

$$\begin{aligned}\nu(B) &= \mu_2(B) = \mu(\{1,2\} \times B) \\ &= \int_{\{1,2\}\times B} f(x,y)(\tau \otimes \nu)(d(x,y)) \\ &= \int_{\{1,2\}} \int_B f(x,y)\tau(dx)\nu(dy) \\ &= \int_B \left(\frac{1}{2}f(1,y) + \frac{1}{2}f(2,y)\right)\nu(dy).\end{aligned}$$

From this, the proposition follows:

PROPOSITION 2. *Let $\mu \in M_\nu(TP_2)$ and f a version of the Radon-Nikodym derivative of μ with respect to $\tau \otimes \nu$. Then*

(i) $f(1,y)$ is a decreasing function of y almost surely $[\nu_U]$

and

(ii) $f(2,y)$ is an increasing function of y almost surely $[\nu_U]$.

PROOF. Since $\mu \in M_\nu(\mathrm{TP}_2)$, $f(1,y)f(2,y') \geq f(1,y')f(2,y)$ a.s. $[\nu_U]$. Thus,

$$f(1,y)[1 - \frac{1}{2}f(1,y')] \geq f(1,y')[1 - \frac{1}{2}f(1,y)] \text{ a.s. } [\nu_U]$$

and so, $f(1,y) \geq f(1,y')$ a.s. $[\nu_U]$. Thus, (i) follows. (ii) is a consequence of (i) and Proposition 1.

THEOREM 2. *The set $M_\nu(TP_2)$ is a compact convex set. (Compactness is in the topology of weak* convergence.)*

PROOF. We first settle convexity. Let μ and λ belong to $M_\nu(\mathrm{TP}_2)$ and $0 \leq \alpha \leq 1$. Let f and g be versions of Radon-Nikodym derivatives of μ and λ, respectively, with respect to $\tau \otimes \nu$. Observe that $\alpha f + (1-\alpha)g$ is a version of the Radon-Nikodym derivative of $\alpha\mu + (1-\alpha)\lambda$ with respect to $\tau \otimes \nu$. Then a.s. $[\nu_U]$,

$$\begin{aligned}&[\alpha f(1,y) + (1-\alpha)g(1,y)][\alpha f(2,y') + (1-\alpha)g(2,y'] \\ - &[\alpha f(1,y') + (1-\alpha)g(1,y')][\alpha f(2,y) + (1-\alpha)g(2,y)] \\ &= [\alpha f(1,y) + (1-\alpha)g(1,y)][\alpha\left(1 - \frac{1}{2}f(1,y')\right) + (1-\alpha)\left(1 - \frac{1}{2}g(1,y')\right)] \\ &\quad -[\alpha f(1,y') + (1-\alpha)g(1,y')][\alpha\left(1 - \frac{1}{2}f(1,y)\right) + (1-\alpha)\left(1 - \frac{1}{2}g(1,y)\right)] \\ &= [\alpha f(1,y) + (1-\alpha)g(1,y)][1 - \frac{1}{2}\left(\alpha f(1,y') + (1-\alpha)g(1,y')\right)]\end{aligned}$$

$$-[\alpha f(1,y') + (1-\alpha)g(1,y')][1 - \frac{1}{2}(\alpha f(1,y) + (1-\alpha)g(1,y))]$$
$$= \alpha f(1,y) + (1-\alpha)g(1,y) - [\alpha f(1,y') + (1-\alpha)g(1,y')]$$
$$= \alpha[f(1,y) - f(1,y')] + (1-\alpha)[g(1,y) - g(1,y')]$$
$$\geq 0, \text{ by Proposition 2.}$$

For the compactness of $M_\nu(\mathrm{TP}_2)$, let $M_\nu = \{\mu : \mu_2 = \nu\}$. Since $\mu_1(\{1,2\}) = 1$ and $\mu(\{1,2\} \times A) = \mu_2(A)$, it is immediate that M_ν is compact. Thus, since $M_\nu(\mathrm{TP}_2) \subset M_\nu$, it suffices to show that $M_\nu(\mathrm{TP}_2)$ is closed. But, this is immediate from Block, Savits and Shaked's Remark (vii) (1982).

Extreme Points. Now we embark on determining the extreme points of the compact convex set $M_\nu(\mathrm{TP}_2)$ and obtain a representation of μ in $M_\nu(\mathrm{TP}_2)$ in terms of extreme points of $M_\nu(\mathrm{TP}_2)$. We are not entirely successful.

Let D be the support or spectrum of ν. D is the smallest closed subset of R with $\nu(D) = 1$. An equivalent description is: $x \in D$ if and only if $\nu\{(x-\epsilon, x+\epsilon)\} > 0$ for every $\epsilon > 0$.

For each $u \in D$, define μ_u on $\mathcal{B}$ by

$$\mu_u(\{1\} \times B_1 \cup \{2\} \times B_2) = \nu((-\infty, u] \cap B_1) + \nu((u, \infty) \cap B_2),$$
$$\text{for } B_1, B_2 \in \mathcal{C}.$$

It is easy to check that μ_u is a probability measure on $\mathcal{B}$ and $(\mu_u)_2 = \nu$. In an intuitive way, μ_u is built up on $\mathcal{B}$ by splitting ν into 2 parts: $\{1\} \times (-\infty, u]$ and $\{2\} \times (u, \infty)$. The compression of μ_u to a 2×2 table gives the following picture.

Description of μ_u			
$X \backslash Y$	$Y \leq u$	$Y > u$	Marginal sum
1	$\nu((-\infty, u])$	0	$\nu((-\infty, u]) = (\mu_u)_1(\{1\})$
2	0	$\nu((u, \infty))$	$\nu((u, \infty)) = (\mu_u)_1(\{2\})$
			1

Let the function $f_u : \{1,2\} \times R \to R$ be defined by

$$f_u(1,y) = 2 \text{ if } -\infty < y \leq u, \quad = 0 \text{ if } u < y < \infty$$

and

$$f_u(2,y) = 0 \text{ if } -\infty < y \leq u, \quad = 2 \text{ if } u < y < \infty.$$

It can be checked that f_u is a version of the Radon-Nikodym derivative of μ_u with respect to $\tau \otimes \nu$. From the description of f_u, it is clear that $\mu_u \in M_\nu(\mathrm{TP}_2)$.

Further, for distinct u_1 and u_2 in D, μ_{u_1} and μ_{u_2} are distinct. We now show that each μ_u is an extreme point of $M_\nu(\mathrm{TP}_2)$. Suppose $\mu_u = \alpha\mu + (1-\alpha)\lambda$ for some $\mu, \lambda \in M_\nu(\mathrm{TP}_2)$ and $0 \le \alpha \le 1$. Let f and g be versions of Radon-Nikodym derivatives of μ and λ, respectively, with respect to $\tau \otimes \nu$. Then

$$f_u(1,y) = \alpha f(1,y) + (1-\alpha)g(1,y) \text{ for almost all } y\ [\nu]$$

and

$$f_u(2,y) = \alpha f(2,y) + (1-\alpha)g(2,y) \text{ for almost all } y\ [\nu].$$

From Proposition 1 and the description of f_u, it follows that

$$f_u = f = g \text{ a.e. } [\tau \otimes \nu]$$

and

$$\mu_u = \mu = \lambda.$$

Now we come to the representation theorem. We need to distinguish several cases of D.

Case 1. D is bounded.

Let a and b be the left and right extremities of D, respectively. Note that a, $b \in D$. We distinguish two cases. Suppose a is an atom of ν, i.e., $\nu(\{a\}) > 0$. Define μ_{a^*} on $\mathcal{B}$ by

$$\mu_{a^*}(\{1\} \times B_1 \cup \{2\} \times B_2) = \nu(B_2) \text{ for all } B_1 \text{ and } B_2 \text{ in } \mathcal{C}.$$

The measure μ_{a^*} spreads the measure ν on line $x = 2$ leaving nothing for the line $x = 1$. Note that the probability measure μ_b spreads ν on the line $x = 1$ leaving nothing for the line $x = 2$. One can check that μ_{a^*} is an extreme point of $M_\nu(\mathrm{TP}_2)$ and distinct from μ_u for every u in D.

We conjecture that these are all the extreme points of $M_\nu(\mathrm{TP}_2)$. If this conjecture is true, then every measure μ in $M_\nu(\mathrm{TP}_2)$ is a mixture of extreme points of $M_\nu(\mathrm{TP}_2)$, i.e., there exists a probability measure λ on an appropriate σ-field on $D^* = \{a^*\} \cup D$ (which depends on μ) such that

$$\mu(A) = \int_{D^*} \mu_u(A)\lambda(du), \text{ for every } A \in \mathcal{B}. \tag{2}$$

The conjecture is true if D is finite. This can be shown as follows. Let $D = \{1, 2, \ldots, n\}$, $\mu \in M_\nu(\mathrm{TP}_2)$, $q_i = \nu(\{i\})$, $i = 1, 2, \ldots, n$, and $\mu(\{i, j\}) = p_{ij}$, $i = 1, 2$, $j = 1, 2, \ldots, n$. The Radon-Nikodym derivative of μ with respect to $\tau \otimes \nu$ works out to be

$$\begin{aligned} f(1,i) &= 2p_{1i}/q_i, \quad i = 1, 2, \ldots, n, \\ f(2,i) &= 2p_{2i}/q_i, \quad i = 1, 2, \ldots, n. \end{aligned}$$

The measure λ on $D^* = \{a^*\} \cup \{1, 2, \ldots, n\}$ is given by

$$\begin{aligned}
\lambda(\{a^*\}) &= 1 - p_{11}/q_1, \\
\lambda(\{1\}) &= p_{11}/q_1 - p_{12}/q_2, \\
\lambda(\{2\}) &= p_{12}/q_2 - p_{13}/q_3, \ldots, \\
\lambda(\{n-1\}) &= p_{1n-1}/q_{n-1} - p_{1n}/q_n, \\
\lambda(\{n\}) &= p_{1n}/q_n.
\end{aligned}$$

The representation (2) is then valid. See Subramanyam and Bhaskara Rao (1988).

The other possibility under Case 1 is that a is not an atom of ν. In this case, μ_{a^*} and μ_a are identical. There is no need to introduce μ_{a^*}. We again conjecture that the set of extreme points of $M_\nu(\mathrm{TP}_2)$ is precisely $\{\mu_u : u \in D\}$.

Case 2. D is unbounded.

Assume that D is unbounded on both sides. Introduce two new measures $\mu_{-\infty}$ and μ_∞ by

$$\mu_{-\infty}(\{1\} \times B_1 \cup \{2\} \times B_2) = \nu(B_1)$$

and

$$\mu_\infty(\{1\} \times B_1 \cup \{2\} \times B_2) = \nu(B_2) \text{ for all } B_1, B_2 \in \mathcal{C}.$$

Then $\mu_{-\infty}, \mu_\infty \in M$ (TP_2), $\mu_{-\infty}$ is concentrated on the line $x = 1$, μ_∞ on the line $x = 2$, and $\mu_{-\infty}$ and μ_∞ are extreme points of $M_\nu(\mathrm{TP}_2)$. We again conjecture that $\mu_{-\infty}$, μ_∞, μ_u, $u \in D$ are the only extreme points of $M_\nu(\mathrm{TP}_2)$.

The last case that D is unbounded on one side only can be discussed in a similar vein.

REFERENCES

BARLOW, R.E. and PROSCHAN, F. (1981). *Statistical Theory of Reliability and Life Testing.* To Begin With, Silver Spring, MD.

BLOCK, H.W., SAVITS, T.H., and SHAKED, M. (1982). Some concepts of negative dependence. *Ann. Prob.* **3** 765–772.

HALMOS, P.R. (1950). *Measure Theory.* Van Nostrand, New York.

KARLIN, S. (1968). *Total Positivity.* Stanford University Press, Stanford.

LEHMANN, E.L. (1966). Some concepts of dependence. *Ann. Math. Statist.* **43** 1137–1153.

SUBRAMANYAM, K. and BHASKARA RAO, M. (1988). Analysis of odds ratios in $2 \times n$ ordinal contingency tables. *J. Multiv. Anal.* **27** 478–493.

DEPARTMENT OF MATHEMATICAL SCIENCES
UNIVERSITY OF NORTH CAROLINA
WILMINGTON, NC 28403

DEPARTMENT OF STATISTICSS
NORTH DAKOTA STATE UNIVERSITY
FARGO, ND 58105

ON STOCHASTIC DEPENDENCE AND A CLASS OF DEGENERATE DISTRIBUTIONS

By Richard A. Vitale[1]

University of Connecticut

We investigate the approximation of stochastic dependence by functional relationships involving so-called cyclic permutations of the interval.

1. Introduction. Dependence between two random variables can take, of course, a variety of forms, of which stochastic independence and functional dependence can be argued to be most opposite in character. In the one case, neither variable provides any information about the other, whereas in the second case there is complete determination (or *complete dependence*: Lancaster, 1963). By means of a direct construction for uniform variables, Kimeldorf and Sampson (1978) showed, however, that one can pass continuously from one to the other of these situations in the natural sense of weak convergence. This obviously weakens complete dependence as a foil for independence (and led Kimeldorf and Sampson, 1978, to the fruitful concept of monotone dependence). Indeed, couched in somewhat different language, Theorem 1 of Brown (1966) can be read to state that *any* form of dependence between uniform random variables can be approximated in the weak sense by functionally related random variables.

On the other hand, this raises the question, of theoretical and obvious computational interest, of the extent to which complete dependence can be used to approximate forms of stochastic dependence. We pursue this in several directions. First we show that functional dependence can be specified to a highly stylized class of invertible functions, the so-called cyclic permutations of the interval. Second, we show that it is possible to move from two to an arbitrary finite number of random variables. Finally we extend to arbitrary (continuous) marginals. Regarding the last point, we systematically take the viewpoint of fixing marginals and consider dependence within this constraint; for the narrower question of regression in this context, see Vitale and Pipkin (1976) and Vitale (1979). We make extensive use of the *uniform representation* of random variables (Kimeldorf and Sampson, 1975).

[1]Research supported in part by NSF Grant DMS 8603944.

AMS 1980 subject classifications (1985 revision). Primary 60E05; secondary 28D05, 60B10, 62E10, 62H05.

Key words and phrases. Complete dependence, functional dependence, maximally dependent collection, measure-preserving transformation, permutation, stochastic independence.

For various references I am indebted to P. Shields and R. Sine. M. Klass offered stimulating comments.

In the next section, we set out some definitions and make some preliminary comments. Theorem 1 of Section 3 states that any pair of uniform random variables can be approximated in distribution by a second pair which exhibits (invertible) functional dependence. Although Theorem 2 of Section 5 strictly includes this result, we present a proof in detail to give an idea of what the hands-on analysis looks like and, in particular, to present a version of the construction of Brown (1966). The interested reader may like to compare Theorem 1 to results of Garsia (1976) and Holbrook (1981) which give exact, but generally non-invertible, relationships in the case of an independent pair of random variables. Section 4 contains remarks on Theorem 1. Using an approach that differs from those of both Brown (1966) and Kimeldorf and Sampson (1978), Section 5 takes up the general case of approximating a collection of random variables with continuous marginals but otherwise arbitrary joint distribution. Section 6 relates the foregoing to the extremal distributions of Hoeffding. Discrete approximating distributions occupy Section 7, and a result of Fairley, Pearl, and Verducci (1987) is sharpened in Section 8. Section 9 concludes with a construction of a canonical sequence (Lai and Robbins, 1976, 1978) which is degenerate in a stronger sense than previous examples.

2. Notation and Preliminaries. We shall deal with Borel maps of the line (or the interval) equipped with Lebesgue measure $m(\cdot)$. A Borel map T from the interval to itself such that $m(T^{-1}(B)) = m(B)$ for any Borel B will be called *measure-preserving* and the entire collection denoted T. Within T, we consider the class $\mathrm{T}_{\mathrm{inv}}$ of measure-preserving maps T which are invertible, i.e., T is $1-1$ and T^{-1} is Borel and measure-preserving as well.

An interval of the form $((j-1)/n, j/n)$ for some $n \geq 1$ and $1 \leq j \leq n$ will be called a *dyadic interval of rank n*. Among invertible measure-preserving maps, we call T a *permutation of rank n* if it maps by a translation each dyadic interval of rank n onto a dyadic interval of rank n. This specifies T except for its values at the end-points of the subintervals; we allow these values to be assigned in any way that makes T one-to-one. We shall also refer to usual permutations $\pi : \{1, \ldots, n\} \to \{1, \ldots, n\}$, but there should be no confusion. Recall that π is *cyclic* if it has a single closed cycle (i.e., of length n, which need not be traversed in the natural order). We call a permutation T *cyclic*, and denote the entire class by $\mathrm{T}_{\mathrm{cyc}}$, if as a map of dyadic intervals it similarly has a single cycle. Our aim in focusing on the last class of maps is two-fold. One is to produce a simple universal model for (approximate) dependence among random variables; the second is to prepare the way for future work of a computational nature (see Section 7). These definitions are drawn from Halmos (1956) to which we refer later.

The following is standard.

PROPOSITION 2.1. *Suppose that* $\{E_1, \ldots, E_k\}$ *is a Borel partition of* $[0,1]$ *and that* $\delta > 0$ *is given. There is a partition* $\{F_1, \ldots, F_k\}$ *of* $[0,1]$ *in which each* F_i *is a finite union of intervals,* $m(F_i) = m(E_i)$, *and* $m(E_i \triangle F_i) < \delta$, $i = 1, \ldots, k$. *(Here* $\triangle$ *denotes symmetric difference.)*

3. Dependence in the Square.

THEOREM 1. *Let U and V be uniformly distributed variables. There is a sequence of cyclic permutations $T_1, T_2, \ldots$ such that (U, T_nU) converges in distribution to (U, V) as $n \to \infty$.*

PROOF. The proof consists of two parts. First we divide $[0,1] \times [0,1]$ into subsquares and adapt an argument of Brown (1966) to find a permutation T such that the distributions of (U, V) and (U, TU) coincide on subsquares. If T is not a cyclic permutation, then we proceed to find such an approximation to it; for this, Halmos (1956, p. 56) would suffice, but we supply a direct proof for completeness. Finally, we note that reducing the size of the sub-squares finishes the proof.

PART ONE. Let n, a power of 2, be fixed, and let $I_j = ((j-1)/n, j/n)$, $j = 1, \ldots, n$. We first produce a $T \varepsilon \text{T}_{\text{inv}}$ such that

$$P(U \varepsilon I_j, TU \varepsilon I_k) = P(U \varepsilon I_j, V \varepsilon I_k) \equiv p_{jk}, \quad j, k = 1, \ldots, n. \tag{1}$$

Define two systems of subintervals

$$\begin{aligned} I_{j1} &= \left(\frac{(j-1)}{n}, \frac{j-1}{n} + p_{j1}\right) \\ I_{j2} &= \left(\frac{j-1}{n} + p_{j1}, \frac{j-1}{n} + p_{j1} + p_{j2}\right) \quad 1 \le j \le n \\ &\vdots \\ I_{jn} &= \left(\frac{j-1}{n} + p_{j1} + \ldots + p_{j,n-1}, j/n\right) \end{aligned}$$

and

$$\begin{aligned} \tilde{I}_{j1} &= \left(\frac{j-1}{n}, \frac{j-1}{n} + p_{1j}\right) \\ \tilde{I}_{j2} &= \left(\frac{j-1}{n} + p_{1j}, \frac{j-1}{n} + p_{1j} + p_{2j}\right) \quad 1 \le j \le n \\ &\vdots \\ \tilde{I}_{jn} &= \left(\frac{j-1}{n} + p_{1j} + \ldots + p_{n-1,j}, j/n\right) \end{aligned}$$

Note that there is a coincidence of Lebesgue measure, $m(I_{jk}) = m(\tilde{I}_{kj})$. The invertible map T which sends each I_{jk} onto $\tilde{I}_{jk}$, $j, k = 1, \ldots, N$ by a translation is what we need (a pencil sketch will help to see this).

Observe that if each p_{jk} is a dyadic rational, then T is a permutation of the interval. If not, we approximate: given $\varepsilon > 0$, find dyadic rationals $\tilde{p}_{jk}$ such that

$$| p_{jk} - \tilde{p}_{jk} | < \varepsilon, \quad j, k = 1, \ldots, n \tag{2}$$

under the constraints (for eventual uniform marginals)

$$\sum_{j=1}^{n} \tilde{p}_{jk} = 1/n, \quad k = 1, \ldots, n \tag{3}$$

and

$$\sum_{k=1}^{n} \tilde{p}_{jk} = 1/n, \quad j = 1, \ldots, n. \tag{4}$$

This is a tricky problem to solve directly, but it yields to an appeal to Birkhoff's (1946) theorem (see, for example, Roberts and Varberg, 1973): if P is the $n \times n$ matrix with (j,k) entry $p_{j,k}$, then it can be written as a convex combination of permutation matrices

$$P = \sum_{i=1}^{n!} \theta_i \Pi^{(i)}, p_{jk} = [P]_{jk} = \sum_{i=1}^{n!} \theta_i [\Pi^{(i)}]_{jk} \,.$$

Now the θ_i's can be varied at will (subject to $\theta_i \geq 0$ and $\sum_{i=1}^{n!} \theta_i = 1$) and the constraints (3), (4) will still hold. Accordingly, find dyadic rationals $\tilde{\theta}_i$, $i = 1, \ldots, n$, so that $\tilde{\theta}_i \geq 0$ and $| \theta_i - \tilde{\theta}_i | < \varepsilon/n!$, $i = 1, \ldots, n!$. It follows that for each j, k

$$\tilde{p}_{jk} = \sum_{i=1}^{n!} \tilde{\theta}_i [\Pi^{(i)}]_{jk}$$

is within ε of p_{jk}, and the marginal constraints are satisfied. With this done, we proceed as before to get a permutation $\tilde{T}$ which satisfies

$$| P(U \varepsilon I_j, V \varepsilon I_k) - P(U \varepsilon I_j, \tilde{T} U \varepsilon I_k) | < \varepsilon, \quad j,k = 1, \ldots, n. \tag{5}$$

Part Two. We next want to arrange for a *cyclic* permutation. Note that if $\tilde{T}$ is a permutation of rank N, then it is also a permutation of rank $\hat{N}$ for any $\hat{N} > N$. We exploit this by choosing $\hat{N}$ very large and modifying $\tilde{T}$ on a small number of dyadic intervals of rank $\hat{N}$. More precisely, given $\varepsilon > 0$, we shall find a cyclic permutation $\tilde{\tilde{T}}$ such that

$$P(\tilde{T} U \varepsilon I_j, \tilde{\tilde{T}} \not\varepsilon I_j) + P(\tilde{T} U \not\varepsilon I_j, \tilde{\tilde{T}}\, U \varepsilon I_j) < \varepsilon, \quad j = 1, \ldots, n. \tag{6}$$

The details follow, but first we must modify $\tilde{T}$ so as to leave no dyadic interval invariant. Suppose that $\tilde{T}$ leaves $(0, k/2^N)$ invariant (i.e., it acts as the identity). Then redefine $\tilde{T}$ on this interval to be $\tilde{T}x = x + r \bmod k/2^N$ for some small dyadic rational r. A similar procedure can be done for other invariant intervals. It is clear that the new $\tilde{T}$ can be constructed so as to satisfy (6) in place of $\tilde{\tilde{T}}$ and with ε replaced by, say, 2ε.

Assume that this adjustment has been done and the modified function is $\tilde{T}$. For ease of notation, we retain N as the rank of $\tilde{T}$. Suppose that $\tilde{T}$ has p cycles of length $h_1, h_2, \ldots, h_p$ ($p \leq 2^{N-1}$ since each cycle contains at least two intervals). If we instead consider $\tilde{T}$ as of rank $\hat{N} > N$, then it has p cycles of length $h_1 2^{\hat{N}-N}, h_2 2^{\hat{N}-N}, \ldots, h_p 2^{\hat{N}-N}$.

We snip these cycles and patch them together. In cycle #1, choose one of the $h_1 2^{\hat{N}-N}$ intervals to be the "out-interval" and its image under $\tilde{T}$ to be the "in-interval." Now define $\tilde{T}$ on in- and out-intervals so that it maps the out-interval of cycle $\#j$ onto the in-interval of cycle $\#j+1$, $j = 1, \ldots, p-1$, and the out-interval of cycle $\#p$ onto the in-interval of cycle #1. The resulting permutation $\approx{T}$ is cyclic and differs from $\tilde{T}$ on at most a set of measure $p2^{-\hat{N}}$. It follows that, with $\hat{N}$ suitably large, (6) holds as does (5) with $\tilde{T}$, ε replaced by $\approx{T}$, 2ε respectively.

Finally, we note that the required sequence $\{T_n\}$ is obtained by effecting the construction for $\approx{T}$ for successive values $n = 1, 2, 4, 8, \ldots, 2^j, \ldots$ so that (5) and (6) are satisfied with $\varepsilon = \varepsilon_n = o(1/n^2)$ at each step.

4. Discussion. Especially in the case when U and V are independent, Theorem 1 calls for some explanation. How is it possible, after all, to arrive at a limiting pair of independent variables when at each stage the components of each approximating pair stand in an invertible, functional relationship to one another? We refer the reader to the discussion of Kimeldorf and Sampson (1978) and provide some other remarks here.

One ingredient which might be questioned is the mode of convergence. Convergence in distribution may be insufficiently stringent. If, for instance, convergence in the variation metric is substituted then the theorem obviously fails. On the other hand, so important a mode as convergence in distribution ought not to be easily dismissed. It is after all central to questions of sampling. In this context, given any sample (u_i, v_i), $i = 1, \ldots, N$ in the square with the u_i's distinct there is a $T \varepsilon \mathrm{T}_{\mathrm{inv}}$ which fully "explains" the sample in the sense that $Tu_i = v_i$, $i = 1, \ldots, N$.

This leads to a related comment. Granted that the components of each approximating pair stand in a functional relationship to one another, the function itself may be so wild that in practical terms it is not feasible as a predictive device. That is, small errors in U may lead to large errors in $T_n U$. To combat this, one may try to refine the process. But this seems to butt up against an inevitable problem of computational complexity. In this sense, T_n cannot be used for prediction in any meaningful way. A possible rejoinder to these comments is that rather than using T_n pointwise one should smooth it slightly so as to get, in effect, a smoothed regression function. We take up an aspect of regression in a later section.

A different, and provocative, comment has been offered by M. Klass. As we have said, the puzzling note is that the theorem makes an asymptotic statement in which there is an abrupt change of behavior *at* the limit. This type of phenomenon often occurs when one is dealing with a "large" space where there is flexibility for discontinuous behavior to occur. The key here is the roominess of $[0, 1]$. A smaller domain, or what is the same, random variables with "fewer" values illustrates the point. Take, for instance, U and V to be independent, $p = 1/2$, Bernoulli variables. Then the analogous theorem fails. It is clear that there is no sequence, convergent in distribution to (U, V), of pairs $(U, T_n U)$ where $T_n : \{0,1\} \to \{0,1\}$. This is true more generally for any pair of discrete random variables which are not already in a relationship of functional dependence. The case of random variables U and V, each uniformly distributed on $\{1, 2, \ldots, N\}$, but with otherwise arbitrary joint distribution, is interesting to consider. There is generally *no* approximating $(U, T_n U)$, as we have said, but we are always within a randomization of matching the joint distribution exactly. Consider that Birkhoff's theorem provides a set $\{\Pi_1, \ldots, \Pi_M\}$ of permutations of $\{1, \ldots, N\}$ and probabilities $\theta_1, \ldots, \theta_M$, $\theta_i > 0$, $\Sigma\theta_i = 1$ such that if X is chosen at random in $\{1, 2, \ldots, N\}$ and if J is chosen equal to j with probability θ_j, $j = 1, \ldots, M$, then the pair $(U, \Pi_j U)$ is distributed like (U, V).

5. General Form. Theorem 1 can be generalized in two ways: relaxation of the condition of uniform marginals and, what we take up first, treatment of an arbitrary number of uniform random variables. This generalizes Theorem 1; now we take some shortcuts in the proof.

THEOREM 2. *Let $U_1, \ldots, U_d$ be a collection of uniform random variables. There is a sequence of vectors converging in distribution to $(U_1, \ldots, U_d)$ of the form $(T_1 U, T_2 U, \ldots, T_d U)$ where U is uniform and $T_1, \ldots, T_d$ are cyclic permutations.*

PROOF. It is well-known that there are $S_1, S_2, \ldots, S_d \varepsilon$ T such that $(S_1 U, S_2 U, \ldots, S_d U)$ is distributed like $(U_1, \ldots, U_d)$ (see, for example, Billingsley, 1971, Theorem 3.2).

We proceed to approximate each S_i by an invertible map. Consider S_1. Let $E_j = S_1^{-1}((j-1)/n, j/n)$, $j = 1, \ldots, n$. Proposition 2.1 provides a partition $\{F_1, \ldots, F_n\}$ (each F_j a finite union of intervals) such that $m(E_j \triangle F_j) < \delta$. Create $\tilde{S}_1$ by taking a $1-1$ piecewise linear map of the interior of F_j onto $((j-1)/n, j/n)$, $j = 1, \ldots, n$. If $\max\{\delta, 1/n\}$ is small, then $\tilde{S}_1 U$ is close to $S_1 U$ in probability. Since $\tilde{S}_1$ is invertible, Halmos (1956, p. 65) applies and ensures a cyclic permutation T_1 such that $T_1 U$ is close to $\tilde{S}_1 U$, and, from our construction, to $S_1 U$ in probability. Proceeding similarly for $S_2, S_3, \ldots, S_n$ completes the argument.

COROLLARY. *The approximating random vectors in the theorem may also be taken of the form $(U, \tilde{T}_2 U, \tilde{T}_3 U, \ldots, \tilde{T}_d U)$ where $\tilde{T}_2, \tilde{T}_3, \ldots, \tilde{T}_d$ are cyclic permutations.*

PROOF. $(T_1 U, T_2 U, \ldots, T_d U) \stackrel{d}{=} (U, T_2 T_1^{-1} U, \ldots, T_d T_1^{-1} U)$ and define $\hat{T}_j = T_j T_1^{-1}$, $j = 2, \ldots, d$. If $\hat{T}_j$ is a cyclic permutation, then set $\tilde{T}_j = \hat{T}_j$. Otherwise, let $\tilde{T}_j$ be a cyclic permutation such that $\tilde{T}_j U$ and $\hat{T}_j U$ are close in probability as argued above.

We turn next to our central result, which treats more general marginals. In view of the discussion in Section 4, we fix attention on continuous marginals. We recall a standard definition.

DEFINITION. The INVERSE of a distribution function F is given by $F^{-1}(u) = \inf\{x \mid u \leq F(x)\}$.

THEOREM 3. *Let $X_1, \ldots, X_d$ be random variables with continuous marginal distributions $X_j \sim F_j$, $j = 1, \ldots, d$. There is a sequence of random vectors, converging in distribution to $(X_1, \ldots, X_d)$, of the form*

$$(F_1^{-1}(T_1U), F_2^{-1}(T_2U), \ldots, F_d^{-1}(T_dU))$$

where $T_1, \ldots, T_d$ are cyclic permutations.

PROOF. Set $U_1 = F_1(X_1), \ldots, U_d = F_d(X_d)$ as uniform random variables and note that

$$(X_1, \ldots, X_d) = (F_1^{-1}(U_1), \ldots, F_d^{-1}(U_d)) \quad \text{a.s.}$$

The event $(F_1^{-1}(U_1) \leq x_1, \ldots, F_d^{-1}(U_d) \leq x_d)$ is the same (up to an event of probability zero) as $(U_1 \leq F_1(x_1), \ldots, U_d \leq F_d(x_d))$ whose image in $[0,1]^d$ is a continuity set of the joint measure of $(U_1, \ldots, U_d)$ (e.g., Billingsley, 1971, p. 3). It follows that if a sequence of random vectors of the form $(T_1U, \ldots, T_dU)$ converges in distribution to $(U_1, \ldots, U_d)$, then the same holds for $(F_1^{-1}(T_1U), \ldots, F_d^{-1}(T_dU))$ and $(X_1, \ldots, X_d) = (F_1^{-1}(U_1), \ldots, F_d^{-1}(U_d))$. Together with Theorem 2 this concludes the proof.

The variable U serves to parameterize each approximating vector. It can be removed by observing that if we set $Y_j = F_j^{-1}(T_jU)$, then $U = T_jF_j(Y_j)$ a.s. and hence $Y_k = F_k^{-1} \circ T_k \circ T_jF_j(Y_j)$ a.s. As in the corollary to Theorem 2, $T_k \circ T_j$ may be adjusted to be a cyclic permutation.

COROLLARY. *The approximating vectors in the theorem may be taken to be of the form*

$$(X_1, F_2^{-1} \circ T_2 \circ F_1(X_1), \ldots, F_d^{-1} \circ T_d \circ F_1(X_1))$$

where $T_2, \ldots, T_d$ are cyclic permutations.

6. Remarks. It is interesting to link up Theorem 3 with the extremal distributions of Hoeffding (1940) (see also, Frechet, 1951; Whitt, 1976; Tchen, 1980). They relate to the following question: Among all random vectors (X, Y) with $X \sim F$ and $Y \sim G$, when is $\text{Cov}(X, Y)$ smallest and largest? (F and G are assumed to yield finite second moments). The answers use $T_{\min} \varepsilon\ \mathrm{T}_{\text{inv}}$ and $T_{\max} \varepsilon\ \mathrm{T}_{\text{inv}}$ respectively:

$$(F^{-1}(U), G^{-1}(T_{\min}U)),\ T_{\min}u = 1 - u,$$

and

$$(F^{-1}(U), G^{-1}(T_{\max}U)),\ T_{\max}u = u,\ \ 0 \leq u \leq 1.$$

Thus Theorem 3 can be thought of as embedding the (degenerate) distributions of Hoeffding in a class dense in the collection of all distributions with the specified marginals.

An amusing note regarding Theorem 3 is a quick answer to the ancient classroom problem of displaying a random vector (X, Y) which has normal marginals but is not bivariate normal (Vitale, 1978). It is enough to take $X \sim N(0,1)$ and $Y = N^{-1} \circ T \circ N(X)$ where $T \varepsilon$ T is *anything but* $T_{\min}$ or $T_{\max}$, say, $Tu = u + 1/2$ mod 1.

7. Discretization. While it is true, as discussed before, that Theorem 1 (and hence Theorem 3) has no general analogue for discrete marginal distributions, the result itself can be discretized.

THEOREM 4. *Suppose that* $(X_1, \ldots, X_d)$ *is a random vector with continuous marginals* $F_1, \ldots, F_d$ *respectively. Then there is a sequence of random vectors converging in distribution to* $(X_1, \ldots, X_d)$ *of the form*

$$(F_1^{-1}(\Pi_1(J)/n), \ldots, F_d^{-1}(\Pi_d(J)/n)$$

where $n \geq 1$, J *is uniform on* $\{1, 2, \ldots, n\}$ *and* $\Pi_1, \ldots, \Pi_d$ *are cyclic permutations of* $\{1, 2, \ldots, n\}$.

PROOF. We adapt the argument of Theorem 3. With the uniform variable U given, define J via $J = j$ if $U \varepsilon ((j-1)/n, j/n)$. Also, let $\Pi_\ell(r) = s$ if T_ℓ takes the rth dyadic subinterval onto the sth dyadic subinterval. It follows that $((\Pi_1(J)/n), \ldots, (\Pi_d(J)/n))$ approximates $(U_1, \ldots, U_d)$ in distribution and this suffices.

COROLLARY. *The approximating vectors may be taken of the form*

$$(F_1^{-1}(J/n),\ F_2^{-1}(\Pi_2(J)/n), \ldots, F_d^{-1}(\Pi_d(J)/n)$$

with n, J, $\Pi_2, \ldots, \Pi_d$ *as given in the theorem.*

An interesting special case is that of a Markov chain.

COROLLARY. *Let* $X_1, X_2, \ldots$ *be a discrete time Markov chain in equilibrium with stationary continuous density* F. *Then, for any* d, $(X_1, \ldots, X_d)$ *can be approximated in distribution by*

$$(F^{-1}(J/n),\ F^{-1}(\Pi(J)/n),\ F^{-1}(\Pi^2(J)/n), \ldots, F^{-1}(\Pi^{d-1}(J)/n))$$

with n, J *as in the theorem and* Π *a cyclic permutation of* $\{1, 2, \ldots, n\}$.

These results suggest a practical method for generating random vectors with arbitrary dependence structure among components. Aside from evaluation of inverse distribution functions and a single uniform variate, all that is needed are the appropriate permutations of $\{1, 2, \ldots, n\}$. How these permutations are determined seems to be an interesting question.

8. The Penalty for Using Linear Regression. In an interesting study, Fairley, Pearl, and Verducci (1987) look at the penalty incurred using various forms of constrained regression. In particular, the expected unexplained variation left from linear prediction can be partitioned into an "intrinsic variation" component and an "extra-linear variation" component

$$(7) \qquad E(Y - \rho X)^2 = E(Y - \phi(X))^2 + E(\phi(X) - \rho X)^2, \; \phi(X) = E[Y \mid X].$$

They point out that it is of interest to bound the first, $\eta = E(Y - \phi(X))^2$, and note that it can be seen arbitrarily close to 0 under certain conditions. Their technique uses data (i.e., point-mass) distributions as approximants. The machinery we have developed provides a stronger form of their result.

THEOREM 5. *Suppose that a bivariate distribution is given with marginals F and G and correlation ρ. Then there is another bivariate distribution with marginals uniformly close to those of the first, correlation arbitrarily close to ρ, and for which $\eta = 0$.*

PROOF. First slightly smooth the given distribution so as to have continuous marginals and then truncate it to a large rectangle. By the corollary to Theorem 3, the resulting distribution has a degenerate distribution close to it and for which $\eta = 0$.

9. Maximally Dependent Random Variables. Suppose that $X_1, X_2, \ldots, X_n$ are independent, identically distributed random variables with common distribution F. They are said to be *maximally dependent* if

$$(8) \qquad P(M_n > x) = \min\{1, n(1 - F(x))\}$$

where $M_n = \max\{X_1, \ldots, X_n\}$. The expression on the right in (8) is clearly the largest conceivable value for $P(M_n > x)$. The study of maximally dependent random variables was inaugurated in Lai and Robbins (1976) and continued in Lai and Robbins (1978). Elaborations and generalizations appear in Tchen (1980) and Rüschendorf (1981).

An important question is the existence of a *canonical* sequence; that is, $X_1, X_2, \ldots, (X_i \sim F \forall i)$ such that, for each n, $X_1, \ldots, X_n$ are maximally dependent. Lai and Robbins (1978) establish existence (and non-uniqueness) in the case when F is continuous (see Tchen, 1980, for the case F discontinuous). They argue that an application of the inverse distribution function renders it sufficient to consider the case of uniform variables and then they provide a direct construction. We show that it is possible to construct a different canonical sequence in which the joint distributions completely degenerate in the sense of our foregoing discussion.

PROPOSITION 9.1. *There is a canonical sequence $U_1, U_2, \ldots$ of uniform random variables such that for each $i \geq 2$*

$$U_i = T_i U_1 \text{ where } T_i \varepsilon \text{ T}_{\text{inv}} \text{ is piecewise linear.} \tag{9}$$

PROOF. We use Lai and Robbins' (1978) observation that a collection $U_1, \ldots, U_n$ of uniform random variables is maximally dependent if and only if M_n is uniformly distributed on $(1 - 1/n, 1)$.

We proceed by induction. The case $n = 1$ is trivial. Assume that $U_1, \ldots, U_n$ have been constructed in the form (9). Note that

$$B = \{u\varepsilon[0,1] \mid \max\{u, T_2 u, \ldots, T_n u\} < 1 - 1/(n+1)\}$$

is a union of intervals and that, by Lai and Robbins' criterion, $m(B) = \frac{1}{n+1}$. Construct U_{n+1} by mapping the interior of B in a piecewise linear manner onto $\left(1 - \frac{1}{n+1}, 1\right)$. Elsewhere define U_{n+1} to be piecewise linear so as to be finally $1-1$ and measure-preserving. It follows that M_{n+1} is uniform on $\left(1 - \frac{1}{n+1}, 1\right)$ and hence $U_1, \ldots, U_{n+1}$ is a maximally dependent collection.

REFERENCES

BILLINGSLEY, P. (1971). Weak Convergence of Measures: Applications in Probability. Regional Conference Series in Applied Mathematics No. 5, SIAM, Philadelphia.

BIRKHOFF, G. (1946). Tres observaciones sobre et lineal. *Univ. Nac. Tucumán Rev. Ser. A* **5** 147–150.

BROWN, J.R. (1966). Approximation theorems for Markov operators. *Pac. J. Math.* **16** 13–23.

FAIRLEY, D., PEARL, D.K., and VERDUCCI, J.S. (1987). The penalty for assuming that a monotone regression is linear. *Ann. Statist.* **15** 443–448.

FRÉCHET, M. (1951). Sur les tableaux de corrélation dont les marges sont données. *Ann. Univ. Lyon Sect. A* **14** 53–77.

GARSIA, A.M. (1976). Combinatorial inequalities and smoothness of functions. *Bull. Amer. Math. Soc.* **82** 157–170.

HALMOS, P. (1956). *Lectures on Ergodic Theory.* Chelsea, New York.

HARDY, G.H., LITTLEWOOD, J.E., and PÓLYA, G. (1952). *Inequalities*, 2nd ed. Cambridge University Press, Cambridge.

HOEFFDING, W. (1940). Masstabinvariante Korrelationstheorie. *Schriften des Mathematischen Instituts und des Instituts fur Angewandte Mathematik der Universität Berlin* **5** 179–233.

HOLBROOK, J.A.R. (1981). Stochastic independence and space-filling curves. *Amer. Math. Monthly* **88** 426–432.

KIMELDORF, G. and SAMPSON, A.R. (1975). Uniform representation of bivariate distributions. *Comm. Statist.* **4** 617–627.

KIMELDORF, G. and SAMPSON, A.R. (1978). Monotone dependence. *Ann. Statist.* **6** 895–903.

LAI, T.L. and ROBBINS, H. (1976). Maximally dependent random variables. *Proc. Nat. Acad. Sci. USA* **73** 286–288.

LAI, T.L. and ROBBINS, H. (1978). A class of dependent random variables and their maxima. *Z. Wahrsch. verw. Geb.* **42** 89–111.

LANCASTER, H.O. (1963). Correlation and complete dependence of random variables. *Ann. Math. Stat.* **34** 1315–1321.

ROBERTS, A.W. and VARBERG, D.E. (1973). *Convex Functions.* Academic Press, New York.

RÜSCHENDORF, L. (1981). Sharpness of Fréchet-bounds. *Z. Warsch. verw. Geb.* **57** 293–302.

TCHEN, A. (1980). Inequalities for distributions with given marginals. *Ann. Statist.* **8** 814–827.

VITALE, R.A. (1978). Joint vs. individual normality. *Math. Mag.* **51** 123.

VITALE, R.A. (1979). Regression with given marginals. *Ann. Statist.* **7** 653–658.

VITALE, R.A. and PIPKIN, A.C. (1976). Conditions on the regression function when both variables are uniformly distributed. *Ann. Prob.* **4** 869 873.

WHITT, W. (1976). Bivariate distributions with given marginals. *Ann. Statist.* **4** 1280–1289.

DEPARTMENT OF STATISTICS
UNIVERSITY OF CONNECTICUT
STORRS, CT 06269

ESTIMATING A DISTRIBUTION FUNCTION BASED ON MINIMA-NOMINATION SAMPLING

BY MARTIN T. WELLS AND RAM C. TIWARI

Cornell University and University of North Carolina

The nonparametric maximum likelihood estimator of a distribution function based on a maxima-nomination sample has been derived recently by Boyles and Samaniego (1986). In this article we study minima-nominations for the case of censored data.

1. Introduction. Let $X_{i1}, \ldots, X_{iK_i}$, $i = 1, \ldots, n$ be independent identically distributed (i.i.d.) random variables (r.v.'s) having a common continuous distribution function F with support $(0,\infty)$. Denote the vector $(X_{i1}, \ldots, X_{iK_i})$ by $\mathbf{X}_i$, $i = 1, \ldots, n$. Define the map $\Pi_i : \mathbb{R}^{K_i} \to \mathbb{R}$ such that Π_i maps $\mathbf{X}_i$ into a particular element in $\mathbf{X}_i$, say X_i $(i = 1, \ldots, n)$. We shall call X_i the nominee of $\mathbf{X}_i$ and the collection $\{X_i : i = 1, \ldots, n\}$ is called the nomination sample. The case when $\Pi_i(\mathbf{X}_i) = \max_{1 \le j \le K_i} X_{ij}$ has been studied by Willemain (1980) and Boyles and Samaniego (1986). Another important case is where $\Pi_i(\mathbf{X}_i) = \min_{1 \le j \le K_i} X_{ij}$; that is, when the nominee of $\mathbf{X}_i$ is the minimum. As an example of such a data generating process suppose that a factory has n identical machines; the ith machine having K_i components $(i = 1, \ldots, n)$. Suppose also that each machine is set up as a series system of i.i.d. components with common d.f. F. Let $\mathbf{X}_i$ be the life lengths of the components in the ith machine. As soon as the first component fails the entire machine fails, these first failure times for the entire factory are $(X_1, \ldots, X_n)$, the nomination sample. A reliability engineer may be interested in inference about the components of the machines, that is about F, rather than the machines itself.

Another example of such a data generating process is the following. Suppose a consumer has a known number of options from which he/she has to make a single decision. The wise consumer will usually choose the option that costs the least and hence the nominee will be the option of minimal cost. Although the distribution of all option costs is unknown, one would like to be able to draw some inference about this distribution from the nomination sample.

In this note we consider the estimation of the distribution function with a nomination sample in the presence of random censoring. This estimator is derived

AMS 1980 subject classifications. Primary 62G05; secondary 60G15, 60G55.

Key words and phrases. Martingale limit theorems, nomination sampling, nonparametric maximum likelihood, censored data, Kaplan-Meier estimator.

in Section 2. The asymptotic theory of this estimator and functionals of this estimator is studied in Section 3.

2. Estimation. Let K be a positive integer valued r.v. with probability mass function $p(\cdot)$ and the probability generating function (p.g.f.) $\psi(\cdot)$. Assume $E|K| < \infty$. Let F and G be continuous d.f.'s on $(0,\infty)$. Given $K = K_i$, let X_i be the minimum of the sample $\mathbf{X}_i$ of size K_i, $i = 1,\ldots,n$. Then X_i has conditional d.f. $1-(1-F)^{K_i}$, $i=1,\ldots,n$. Let $Z_1,\ldots,Z_n$ be i.i.d. G. Define

$$Y_i = \min\{X_i, Z_i\} = X_i \wedge Z_i \text{ and } \delta_i = 1[X_i \le Z_i], \quad i = 1,\ldots,n, \tag{1}$$

where $1[\cdot]$ is the indicator function of the event $[\cdot]$. One can see that Y_i is the nominee of the ith sample if there is no censoring, otherwise we observe the censoring variable Z_i. Hence, if there is no nominee from the ith sample, $\delta_i = 0$. In the reliability example discussed above, this would correspond to no failure in the ith series system, surely this information should be accounted for when estimating F.

Let $Y_{1:n} \le,\ldots,\le Y_{n:n}$ denote the ordered values of observed Y_i's. Denote the $\{(Y_{i:n}, K_{i:n}, \delta_{i:n}); i = 1,\ldots,n\}$ by $\mathcal{D}_c$, where $K_{i:n}$ and $\delta_{i:n}$ are the values of K_i and δ_i that correspond to $Y_{i:n}$, $i = 1,\ldots,n$. Proceeding as in Boyles and Samaniego (1986) (hereafter denoted B–S (1986)) we can obtain the nonparametric maximum likelihood estimator (NPMLE) of F by finding the d.f. F that maximizes

$$\begin{aligned} L(F|\mathcal{D}_c) &= \Pi_{i=1}^n\{[1-(1-F(Y_{i:n}))^{K_{i:n}}] - [1-(1-F(Y_{i-1:n}))^{K_{i:n}}]\}^{\delta_{i:n}} \\ &\quad \times\{\{1-F(Y_{i:n})\}^{K_{i:n}}\}^{(1-\delta_{i:n})} \\ &= \Pi_{i=1}^n\{\bar{F}^{K_{i:n}}(Y_{i-1:n}) - \bar{F}^{K_{i:n}}(Y_{i:n})\}^{\delta_{i:n}}\{\bar{F}(Y_{i:n})\}^{K_{i:n}(1-\delta_{i:n})}, \end{aligned} \tag{2}$$

where $\bar{F} \equiv 1 - F$ is the survival function.

Now, letting $p_i = \bar{F}(Y_{i:n})/\bar{F}(Y_{i-1:n})$ we have from (2) that

$$\begin{aligned} L(F|\mathcal{D}_c) &= \Pi_{i=1}^n(1-p_i^{K_{i:n}})^{\delta_{i:n}} p_i^{K_{i:n}(1-\delta_{i:n})}\Pi_{i'<i}\, p_{i'}^{K_{i':n}} \\ &= \Pi_{i=1}^n(1-p_i^{K_{i:n}})^{\delta_{i:n}} p_i^{\gamma_{[i]}-K_{i:n}\delta_{i:n}} \\ &= L(\mathbf{p}), \text{ say,} \end{aligned} \tag{3}$$

where

$$\gamma_{[i]} = \Sigma_{j=1}^n K_{j:n}, \quad i = 1,\ldots,n.$$

We maximize $L(\mathbf{p})$ in (3) by separate maximization of each factor. One can verify that the function $x^a(1-x^b)^c$ is concave and is uniquely maximized by $\hat{x} = (a/(a+bc))^{1/b}$. It follows that $L(\mathbf{p})$ is maximized by

$$\begin{aligned} \hat{p}_i &= ([\gamma_{[i]} - K_{i:n}\delta_{i:n}]/\gamma_{[i]})^{1/K_{i:n}} \\ &= ([\Sigma_{j=1}^n K_{j:n} - K_{i:n}\delta_{i:n}]/\Sigma_{j=1}^n K_{j:n})^{1/K_{i:n}}, \quad i = 1,\ldots,n. \end{aligned} \tag{4}$$

Therefore the NPMLE of F is given by

$$\text{(5)} \qquad \hat{F}_n(x) = \begin{cases} 0 & , \text{ if } x < Y_{1:n} \\ 1 - \Pi_{j=1}^{i} \hat{p}_j & , \text{ if } Y_{i:n} \leq x < Y_{i+1:n},\ i = 1, \ldots, n-1 \\ 1 & , \text{ if } Y_{n:n} \leq x. \end{cases}$$

Note that $\hat{F}_n(x)$ given by (5) is closely related to the estimate developed in B–S (1986) where the nomination function Π_i was the maximum and there was no censoring. Note also that if $K_i = 1$, for all i, $\hat{F}_n$ reduces to the Kaplan-Meier estimate. Hence if $K_i = 1$ and $\delta_i = 1$, for all i, $\hat{F}_n$ reduces to the empirical distribution function of $(X_1, \ldots, X_n)$. These analogies will become more apparent in the next section when we discuss the asymptotic theory of the process $\sqrt{n}\,(\hat{F}_n - F)$.

3. Asymptotic Theory. The weak convergence results of $\hat{F}_n$ presented here are based on the methods of martingale based inference. The approach is to propose an estimator which is asymptotically equivalent to $\hat{F}_n$ and to demonstrate its limiting distribution.

Note that the stochastic intensity (failure rate) of X_i given $K = K_i$ is given by $\lambda_i(t) = \lambda_0(t)K_i\delta_i$, where $\lambda_0(t)$ is the intensity of the distribution F. Let $\mathcal{H}_i$ be a history which satisfies "the usual conditions" (see Dellacherie (1972)). Embed K_i into an $\mathcal{H}_i$-predictable and locally bounded process $K_i(t)$. Also, embed δ_i into an $\mathcal{H}_i$-predictable process $\delta_i(t)$ taking values in {0,1}, indicating (by the value one) when the ith sample is under observation; thus $\delta_i(\cdot)$ is the censoring process. Now, define the multiplicative intensity model $N_i(t)$ as the point process having stochastic intensity $\lambda_i(t)$. Also, define $N(t) = \Sigma_{i=1}^{n} N_i(t)$ with stochastic intensity $\lambda_0(t)\Sigma_{i=1}^{n} K_i(t)\delta_i(t)$ and history $\mathcal{H} = \vee_{i=1}^{n} \mathcal{H}_i$. Hence, this is related to the Cox regression model as studied by Anderson and Gill (1982) (hereafter denoted by AG (1982)). This is also the approach of B–S (1986), however, we will incorporate the censoring process in our development.

The theory of martingale based inference will give us an estimate of $\Lambda_0(t) = \int_0^t \lambda_0(s)ds$, the integrated hazard, and hence an estimate of $\bar{F}(t) = \exp(-\Lambda_0(t))$. From the results of Section 5.2 of Karr (1986) it can be seen that the estimator of $\Lambda_0(t)$ is given by

$$\text{(6)} \qquad \hat{\Lambda}_{0n}(t) = \int_0^t \frac{dN(s)}{\sum_{i=1}^{n} K_i(t)\delta_i(t)} = \sum_{i:X_i \leq t} \delta_i \left[\sum_{j:X_j > X_i} K_j \right]^{-1}.$$

Define $H_n = -\log(1 - \hat{F}_n)$ as the NPMLE of the integrated hazard rate of F. A simple modification of B–S (1986) to allow for censoring yields

LEMMA 3.1. $$\sup_{t \in \mathbb{R}} \sqrt{n}\,|H_n(t) - \hat{\Lambda}_{0n}(t)| \xrightarrow{P} 0 \text{ as } n \to \infty.$$

This lemma implies that $\sqrt{n}(\hat{\Lambda}_{0n} - H)$ and $\sqrt{n}(H_n - H)$ will have the same asymptotic distribution where $H = -\log \bar{F}$.

The weak convergence of $\sqrt{n}(\hat{\Lambda}_{0n} - H)$ will follow from Rebolledo's (1980) martingale central limit theorem as found in the appendix of AG (1982). We need to verify their conditions (I.3) and (I.4) employed in its proof. Recall that $\psi(\cdot)$ is the p.g.f. of the r.v. K and that $\bar{G} \equiv 1 - G$ is the survival function of the censoring distribution. Define

$$\xi_n(t) = \sum_{i:X_i>t} K_i\delta_i. \tag{7}$$

Hence we have that $\hat{\Lambda}_{0n}(t) = \int_0^t \frac{dN(s)}{\xi_n(s)}$.

The following result will be used in verifying the conditions of Rebolledo's (1980) martingale central limit theorem.

LEMMA 3.2.
(a)

$$\begin{aligned}\int_0^t \frac{\lambda_0(s)ds}{\xi_n(s)/n} &\xrightarrow{P} \int_0^t \frac{\lambda_0(s)ds}{\bar{F}(s)\bar{G}(s)\psi'(\bar{F}(s))} \\ &= \int_0^t \frac{dF(s)}{\bar{F}^2(s)\bar{G}(s)\psi'(\bar{F}(s))},\end{aligned}$$

where ψ' is the first derivative of ψ.

(b) *For any $\epsilon > 0$*

$$\int_0^t \frac{n\lambda_0(s)}{\xi_n(s)} 1[\frac{\sqrt{n}}{\xi_n(s)} > \epsilon]\, ds \xrightarrow{P} 0.$$

PROOF. We will prove that

$$\Delta_n = \sup_{t\in\mathbb{R}} \left| \frac{\lambda_0(t)}{\xi_n(t)/n} - \frac{\lambda_0(t)}{\bar{F}(t)\bar{G}(t)\psi'(\bar{F}(t))} \right| \xrightarrow{P} 0 \text{ as } n \to \infty.$$

Note that we have the identity

$$\Delta_n = \sup_{t\in\mathbb{R}} \left| \frac{\lambda_0(t)\bar{F}(t)\bar{G}(t)\psi'(\bar{F}(t)) - \lambda_0(t)\xi_n(t)/n}{\bar{F}(t)\bar{G}(t)\psi'(\bar{F}(t))\xi_n(t)/n} \right|. \tag{8}$$

The strong law of large numbers yields

$$\begin{aligned}\frac{1}{n}\xi_n &\xrightarrow{\text{a.s.}} E\{K1[X \geq t]1[Z \geq X]\} \\ &= \bar{G}(t)\sum_{k=1}^{\infty} k\bar{F}^k(t)p(k) = \bar{G}(t)\bar{F}(t)\psi'(\bar{F}(t)).\end{aligned} \tag{9}$$

Note that the expression in (9) is positive for all $t \epsilon \mathbb{R}$. Hence the numerator in (8) tends to zero and the denominator of (8) is $O_p(1)$. Therefore $\Delta_n \xrightarrow{P} 0$. The second equality follows since $\lambda_0(t) = \frac{f(t)}{\bar{F}(t)}$, where f is the density of F.
(b) The proof follows from a minor modification of Lemma 2.3b in B–S (1986). ||

Let θ be a positive real number such that $\theta < \tau^{-1}(1)$, where $\tau = (1-F)(1-G)$. By applying Rebolledo's (1980) theorem we will show that following result holds.

THEOREM 3.3. *The process* $\beta_n = \sqrt{n}\,(\hat{\Lambda}_{0n} - H)$ *converges weakly in* $D[0,\theta]$ *as* $n \to \infty$ *to a mean zero Gaussian martingale* B *with covariance function*

$$\text{(10)} \qquad \operatorname{Cov}(B(s), B(t)) = \int_0^s \frac{dF(u)}{\bar{F}^2(u)\bar{G}(u)\psi'(\bar{F}(u))} \quad 0 \le s \le t \le \theta.$$

PROOF. By the Doob-Meyer decomposition for submartingales (and hence for counting processes) we have

$$\text{(11)} \qquad dN_i(t) = \lambda_0(t)K_i(t)\delta_i(t) + dM_i(t)$$

$$\text{(12)} \qquad dN(t) = \lambda_0(t)\xi_n(t) + dM(t)$$

where $M(t) = \Sigma_{i=1}^n M_i(t)$ is an $\mathcal{H}$-martingale. Therefore the process in the theorem may be expressed as

$$\text{(13)} \qquad \sqrt{n}\,\{\hat{\Lambda}_{0n}(t) - H(t)\} = \bar{M}_n(t) + R_n(t),$$

where

$$\bar{M}_n(t) = \frac{1}{\sqrt{n}} \int_0^t \frac{dM(s)}{\xi_n(s)/n} \text{ and } R_n(t) = \int_0^t I(\xi_n(s) = 0)dH(s).$$

Note that (9) implies

$$\sup_{t \epsilon [0,\theta]} |R_n(t)| \xrightarrow{P} 0 \text{ as } n \to \infty.$$

Also, note that $\bar{M}_n(t)$ is a square integrable martingale. Therefore to deduce that the process β_n converges to a Gaussian martingale, we will apply Rebolledo's (1980) martingale central limit theorem to the martingale $\bar{M}_n$. The version of Rebolledo's theorem we will use is found in AG (1982) with $p = 1$ and $H_{1\ell}(t) = n^{-1/2}(\xi_n(t)/n)^{-1}$. By Lemma 3.2a and (12) we have that

$$< \bar{M}_n, \bar{M}_n > (t) = \int_0^t \frac{\lambda_0(s)}{\xi_n(s)/n} ds \xrightarrow{P} \int_0^t \frac{\lambda_0(s)ds}{\bar{F}(s)\bar{G}(s)\psi'(\bar{F}(s))} =< B, B > (t).$$

The Lindeberg condition of AG (1982) may be verified by applying Lemma 3.2b. Therefore the conditions of the theorem have been met and we have the desired result. ||

THEOREM 3.4. *The process* $\chi_n(t) = \sqrt{n}\ \{\hat{F}_n(t) - F(t)\}$ *converges weakly in* $D[0,\theta]$ *as* $n \to \infty$ *to* $\chi(t) = \bar{F}(t)B(t)$, *where* $B(t)$ *is the Gaussian martingale in Theorem 3.3. The covariance kernel of* χ *is given by*

$$K(s,t) = \bar{F}(s)\bar{F}(t)\int_0^s \frac{dF(u)}{\bar{F}^2(u)\bar{G}(u)\psi'(\bar{F}(u))} \quad 0 \le s \le t \le \theta. \tag{14}$$

PROOF. By applying the Doléans-Dade exponential, $\mathcal{E}(\cdot)$ (see Liptser and Shiryayev, 1978, pp. 255–256) it is immediate that $\bar{F} = \mathcal{E}(-H)$ since $\bar{F}$ satisfies

$$\bar{F}(t) = 1 + \int_0^t \bar{F}(s)\, d(-H(s)).$$

Therefore,

$$\mathcal{E}(H(t) - \hat{\Lambda}_{0n}(t)) = (1 - \hat{F}_n(t))/\bar{F}(t)$$

so that

$$(1 - \hat{F}_n(t))/\bar{F}(t) = 1 + \int_0^t \frac{(1 - \hat{F}_n(s))}{\bar{F}(s)}\, d(H(s) - \hat{\Lambda}_{0n}(s)).$$

Using the decomposition in (13) it is clear that

$$\chi_n(t) = \bar{F}(t)\int_0^t \frac{(1 - \hat{F}_n(s))}{\bar{F}(s)}\, d\bar{M}_n(s) + R_n^*(t) \tag{15}$$

for some remainder term R_n^* which tends to zero in probability uniformly in $t \epsilon [0,\theta]$. Define $L_n(t)$ to be the integral in first term on the right hand side of (15). Note that $L_n(t)$ is a square integrable martingale with predictable variation process

$$< L_n, L_n > (t) = \int_0^t \left[\frac{1 - \hat{F}_n(s-)}{(1 - F(s))}\right]^2 \frac{\lambda_0(s)ds}{\xi_n(s)/n}. \tag{16}$$

By an application of Lenglart's (1977) inequality and the decomposition in (15) we have that the term on the right hand side of (16) in the square bracket tends to one. Therefore, it follows that

$$< L_n, L_n > (t) \xrightarrow{P} < B, B > (t). \tag{17}$$

As in the proof of Theorem 3.3, for $\bar{M}_n$, the Lindeberg condition for L_n may be verified. The result then follows by an application of Rebolledo's martingale central limit theorem. ||

As to be expected, if $\psi(u) \equiv u$, that is $K_i \equiv 1$ for all $i = 1, \ldots, n$ the covariance function in (11) reduces to the limiting covariance function of the Kaplan-Meier estimate. Similarly, if $\psi(u) \equiv u$ and $\bar{G}(u) \equiv 1$, that is, there is no censoring, the result reduces to the classical result for the empirical distribution function.

To apply Theorem 3.4 it is necessary to estimate the variance in (14), but this can also be done using martingale based methods. The process

$$W_n(t) = \int_0^t \frac{\lambda_0(s)ds}{\Sigma_{i=1}^n K_i(s)\delta_i(s)},$$

which converges to the variance in (10), has a martingale based estimator

$$\hat{W}_n(t) = \int_0^t \frac{1}{(\Sigma_{i=1}^n K_i(s)\delta_i(s))^2}\, dN(s) = \sum_{i:X_i \le t} \delta_i \left[\sum_{j:X_j > X_i} K_j \right]^{-2}.$$

By applying the results of Theorem 5.12 of Karr (1986) the estimate $\hat{W}_n(t)$ may be shown to be a consistent estimate of the variance in (10). Therefore,

$$\hat{U}_n(t) = (1 - \hat{F}_n(t))\hat{W}_n(t) \tag{18}$$

will consistently estimate the variance in (14). This is the type of estimator introduced by Tsiatis (1981) in the context of Cox regression models.

Using the results of Theorem 3.4 one may study the asymptotic behavior of the estimated quantiles. Define the quantiles of $\hat{F}_n$ as $\hat{F}_n^{-1}(t) = \inf\{x : \hat{F}_n(x) \ge t\}$. Applying the general results for the asymptotic behavior of quantiles by Tiwari and Wells (1988) we have

THEOREM 3.5. *Under the conditions of Theorem 3.4 the process* $\sqrt{n}\,[\hat{F}_n^{-1}(t) - F^{-1}(t)]f(F^{-1}(t))$ *converges weakly on* $D[0, F(\theta)]$ *to the mean zero Gaussian process* $\chi(F^{-1})$, *where* $\chi(\cdot)$ *has covariance function given by (14).*

The above theorems are stated under the assumptions that K is a random variable with finite expectation. In some applications the assumption that K is random may not be appropriate. B–S (1986) suggest a possible modification when K is not random. For further details see Section 3 of B–S (1986).

Many statistical procedures under the nomination sampling scheme can be viewed as a functional of the process $\sqrt{n}(\hat{F}_n - F)$ and the asymptotic properties of such procedures can be inferred from the process itself. In what follows, in the remainder of this section, we will consider the problem of estimation of a parameter of the unknown distribution F. Specifically, we will examine the properties of linear combinations of functions of estimated quantiles (lcfeq) under the nomination sampling scheme. In the case of simple random sampling the estimated quantiles are the order statistics and in that case parameter estimates are based on linear combinations of functions of order statistics (lcfos). In the case of nomination sampling we do not record the order statistics of the individual samples, thus we will use the estimated quantiles discussed in Theorem 3.5.

Let J_n be some known score generating function and let $h(\cdot)$ denote a known function of the form $h = h_1 - h_2$ with $h_i (i = 1, 2)$ increasing and left continuous. Consider the lcfeq

$$T_n = \int_0^t h(\hat{F}_n^{-1}(s))J_n(s)\,ds.$$

If $J_n \to J$, in some sense, T_n can be used as an estimate for the functional

$$\theta = \int_0^t h(F^{-1}(s))J(s)\,ds.$$

See Serfling (1980) for an extensive survey of functionals of this type.

Associated with the function $g = h(F^{-1})$ is a Lebesgue-Stieltjes signed measure; let $|g|$ denote the total variation measure of this measure. We shall need the following assumptions to demonstrate the asymptotic normality of $\sqrt{n}\,(T_n - \theta)$.

Assumption 1: (i) Suppose $|J| < \alpha(t)$ and, for all n, $|J_n| < \alpha(t)$ on (0,1) where $\alpha(t) = Mt^{-b_1}(1-t)^{-b_2}$ for $0 < t < 1$ with $M > 0$ and $(b_1 \wedge b_2) < 1$.

(ii) Suppose $h = h_1 - h_2$, with h_i increasing and left continuous on $\mathbb{R}$ with $|h_i(F^{-1})| < D(t)$ for $i = 1, 2$, where $D(t) = Mt^{-d_1}(1-t)^{-d_2}$, for $0 < t < 1$, with $M > 0$ and any fixed d_1, d_2.

Assumption 2: Except on a set of t's of $|g|$-measure zero we have both J is continuous at t and $J_n \to J$ uniformly in some small neighborhood of t as $n \to \infty$.

Under the above assumption we have a theorem which is an analog of the result of Shorack (1972) concerning lcfos in the simple random sampling set up. The proof of our result is quite similar to Shorack's and will be omitted.

THEOREM 3.6. *If* $(b_1 + d_1) \vee (b_2 + d_2) < \frac{1}{2}$, *then*

$$\sqrt{n}\,(T_n - \theta) \xrightarrow{d} N(0, \sigma^2) \text{ as } n \to \infty,$$

where

$$\sigma^2 = \int_0^1 \int_0^1 K(F^{-1}(s), F^{-1}(t))\, J(s)J(t)\, dg(s)dg(t)$$

with $K(\cdot,\cdot)$ *being the covariance kernel given by (14).*

REFERENCES

ANDERSON, P.K. and GILL, R.D. (1982). Cox's regression model for counting processes: A large sample study. *Ann. Statist.* **10** 1100–1120.

BOYLES, R.A. and SAMANIEGO, F.J. (1986). Estimating a distribution function based on nomination sampling. *J. Amer. Statist. Assoc.* **81** 1039–1045.

DELLACHERIE, C. (1972). *Capacites' et Processus Stochastiques.* Springer Verlag, Berlin.

KARR, A.F. (1986). *Point Processes and Their Statistical Inference.* Marcel Dekker, Inc., New York.

LENGLART, E. (1977). Relation de domination entre deux processus. *Ann. Inst. Henri Poincare* **13** 171–179.

LIPTSER, R.S. and SHIRYAYEV, A.N. (1978). *Statistics of Random Processes II: Applications.* Springer-Verlag, New York.

REBOLLEDO, R. (1980). Central limit theorems for local martingales. *Z. Wahrsch. verw. Geb.* **51** 269–286.

SHORACK, G.R. (1972). Functions of order statistics. *Ann. Math. Statist.* **43** 412–427.

SERFLING, R. (1980). *Application Theorems of Mathematical Statistics.* Wiley, New York.

TIWARI, R.C. and WELLS, M.T. (1988). On the asymptotic theory for quantile estimation for various sampling schemes. Submitted for publication.

TSIATIS, A.A. (1981). A large sample study of Cox's regression model. *Ann. Statist.* **9** 93–108.

WILLEMAIN, T.R. (1980). Estimating the population median by nomination sampling. *J. Amer. Statist. Assoc.* **75** 908–911.

DEPARTMENT OF ECONOMIC
AND SOCIAL STATISTICS
CORNELL UNIVERSITY
ITHACA, NY 18451–0952

DEPARTMENT OF MATHEMATICS
UNIVERSITY OF NORTH CAROLINA
CHARLOTTE, NC 28223

A STRONG LIMIT THEOREM FOR PROCESSES WITH ASSOCIATED INCREMENTS

BY A. LARRY WRIGHT

University of Arizona

A finite collection of random variables $X_1, \ldots, X_m$ is said to be associated if for any two coordinatewise nondecreasing functions f, g on R^m

$$\mathrm{Cov}(f(X_1, \ldots, X_m), g(X_1, \ldots, X_m)) \geq 0$$

whenever the covariance is defined; a stochastic process $X(t)$ is said to have associated increments if for each t, $X(t)$ and the increments of the process in (t, ∞) are associated.

THEOREM. *If $X(t)$ is a separable, mean zero stochastic process with associated increments, and $H(t) \uparrow \infty$ is positive and continuous and such that $\int_0^\infty \frac{d\sigma(t)}{H(t)} < \infty$, where $\sigma(t)$ = standard deviation $(X(t))$, then $\frac{X(t)}{H(t)} \to 0$ a.s.*

1. Introduction. Let $\{X_j, 1 \leq j \leq m\}$ be a collection of random variables. The collection is said to be associated if for any coordinatewise non-decreasing functions f, g on R^m, $\mathrm{Cov}(f(X_1, \ldots, X_m), g(X_1, \ldots, X_m)) \geq 0$ whenever the covariance is defined; an infinite collection is associated if every finite subcollection is associated. In Newman and Wright (1982) it was shown that associated random variables satisfy several of the classical martingale inequalities.

In this paper we consider continuous time stochastic processes. A stochastic process $X(t)$ is said to have associated increments if for each t, $X(t)$ and the increments of the process in (t, ∞) are associated. Such processes have many of the sample function properties that a separable submartingale has. Many of these properties are discussed in Wood (1983), in the more general case of (continuous time) demimartingales. These properties also hold for separable and centered processes with independent increments. In our case we assume the process is separable, mean zero, with *associated* increments. One interesting consequence shown here is that a separable, mean zero process with associated increments is automatically centered.

The properties above allow us to prove the following strong limit theorem: Let $\{X(t), t \geq 0\}$ be a separable, mean zero process with associated increments. If

AMS 1980 subject classifications. 60G17, 60G44, 60J30.

Key words and phrases. Associated random variables, processes with independent increments, infinite divisibility.

$\int_0^\infty \frac{d\sigma(t)}{H(t)} < \infty$, where $\sigma(t)$ = standard deviation $(X(t))$, and $H(t) \uparrow \infty$ is positive and continuous, then $\frac{X(t)}{H(t)} \overset{\text{a.s.}}{\to} 0$.

2. Associated Processes. Let $\{X(t), t \geq 0\}$ be a stochastic process. We say that the process has associated increments if for any $n \geq 1$ and $0 \leq t_1 < t_2 < \cdots < t_n$ the random variables $X(t_1), X(t_2) - X(t_1), \ldots, X(t_n) - X(t_{n-1})$ are associated.

We first discuss almost sure sample function properties of such processes. Outside a set of probability zero, the sample functions are bounded on every bounded interval, have finite left and right-hand limits at each point, and their discontinuities are jumps, outside a fixed and countable set T^*. These properties are known to hold for separable submartingales and separable centered processes with independent increments (Doob, 1953, pages 361, 422). In Wood (1983) it is mentioned that the first two properties above hold in the demimartingale case.

We will need the following two lemmas:

LEMMA 1. *Let $\{X_j, 1 \leq j \leq n\}$ be a sequence of associated, mean zero random variables, and set $S_j = \sum_{j=1}^n X_j$. Then for $\lambda > 0$*

$$P\left\{\max_{1\leq j\leq n} S_j \geq \lambda\right\} \leq \frac{E|S_n|}{\lambda} \leq \frac{\sigma(S_n)}{\lambda}.$$

The proof may be found in Newman and Wright, 1982, Theorem 3, page 363.

For the next lemma, we first recall a version of Doob's upcrossing inequality which holds for demimartingales:

Let $\{X_j, 1 \leq j \leq n\}$ be as in Lemma 1, and define a sequence of stopping times $J_0 = 0, J_1, J_2 \ldots$ as follows (for $k = 1, 2, \ldots$):

$$J_{2k-1} = \begin{cases} n+1, \text{ if } \{j : J_{2k-2} < j \leq n \text{ and } S_j \leq a\} \text{ is empty} \\ \min\{j : J_{2k-2} < j \leq n \text{ and } S_j \leq a\}, \text{ otherwise} \end{cases}$$

$$J_{2k} = \begin{cases} n+1, \text{ if } \{j : J_{2k-1} < j \leq n \text{ and } S_j \geq b\} \text{ is empty} \\ \min\{j : J_{2k-1} < j \leq n \text{ and } S_j \geq b\}, \text{ otherwise} \end{cases}$$

We define the number of complete upcrossings of [a, b] by $S_1, \ldots, S_n$ by $U_{a,b} = \max\{k : J_{2k} < n+1\}$. We have the following:

LEMMA 2. *If $\{X_j, 1 \leq j \leq n\}$ is a sequence of mean zero, associated random variables, then for any $a < b$,*

$$E(U_{a,b}) \leq \frac{E((S_n - a)^+) - E((S_1 - a)^+)}{b - a}$$

The proof may be found in Newman and Wright, 1982, Theorem 7.

We now prove the theorem of this section.

THEOREM 3. *Let* $\{X(t), t \geq 0\}$ *be a separable,* mean zero *process with associated increments. Then outside a set of probability zero, the sample functions have the following properties:*

(a) they are bounded on every finite interval,

(b) they have finite right and left-hand limits at every point, and

(c) their discontinuities are jumps, except possibly for some points $t \epsilon T^*$ *where* T^* *is a fixed countable set.*

PROOF. For $T > 0$ finite, $0 = t_1 < t_2 < \cdots < t_n = T$, and $\lambda > 0$ we have, by Lemma 1 and the definition of a process with associated increments,

$$\lambda P\left\{\max_{1\leq j\leq n} X_{t_j} \geq \lambda\right\} \leq \int_\Omega |X_T| dP < \infty.$$

Taking successively finer partitions of $[0, T]$ and using separability,

$$\lambda P\left\{\sup_{t\epsilon[0,T]} X_t > \lambda\right\} \leq \int_\Omega |X_T| dP < \infty.$$

It follows that the process is bounded from above with probability one. Also, from the definition of separability, with probability one, for $\lambda > 0$,

$$\lambda P\left\{\inf_{t\epsilon[0,T]} X_t < -\lambda\right\} = \lambda P\left\{\inf_{t_j\epsilon[0,T]} X_{t_j} < -\lambda\right\} = \lambda P\left\{\sup_{t_j\epsilon[0,T]} -X_{t_j} > \lambda\right\},$$

where $\{t_j, j = 1, 2, \ldots\}$ is a separating set. But $\{-X_t, t \geq 0\}$ is a mean zero process with associated increments, so it follows from Lemma 1 that

$$\lambda P\left\{\sup_{t\epsilon[0,T]} -X_t > \lambda\right\} \leq E|X_T| < \infty,$$

or

$$\lambda P\left\{\inf_{t\epsilon[0,T]} X_t < -\lambda\right\} \leq E|X_T| < \infty.$$

It follows that the process is bounded from below with probability one, completing the proof of (a).

The proof of (b) follows directly from Doob, 1953, page 361, using our Lemma 2 in place of the usual upcrossing inequality.

For (c), we note that for a sample function satisfying (a) and (b) above, if $\{t_j, j \geq 1\}$ is a separating set for the process, any discontinuity at a point other than a t_j must be a jump. Since the separating set is denumerable, we have proved (c).

3. Centered Processes. Recall that a stochastic process $\{Z_t, t \geq 0\}$ is said to be centered if:

(a) For each $t > 0$, and sequence $s_n \to t$ with $s_n < t, \lim_{n\to\infty} Z_{s_n} = Z_{t-}$ exists with probability one, and for each $t \geq 0$ and sequence $s_n \to t$ with $s_n > t$, $\lim_{n\to\infty} Z_{s_n} = Z_{t+}$ exists with probability one.

(b) If any difference $Z_t - Z_s$ (or $Z_{t+} - Z_s, Z_t - Z_{t-}$, etc.), is identically constant with probability one, the constant is equal to zero.

(c) Except (possibly) for the points of an enumerable set $S \subset [0, \infty)$, the following holds with probability one: $Z_{t-} = Z_t = Z_{t+}$.

We have shown that any separable, mean zero process with associated increments satisfies (a) and (b). That such a process also satisfies (c), and hence is centered, follows from the following theorem:

Let $\{X(t), t \geq 0\}$ be a stochastic process. If for every $t > 0$ at least one of the limits in probability $\lim_{s\uparrow t} X_s = X_{t-}, \lim_{s\downarrow t} X_s = X_{t+}$ exists, then there is an at most enumerable subset $T \subset (0, \infty)$ such that for all $t \epsilon (0, \infty) \setminus T$, both stochastic limits X_{t-} and X_{t+} are defined, and $X_{t-} = X_{t+} = X_t$ with probability 1 (Doob, 1953, Theorem 11.1, page 356).

In particular, a separable, mean zero process with independent increments is centered. It is easy to show that any mean zero, L^2 process with independent increments satisfies (a) and (b) of the definition above (Wright, 1982, Theorem 2, page 110), but may not be centered, if it is not separable.

A trivial example of an L^2, separable and infinitely divisible process with independent increments which is *not* centered is the following:

$$Z(t) = \begin{cases} 0, & 0 \leq t < 1 \\ 1, & 1 \leq t < \infty. \end{cases}$$

4. A Strong Limit Theorem. In this section we prove a strong limit theorem for stochastic processes with associated increments. For its proof we make use of separability, Lemma 1, Theorem 3, and the easily proven fact that for a process $\{X(t), t \geq 0\}$ with associated increments, $\sigma(X(t) - X(s)) \leq \sigma(t)$ for $s < t$.

THEOREM 4. *Let $\{X(t), t \geq 0\}$ be an L^2, mean zero and separable process with associated increments. If $H(t) \uparrow \infty$ is positive and continuous and such that*

$$\int_0^\infty \frac{\sigma(t)}{H(t)} < \infty, \textit{ then } \frac{X(t)}{H(t)} \xrightarrow{a.s.} 0.$$

PROOF. It is easy to show that we may choose points $0 \leq t_0 < t_1 < \cdots$ such that $\frac{\sigma(t_j)}{H(t_j)} < \frac{1}{2^{2j}}$ and $\int_{t_j}^\infty \frac{\sigma(t)}{H(t)} < \frac{1}{2^{2j}}$ for $j = 1, 2, \ldots$. For such a sequence we have for $t_{j-1} \leq t < t_j, j = 1, 2, \ldots$

$$\frac{|X(t)|}{H(t)} < \frac{1}{H(t)}\{|X(t_0)| + \sum_{i=1}^{j-1}\{\sup_{t_{i-1}<s\le t_i}|X(s)-X(t_{i-1})|\}$$

(1) $$+ |X(t)-X(t_{j-1})|\}.$$

Trivially, $\frac{|X(t_0)|}{H(t)} \overset{\text{a.s.}}{\to} 0$. As for the second term on the right-hand side, we note that if $t_{j-1} = t_{j,0} < t_{j,1} < \cdots < t_{j,N_j} = t_j$ is any partition of $[t_{j-1}, t_j]$, then by Lemma 1,

$$P\left[\max_{1\le k\le N_j}\frac{|X(t_{j,k})-X(t_{j,0})|}{H(t_j)} \ge \frac{1}{2^j}\right] \le \frac{\sigma(X(t_j)-X(t_{j-1}))}{H(t_j)2^{-j-1}}.$$

Since $\{-X(t), t\ge 0\}$ has the same properties as $\{X(t), t\ge 0\}$ it follows that

$$P\left[\max_{1\le k\le N_j}\frac{|X(t_{j,k})-X(t_{j,0})|}{H(t_j)} \ge \frac{1}{2^j}\right] \le \frac{\sigma(X(t_j)-X(t_{j-1}))}{H(t_j)2^{-j-1}}$$

$$\le \frac{\sigma(t_j)}{H(t_j)2^{-j-1}} \le \frac{1}{2^{j-1}}.$$

Using Theorem 3 we see that

$$P\left[\sup_{t_{j-1}<s\le t_j}\frac{|X(s)-X(t_{j-1})|}{H(t_j)} > \frac{1}{2^j}\right] \le \frac{1}{2^{j-1}}.$$

By the Borel-Cantelli theorem,

$$\sum_{j\ge 1}\sup_{t_{j-1}<s\le t_j}\frac{|X(s)-X(t_{j-1})|}{H(t_j)} < \infty \text{ a.s.}$$

From this and (1) it follows that we only need to show

$$X_j \equiv \sup_{t_{j-1}<s\le t_j}\frac{|X(s)-X(t_{j-1})|}{H(t)} \overset{\text{a.s.}}{\to} 0.$$

We proceed by picking points $\{s_{j,k} : 1\le j<\infty, 0\le k\le N_j\}$ as follows:

$$t_{j-1} = s_{j,0} < s_{j,1} < \cdots < s_{j,N_j} \quad \text{with} \quad s_{j,N_j} - 1 < t_j \le s_{j,N_j}$$

(2) $$\text{and} \quad H(s_{j,i}) = 2H(s_{j,i-1})$$

It follows that

$$|X_j| \le \max_{0\le i\le N_j-1}\left\{\sup_{s_{j,i}<s\le s_{j,i+1}}\frac{|X(s)-X(s_{j,0})|}{H(s_{j,i})}\right\}.$$

From (1), (2), and Lemma 1,

$$P\left\{|X_j| > \frac{1}{2^j}\right\}$$

$$\begin{aligned}
&\le \sum_{i=0}^{N_j-1} P\left\{\sup_{s_{j,i}<s\le s_{j,i+1}} \frac{|X(s)-X(s_{j,0})|}{H(s_{j,i})} > \frac{1}{2^j}\right\} \\
&\le 2^{j+1}\sum_{i=0}^{N_j-1}\left[\frac{\sigma(X(s_{j,i+1})-X(s_{j,0}))}{H(s_{j,i})}\right] \\
&\le 2^{j+1}\sum_{i=0}^{N_j-1}\left\{\frac{\sigma(s_{j,i+1})-\sigma(s_{j,0})}{H(s_{j,i})}\right\} + 2^{j+1}\sum_{i=0}^{N_j-1}\frac{\sigma(s_{j,0})}{H(s_{j,i})} \\
&\le 2^{j+1}\sum_{i=0}^{N_j-1}\{\sigma(s_{j,i+1})-\sigma(s_{j,i})\}\left[\frac{1}{H(s_{j,i})}+\frac{1}{H(s_{j,i+1})}+\cdots\right] + \frac{2^{j+2}\sigma(s_{j,0})}{H(s_{j,0})} \\
&\le 2^{j+3}\sum_{i=0}^{N_j-1}\frac{\{\sigma(s_{j,i+1})-\sigma(s_{j,i})\}}{H(s_{j,i+1})} + \frac{2^{j+2}\sigma(s_{j,0})}{H(s_{j,0})} \\
&\le 2^{j+3}\int_{t_{j-1}}^{\infty}\frac{d\sigma(t)}{H(t)} + \frac{2^{j+2}\sigma(s_{j,0})}{H(s_{j,0})} < 2^{-j+6}.
\end{aligned}$$

From the Borel-Cantelli Theorem we get $X_j \xrightarrow{\text{a.s.}} 0$, completing the proof of the theorem.

REMARKS. The converse of Theorem 4 is not true. For a counterexample, let $\{X(t), t \ge 0\}$ be standard Brownian motion. In this case

$$\int_0^\infty d\sigma(t) = \frac{1}{2}\int_0^\infty t^{-1/2}dt = \infty,$$

and it is easy to construct $H(t) \uparrow 0$, such that $H(t) = b(t)t^{1/2}$ for $t \ge 1$ with $b(t) \uparrow \infty$ and $\int_0^\infty \frac{d\sigma(t)}{H(t)} = \infty$. It follows that $\frac{X(t)}{H(t)} \to 0$.

If $\{X(t), t \ge 0\}$ is as above except that $EX(t) \equiv 0$ does not necessarily hold, we would like to assert that $\frac{X(t)-EX(t)}{H(t)} \xrightarrow{\text{a.s.}} 0$.

Since the process $\{X(t) - EX(t), t \ge 0\}$ has associated increments if $\{X(t), t \ge 0\}$ does, what is needed are conditions that imply that $\{X(t) - EX(t), t \ge 0\}$ is a separable process.

It is easy to show (Wood, 1983, page 4), that if $\{X(t), t \ge 0\}$ is a separable process and f is a function that has finite right and left-hand limits at every point $t > 0$, then $\{X(t) + f(t), t \ge 0\}$ is also a separable process. This observation leads to the following:

COROLLARY 5. *Let* $\{X(t), t \ge 0\}$ *be a separable and centered stochastic process with associated increments. If* $\sup_{t\in[0,T]} EX^2(t) < \infty$ *for all* $T > 0$ *and* $H(t) \uparrow \infty$

is positive and continuous and such that $\int_0^\infty \frac{d\sigma(t)}{H(t)} < \infty$, *then* $\frac{X(t)-EX(t)}{H(t)} \to 0$.

The proof follows from noting that $\{X(t), 0 \leq t \leq T\}$ is uniformly integrable over $[0,T]$ for each $T > 0$, and that $\lim_{s\uparrow t} X(s) = X(t-)$ and $\lim_{s\downarrow t} X(s) = X(t+)$ exist a.s. for each $t > 0$, implying that $\lim_{s\uparrow t} EX(t)$ and $\lim_{s\uparrow t} EX(t)$ exist for each $t > 0$.

REFERENCES

DOOB, J. (1953). *Stochastic Processes.* Wiley, New York.

HAHN, N. (1948). *Set Functions.* Univ. of New Mexico Press, Albuquerque.

LOEVE, M. (1963). *Probability Theory.* Van Nostrand, New York.

NEWMAN, C. and WRIGHT, A. (1982). Associated random variables and martingale inequalities. *Z. Wahrsch. verw. Gebrete.* **59** 361–371.

PYKE, R. (1961). On centering infinitely divisible processes. *Ann. Math. Statist.* **31** 797–800.

WOOD, T. (1983). Sample paths of demimartingales. *Probability Theory on Vector Spaces III*, Springer's LNM.

WRIGHT, A. (1982). Two strong limit theorems for processes with independent increments. *J. Mult. Anal.* **12** 178–185.

DEPARTMENT OF MATHEMATICS
UNIVERSITY OF ARIZONA
TUCSON, AZ 85721

DEPENDENCE ORDERING IN STATISTICAL MODELS AND OTHER NOTIONS

By Takemi Yanagimoto

Institute of Statistical Mathematics

The relation of notions of dependence ordering with other notions in statistics such as heaviness of tail and largeness of dispersion are reviewed and developed. The problem of overdispersion in reproductive studies is discussed as a practical, attractive example.

1. Introduction. The notion of positive dependence is closely associated with other various statistical notions. This has been emphasized by many authors including Yanagimoto and Sibuya (1972) and Karlin and Rinott (1980a). Recent theoretical development of the notion of positive dependence permits us better understandings of statistical models. However, as Kimeldorf and Sampson (1987) stressed, the research on dependence orderings does not appear fully developed. We often assume a family of distributions having monotone dependence. To state this clearer, consider a distribution function, $F_\alpha(x)$, of a random variable X on R^n. The suffix α, representing the degree of largeness of dependence, is considered to be favorably parameterized, if $\alpha = 0$ stands for independence and dependence ordering is monotone increasing in α; as a result $\alpha > 0$ means positive dependence of $F_\alpha(x)$.

The situation is largely different according to the value of n. In the bivariate case negative dependence of (X_1, X_2) is reasonably recognized as positive dependence of $(X_1, -X_2)$. The notion of negative dependence is much more complicated in the multivariate case than in the bivariate case. The aim of the present paper is to review and develop dependence orderings with emphasis on the relation with other statistical notions and practical models. In Sections 2 and 4 definitions of dependence orderings are studied in the bivariate case and in the multivariate case. Some relations are discussed in Sections 3 and 5.

As usual conventions, we will employ simple descriptions unless any confusion is anticipated. Therefore, for example, dependence of a distribution, that of a random variable with the distribution and that of a distribution function of the distribution are not distinguished, and an increasing function means a nondecreasing function.

2. Dependence Ordering: Bivariate Case. A systematic definition of notions of positive dependence in the bivariate case was developed by Lehmann

AMS 1980 subject classifications. 62E10.

Key words and phrases. Heaviness of tail, multivariate distribution, overdispersion, spread.

The author expresses his thanks to a referee and the editors for their helpful comments.

(1966), which was followed by many works such as Esary, Proschan, and Walkup (1967), Yanagimoto (1972), and Shaked (1977a). Among these notions, that of association introduced by Esary et al. is inconvenient to be extended to a dependence ordering.

Unified definitions of notions of positive and negative dependence given in Yanagimoto (1972) permit us a straightforward extension to those of dependence orderings. Let $F(x, y)$ and $G(x, y)$ be cumulative distribution functions both having common marginal distribution functions. Following the idea and the notations of Yanagimoto (1972) and Kimeldorf and Sampson (1987), we give the following definition.

DEFINITION 1. Let I_1 and I_2 be real intervals. We say $I_1 < I_2$ if and only if $x_1 \in I_1$ and $x_2 \in I_2$ imply $x_1 < x_2$. Define the four families of products of interval as

$$\begin{aligned} S(1) &= \{(-\infty, x_1] \times (x_1, \infty) \mid -\infty < x_1 < \infty\} \\ S(2') &= \{(x_1, x_2] \times (x_2, \infty) \mid -\infty < x_1 < x_2 < \infty\} \\ S(2'') &= \{(-\infty, x_1] \times (x_1, x_2] \mid -\infty < x_1 < x_2 < \infty\} \\ S(3) &= \{(x_1, x_2] \times (x_2, x_3] \mid -\infty < x_1 < x_2 < x_3 < \infty\}. \end{aligned}$$

Then $G(x, y)$ is said to have larger dependence $F(x, y)$ in the sense of $P(i, j)$, if it holds

$$\begin{aligned} &P_F(I_1, J_1)P_F(I_2, J_2)P_G(I_1, J_2)P_G(I_2, J_1) \\ &\leq P_G(I_1, J_1)P_G(I_2, J_2)P_F(I_1, J_2)P_F(I_2, J_1) \end{aligned}$$

for any $I_1 < I_2$ and $J_1 < J_2$ satisfying $I_1 \times I_2 \in S(i)$ and $J_1 \times J_2 \in S(j)$, where $P_F(I, J)$ represents the probability assigned by F to the rectangle $I \times J$.

The above definition provides the 16 strictly different notions of dependence orderings. The implication scheme among these orderings holds parallel to that of $P(i, j)$'s. Among them two orderings coincide with ones in existing literature. Yanagimoto and Okamoto (1969) called $G(x, y)$ has larger quadrant dependence than $F(x, y)$ when (i, j) takes (1,1). The ordering was also discussed in Tchen (1980). Kimeldorf and Sampson (1987) introduced the ordering called that $G(x, y)$ is more TP_2 than $F(x, y)$. This ordering is strictly stronger than that corresponding $(i, j) = (3, 3)$.

Yanagimoto and Okamoto (1969) present another extension of the positive regression dependence ordering, which corresponds to $P(3, 1)$ in Yanagimoto (1972).

DEFINITION 2. Suppose that $G(x, y)$ and $F(x, y)$ have both the common marginal distribution functions. $G(x, y)$ is called to have larger regression dependence on x than $F(x, y)$, if it holds

$$F^{-1}(u \mid x') \geq F^{-1}(v \mid x) \text{ implies } G^{-1}(u \mid x') \geq G^{-1}(v \mid x)$$

for any $x' > x$ and $0 < u, v < 1$, where $F^{-1}(u \mid x) = \inf\{y \mid F(y \mid x) \geq u\}$.

Schriever (1987) introduced a new dependence ordering, weaker than the above. We assume common marginal distributions to compare dependence orderings of random variables X and Y. Obviously, the assumption is inessential.

Examples of distributions $F_\alpha(x, y)$ whose dependence ordering is monotone in α were presented in Yanagimoto and Okamoto (1969), Yanagimoto (1971) and Kimeldorf and Sampson (1987). Examples cover the normal distribution and families due to Farlie (1960) or Plackett (1965).

3. Some Relations – I. Dependence ordering is expected to possess a close relation with a measure of dependence and the distribution of a test statistic for independence. When $G(x, y)$ has larger dependence than $F(x, y)$, we expect $\mathrm{Cov}_G(Y_1, Y_2) \geq \mathrm{Cov}_F(X_1, X_2)$, if they exist. This is true if $G(x, y)$ has larger quadrant dependence, which was essentially given in Lehmann (1966). Yanagimoto and Okamoto (1969) obtained monotonic properties of the distribution of a test statistic under dependence orderings. The statement (ii) was improved in Schriever (1987).

PROPOSITION 1. *(Yanagimoto and Okamoto, 1969) (i) Suppose $G(x, y)$ has larger quadrant dependence than $F(x, y)$. Let Q be Blomqvist's statistic. Then the distribution of Q under $G(x, y)$ is stochastically larger than that under $F(x, y)$.*

(ii) Suppose $G(x, y)$ has larger regression dependence on x than $F(x, y)$. Let T be a test statistic in a family satisfying regularity conditions, which includes Kendall's and Spearman's statistics. Then the distribution of T under $G(x, y)$ is stochastically larger than that under $F(x, y)$.

Next, we consider the notion of heaviness of tail of a positive distribution function $F(x)$. $F(x)$ is said to have increasing hazard rate, when $-\log(1 - F(x))$ is convex. This notion is regarded as describing lighter tails for $F(x)$ than that of the exponential distribution. Let $X_{(1)}$ and $X_{(2)}$ be order statistics in ascending order from independent identically distributed random variables having a distribution function $F(x)$. If $F(x)$ is exponential, $T = (X_{(1)}, X_{(2)} - X_{(1)})$ is independent. Conversely Shanbhag (1970) showed independency of T yields that $F(x)$ is exponential or geometric. Yanagimoto (1972) showed that negative dependence of T can characterize increasing hazard rate.

The notion of heaviness of tail is extended to that of heavier tail. $G(x)$ is said to have heavier HR tail than $F(x)$, if $G^{-1}(F(x))$ is convex.

PROPOSITION 2. *Suppose $G(x)$ and $F(x)$ are continuous. $G(x)$ has heavier HR tail than $F(x)$, if and only if the distribution of T under $G(x)$ has smaller regression dependence on x than that under $F(x)$.*

Results concerning positive dependence and heaviness of tail were explored in Shaked (1977b) and Shaked and Tong (1985). Yanagimoto and Sibuya (1976) showed the close relationship of the notion of heaviness of tail and that of dispersion.

4. Dependence Ordering – Multivariate Case. Only a few number of papers on positive or negative dependence in the multivariate case are found in existing literature. For our purpose, it appears that MTP_2 by Karlin and Rinott (1980a,b), negative association by Joag-Dev and Proschan (1983) and *LOD* and *UOD* by Shaked (1982) are attractive. Suppose $G(x)$ and $F(x)$ have all the common one dimensional marginal functions. $G(x)$ is said to more *LOD* than $F(x)$, if it holds that

$$P_G(Y_i \le c_i,\ i = 1, \dots, n) \ \ge\ P_F(X_i \le c_i,\ i = 1, \dots, n) \tag{1}$$

for any c_i. The dependence ordering more *UOD* is given by replacing $\le$ by $\ge$ in (1). Karlin and Rinott (1980a,b) defined positive TP_2 dependence by

$$f(x \vee y) f(x \wedge y) \ \ge\ f(x) f(y) \tag{2}$$

for any $x,\ y \in R^n$, where $x \vee y$ denotes the vector with each component having the greater one of corresponding components of x and y, and $x \vee y + x \wedge y = x + y$. They defined negative TP_2 dependence by reversing the inequality in (2) and adding additional requirements. Since TP_2 dependence of $f(x)$ does not necessarily mean that of a marginal density of $f(x)$, we require additional assumptions. We write a marginal density of $f(x)$ as $\tilde{f}(y)$. Analogous with Definition 1, we present a definition of larger TP_2 dependence ordering.

Definition 3. Let $G(x)$ and $F(x)$ be distribution functions having density functions on a common measure. $G(x)$ is said to have larger TP_2 dependence than $F(x)$, if it holds that

$$\tilde{g}(y_1 \vee y_2)\tilde{g}(y_1 \wedge y_2)\tilde{f}(y_1)\tilde{f}(y_2) \ \ge\ \tilde{g}(y_1)\tilde{g}(y_2)\tilde{f}(y_1 \vee y_2)\tilde{f}(y_1 \wedge y_2)$$

for any marginal density $\tilde{g}(y)$ and $\tilde{f}(y)$ and any $y_1,\ y_2 \in R^m$.

The above definition is not an extension of positive or negative TP_2 dependence in a strict sense, while it is intended to extend negative TP_2 dependence in a simpler way. In the bivariate case the larger TP_2 dependence ordering is weaker than the ordering in Kimeldorf and Sampson (1987), when the density functions exist.

An attractive notion of multivariate negative dependence is negative association (Joag-Dev and Proschan, 1983). The random variable X is called negatively associated, if

$$\mathrm{Cov}(\varphi(X_1),\ \phi(X_2)) \ \le\ 0, \tag{3}$$

provided it exists, where X_1 and X_2 are any partitions of X, and φ and ϕ are increasing functions. The extension of this notion to positive dependence is straightforward, though the corresponding notion of positive dependence is strictly weaker than well known association (Esary et al. 1967). To avoid confusion of nomenclature, we will call negative association negative weak-association (w-association).

A difficulty arises from the fact that the distributions of $\varphi(X_1)$ and $\phi(Y_1)$ are different in general, even though all the one dimensional marginals of X and Y are common. Note that the condition (3) can be replaced by that $(\varphi(X_1), \phi(X_2))$ has negative quadrant dependence. This fact permits us a definition of w-association ordering.

DEFINITION 4. Let X and Y be random variables having all the common one dimensional marginal distributions. Y is said to have larger w-association than X, if it holds that $(\varphi(Y_1), \phi(Y_2))$ has larger quadrant dependence than $(\varphi(X_1), \phi(X_2))$ for any corresponding partitions X_1, X_2 and Y_1, Y_2 and increasing functions φ and ϕ.

5. Some Relations – II. In this section we will review relations of the notion of multivariate dependence with other notions, with emphasis paid to the problem of overdispersion arising in the multi-generation experiments for reproductive toxicology.

Consider an n dimensional exchangeable random variable $X = (X_1, \ldots, X_n)$. Shaked and Tong (1985) suggested that larger dependence of X is associated with "hanging together" within components of X. They studied the joint distribution of order statistics from components of X, and obtained several relations of larger dependence with larger dispersion and majorization properties. A simple fact of their results is that larger dependence of X is associated with larger dispersion of $T = \Sigma X_i$. Recall that $V(T) = nV(X_1) + n(n-1)\mathrm{Cov}(X_1, X_2)$. Therefore, when we fix all the one dimensional marginal distributions, larger covariance of X means larger variance of T. In other words, positive dependence of X results in overdispersion of T.

The problem of overdispersion is important for a model used for analyzing reproductive experiment data (for example, Krewski, Colin, Hogan, and Yanagimoto, in press). Assume that X represents an underlying (tolerance) distribution of n fetuses in a litter, and that for a critical point c, say $c = 0$, a pathological finding is observed in the ith fetus when $X_i > c$. Define the ith component of $Z(X)$, a statistic, to be 1 for $X_i > c$, and as 0, otherwise. Note $Z(X)$ is expressed as an increasing function of X. It is widely accepted that the random variables X_i, are not necessarily independent and consequently $Z(X)$ is not a vector of independent elements. It is likely to be positively dependent (and possibly negatively dependent). Positive dependence is reasonably interpreted by the fact that fetuses within a litter share common genetic and environmental factors. Potential negative dependence may be interpreted as unequal distribution of toxic substances to fetuses in a pregnant animal.

These two interpretations can be formulated by a mixture distribution. Let $H_\alpha(x)$ be a one dimensional distribution function, which is stochastically monotone in α. Let $K(\alpha)$ be a mixing distribution. The positively dependent distribution function $F(x)$ of X can be expressed by

$$F(x) = \int \Pi_{i=1}^n H_\alpha(x_i) dK(\alpha). \tag{4}$$

When $H_\alpha(x)$ is expressed as $H(x-\alpha)$, we can give another expression,

$$X \sim U + Ve, \tag{5}$$

where $U = (U_1, \ldots, U_n)$, $e = (1, \ldots, 1)$, and V and U_i, $i = 1, \ldots, n$ are random variables having distribution functions $K(x)$ and $H(x)$, respectively. The negatively dependent distribution of X may be expressed by

$$X \sim (U_1, \ldots, U_n) \mid_{\Sigma U_i = u_0} \tag{6}$$

for a constant value u_0.

All the distributions represented in (4)–(6) have been studied as typical multivariate positive or negative distributions. A family of distribution functions given by (4) was pursued in depth by Dykstra, Hewett, and Thompson (1973) and Shaked (1975). A family of distributions given by (6) was discussed by Block, Savits, and Shaked (1982) in a more general manner. They generalized conditioning to $U_0 + \Sigma U_i$ by adding an independent random variable, and showed that such a distribution is *UOD* and also *LOD* if each random variable has a TP_2 density.

In practical applications the distribution function $K(\alpha)$ in (4) contains a parameter representing dispersion. We can expect that larger dispersion of a mixing distribution results in larger dependency of the mixture distribution, though we need marginal adjustments for comparison. The generalization of (6) by Block et al. (1982) looks appealing, and larger dispersion of U_0 is expected to result in smaller dependency of X. Finally, we present examples of the normal case (Ochi and Prentice, 1984) and the beta-binomial case (Skellam, 1948, Williams, 1975).

EXAMPLE 1. As usual the normal case presents us a simple, clear example. Let $U_i \sim N(0, \sigma^2)$, $\sigma^2 \geq 0$, $V \sim N(0, \delta^2)$, $\delta^2 \geq 0$, $U_0 \sim N(0, \tau^2)$, $\tau^2 \geq 0$ and $u_0 = 0$. Then the random variable X in (5) belongs to $N(0, \sigma^2 I + \delta^2 e'e)$ and that in (6) belongs to $N(0, \sigma^2 I - (\sigma^4/(n\sigma^2 + \tau^2))e'e)$. By adjusting all the one dimensional marginal distributions as $N(0,1)$, the former family is written as $N(0, (1-p)I + pe'e)$ for $1 \geq p \geq 0$ and the latter is written as $N(0, (1-p)I + pe'e)$ for $0 > p \geq -1/(n-1)$. Note that both the families superficially share the same form of the normal distribution. It is easily checked that the larger TP_2 dependence ordering of the above families of distributions is monotone increasing in p, that is, monotone increasing in δ^2 and monotone decreasing in τ^2. As a result the more *LOD* or the more *UOD* ordering also is monotone increasing in p.

EXAMPLE 2. The beta-binomial distribution is the most familiar distribution employed in the model for reproductive studies because of its simple form as a mixture distribution of the binomial distribution. The probability function of the beta-binomial distribution is written as

$$(7)\qquad p(x;n,\alpha,\beta) = {}_nC_t\Pi_{r=0}^{x-1}(\mu + r\theta)\Pi_{r=0}^{n-x-1}\{(1-\mu)+r\theta\}/\Pi_{r=0}^{n-1}(1+r\theta).$$

Note that (7) makes sense to some extent, even when θ is less than 0 (Prentice, 1986).

Let W be a random variable having the beta distribution, $Be(\alpha, \beta)$ and $V = \log W/(1-W)$. Let U_i be a random variable having the distribution function $\exp x/(1+\exp x)$. To define $Z(X)$ having the multivariate beta-binomial distribution we set $X = U + Ve$ and $c = 0$. The probability function of $t = \Sigma X_i$ is given by $p(t;n,\theta,\mu)/{}_nC_t$ with $\mu = \alpha/(\alpha+\beta)$ and $\theta = 1/(\alpha+\beta)$. It follows that this distribution is associated if $\theta > 0$. Straightforward calculations yield that the larger TP_2 dependence ordering of the multivariate beta-binomial distribution is monotone increasing in θ for a fixed μ as far as (7) makes sense.

REFERENCES

BLOCK, H.W., SAVITS, T.H. and SHAKED, M. (1982). Some concepts of negative dependence. *Ann. Prob.* **10** 765–772.

DYKSTRA, R.L., HEWETT, J.E. and THOMPSON, W.A. (1973). Events which are almost independent. *Ann. Statist.* **1** 674–681.

ESARY, J., PROSCHAN, F. and WALKUP, D. (1967). Association of random variables with application. *Ann. Math. Statist.* **38** 1466–1474.

FARLIE, D.J.G. (1960). The performance of some correlation coefficients for a general bivariate distribution. *Biometrika* **47** 307–323.

JOAG-DEV, K. and PROSCHAN, F. (1983). Negative association of random variables, with applications. *Ann. Statist.* **11** 286–295.

KARLIN, S. and RINOTT, Y. (1980a). Classes of orderings of measures and related correlation inequalities–I. *J. Multivar. Anal.* **10** 467–498.

KARLIN, S. and RINOTT, Y. (1980b). Classes of orderings of measures and related correlation inequalities–II. *J. Multivar. Anal.* **10** 499–516.

KIMELDORF, G. and SAMPSON, A.R. (1987). Positive dependence orderings. *Ann. Inst. Statist. Math.* **39** 113–118.

KREWSKI, D., COLIN, D. and YANAGIMOTO, T. (1989). Statistical methods for developmental toxicity studies. In *Interpretation and Extrapolation of Reproduction Data to Establish Human Safety Standards*, H.C. Grice and K.S. Khera, eds., Springer, New York.

LEHMANN, E.L. (1966). Some concepts of dependence. *Ann. Math. Statist.* **37** 1137–1153.

OCHI, Y. and PRENTICE, R.L. (1984). Likelihood inference in a correlated probit regression model. *Biometrika* **71** 531–543.

PLACKETT, R.L. (1965). A class of bivariate distributions. *J. Amer. Statist. Assoc.* **60** 516–522.

PRENTICE, R.L. (1986). Binary regression using an extended beta-binomial distribution, with discussion of correction induced by covariate measurement errors. *J. Amer. Statist. Assoc.* **81** 321–327.

SCHRIEVER, B.F. (1987). An ordering of positive dependence. *Ann. Statist.* **15** 1208–1214.

SHAKED, M. (1977a). A family of concepts of dependence for bivariate distributions. *J. Amer. Statist. Assoc.* **72** 642–650.

SHAKED, M. (1977b). A concept of positive dependence for exchangeable random variables. *Ann. Statist.* **5** 505–515.

SHAKED, M. (1982). A general theory of some positive dependence. *J. Multivar. Anal.* **12** 219–229.

SHAKED, M. and TONG, Y.L. (1985). Some partial orderings of exchangeable random variables by positive dependence. *J. Multivar. Anal.* **17** 339–349.

SHANBHAG, D.N. (1970). The characterizations for exponential and geometric distributions. *J. Amer. Statist. Assoc.* **65** 1256–1259.

SKELLAM, J.G. (1948). A probability distribution derived from the binomial distribution by regarding the probability of success as variable between the sets of trials. *J. Roy. Statist. Soc.* Ser. B **10** 257–261.

TCHEN, A. (1980). Inequalities for distributions with given marginals. *Ann. Prob.* **8** 814–827.

WILLIAMS, D.A. (1975). The analysis of binary responses from toxicological experiments involving reproduction and teratogenicity. *Biometrics* **31** 949–952.

YANAGIMOTO, T. (1971). Examples of families of distributions with positive dependence (in Japanese). Conference Proc. (limited circulation).

YANAGIMOTO, T. (1972). Families of positively dependent random variables. *Ann. Inst. Statist. Math.* **24** 559–573.

YANAGIMOTO, T. and OKAMOTO, M. (1969). Partial orderings of permutations and monotonicity of a rank correlation statistic. *Ann. Inst. Statist. Math.* **21** 489–506.

YANAGIMOTO, T. and SIBUYA, M. (1972). Stochastically larger component of a random vector. *Ann. Inst. Statist. Math.* **24** 259–269.

YANAGIMOTO, T. and SIBUYA, M. (1976). Isotonic test for spread and tail. *Ann. Inst. Statist. Math.* **28** 329–342.

INSTITUTE OF STATISTICAL MATHEMATICS
4-6-7 MINAMI-AZABU
MINATO-KU
TOKYO 106, JAPAN

SYMPOSIUM ON DEPENDENCE IN

STATISTICS AND PROBABILITY

AUGUST 1-5, 1987

• • •

Hidden Valley Conference Center

Laurel Highlands

Somerset, Pennsylvania 15501

PROGRAM

Sponsored by: AFOSR, NSF, ARO and the University of Pittsburgh

- 2 -

SYMPOSIUM ON DEPENDENCE IN STATISTICS AND PROBABILITY

August 1-5, 1987

This research symposium focuses on the theory and applications of positive and negative dependence. The four basic areas to be examined are: (a) modeling foundations in monotone dependence; (b) applications to probability theory, reliability theory, mathematical statistics and other areas; (c) statistical issues arising from monotone dependence considerations; and (d) relationships of monotone dependence concepts to other statistical concepts.

CONFERENCE ORGANIZERS

H. W. Block
A. R. Sampson
T. H. Savits

LOCAL ARRANGEMENTS

D. Stoffer
D. Chhetry
Z. Fang

- 3 -

SATURDAY, AUGUST 1

1:00-4:00 p.m. Reception (Room 4)

4:00-5:30 p.m. *Dependence Concepts, I* (Rooms 1 and 2)
Chair: **Thomas H. Savits,** University of Pittsburgh,

Henry W. Block, University of Pittsburgh, "Dependence in probability and statistics."

Devendra Chhetry and **Allan R. Sampson*,** University of Pittsburgh, "A projection decomposition for bivariate discrete probability distributions."

Henry W. Block, Devendra Chhetry*, Zhaoben Fang*, and **Allan R. Sampson,** University of Pittsburgh, "Partial orderings on permutations and dependence orderings on bivariate distributions."

6:30-8:00 p.m. DINNER (Room 4)

8:00 p.m. CONFERENCE PARTY (Four Seasons Lounge)

SUNDAY, AUGUST 2

8:15 - 9:15 a.m. CONTINENTAL BREAKFAST (Balcony off Rooms 1 and 2)

9:15 - 9:30 a.m. OPENING (Rooms 1 and 2)
Jack E. Freeman, Executive Vice President, University of Pittsburgh

9:30 -11:00 a.m. *Modeling Multivariate Distributions* (Rooms 1 and 2)
Chair: **Frank Proschan,** Florida State University

Albert Marshall, University of British Columbia and **Ingram Olkin*,** Stanford University, "Modeling multivariate distributions."

Albert Marshall*, University of British Columbia and **Ingram Olkin,** Stanford University, "Families of multivariate distributions."

11:00-12:00 Noon *Multivariate Statistical Models* (Rooms 1 and 2)
Chair: **Naftali Langberg,** University of Pittsburgh and Haifa University, Israel

Takemi Yanagimoto, Institute of Statistical Mathematics, Toyko, "Positive dependence in statistical models and other notions."

G. P. Patil, Pennsylvania State University and Harvard University, "Dependence perspectives of bivariate distributions arising in encountered data, joint toxicity assessment, and bivariate exponential extensions."

12:00-1:30 p.m. LUNCH (Room 4)

1:30-3:00 p.m. *Life Testing* (Rooms 1 and 2)
Chair: **Morris DeGroot,** Carnegie-Mellon University

Nozer Singpurwalla*, and **Mark Youngren,** George Washington University, "Multivariate distributions for the life lengths of items sharing a common (fixed or varying) environment."

- 4 -

Myles Hollander*, Florida State University, **Frank Proschan,** Florida State University, and **James Sconing,** University of Iowa. "Information, censoring, and dependence."

Nancy Flournoy, University of Washington, "Bivariate Markov chains."

3:00-3:30 p.m. BREAK (Balcony off Rooms 1 and 2)

3:30-5:30 p.m. *Dependence and Bivariate Discrete Distributions* (Rooms 1 and 2)
Chair: **Allan R. Sampson,** University of Pittsburgh

Harry Joe, University of British Columbia, "Relative entropy measures of multivariate dependence."

Marco Scarsini, Stanford University and Universitá di Parma, "An ordering of dependence."

Truc T. Nguyen, Bowling Green State University, "Complete dependence and a characterization of random variables with normal distributions."

Mei-Ling Ting Lee, Boston University, "Some cross-product difference statistics and a test for trends in ordered contingency tables."

Allan R. Sampson, University of Pittsburgh, and **Lyn Whitaker***, University of California, Santa Barbara, "Estimation of multivariate distributions under stochastic ordering."

3:30-5:30 p.m. *Probability and Dependence*
Chair: **Joel Greenhouse,** Carnegie-Mellon University

A. Larry Wright, University of Arizona, "Associated processes: results and conjectures."

E. G. Enns, University of Calgary, "Dependence relationships for chords through a convex body generated from within an embedded convex body."

Jürg Hüsler, University of Bern, "Multivariate extreme values of nonstationary sequences."

Nicholas Longford, Educational Testing Service, "Multivariate exponential and geometric distributions derived from hitting times in Markov processes."

7:00-9:30 p.m. BANQUET (Family Entertainment) (Room 4)

9:30-12:00 p.m. INFORMAL GATHERING (Four Seasons Lounge)

- 5 -

MONDAY, AUGUST 3

7:30-8:30 a.m. CONTINENTAL BREAKFAST (Patio outside of Room 4)

8:30-9:45 a.m. *Coincidence Probabilities and Inequalities* (Rooms 1 and 2)
Chair: **Henry W. Block,** University of Pittsburgh

Samuel Karlin, Stanford University, "Coincidence probabilities and applications to combinatorics."

L. D. Brown, Cornell University and **Yosef Rinott,*** Hebrew University, "Inequalities for multivariate infinitely divisible processes."

10:15-11:45 a.m. *Processes in Reliability Maintenance* (Rooms 1 and 2)
Chair: **Brian Woodruff,** AFOSR

Henry W. Block, University of Pittsburgh, **Naftali A. Langberg,** Haifa University, and **Thomas H. Savits*,** University of Pittsburgh, "Stochastic comparisons for some repair models."

E. Arjas*, University of Oulu and **I. Norros,** University of Helsinki, "Should minimal repair depend on information?"

M. Abdel-Hameed, Kuwait University, "Inspection and maintenance policies of devices subject to deterioration."

10:00-11:45 a.m. *Sequential and Nonparametric Analysis* (Room 3)
Chair: **Bimal Sinha,** University of Maryland, Baltimore

Joseph Glaz, University of Connecticut, "Approximating the distribution of sums of random variables with applications to group sequential test and scan statistics.

James R. Kenyon, University of Southern Maine, "Calculating improved bounds and approximations for sequential testing procedures."

Z. Govindarajulu, University of Kentucky, "Robustness of Mann-Whitney-Wilcoxon test to dependence in the variables."

Pablo Salzberg, University of the Sacred Heart, "On the independence between polynomials in a multinormal random variable."

11:45-1:45 p.m. POOLSIDE BARBEQUE

1:45-3:15 p.m. *Majorization and Orderings* (Rooms 1 and 2)
Chair: **Robert L. Launer,** ARO

Persi Diaconis, Stanford University, and **Michael D. Perlman*,** University of Washington, "Bounds for tail probabilities of linear combinations of independent gamma random variables."

Wai Chan, Ohio State University, **Frank Proschan,** Florida State University, and **Jayaram Sethuraman*,** Florida State University, "Convex ordering among functions with applications to reliability and mathematical statistics."

Arthur Cohen* and **Harold B. Sackrowitz,** Rutgers University, "Unbiasedness of tests of homogeneity when alternatives are ordered."

3:15-3:45 p.m. BREAK (Balcony outside of Rooms 1 and 2)

- 6 -

3:45-5:15 p.m. *Dependence Concepts, II* (Rooms 1 and 2)
Chair: **Moshe Shaked,** University of Arizona

Barry C. Arnold, University of California, Riverside, "Dependence in conditionally specified distributions."

D. R. Jensen, V.P.I., "Invariance properties of arrays through dependence by mixing."

Tadeusz Bromek, Polish Academy of Sciences. Title to be announced.

6:30-8:00 p.m. DINNER (Room 4 and Patio outside)

8:00-9:00 p.m. *Special Interest Group Discussions* (Boardroom), "Contingency tables and positive dependence"

8:00-12:00 p.m. INFORMAL GATHERING (Four Seasons Lounge)

TUESDAY, AUGUST 4

7:30-8:30 a.m. CONTINENTAL BREAKFAST (Patio outside of Room 4)

8:30-9:30 a.m. *Topics in Reliability* (Rooms 1 and 2)
Chair: **Brian Woodruff,** AFOSR

Mark Brown, City College, CUNY, "Bivariate constructions, coupling and error bounds."

Moshe Shaked,* University of Arizona and **J. George Shanthikumar,** University of California, Berkeley, "Multivariate conditional hazard rate functions."

8:45-10:00 a.m. *Dependence Concepts, III* (Room 3)
Chair: **David Stoffer,** University of Pittsburgh

K. Subramanyam, University of Pittsburgh, and **M. Bhaskara Rao*,** University of Sheffield, "On the structure of $2 \times \infty$ bivariate distributions which are totally positive of order two."

K. Subramanyam, University of Pittsburgh, "Some comments on positive quadrant dependence."

Sanat K. Sarkar, Temple University, "On quasi- independence in ordinal contingency tables."

10:00-11:30 a.m. *Inequalities and Orderings* (Rooms 1 and 2)
Chair: **Ingram Olkin,** Stanford University

J.H.B. Kemperman, Rutgers University, "Inequalities for the Borda ranking of random variables."

Y. L. Tong, Georgia Institute of Technology, "Inequalities for a class of positively dependent random variables with a common marginal."

Phillip Boland*, University College, Dublin, and **Frank Proschan,** Florida State University, "Multivariate arrangement increasing functions with applications in probability and statistics."

11:30-12:45 p.m. Title to be announced.
Chair: **Michael Meyer,** Carnegie-Mellon University

- 7 -

Leo A. Goodman, University of California, Berkeley (tentative). Title to be announced.

Ruth Douglas and **Stephen E. Fienberg*,** Carnegie-Mellon University, "An overview of dependency models for cross-classified categorical data involving ordinal variables."

12:45-1:30 p.m.	LUNCH (Room 4)
1:30-3:00 p.m.	*Ising Models and Dependence Inequalities* (Rooms 1 and 2) *Chair:* **Y. Mittal,** NSF and VPI

Charles M. Newman, University of Arizona, "Ising models as dependent percolation models."

Kumar Joag-Dev, EPFL Lausanne and University of Illinois, "A version of Kelly-Sherman inequality for Ising model with an application."

Loren D. Pitt, University of Virginia, "Semi-groups and positive correlations."

1:30-3:00 p.m.	*Reliability and Dependence* (Room 3) *Chair:* **Satish Iyengar,** University of Pittsburgh

A. P. Basu, University of Missouri, Columbia, "Some inference problems for dependent systems."

William S. Griffith, University of Kentucky, "On the role of dependence in the analysis of consecutive k-out-of-n systems."

Jerzy Filus, Illinois Institute of Technology, "On multivariate probability distributions as models of reliability for systems with dependent life times of their components."

3:00-6:00 p.m.	FALLINGWATER TRIP (Sign-up required.)
7:00-8:30 p.m.	DINNER (Hearthside Resturant - reservations required)
8:00-9:00 p.m.	*Special Interest Group Discussions* (Boardroom) - "Topics in Dependence"
8:30-12:00 p.m.	INFORMAL GATHERING (Four Seasons Lounge)

- 8 -

WEDNESDAY, AUGUST 5

7:00-8:30 a.m. CONTINENTAL BREAKFAST (Patio outside of Room 4)

8:30-10:00 a.m. *Time Series and Bayesian Statistics* (Rooms 1 and 2)
Chair: **Devendra Chhetry,** University of Pittsburgh

Richard A. Johnson* and **Thore Langeland,** University of Wisconsin, "A linear combination test for detecting serial correlation in multivariate samples."

Henry W. Block, University of Pittsburgh, **Naftali Langberg,** Haifa University, and **David S. Stoffer*,** University of Pittsburgh, "Time series models for non-Gaussian processes: some recent developments."

Richard E. Barlow and **Carlos A. de B. Pereira*,** University of California, Berkeley, "Probabilistic influence diagrams."

10:00-11:00 a.m. *Miscellaneous Topics* (Rooms 1 and 2)
Chair: **Zhaoben Fang,** University of Pittsburgh

R. D. Gupta, University of New Brunswick, "Estimation of reliability in the case of normal distributions."

Richard A. Vitale, The Claremont Graduate School, "A differential version of the Efron-Stein inequalities."

11:30-12-30 p.m. LUNCH (Hearthside Restaurant or Box Lunch)

CONFERENCE PARTICIPANTS

Abdel-Hameed, Mohamed
Arjas, Elja
Arnold, Barry C.
Basu, Asit P.
Benedict, Jeffrey P.
Block, Henry W.
Boland, Philip
Bromek, Tadeusz
Chen, C.S.
Chhetry, Devendra
Cohen, Arthur
Constantine, Gregory
Costigan, Timothy M.
DeGroot, Morris H.
DiFranco, Donna M.
Douglas, Ruth
Enns, Ernest
Fang, Zhaoben
Fienberg, Stephen E.
Filus, Jerzy
Flournoy, Nancy
Gatsonis, Constantine
Glaz, Joseph
Govindarajulu, Z.
Greenhouse, J.
Gupta, R.D.
Heldal, Johan
Hollander, Myles
Hüsler, Jürg
Iyengar, Satish
Jensen, Donald R.
Joag-Dev, Kumar
Joe, Harry
Johnson, Richard A.
Karlin, Samuel
Kemperman, J.H.B.
Kenyon, James R.
Kim, Song Ho
Langberg, Naftali
Launer, Robert
Lee, Mei-Ling
Longford, Nicholas T.
Marshall, Albert W.
McCloskey, Joseph P.
Metry, Magdy H.
Meyer, Michael
Mi, Jie
Mittal, Yashaswini, D.
Nassar, Manal M.
Neuenschwander, Beat
Newman, Charles M.
Nguyen, Truc Truong
Olkin, Ingram
Patil, G.P.
Pereira, Carlos
Perlman, Michael D.
Perrugia, Mario
Pitt, Loren D.
Proschan, Frank
Qian, Shixian
Quesada, Jose Juan
Quinn, Joseph E.
Rao, M.B.
Rinott, Yosef
Sampson, Allan R.
Santi, Mary
Sarkar, Sanit
Savits, Thomas H.
Scarsini, Marco
Sethuraman, Jayaram
Shaked, Moshe
Sheu, Shey-Huei
Singpurwalla, Nozer D.
Sinha, Bimal K.
Stoffer, David
Subramanyam, Kassala
Tiwari, Ram Chandra
Tong, Yung L.
Vitale, R.A.
Wang, Jane-Ling
Whitaker, Lyn
Woodruff, Brian
Wright, A. Larry
Yanagimoto, Takemi
Zeng, Weibin

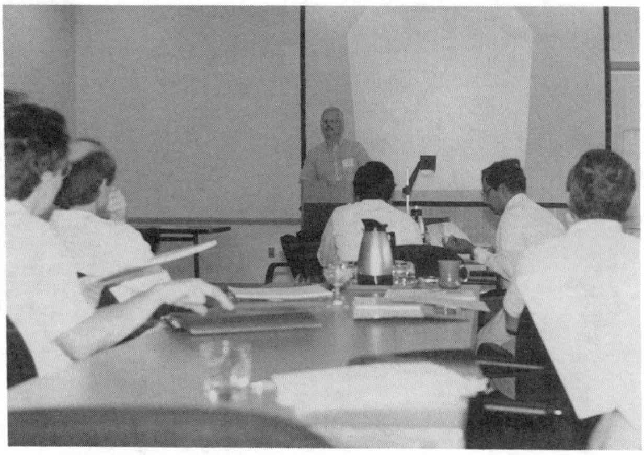

Henry Block

Tom Savits, Allan Sampson
and Henry Block

Hidden Valley

Z. Govindarajulu and
Mei-Ling Lee

Myles Hollander,
Frank Proschan
and Morrie DeGroot

Yosh Mittal and Asit Basu

Nancy Flournoy and
Ingram Olkin

Foreground (Facing camera):
Moshe Shaked, Al Marshall
and Rich Johnson

Lyn Whitaker

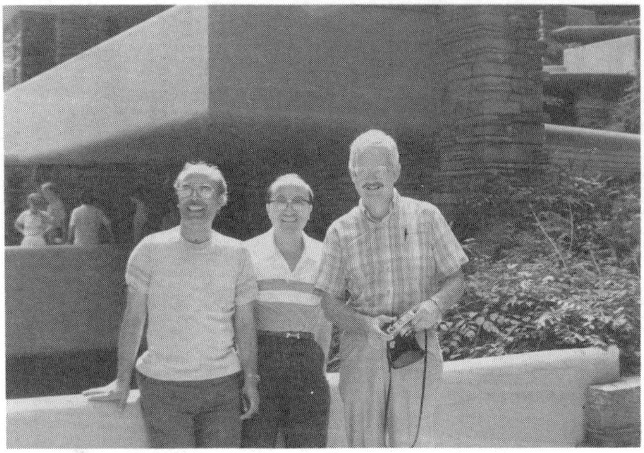

Dev Chhetry, Zhaoben Fang
and Henry Block

Mo Abdel-Hameed
and S.-H. Sheu

Facing Camera:
Ingram Olkin and
Tadeusz Bromek

Lunch on the patio

Henry Block

Allan Sampson,
Josh Sampson
and Mike Sampson

Dev Chhetry and C.S. Chen

Joyce Fienberg and
Steve Fienberg

Zhaoben Fang and
Weibin Zeng

Author Citation Index

* The cited page is the first page of the article in which the author's name appears.

Key Words and Phrases Index

* This index utilizes authors' key words and phrases. The cited page is the first page of the article in which that word or phrase appears.